Profitable Partnering in Construction Procurement

Profitable Partnering in Construction Procurement

CIB W92 (Procurement Systems)
and
CIB TG 23 (Culture in Construction)

Joint Symposium

Edited by
Dr. Stephen O. Ogunlana

SPONSORED BY

E & FN SPON
An imprint of Routledge
London and New York

This edition published 1999
by E & FN Spon, an imprint of Routledge
11 New Fetter Lane, London EC4P 4EE

Simultaneously published in the USA and Canada
by Routledge
29 West 35th Street, New York, NY 10001

© 1999 E & FN Spon

Printed and bound in Great Britain by
St Edmundsbury Press, Bury St Edmunds, Suffolk

Publisher's Note
This book has been prepared from camera-ready copy provided by the editor.

British Library Cataloguing in Publication Data
A catalogue record for this book is available
from the British Library

ISBN 0 419 24760 2

CONTENTS

FOREWORD

Working Commission 92 (W92) of the Conseil International du Bâtiment was established in 1989 W92's aims and objectives were established as follows:

- "to research into the **social, economic and legal** aspects of contractual arrangements that are deployed in the procurement of construction projects;
- to establish the practical aims and objectives of **contractual arrangements** within the context of procurement;
- to report on and to evaluate areas of commonality and difference;
- to formulate recommendations for the **selection and effective implementation** of project procurement systems;
- to recommend standard conventions."

These aims and objectives still hold today although it was recognised from the outset that some of this subject matter had been included in the broader remits of other Working Groups, in particular W65 (Organisation and Management of Construction) and W55 (Building Economics). However, it was argued that there was a need for a working group which would serve the needs of the international research community in times of great change. In addition to long-standing concerns of international comparability and standardisation of contracts and contract procedures, the 1980's had brought significant changes in the legal, economic, and social structures of states in both Developing and Developed Countries. Private finance initiatives had been effected not only in Europe and North America, but also in eastern Europe (through the transitions from socialist to capitalist systems), in Africa (through Structural Adjustment Programs) and Asia.

The contributions to W92 Symposia have been made from very different theoretical and cultural perspectives; the contributors have come from all continents, and from different economic, legal, and political systems. This is one of the main strengths of the working commission. The significant strategic issues prevalent in the work of the commission have been identified as follows: - competition, privatisation, development, culture, trust and institutions all within the context of procurement strategies, contractual arrangements and forms of contract. Emergent themes in W92's work - those which have begun to emerge in recent years and are expected to grow significantly have been identified as - organisational learning, culture, developmentally orientated procurement systems, and sustainable procurement.

The volume and quality of the published work of the commission has grown throughout the decade and it is pleasing to note that this book and "Procurement Systems: A guide to best practice in construction" are both being published in 1999 by E & FN Spon as the output from the commission.

The theme of this symposium is harmony and profit. This theme aims to reflect not only the cultural values of Asia, where the symposium takes place, but also the cultural changes taking place in the construction industry where the concepts of partnering, collaboration and mutual benefit are coming to the fore. This new direction is a positive response to the past when disputes and dispute avoidance were key issues in procurement systems.

The theme of the symposium has been chosen on the basis that W92 has had a rolling programme of symposia which have dealt in recent years with the transfer of knowledge and experience between regions, such as East meets West in Hong Kong in 1994 and North meets South in South Africa in 1996. This attitude of learning and of organisational change was taken up in Montreal in 1997 where the theme was Procurement systems: A key to innovation.

W92 has been one of the most active of the CIB working commissions over the past five years and the activity is underpinned through strong links to other task groups such as those working on the topics of culture, TG23, and dispute resolution, TG15. There are also strong links with W65, organisation and management, and W55, construction economics.

There are over 65 papers presented in these proceedings and the range of topics relating to procurement systems and having a theme relating to harmony and profit is remarkable. Also, what makes these proceedings important is the fact that the contributors come from around the world and an ideal opportunity is thus provided for knowledge transfer. TG23 was established to facilitate this, as the issue of culture is becoming increasingly significant, especially where multi-national organisations are involved in the procurement process.

Hence the following terms of reference were established for the Task Group:

- To identify and define concepts of culture in the international construction industry and to carry out research into their manifestations and effects.
- To discuss and develop appropriate methodologies for the study of culture in construction.
- To examine and, where appropriate, adopt methodologies used in other disciplines, with special reference to the Social Sciences, for researching culture in the construction industry worldwide.

These aims are being met through this joint symposium and the publication of this book.

Finally we would like to thank AIT most sincerely for making this publication and symposium possible through its generosity and enthusiasm. We would also like to thank the scientific committee and especially Dr. Stephen Ogunlana for all their endeavours in making all this happen.

CIB W92 Coordinators: Professor David Jaggar and Dr. Steve Rowlinson

CIB TG23 Coordinators: Dr. Richard Fellows and Dr. David Seymour

PREFACE

Profitable Partnering in Construction Procurement contains papers concerned with nine themes:

- Partnering
- Harmonious relations
- Contracts and contractual relations
- Procurement systems and strategies
- Designing and building
- Private finance projects
- Culture
- Information and decision systems
- Change, technology and value management

Papers dealing with relationships and how they are managed have been grouped under the three themes: partnering; harmonious relations; and contracts and contractual relationships. The three themes 'procurement systems and strategies'; 'private finance projects'; and 'designing and building' contain papers dealing with traditional and emerging forms of relationships in the construction industry. Culture is the focus of CIB's Task Group 23. As such, papers specific to culture have been grouped into one theme. The rest of the papers in the book address systems and technologies used for relationship management. They have been grouped under the two themes: information systems and technologies; and change, technology and value management.

Maintaining standards

There is growing interest in the work of CIB W92 and TG23. In line with the growing maturity of our discipline, the papers contained in this book have been subjected to very high standards of refereeing. Of the 96 abstracts received, 85 were selected for development into full papers. All of the full papers were then refereed internationally and authors invited to incorporate the referee's comments into their papers. At the time of going to press, 67 papers had been accepted. The papers presented in this book are of very high quality.

Acknowledgements

I would like to record my special thanks to members of the organising committee, the international scientific committee. Particular thanks are due to Dr. Akintola Akintoye, Dr. Steve Rowlinson, Dr. Richard Fellows and Professor David Jaggar for working extremely hard to bring the symposium to the attention of the contributors. I would also like to thank Mr. Tim Robinson and Mr. Michael Doggwiler of E & FN Spon for the support given in publishing this book. Finally, I would like to thank Mr. Thomas Aduloju, Mr. Arun Bajracharya, and Mr. Abdul Samad Kazi for putting together all the papers presented at the CIB W92 and TG23 Joint Symposium (Harmony and Profit) in Chiang Mai, Thailand.

Stephen O. Ogunlana
Pathum Thani, Thailand. November, 1998

1. Partnering

PARTNERING: THE PROPAGANDA OF CORPORATISM?

S. D. GREEN
Department of Construction Management & Engineering, The University of Reading,
Whiteknights, PO Box 219, Reading, RG6 6AW, UK. s.d.green@reading.ac.uk

Abstract

A critical perspective on partnering is developed with reference to current concerns regarding the increasingly corporatist nature of global capitalism. Partnering is advocated by many leading clients as a means of improving customer responsiveness and ensuring continuous improvement. The seductive rhetoric of partnering too often serves only to disguise the crude exercise of buying power. In the UK, the four largest supermarket chains are all leading advocates of partnering. Ironically, all are currently under investigation by the Office of Fair Trading for failing to pass on savings to their customers. It is suggested that the doctrine of customer responsiveness ultimately owes more to corporatist propaganda than to a coherent management policy. The buying power of the industry's major clients continually discourages dissent to the partnering ideal. Construction companies which do not appear similarly committed risk being denied access to a substantial proportion of the UK market. The increasing influence of industry on the construction research agenda also discourages academics from challenging the legitimacy of partnering discourse. There is an urgent need for research which is independent of commercial vested interests.

Keywords: Continuous improvement, corporatism, critical theory, customer responsiveness, partnering, technocratic totalitarianism.

Introduction

In common with many other countries, partnering has received widespread endorsement in the UK construction industry (DETR, 1998; Bennett and Jayes, 1995; Bennett and Jayes, 1998; Construction Industry Board, 1997). The adoption of partnering supposedly improves customer-responsiveness and ensures continuous improvement. To date, there has been an almost total absence of any counter-argument. The purpose of this paper is to overcome this deficiency by developing a critical perspective on partnering. The discussion draws from the traditional of critical theory to challenge the legitimacy of commonly accepted practices as advocated and imposed by powerful vested interests. The intention is to encourage critical debate within the international academic community and to demonstrate the need for research which is independent of commercial vested interests.

The broad context is provided by current concerns that modern society is increasingly characterised by the ideology of 'corporatism'. In the knowledge that such concerns may be perceived as overly political, it is important initially to assert the author's commitment to a market-based economy subject to appropriate regulation. Given that the adopted political position is so moderate, the arguments presented should not be dismissed on the basis of being 'radical'. Indeed, it will be contended that it is the propaganda of partnering that has become dangerously radical by replacing the central tenets of a regulated market economy with an ideology uncomfortably close to naked corporatism. Also of concern is the way in which the assumptions of partnering are potentially in conflict with long-established notions of pluralism and the associated commitment to the principles of humanism.

Corporatism

The ideology of corporatism is often linked with the fascism of Mussolini's Italy. It is based on the belief that business and labour share the same interests (Heywood, 1992). Different interest groups are therefore seen to be bound together by duty and mutual obligations. In essence, a corporatist society is one where interest groups triumph over individuals. Examples of such interest groups include not only corporations, but also owners' associations, trade unions and professional associations. The logic of corporatism dictates that such groups are not in conflict with each other, but seek non-confrontational relationships. Whilst usually associated with past dictators such as Mussolini, Peron and Salazar, corporatism is also evident in the industrial West in the form of neo-corporatism. This describes the tendency of governments to govern in consultation with economic interests such as business and trade unions (Heywood, 1992). Such neo-corporatist tendencies are even more observable in heavily institutionalised Asian economies such as Japan, Korea and Singapore.

Any advance of corporatism is inevitably at the expense of democracy, in that the interests of groups are given preference over the interests of individuals. Whilst today's neo-corporatists do not like to be confused with the unpleasant dictators of the past, there is a convincing argument that humanism is increasingly in retreat in the face of corporate vested interests (Saul, 1997). With the collapse of communism there is now a danger that capitalism runs unchecked. Unfortunately, for those who adhere to the humanist tradition, there is no historical evidence that capitalism automatically results in democracy (Wood, 1995). The various checks and balances against unbridled capitalism achieved since the Industrial Revolution were hard fought for gains which were only attained as a result of prolonged political and social protest. The current concern is that these gains are being lost as Western society becomes increasingly corporatist. The trend is accentuated by the global nature of modern business. Many corporations play-off one government against another, re-deploying their production facilities to best commercial advantage (Greider, 1996). The economic muscle of such corporations often outweighs that of democratically elected governments. Many important checks and balances against the excesses of capitalism are currently being dismantled in the cause of 'de-regulation'. This interpretation of current economic trends is in direct conflict with the propaganda of the free marketplace and the assumption that commerce inevitably leads to democracy. It also provides a radically different starting point for a critique of partnering in construction.

Partnering: defining characteristics

There are numerous definitions of partnering which are currently in circulation. One of the most comprehensive is that offered by the Construction Industry Institute (1989):

> "*A long term commitment between two or more organisations for the purposes of achieving specific business objectives by maximising the effectiveness of each participant's resources. This requires changing traditional relationships to a shared culture without regard to organisational boundaries. The relationship is based on trust, dedication to common goals, and on an understanding of each others individual expectations and values. Expected benefits include improved efficiency and cost effectiveness, increased opportunity for innovations, and the continuous improvements of quality products and services.*"

From this point of view, partnering is primarily concerned with 'maximising effectiveness', thereby reflecting the purpose of countless other management improvement techniques. Emphasis is also given to 'culture' and the need to base relationships on trust and understanding. The influence of the rhetoric of Total Quality Management (TQM) is readily apparent in the reference to 'continuous improvement'. The tone of the definition offered by Bennett and Jayes (1995) reflects similar themes:

> "*Partnering is a management approach used by two or more organisations to achieve specific business objectives by maximising the effectiveness of each participant's resources. The approach is based on mutual objectives, an agreed method of problem resolution, and an active search for continuous measurable improvements.*"

It is notable that improvements must not only be continuous, they must also be 'measurable'. The definition recently offered by the recent 'Egan Report' provides a similar emphasis on continuous, measurable improvement:

> "*Partnering involves two or more organisations working together to improve performance through agreeing mutual objectives, devising a way for resolving disputes and committing themselves to continuous improvement, measuring progress and sharing the gains.*" (DETR, 1998)

The DETR further consider partnering to be a 'tool to tackle fragmentation' which is increasingly used by the best firms in place of traditional contract-based procurement and project management. According to the Construction Industry Board (1997), partnering has three essential components:

- establishment of agreed and understood mutual objectives;
- methodology for quick and co-operative problem resolution;
- culture of continuous, measured improvement.

The achievement of the appropriate 'culture' is almost universally held to be of vital importance to the success of partnering. The Construction Industry Board emphasises that the first step towards partnering is to ensure that the culture of the company is conducive to a 'whole-team co-operative approach'. It is further recommended that a champion should be appointed to promote the partnering concept and that senior management must act as exemplars of the required culture.

Whilst partnering is often equated with long-term relationships, the two terms are not synonymous. Several sources make a distinction between 'project specific' partnering and 'strategic' partnering, where the partners work together on several projects. (Construction Industry Board, 1997). Strategic partnering supposedly allows the benefits of improved understanding to be carried forward to subsequent projects. The philosophy of continuous, measured improvement however demands that each project exceeds the performance of the previous one. Despite the seductive discourse on 'empowerment', 'working together' and 'relationships', the ultimate measure of success seems to hinge on cost improvement.

Underlying influences

McGeorge and Palmer (1997) suggest that formal partnering as a construction management concept dates from the mid-1980s. Several early partnering arrangements were apparently established in the process engineering sector. Specific examples include Union Carbide with Bechtel and Du Pont with Fluor Daniel. Notwithstanding these early American examples, it would seem that the currently advocated philosophy of partnering is heavily influenced by the collaborative practices of Japanese supply-chain management. Indeed, it is difficult to separate partnering from the principles of TQM, from where the emphasis on continuous improvement has been borrowed. Within the UK, the cause of partnering has been championed by a number of powerful clients who have become dissatisfied with the supposed under-performance of the construction industry. The large UK supermarkets have numbered amongst the most enthusiastic advocates of partnering. As regular clients of construction, they understandably wish to extend the control that they exert over the grocery supply-chain to the construction sector. It is by no coincidence that Bennett and Jayes (1998) include exemplar case studies of Sainsbury's and Asda. The Egan Report (DETR, 1998) also cites the case of Tesco, who have apparently:

> "...reduced the capital cost of their stores by 40% since 1991 and by 20% in the last two years, through partnering with a smaller supply base with whom they have established long term relationships. Tesco is now aiming for a further 20% reduction in costs in the next two years and a further reduction in project time".

If true, the benefits achieved by Tesco though partnering are indeed significant. Strangely, the Egan Report says nothing about the corresponding increase in profitability achieved by Tesco's partners. Other large clients who are strong advocates of partnering include BAA and Whitbread, both of whom were also members of the Egan Construction Task Force. Other notable top-twenty UK clients who claim to be committed to partnering include Rover Cars and John Lewis Partnership.

Buying power and the rhetoric of seduction

Given the collective buying power of the aforementioned clients, it is unsurprising to find that many leading contractors also claim to be committed to partnering. To do otherwise would be to risk attracting the label of 'adversarial', thereby denying themselves access to a significant part of the UK market. This exercise of buying power is made especially clear by the Construction Clients' Forum (1998), who collectively account for some 80% of the construction market. The CCF document commits its members to promoting relationships based on teamwork and trust, and to working jointly with their partners to reduce costs. They also promise not to unfairly exploit their buying power, but to look to form lasting relationships with the supply side. The overall tone is one of barely-disguised seduction. However, they then issue a unveiled threat to those who may still be unconvinced:

> *"The message from the Construction Client's Forum is clear. If this Pact is concluded, clients represented on the CCF will seek to place their £40bn of business with companies that are seen to follow the approach described in this document....."*

The message is indeed clear. The CCF is saying to the construction industry that in order to qualify for £40bn worth of work then their ideas must be accepted. Dissent will not be tolerated. An adherence to the language of partnering is an essential pre-requisite of doing business. This is made equally clear by the Construction Industry Board (1997):

> *"If it becomes clear that anyone at the workshop is unable to adopt the spirit of partnering, that person should be replaced in the team."*

It would therefore seem that lurking behind the rhetoric of seduction is an 'iron fist'. The same implied threat lies behind the 'Egan Report' (DETR, 1998). Little wonder that dissenters to partnering are so few and far between.

Living up to the rhetoric

The contrast between the rhetoric of seduction and the enforcing iron fist raises the question of whether large clients live up to their own rhetoric. The influence of the big supermarkets on the propagation of partnering in construction has already been noted, as has their understandable desire to exercise increased control over the construction supply chain. Given this influence, it is interesting to read *The Daily Telegraph's* report on 31st July 1998 that Sainsbury's, Asda, Tesco and Safeway, are all currently under investigation by the Office of Fair Trading (OFT) after complaints from farmers and growers that consumers were not benefiting from low farm prices. The investigation will apparently decide whether supermarkets are unfairly wielding monopoly power in the grocery trade. It will also investigate pricing policies for cleaning products, toiletries and household goods. Sainsbury's and others are very fond of preaching 'customer-responsiveness' to the construction industry. The suggestion that their innovations in supply-chain management are directed towards earning super-normal profits, rather than serving the interests of their customers, is therefore interesting. Collectively, the four above named supermarkets account for 47% of all grocery sales in the UK. According to the report in *The Sunday Times* on 23rd August 1998, Sainsbury's has increased its annual profits to £728m from £279m ten years ago. In the circumstances, it does not seem unreasonable to question their motives in seeking to introduce partnering into the construction supply chain.

Shaping the research agenda

Of further concern is the control that clients such as Sainsbury's exert over the UK's construction research agenda. Throughout the 1990s, a number of powerful industrialists have exercised an increasing influence over the allocation of public funds to construction research (Lansley, 1997). Such powerful industrialists act as spokespersons for commercial interest groups and therefore provide a further example of corporatism in operation. Their legitimacy to talk on behalf of the public good is apparently taken for granted. In the case of partnering, it is especially ironic that the industry task forces behind the two partnering reports published in conjunction with the Reading Construction Forum (Bennett and Jayes, 1995; Bennett and Jayes, 1998) were both chaired by Charles Johnston of Sainsbury's. Given the profits realised by Sainsbury's through partnering in the grocery supply chain, the overwhelmingly positive tone of these two reports is hardly surprising. Not only does the buying power of the supermarkets seemingly deter any critical comment from their supposed 'partners' in the supply chain, it also appears to enable them to influence research outputs. It is surely time that academics developed a construction management research agenda which is independent of commercial vested interests.

Success requires faith and commitment

A critical reading of the existing literature on partnering serves to reinforce the suspicion that it is primarily driven by the propaganda of corporatism. Of particular note is the way in which success is continually linked to 'faith' and 'commitment'. For example, Bennett and Jayes (1995) attach considerable importance to the commitment of top management. This is further reinforced by the Construction Industry Board (1997), who also see commitment to be an essential element of partnering:

> *"to succeed requires fundamental belief, faith and stamina. The commitment must start at the top and it must be shared by the senior management."*

The emphasis of the above quotation is interesting. The plea for 'fundamental belief, faith and stamina' echoes the language of previous manifestations of corporatism. The repeated use of the word 'must' is especially notable. The manager *must* have 'fundamental belief'. The inference is that the manager must not question. The manager must not think. 'True knowledge' is seemingly held by a small technocratic elite, and the rest of the construction industry is required to act on faith. Dissenters are marginalised as deviants. According to the Construction Industry Board (1997), 'cynicism and lack of commitment by the few will destroy the efforts of many'. Such a script could easily have been used to justify the Spanish Inquisition, or Mussolini's Italy, or any other rendition of corporatism. The overwhelming tone of the partnering literature is that of technocratic totalitarianism.

Partnering is a regime which is to be *imposed* on the construction industry. The 'partners' involved are ostensibly limited to commercial entities. There is no role for independent trade unions or for any individual 'deviants' whose objectives do not accord with those of 'top management'. Pluralist models of organisation are sacrificed in favour of crude corporatism.

The associated propaganda is provided by the 'management-speak' of partnering and customer-responsiveness. The ultimate argument rests with mystical appeals to the 'customer' and the imperatives of the global market. What is frightening is that this propaganda is also continually advocated by universities and business schools which are increasingly dependent upon the business sector for funding (Saul, 1997). It would seem that education and research have already been recruited to the cause of corporatism. The doctrines of customer-responsiveness and continuous improvement must seemingly be accepted on faith rather than on rational argument. It is almost as if the Enlightenment had never happened.

Customer responsiveness

It is only when the shackles of corporatist propaganda are removed that customer-responsiveness and continuous improvement become susceptible to critique. Of course, the rhetoric of customer-responsiveness is by no means unique to partnering. It currently pervades the whole range of management improvement techniques from TQM through to process re-engineering and 'lean thinking'. The following quote from the 'Egan Report' is by no means untypical:

> *"...in the best companies, the customer drives everything. These companies provide precisely what the end customer needs, when the customer needs it and at a price that reflects the product's value to the customer."* (DETR, 1998)

Given Tesco's representation on Egan's Construction Task Force, it is difficult to take this advice seriously. However, final judgement must be suspended until the results of the current OFT investigation become known. The results of the investigation will not of course be influenced by the fact that Tesco, Sainsbury's and Safeways are listed amongst the ruling Labour party's biggest financial donors. Also of note is the recent appointment of the unelected Lord Sainsbury as an industry minister. The trends towards an increasingly corporatist society could hardly be more conspicuous. It is further notable that Egan's BAA (formerly the British Airports Authority) operates as a privatised quasi-monopoly which is by no means subject to the same competitive pressures which face the majority of firms in the construction industry. Other participants in Egan's Task Force included Nissan UK, British Steel, Whitbread and the Housing Corporation. Perhaps the legitimacy of the Egan Report owes more to buying power than to rational argument.

The rhetoric of customer-responsiveness creates a cosy image of producers striving honestly to meet 'customer needs'. The assumption is that such needs exist 'out there' waiting to be identified. The possibility that needs are socially ascribed and negotiated by means of corporate marketing is not considered (Alvesson and Willmott, 1996). An extreme example is provided by the tobacco companies, who are arguably as justified as anyone in claiming to be customer-responsive. Leiss (1983) illustrates the way in which corporate advertising relies on the use of symbolism to encourage individuals to purchase particular goods. Many such purchases serve only to satisfy a customer's needs for status, identify confirmation and self-esteem, all of which are socially constructed by a complex process in which marketing is directly implicated. It should also be remembered that the origins of such techniques of mass persuasion lie within the propaganda machines developed in Germany and Italy during the 1930s and 1940s (Saul, 1997).

It is not necessary to take this last point completely seriously to accept that the doctrine of 'customer-responsiveness' owes more to propaganda than to a coherent management policy. It has long been recognised that organisations must satisfy a wide range of stakeholders if they are to be successful (Kast and Rosenzweig, 1985). Such stakeholders include not only customers, but also shareholders, employees, trade associations, unions, suppliers, and public interest groups. Unfortunately, corporatism does not recognise pluralistic models which require management to maintain a *balance* between potentially conflicting interests. It is also tempting to suggest that the stronger a company's propaganda of customer-responsiveness, the greater is the actual emphasis given to the interests of shareholders. Of course, the linkage between managerial salaries and the interests of shareholders is much more direct than that between managerial salaries and the interests of customers.

Further insights into customer-responsiveness can be gained from the burgeoning critical literature relating to the various 'quality' initiatives. The main argument in support of TQM is that it encourages employees to identify themselves as parts of a supply chain which comprises a sequence of relationships between suppliers and customers (Tuckman, 1995). Kerfoot and Knight (1995) suggest that this provides employees with a sense of self-esteem from serving the next person in the chain, rather than having to derive satisfaction from the task itself. Metaphors such as 'teamwork' and 'customer' are therefore intentionally used to mask the reality that most employees are required to act as mindless cogwheels in a remorseless machine. If this critical interpretation is accepted, it would seem that the rhetoric of customer responsiveness is primarily used as hollow propaganda to justify management regimes which are increasingly based on domination and control.

Continuous improvement

There is an obvious paradox in the ethos of continuous improvement as measured in terms of cost improvement. Organisations such as Tesco cannot continue to claim that partnering secures 20% savings every two years for too much longer. If continuous improvement was advocated in the sense of continuously examining how things are done in the context of an ever-changing environment, then it would make some sort of sense. It is the link to *measurable* cost improvement which renders the logic ultimately unsustainable. The Egan Report also places great emphasis on the need for objective measurement, whilst calling for a 10% annual cost reduction. There is no guidance on how such measurements are to be achieved in isolation of varying interest rates, price inflation and currency fluctuations.

Several authors have suggested a link between continuous improvement, or *kaizen*, and 'management by stress'. For example, Garrahan and Stewart (1992) observe that the regime of *kaizen* as implemented by Nissan UK creates the expectation of a continuous flow of ideas for improvement, thereby providing a source of additional stress for their employees. The associated ideas of 'teamworking' can also be interpreted as a means of self-policing through peer surveillance and control. In other words, employees are encouraged to identify defects caused by others and to ensure that the 'guilty' are identified (Legge, 1995). Peer pressure is therefore bought to bear to ensure that workers do not 'let down' fellow team members. According to Beale (1994), the Nissan system of continuous improvement is directly dependent upon the existence of a single union agreement which to all extent-and-purposes is a 'no-strike' deal. Such collaborative agreements between global corporations and single unions once again echo the tenets of corporatism. The ultimate power of course rests with Nissan UK, who can simply re-locate elsewhere if the work force refuses to conform. Beale (1994) further suggests that the contention that *kaizen* equates with management-by-stress is supported by the reported high labour turnover at Nissan's plant in Sunderland, despite very high levels of local unemployment. In contrast, the Egan report presents Nissan UK as a paragon of good practice.

A critical reading of the case studies provided by Bennett and Jayes (1998) provides ready support for the contention that continuous improvement equates to management-by-stress. Sainsbury's approach to partnering was initially linked to the downsizing of their property division from 240 to 80 staff. The remaining employees were ominously 'instilled' with the culture of TQM. Costs were apparently reduced by 35% and typical construction durations were reduced from 42 weeks to 15. However, these impressive achievements were not enough. Sainsbury's management have since established further 'tough and steadily improving cost, time and quality targets'. The regime of management-by-stress is even more apparent in the case study of Rover Cars, who seem especially proud of imposing a regime of 'relentless pressure' on their supposed partners.

Conclusion

This paper has presented a critical perspective on partnering in construction. A significant credibility gap has been demonstrated between the rhetoric of the major clients and the way in which they behave in practice. The arguments in favour of partnering would seem to owe more to the buying power of its advocates rather than to any independent appraisal. The established literature repeatedly exhorts the construction industry to accept partnering as an act of 'faith', whilst casting dissenters in the role of deviants. Construction firms are deterred from critical comment by the threat of being labelled 'adversarial', thereby denying themselves access to a significant proportion of the UK construction market. It has further been suggested that the seemingly sacred concept of 'customer responsiveness' owes more to propaganda than to a coherent management policy. Likewise, the seductive rhetoric of continuous improvement is too often used to camouflage reactionary management regimes which rely on control, surveillance and stress. Unfortunately, the more that managers' behaviour is governed by propaganda, the less likely they are to engage in risk-taking and entrepreneurial behaviour. Within corporatist regimes, individuals rise to join the technocratic elite on the basis of compliance, rather than deviance.

It would be a mistake to read this interpretation of partnering as some sort of conspiracy theory. Such an interpretation would credit the technocratic elite with too high a level of consciousness. The reality of the situation is more one of mindless compliance caused by intellectual laziness. Despite the alleged prolonged crisis in construction productivity, the industry's technocratic elite continue to award themselves very comfortable salaries and to recruit others in the same mould. In this respect, the parallel with previous corporatist regimes is once again unavoidable. The construction industry would likely be much improved if its managers were less accepting of the propaganda continuously propagated by corporatist bodies such as the Construction Industry Board. It is surely the role of academics to expose the dogma and propaganda which all too often prevails. Unfortunately, critical thought in defence of humanism is seemingly in terminal retreat as universities become just another interest group within the corporatist coalition.

References

Alvesson, M and Willmott, H (1996) *Making Sense of Management: A Critical Introduction*, Sage, London.

Beale, D (1994) *Driven by Nissan?: A Critical Guide to New Management Techniques,* Lawrence & Wishart, London

Bennett, J and Jayes, S (1995) *Trusting the Team: The Best Practice Guide to Partnering in Construction,* Centre for Strategic Studies in Construction, The University of Reading.

Bennett, J and Jayes, S (1998) *The Seven Pillars of Partnering*, Thomas Telford, London.

Construction Clients' Forum (1998) *Constructing Improvement,* Construction Clients' Forum, London.

Construction Industry Board (1997) *Partnering in the Team,* Thomas Telford, London.

Construction Industry Institute (1989) *Partnering: Meeting the Challenges of the Future,* CII, Texas.

DETR (1998) *Rethinking Construction*, Department of the Environment, Transport and the Regions, London.

Garrahan, P and Stewart, P (1992) *The Nissan Enigma: Flexibility at Work in a Local Economy*, Mansell, London.

Greider, W (1997) *One World, Ready or Not: The Manic Logic of Global Capitalism,* Penguin, London.

Heywood, A. (1992) *Political Ideologies*, MacMillan, Basingstoke.

Kast, F E and Rosenzweig, J E (1985) *Organization and Management: A Systems and Contingency Approach*, 4th Edn., McGraw-Hill, New York.

Kerfoot, D and Knight, D (1995) Empowering the 'quality worker'?: the seduction and contradiction of the total quality phenomenon, in *Making Quality Critical*, (eds. A. Wilkinson and H. Willmott) Routledge, London, pp. 219-239.

Lansley, P (1997) The impact of BRE's commercialisation on the research community, *Building Research and Information*, Vol. 25, No. 5, pp. 301-312.

Legge, K (1995) *Human Resource Management: Rhetorics and Realities*, MacMillan, Basingstoke.

Leiss, W (1983) The icons of the market place, *Theory, Culture & Society*, Vol. 1, No. 3, pp. 10-21.

McGeorge, D and Palmer, A (1997) *Construction Management: New Directions*, Blackwell Science, Oxford.

Saul, J R (1997) *The Unconscious Civilisation*, Penguin, London.

Tuckman, A (1995) Ideology, quality and TQM, in *Making Quality Critical*, (eds. A. Wilkinson and H. Willmott) Routledge, London, pp. 54-81.

Wood, E M (1995) *Democracy against Capitalism,* Cambridge University Press, Cambridge.

HARMONY AND PROFIT IN SMEs: THE POSSIBILITIES AND LIMITATIONS OF BUILDING PARTNERSHIPS

C L DAVEY, D J LOWE and A R DUFF
Department of Building Engineering, University of Manchester Institute of Science and Technology, P.O. Box 88, Manchester, M60 1QD, United Kingdom. David. Lowe@umist.ac.uk

Abstract

This paper aims to identify opportunities to assist small and medium sized construction companies work in partnership with clients and increase their effectiveness through the exploitation of communication procedures and technologies. A review of the literature highlighted the importance of partnering approaches. The results showed that small and medium construction companies who participated in the research were reluctant to work for main contractors, but welcomed opportunities to work with and form partnerships with blue chip companies and public sector clients. Construction companies were concerned that the continued use of competitive tendering to establish project partners would undermine the process by preventing them from contributing to the design stage and by rewarding firms who submit low bids, only to claw back profit later. They also objected to being denied opportunities to continue working with clients with whom they had established a strong relationship and understanding of the service required. A strategic partnership combined with involvement in an action learn set prompted one medium sized construction company to produce a detailed report about the management of maintenance defects. We plan to run workshops to promote good practice concerning the management of maintenance defects and the introduction of new technologies.

Keywords: Action research, construction, partnering, workshops.

Introduction

Increasingly, government and private clients are putting pressure on construction companies to produce higher standards of building, meet the needs of the social infrastructure and reduce costs (The Housing Corporation, 1997; West, 1997). Identifying and responding to these pressures requires construction companies to increase their sensitivity to the environment, offer innovative solutions to problems and develop collaborative styles of working (Construction Innovation Forum, 1997; The Housing Corporation, 1996; Weaver, 1997).

'Partnering' is a recognised method of improving communication mechanisms and technologies, responding to innovative construction projects, creating a less stressful working environment and reducing transaction costs resulting from uncertainty, competition and information asymmetry (European, Construction Institute, 1997; Loraine, 1994; Construction Industry Board, 1997). The approach can be used to achieve a range of client objectives including equality, training and employment for local people and services for tenants (Davey *et al*, 1998). The extension to relationships with subcontractors has helped large contractors achieve more compliant bids, less confrontation and lower tendering costs from their subcontractors (Mathews *et al.* 1996).

Research to identify and develop opportunities for partnering has mainly targeted large construction companies and clients involved in large-scale projects (e.g. redesign of bank branches for NatWest). Indeed, Sir John Egan (1998) urges large companies to ensure that they are at the forefront of changes to improve productivity. However, small and medium sized construction companies comprise the bulk of the construction industry and are well positioned to take advantage of new market opportunities arising from collaborative building programmes (Davey *et al*, 1998). The primary aim of our research project was to assist small and medium sized construction companies establish and work in partnership with public and private sector clients based in the North West of England and diversity into new business opportunities. The project also sought to encourage companies to increase competitiveness through more effective exploitation of communication procedures and technologies. The 8-month project called *'Building Partnerships'* was funded by UMIST, The Manchester Federal School of Business and Management and the European Regional Development Fund.

Literature review

Partnering is one of several strategies being proposed by practitioners, academics and managers (Cook and Hancher, 1990) and draws heavily upon lessons learned from Japanese manufacturing. It is defined by the Reading Construction Forum (1995) as:

"a management approach used by two or more organisations to achieve specific business objectives by maximising the effectiveness of both parties. The approach is based upon mutual objectives, an agreed method of problem resolution, and an active search for continuous measurable improvements".

In Europe, there are basically two types of partnering: project partnering, where the parties come together for the duration of the project; and strategic partnering where the parties develop a longer term relationship over a series of projects for which contracts are usually negotiated. The former is recommended for public sector clients who have to use market testing in order to comply with EC procurement regulations, usually through the competitive tendering process. Nevertheless, public sector organisations are allowed to use partnering criteria to select and award contracts (Loraine, 1994; European Construction Institute, 1997).

A partnering relationship is only recommended where the management teams of all parties involved display a fundamental commitment to partnering and where companies share a common culture (Smircich, 1985). The partnering process involves allocating time to agree objectives, establishing an open style of communication, developing a mechanism for problem resolution and identifying measures designed to monitor and help improve performance (CIB, 1997).

While partnerships are an effective method of helping construction companies strengthen links with clients, diversify into new projects and enhance competitiveness, they are potentially undermined by the construction industry's existing 'macho' and adversarial culture and its widespread use of short-term, legalistic approaches to procurement and contracting. Partnerships are also difficult to implement and maintain in a system characterised by indirect linkages between clients, contractors, subcontractors, consultants, suppliers, employees and end-users. As a result, industrialists and academics have found it necessary to adapt partnering methods to specific contexts and to build upon success factors, rather than relying entirely upon prescriptive models. They also warn that the benefits of partnering are not necessarily immediately apparent (Barlow *et al.*, 1997; Mathews *et al.*, 1996).

Methodology

An action research methodology was employed where researchers and participants work together to identify and define problems within the industry, develop solutions and bring about improvements through the implementation of good practice. Action research is used to instigate and learn from the process of change, rather than simply explain problems or provide theoretical insights (Easterby-Smith *et al.*, 1991). It is an educational process capable of changing the researchers, participants and the situation and requires the people being studied to be involved in thinking, planning, implementing and disseminating research as well as a willingness on the part of the researchers to learn and change. The process of gathering data, reflecting upon the findings and forming insights, integrating insights into theories and creating closure on discoveries in order to plan action is undertaken together (Reinhardtz, 1981).

We began the research process by eliciting assistance and participation from contacts gained during previous allied research designed to help senior managers from housing associations (Davey *et al*, 1998). Housing associations are charities (i.e. not for profit organisations) funded by grants/loans from central government. They are responsible for providing housing for people in social need, for rent or sometimes for sale, and for improving the wider social fabric through employment and the purchase of goods and services (Council of Mortgage Lenders, 1997).

So far, empirical data has been collected from in-depth semi-structured interviews with 8 managers from 5 construction companies and 5 managers from 4 public sector organisations. The interviews with construction companies covered the following topics: choice of clients and projects; methods of gaining business; successes and problems; future plans; methods of assessing performance; good practice; and relationships with subcontractors, consultants and suppliers; views on partnering; and details of partnering projects. The clients were asked to give similar information, but related to the business available to contractors and methods of procurement. The format was adapted to the specific interests and needs of the participants, and action taken following the interview to help participants develop solutions. Information was also gained from a seminar run by the housing associations, two large contractors, the Chamber of Commerce, the university and a network for women property professionals.

The comments from the interviews and seminars were classified into categories. The categories were initially similar to the topics covered during the interviews, but were adapted to fit participants' comments, the literature and our analyses of the data. Although informed by our personal experiences and knowledge (Marshall, 1981), our insights into the research process and outcomes were discussed and developed with members of an action learning set comprising 3 academics, 6 managers from housing associations and 2 from a medium sized construction company. The set resulted from collaboration between UMIST, the University of Salford and the Revans Centre for Action Learning. It provided regular information, feedback and practical support.

Results of the research

The research revealed that clients and contractors were interested in more cost-effective procurement, improved design and contractual arrangements, achieving higher standards of quality on projects and getting involved in 'added value' projects. It also found commitment to the principles of partnering or forming partnerships amongst some of the participants. The findings from workshops and further research to help strengthen links between clients and improve competitiveness amongst SMEs is detailed below:

SMEs welcomed opportunities to work in partnership with public sector clients and blue chip companies, but were reluctant to work as subcontractors.

A construction manager survey showed that large contractors were obtaining 10% to 70% of turnover/contracts from partnerships and the majority forecasted increases in revenue from partnering relationships (Walter, 1998). However, some small and medium sized contractors expressed reluctance to partner with main contractors due, in part, to the fact that sub-contractors are unable to increase their profit margins by negotiating favourable rates from suppliers, but mainly due to fear of litigation and non-payment. A marketing executive said that he would not want to work for a main contractor because the company's last contract with a large construction company looked likely to result in litigation. Even a manager from a large construction company acknowledged that large firms delayed payment to increase profit, with the unscrupulous failing to pay within the time agreed in the contract or not paying at all:

"Cash management is the way that [large contractors] make money. SMEs get hammered. Certain contractors are unscrupulous because they delay or don't pay at all" (business development manager, large construction company).

While reluctant to work as a subcontractor for a main contractor, contractors were keen to collaborate with blue chip companies. Blue chip companies were perceived to offer large contracts, reliable payment and high rates of quality work completed within a short period of time, as well as professional conduct and experience at partnering relationships. An alliance with a blue chip company also appeared to enhance a contractor's standing amongst its clients. A development director from a housing association believed such an alliance demonstrated the company's professionalism and trustworthiness.

Although profit margins offered by public sector clients were relatively low, the contractors welcomed opportunities to collaborate on projects run by public sector clients such as housing associations and local authorities. These clients provided a steady source of income, which maintained turnover, along with reliable payment. Indeed, a housing association had introduced a payment scheme that guaranteed payment within one week of receiving the invoice in order to attract contractors and obtain value for money. One marketing manager highlights the benefits of work offered by educational authorities:

"The outcry over education has led to money in building schools. Money is spread fairly thinly amongst local authorities. Some of the work is of interest. It's below average in value, but local [to the company's offices]. The work keeps staff and teams together and keeps management on site" (marketing manager, SME).

Although the housing associations were a significant client group in the North West spending over one billion pound in the last five years (Davey *et al*, 1997; Lunney 1996), a marketing manager of a medium sized construction company pointed out that companies bid for projects within their capabilities in terms of size and often specialised in certain types of work (e.g. New build, refurbishment or maintenance) or contracts (e.g. Design and Build or standard contracts). He therefore found it difficult to determine funds available for refurbishment work from capital spending figures for public sector clients. In his view, the clients were unwilling or unable to share information with contractors, perhaps due to lack of knowledge.

SMEs wanted to get onto public sector clients' approved lists and be invited to tender.

While construction companies wanted to work for public sector clients, they sometimes found it difficult to take advantage of specific opportunities. The public sector clients generally restricted opportunities to tender to contractors from their approved lists who complied with their specific criteria (Commission for Racial Equality, 1995; Nicholson, 1998). The construction companies were able to gain membership of approved lists by either applying directly to client to an independent body responsible for assessing applications. Several managers from housing association and universities admitted that they were reluctant to consider applications from new contractors, however, because they were happy with existing companies and the process of evaluation for initial applications and annual review was time consuming and costly. Nevertheless, managers were prepared to consider applications from companies who presented a professional image and/or offered something special.

Housing associations and university clients said they offered opportunities to tender to contractors with whom they had established good working relationships, developed an understanding of the standards required and were likely to get value for money. A development officer from a housing association said that he considered contractors who had successfully completed construction projects in the past for the company, but that only a limited number of construction companies were eligible for contracts involving the provision of training and employment opportunities for local people. A university maintenance manager offered opportunities to contractors approved by the manufacturers to use their materials. Other contractors had simply stopped undertaking work for the university. He believed that this meant that the contractors working for the university all understood the quality required and could therefore price accordingly.

The contractors attempted to increase their chances of being invited to tender by ensuring that they performed well on current projects and through the use of marketing activities. A marketing manager of a medium sized company said he placed advertisements in trade magazines, though he complained that this was costly and the frequent calls from trade magazines were disruptive. The majority attempted to establish personal links with clients. One large company had appointed a business or marketing manager with budgets to cover expenses, but two small companies relied upon directors to gain business and one said he found it difficult to commit time to establishing and maintaining new contacts.

SMEs welcomed opportunities to partner with public sector clients and one firm wanted assistance with a application for project partnerships.

The construction company managers were generally positive about the prospect of working in partnership with public clients. A contracts manager said that partnering improved the quality of the relationship and enabled his firm to meet the needs of the clients in terms of the quality of the product and budget, whilst still making a profit.

A marketing manager from a medium sized company thought that partnering would help the contractor and client talk through and resolve problems together, without resorting to litigation. Nevertheless, the fact that public sector clients can use partnering criteria to select potential partners and award contracts (Loraine, 1994; European Construction Institute, 1997) required new knowledge and skills from construction companies and some managers welcomed assistance in acquiring them. For example, a marketing manager of a medium sized company had received a letter from the a local authority informing contractors of their intention to enter into a partnership arrangement. The letter included the following:

"This initiative is intended to enable a co-operative style of management for the execution of the works, whereby all the parties to the contract can work together without affecting the contractual requirements and obligations... it will include establishing a forum at the post tender stage for identifying possible cost savings and solving areas of potential difficulty or conflict before they impinge on the programme and/or cost of executing the works" (technical services director, local authority).

The company had not yet formally established a partnering relationship, and therefore wanted help in responding to the selection procedure of the local authority. He pointed out in a letter to us that the company's "experience of the partnering process is very much on a learning curve" and that a "contribution towards understanding the procedures is much appreciated" (marketing manager, SME).

Alternatively, construction companies could identify land and property themselves which could be presented to clients as opportunities for collaboration. Although speculative work yielded higher profit margins and enabled companies to increase their control over the building process by employing more staff on permanent contracts, the risks and the difficulties were considered too great for one medium sized company:

"If a Local Authority has land, but no money our organisation could come in and develop the land. It is very complicated for tax, rebates and grants. Often need something special – commercial input. For example, shops or housing for sale, rather than just social renting I can come up with a site and take it to a housing association. I lose money by chasing sites" (marketing manager, SME).

SME felt that the process of competitive tendering used for project partnering undermined the potential benefits of the approach.

Although SMEs were interested in partnering and often positive about the potential benefits, they were concerned that the process of competitive tendering undermined the potential benefits. Construction managers complained about clients failing to comply with codes of practice governing the length of time allowed to prepare and submit tender documentation. A director of a medium sized company pointed out that they had traditionally been given four weeks to price design and build contracts, but three weeks had apparently become the norm and around 15% of clients were asking contractors to return tender documents within two weeks. In one instance, he telephoned to ask whether the deadline had been extended and was told that he was the only one to complain, but later discovered from other contractors that they had also contacted the client. In one instance, despite allowing only two weeks to submit a tender, the client apparently delayed opening and assessing the bids for several months, by which time the figures were out of date.

The managers pointed out that awarding contracts to the company who offered the lowest price encouraged firms to submit a low bid, but to then claw back profit by claiming for items not specified in the contract or specifying overpriced materials. An interviewee explains how the process works:

"If I give a low bid. I then have to get it back. I say I want to know the colour of the curtains required. They go on holiday. I charge extra. I then offer gold braid curtains" (construction company manager).

The process of clawing back money was perceived to increase the likelihood of litigation and break down trust. The lack of trust was symbolised by the client's appointment of a quantity survey to oversee the project and ensure budgetary control:

"Everyone is suspicious of everyone's motives. The client appoints a surveyor because he does not want to get ripped off. He has to protect his price margins" (contracts manager, SME).

A business development manager pointed out that in project partnerships the design and price for the building have already been fixed which, in turn, reduces opportunities for innovation and cost saving. He would like the opportunity to negotiate contracts during the early stage of the design process:

"We want the opportunity to negotiate on what they want to achieve. We don't want to be given a fixed project with a given price. If they bring this to the table, it's too late. Where is the innovation? Partnering means discussing the site layout, subcontractors and suppliers" (business development manager, large contractor).

The majority of contractors wanted to continue working with clients with whom they had developed a strong relationship, where they understood the service required and had already overcome initial problems. They often objected to procurement mechanisms that broke up relationships. For example, a contracts manager complained about a random systems of selecting contractors to tender used by a local authority, even though it guaranteed work and had previously enabled him to diversify into building schools:

"It's a lottery [the tendering process]. It's perceived as fair, but isn't. The local authority has 20 contractors and randomly selects 6. The client wanted us, but couldn't have us because we weren't selected. What's the incentive to be good? At least we are still on the list! Partnering then goes" (contracts manager, SME).

The construction companies welcomed opportunities offered by public sector clients to circumvent the competitive tendering process. A contracts manager from a medium sized firm extended his contract with a local authority by, for example, remaining on site to complete further work and thus providing better value for money:

"The number of return clients we have is substantial, but we still have to tender. We built a [public building] for the --- ---- Authority. We had already shown our ability. We said that if we could do the two together, we could do it cheaper" (contracts manager, SME).

Nevertheless, a group of managers from the housing associations informed us that they were under pressure from their regulatory body to ensure work was evenly spread amongst contractors and to minimise risk by avoiding over reliance on a small number of contractors. They also felt under pressure to use competitive tendering, rather than negotiated contracts.

SMEs involved in strategic partnerships were under pressure to demonstrate good performance.

Despite being discourage from relying upon a small number of contractors and negotiating contracts by regulatory or governing bodies, three public sector organisation had established a minimum number of three contractors with whom they do business and/or regularly negotiated contracts. Although the housing associations had formally selected partners, the university had simply relied upon a select list of companies who met its requirements. Choosing only a small number of partners enabled a housing association with a relatively small construction budget to offer a significant amount of business to those construction companies and thus increase their likelihood of being considered a valued client. A development director of the housing association said that choosing mainly small and medium-sized contractors enabled him to deal directly with the managing director, guarantee the availability and commitment to their projects and influence their organisational strategies and practices. The use of preferred contractors was also intended to reduce costs, maintain strong relationships and improve the quality of service provided, especially during the post-construction phase of the project.

A contractor who was working in partnership with a client had joined a subcommittee designed to improve quality had been prompted by his involvement in the action learning set to produce a comprehensive report about the management of maintenance defects. The report outlined the problems and potential solutions. It also provided examples of reporting forms, schedules, meeting agenda and quality assurance information (McDonald, 1998).

The allocation of work to partners meant that other construction companies appeared unwilling, however, to bring new business opportunities to the clients. One client was considering paying a fee to companies who identified new opportunities. The difficulties were compounded by pressure from the regulatory body to demonstrate the effectiveness and fairness of partnering relationships compared to standard relationships with contractors who have been selected through competitive tendering. Construction companies were also concerned about the consequences of strategic partnerships both for their firms and the industry. A director of a construction company that prided itself on the quality of its workmanship, but which had not been considered for partnering, was unhappy about not being selected as a partner. Several construction managers were also concerned that by reducing competition partnering would prevent new companies from entering closed markets.

Conclusion

The construction companies who participated in the project welcomed opportunities to meet public sector clients and to address some of the problems within the industry. We plan to run a series of workshops designed to facilitate the process of change both through their format and content. The workshops will involve representatives from public sector clients, construction companies and academia. The workshops will enable the researchers to introduction the research project, present detailed information about the problems encountered, recommended solutions and discuss the role of the client in promoting good practice. Small discussion groups will be used to determine the applicability of the recommendations within the delegates' own organisation and to gain more detailed information. The discussion groups' findings will be presented to the entire group during a plenary session.

The subject of the first workshop will be the management of maintenance defects. Initial recommendations include: the joint inspection of the property by the client, main contractor and maintenance manager prior to its hand-over; setting the date for handover at the beginning of a contract; and the use of a standard form to record information from tenants about problems encountered, the form would then form the basis of a formal maintenance/defect tracking system. It is anticipated that the form should also enable maintenance managers to record instances where problems have arisen not from defects, but from other factors such as tenant damage, tenant lack of knowledge or lack of routine maintenance. The information should be recorded and used for monitoring the performance of all parties. The second workshop on communication procedures and technologies will involve demonstrations of video conferencing equipment, maintenance management software and digital cameras.

References

Barlow, J., Jashapara, A. and Cohen, M. (1997) Organisational Learning and Inter-firm 'Partnering' in the UK Construction Industry. British Academy of Management Conference, London, September 8-10
CIB (1997) *Partnering in the Team.* A report by the Construction Industry Board Working Group 12, London: Thomas Telford Publishing.

Construction Innovation Forum (1997) 8 October. BRE Garston.

Company report (1998) *The Company's approach to partnering*. Personal contact with consultant.

Commission for Racial Equality (1995) *Racial Equality and Council Contractors*. Claxton House Press.

Cook. L., and Hancher, D. E., (1990) Partnering Contracting for the Future. *Journal of Management in Engineering*, Vol. 6, No.4, October, pp431-447.

Council of Mortgage Lenders Research (1997) *Housing and the Economy*. Council of Mortgage Lenders.

Davey, C., Davidson, M., Gale, A., Hopley, A. and Rhys Jones, S. (1998*) Building Equality in Construction: Good practice guidelines*. Partners in Technology project report, UMIST, Manchester

Easterby-Smith, M., Thorpe, R. and Lowe, A. (1991) *Management Research*. Sage; Oxford. UK.

Egan, Sir John (1998) *Rethinking Construction: The report of the Construction Task Force*. The Stationary Office, London

European Construction Institute (1997*) Partnering in the Public Sector: a toolkit for the implementation of post award*. Project specific partnering on construction projects. ECI, Loughborough University.

Loraine. R.K. (1994) Project Specific Partnering. *Engineering, Construction and Architectural Management*, Vol. 1, issue 1, pp5-6.

Lunney, J (1997) *Housing Associations in the lead for change*. Building Equality in Construction: the Business Case. Seminar organised by UMIST, Building Positive Action and Rhys Jones Consultants, UMIST Conference Centre. 23 April.

McDonald, J (1998) *Action Learning for Managers: Maintenance Defects Period* Report for Action Learning Set, Rowlinson Construction Limited

Marshall, J. (1981) *Making sense as a personal process* Chapter 34, pp395-399, in Human Enquiry. Reason, P. and Rowen, J., (Eds.) John Wiley and Sons Ltd, Chichester, UK.

Matthews, J., Tyler, A. and Thorpe, A. (1996) Pre-construction project partnering: developing the *process Engineering, Construction and Architectural Management*, Vol. 3 Nos. 1/2 pp 117-131

Nicholson, N. (1998) *Welcome to Building* Positive Action Hernric Services

Reading Construction Forum (1995*) Trusting the Team., the best practice guide to partnering in construction*, Centre for Strategic Studies in Construction, Reading University.

Reinhardtz, S. (1981) *Implementing New Paradigm Research: A Model for Training and Practice*, Chapter 36, pp415-434, in Human Enquiry. Reason, P. and Rowen, J., (Eds.) John Wiley and Sons Ltd, Chichester, UK.

Smircich, L. (1985) *Is the concept of culture a paradigm for understanding organisations and ourselves*. in Frost, P., et al. Organisation Culture. Sage.

The Housing Corporation (1996) *Innovation and Good Practice Grants available from the Housing Corporation*. The Housing Corporation.

The Housing Corporation (1997) *A Housing Plus Approach to Sustainable Communities*. The Housing Corporation.

Walter, M. (1998) The essential accessory. *Construction Management*. Vol.16, February, pp.1-2.

Weaver, M. (1997) New Chairman, New Ideas. *Housing Today*. 30 October.

West. T . (1997) Quality the aim of £10m project. *Housing Today*. 16 October.

OPERATIONAL RISKS ASSOCIATED WITH PARTNERING FOR CONSTRUCTION

AKINTOLA AKINTOYE
Department of Building and Surveying, Glasgow Caledonian University, Glasgow, G64 3JG, United Kingdom. akin@gcal.ac.uk

CAROLYNN BLACK
Procurement Section, West of Scotland Water, 419 Balmore Road, Glasgow, G22 6NU, United Kingdom

Abstract

Organisations which have used partnering for construction projects are now reporting favourable results, including decreased costs, quality improvement and delivery of project to programme. This paper presents trends in the usage of partnering and risks associated with partnering based on a questionnaire survey of construction clients, contractors and consultants. The paper reviews risks faced by the construction industry from the use of the partnering procurement method. The study reveals an increase in the use of partnering since the Latham report was published. Also, there has been a shift in its use from the post contract stage of construction to the design stage of project development. Since most projects' costs are committed at the design stage, it is reckoned that this shift would assist the partners to make a significant contribution to the achievement of project objectives. Factor analysis of the risks involved in partnering shows that these are dominated by operational risks, and it is reckoned that these could constitute a barrier to the successful adoption of this procurement method in construction. Cost risks are not considered important in the use of partnering procurement method.

Keywords: Conflict, clients, consultants, contractors, factor analysis, partnering, procurement,

Introduction

Partnering is increasingly being used for construction projects since the Latham report "Constructing the Team" was published which recommended partnering as a means of improving inter-firm relations [Latham (1994)]. Partnering involves the parties to a construction project working together in an environment of trust and openness to realise the project efficiently and without conflict. This requires a major change to the construction industry culture. NEDO (1991) describes partnering as "a long term commitment between two or more organisations for the purpose of achieving specific business objectives by maximising the effectiveness of each participant's resources". Baker (1990) has identified conditions conducive to partnering from both the owner's and contractor's perspective. The conditions include the fact that the owner must be willing to change, possesses the ability to transfer some of its responsibilities to the partnership, has a commitment from its executive ranks to the partnering, desires to focus on overall results rather than strictly a singular component, etc.

Organisations which have used partnering for construction projects are now reporting favourable results, including decreased costs, quality improvement and delivery of project to programme. Many claims have been made for the benefits of partnering, in terms of project cost, time, quality, buildability etc. (e.g. Bennett et al, 1996). For example, US research claimed, on average, reduced overall projects costs of 5%, client's costs of 10% and time taken to completion of 6% on partnering projects [McLellan (1995)]. Despite these benefits, there remain to be investigated the risks associated with this mode of construction procurement. This paper presents trends in the use of this procurement route and the risks arising from the use of partnering for construction. The investigation adopts a questionnaire survey to gauge the views of construction clients, contractors and consultants on partnering trends for construction and the risk factors involved.

Overview of risk involved in construction

Procurement now represents a significant risk, possibly the most significant, faced by most organisations [Griffiths (1992)]. Risk is a function of the interaction of uncertainty and the magnitude of the potential loss or gain. Construction work involves considerable risk due to the complex nature and uncertainties inherent in the construction process [Al-Bahar and Crandall (1990)]. Consequently, the construction industry suffers from several factors, which act as barriers to the introduction of a partnering approach to procurement. For example, specifications tend to be imprecise and are firmed up too late in the process; parties often make promises they do not expect to honour and there is a constant focus on cost and programme to the detriment of quality and the client's requirements; construction contracts usually have built in penalties in the form of liquidated damages without providing incentives for contractors, i.e. there is a focus on the punishment of undesirable behaviour rather than the reward of exemplary behaviour etc. [Smit (1995)]. The result of risk exposure, for example from poorly drafted contracts, poor project planning or contractor management, is inflated contract costs, poor workmanship, inferior materials, contractor bankruptcy and programme delays. Risk is also increased simply by the hiring of a contractor to undertake work due to the inevitable loss of employee loyalty and loss of control over sub-contractor activities [Bova (1995)].

Traditionally, most UK construction clients package their construction requirements into one-off projects and use competitive tendering to determine the award of the contract. This results in a short-term and reactive procurement strategy, which requires contractors to respond to fragmented demand [Cox and Townsend (1996)]. Considering all these, it is expected that full benefits from partnering will take time to develop in an industry dominated by a focus on the short-term [Matthews *et al.* (1996)].

Lamming (1993) is of the opinion that the intensity of the partnership relationship and the central philosophy of commitment can lead to a high level of pressure to perform whereby partners under pressure may be encouraged to take unnecessary risks to prove their worth. Saunders (1994) and Ramsay (1996) recognise that the formation of a partnership with a supplier involves considerable risk. They assert that the risk of the transfer of power from buyer to the supplier is significant in a single source relationship. They argued further, that very large buyers will be in a position to overcome this risk by being able to dedicate resources to developing new sources of supply in the event of the original supplier flexing its new found power, while smaller companies will not be in a position to insure themselves against this type of risk. With respect to construction, Baxendale and Greaves (1997) believe that construction firms entering partnering with sub-contractors may limit competition resulting in the remaining firms forming cartels.

Methodology

A postal questionnaire was considered appropriate for the investigation as the total population of organisations involved in construction projects, i.e. consultants, contractors and clients, is extremely large. Random sampling was used to reduce the data to manageable proportions. Following a pilot study which involved 5 people, questionnaires were sent to 290 organisations (comprising consultants, contractors and clients) involved in construction work under cover of a letter explaining the aims of the research.

The survey comprised of closed and open-ended questions. The questionnaire comprises six main sections. Section one covers general information about the respondents. Section two deals with partnering trends and section three with the outcome from the use of partnering. Section four covers the reasons for using partnering and the benefits which result and section five the risks associated with partnering in particular and the construction industry in general. Finally, section six allows respondents to make general comments on the subject matter.

The questionnaire was designed to allow comparisons to be drawn between the organisational categories involved in construction work (clients, consultants and contractors) and to compare the opinions of organisations which have experienced partnering with those with no experience of partnering. This paper reports only the sections of the questionnaire dealing with trends and risks associated with partnering for construction.

Questionnaire response

Responses were received to the questionnaire as follows: 25 consultants responded representing 25.0% response rate, 32 contractors representing 32.0% of contractors approached and 21 clients responded which is 26.7% of the those to which a questionnaire was sent. Overall, response was 78 out of 290 representing 26.7% response rate. This response rate is not unusual for a construction industry survey; for example, Vidogah and Ndekugri (1998) received a 27% response rate to their survey questionnaire and Shash (1993) 28.3%.

Characteristics of responding firms

Most of the contractors (90.6%) that responded to the questionnaire have been involved in partnering for construction work. Less than half of the consultants (48%) and clients (47.6%) have been involved in partnering. Overall, 51 companies out of the 78 that responded to the questionnaire have been involved in construction partnering.

Trend in the use of partnering

The questionnaire examined the trend towards partnering for construction; in particular, when organisations were first involved, which party introduced partnering and the stage of the construction process where partnering is introduced.

The majority of the contractors became involved in partnering in 1991 followed by consultants in 1993. The year of first involvement for clients is 1995. This may indicate that contractors have been involved in partnering arrangements with sub-contractors or that the clients who responded to the questionnaire were not representative of the client population. This later entry into partnering by clients is also reflected in the level of involvement as they have the least experience of partnering compared to consultants and contractors. Although there is significant difference in the timing of first involvement of contractors, clients and consultants in partnering (p = 0.09), there is a general trend of increasing use of partnering which agrees with the view of Chadwick and Rajagopal (1995). The year of first involvement in partnering by the responding groups is shown in Table 1.

Table 1 Year of first involvement in partnering

Year	Total	Consultants	Contractors	Clients
> 1985	7	2	5	0
1986/87	0	0	0	0
1988/89	3	1	2	0
1990/91	4	0	3	1
1992/93	4	1	3	0
1994/95	14	3	9	2
1996/97	21	6	6	9
Total	53	13	28	12

(F Stat = 2.53; p = 0.09)

Table 2 shows that sixty percent of responding firms have recommended the use of partnering for further construction projects since first being involved in the use of the procurement method. On individual organisational categories, the contractors are the most enthusiastic with 84.4%, followed by consultants at 44% and clients at 42.3%. This may reflect that contractors have felt the most benefit from partnering considering that contractors may have suffered most from the conflict which often occurs in traditionally procured projects. These results contradict an American study which found that all 71 construction clients, contractors and consultants which had been involved in at least one partnering project claimed they would be happy to continue working in that way (McLellan, 1995).

Figures 1 to 3 show the trends in the use of partnering procurement method by consultants, contractors and clients, respectively, at the various stages of the construction process. These figures show that most firms were not involved in the use of partnering to a great extent at any stages of the construction process in the 1980s, although this procurement method has long been in use in the petro-chemical sector compared with the building sector. The 1990s are associated with the increasing popularity in the use of the procurement method, which coincides, with publication of the Latham report. Latham (1994) advocates the use of partnering for construction project developments and the government's endorsement of the procurement route in the public sector. Although, the procurement method has been used predominantly by the private sector, it is expected that this will be adopted with time for public sector projects, as far as some conditions are met, namely: it does not create an uncompetitive environment, does not create a monopoly, the partnering arrangement is tested competitively, it is established on clearly defined needs and objectives over a specified period of time and the construction firm does not become over-dependent on the partnering arrangement [McLellan (1995)].

Table 2 Number of firms that have recommended partnering since first used the method

Organisational Category	Number of Firms	% of Firms
Consultants	11	44.0
Contractors	27	84.4
Clients	9	42.3
Total	47	60.3

(F Stat = 0.79; P = 0.46)

The figures also show that while the consultants tend to get involved in partnering at pre-design stage, the most common stages at which contractors are involved are post-design and post-competitive tendering. It would appear that clients prefer to introduce partnering at the post contract award stage; which tends to suggest that clients prefer to introduce partnering as part of the negotiation process following the traditional competitive tendering. A number of clients have used partnering at the pre-design stage, however few introduce partnering at the post-design stage.

Figure 4 presents a summary of the trends in the use of partnering at various stages of construction process. This shows that while the construction industry has been involved in 'partnering' at the post contract and post competitive tendering phases of construction since 1980s, project partnering from pre-design stage is a recent development in the industry.

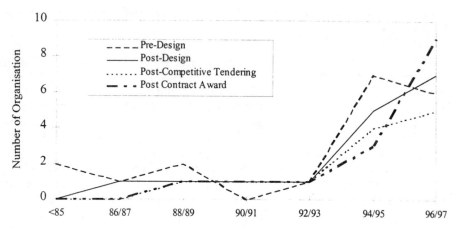

Figure 1 Trends in involvement of consultants in partnering

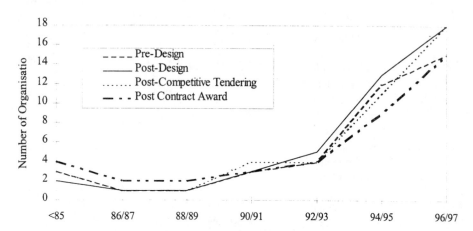

Figure 2 Trends in involvement of contractors in partnering

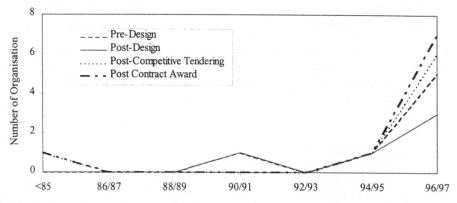

Figure 3 Trends in involvement of clients in partnering

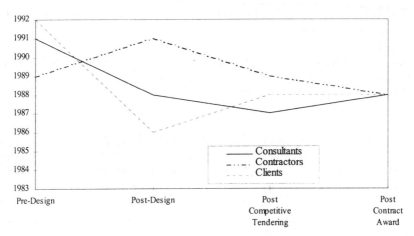

Figure 4 Summary information on the trends in the use of partnering

Construction partnering risks

Harback *et al.* (1994) have identified five pitfalls of partnering: unfulfilled expectations, unfinished business in which some elements or process of the partnering are still in dispute, assumption that all parties involved in the partnering are willing to share personal beliefs and thoughts, one-size-fits-all approach rather than seeing partnering as being specific to a project and conflict between internal (relationship between various departments of the company) and external (relationship with other parties) partnering. There are however some features of the construction industry which make the introduction of partnering more difficult than in other industries.

These features of the construction industry are taken further by providing the respondents with a list of probable risks associated with partnering, which were identified from a literature review, for them to assess the level important to their establishments. The respondents were asked to rate the level of importance on a Likert scale of 1 to 5; 5 denotes most important and 1 denotes least important. As part of the subsequent analysis of the data, the Cronbach alpha reliability is produced. Cronbach alpha reliability (the scale of coefficient) measures or tests the reliability of the five-point Likert-type scale used for study [Norusis/SPSS (1992)]. The Cronbach's coefficient alpha is 0.84 (F statistics = 7.512, $p = 0.000$), indicating that the 5-point Likert scale used for measuring the risk factors is reliable at 5% significant level. Table 3 presents the importance ranking of the risk factors by the contractors, clients and consultants and all the respondents combined.

The five most important operation risk factors identified by the respondents were managers' unwillingness to relinquish control, partners become complacent, increasing dependence on partner, pressure to perform and partner reverts to adversarial relationship. With the exception of the risks of partner reverting to an adversarial approach and limited competition leading to cartels, the opinions of the three parties on the risks associated with partnering were not statistically significant. The clients recognised that partnering may limit competition, which could lead to cartels.

The consistently least important risk factors overall and within the groups are those associated with cost, which tends to suggest that the parties do not see cost as important risk factor in partnering. These opinions on costs associated with partnering contradict the claims made by MacBeth and Ferguson (1994) and Ramsay (1996).

To capture the multivariate relationship between the risk factors, the Factor Analysis technique was used to investigate the cluster of relationship. Various tests are required for the appropriateness of the factor analysis for the factor extraction including Kaiser-Meyer-Olkin (KMO) measure of sampling accuracy, anti-image correlation, measure of sampling activities (MSA) and Barlett Test of Sphericity.

Table 3 Risks associated with partnering

Risk Factors	Code	Total	Consults	Ctrs	Clients	F	Sig.
Managers unwilling to relinquish control	R4	3.58	3.23	3.71	3.81	2.24	0.11
Partners become complacent	R1	3.55	3.23	3.68	3.76	1.99	0.14
Increasing dependence on partner	R2	3.47	3.50	3.29	3.71	0.83	0.44
Pressure to perform	R9	3.33	3.31	3.26	3.48	0.23	0.79
Partner reverts to adversarial approach	R10	3.27	3.08	3.06	3.81	2.74	0.07
Limited competition leading to cartels	R3	3.15	2.96	3.00	3.62	2.79	0.07
Suppliers add contingency to tenders	R11	3.06	2.88	3.13	3.19	0.48	0.62
Downsizing results in inadequate resources	R5	3.01	3.11	2.90	3.05	0.32	0.73
Cost of new procedures/systems	R8	3.00	3.15	2.74	3.19	1.73	0.18
Costs of understanding partner	R6	2.81	3.15	2.61	2.67	2.17	0.12
Cost of increased communication	R7	2.78	2.92	2.61	2.86	0.73	0.48

The eleven risk factors were subject to factor analysis, with principal component analysis and varimax rotation. The first stage of the analysis is to determine the strength of the relationship among the variables based on either correlation coefficient or partial correlation coefficients of the variables. Table 4 shows the partial correlation coefficient (same as the matrix of anti-image correlation) between the variables. The results of the partial correlation matrix show that the variables share common factors, as the partial correlation coefficients between pairs of the variables are small when the effect of the other variables are eliminated. According to Norusis/SPSS (1992) the partial correlations should be close to zero when factor analysis assumptions are met and that if the proportion of large coefficients are high, the use of a factor model should be reconsidered. The table also displays the Measures of Sampling Adequacy (MSA) on the diagonal of the matrix. The value of MSA must be reasonably high for a good factor analysis. In this case, the value of MSA ranged between 0.749 - 0.855 suggesting no need to eliminate any variable from the analysis.

Table 4 Anti-image correlation Matrix (the MSA is shown on the diagonal)

	R1	R2	R3	R4	R5	R6	R7	R8	R9	R10	R11
R1	0.794										
R2	0.140	0.850									
R3	-0.238	0.228	0.820								
R4	-0.338	0.171	0.111	0.809							
R5	-0.057	-0.090	-0.090	0.147	0.795						
R6	0.026	0.022	0.065	-0.061	0.252	0.767					
R7	-0.172	0.056	-0.194	0.119	0.024	-0.376	0.749				
R8	-0.004	-0.220	0.113	-0.038	0.030	0.050	-0.498	0.795			
R9	-0.219	-0.185	-0.150	0.050	0.199	0.038	0.057	-0.187	0.789		
R10	0.103	-0.186	-0.329	0.246	0.012	-0.127	0.034	0.037	0.020	0.855	
R11	-0.253	-0.170	0.000	-0.111	0.025	0.016	0.019	0.095	0.000	0.124	0.910

Barlett's test of spericity tests the hypothesis that the correlation matrix is an identity matrix. In this case the value of the test statistic for spericity is large (Barlett Test of Sphericity = 260.591) and the associated significant level is small (p=0.000), suggesting that the population correlation matrix is an identity. Observation of the correlation matrix of the risk factors shows that they all have significant correlation at 5% level suggesting no need to eliminate any of the variables for the principal component analysis. The value of the KMO statistic is 0.8141, which according to Kaiser (1974) is satisfactory for factor analysis. In essence, these tests show that factor analysis is appropriate for the factor extraction.

Principal component analysis was undertaken, which produced a three-factor solution with eigenvalues greater than 1, which explains 60.8 percent of the variance. Varimax orthogonal rotation of principal component analysis is then used to interpret these factors. An unrotated principal component analysis factor matrix only indicates the relationship between individual factors and the variables, and it is sometimes difficult to interpret the pattern. Rotation techniques, such as varimax method, transform the factor matrix produced from an unrotated principal component matrix into one that is easier to interpret. The factor loading based on varimax rotation is shown in Table 5. Each of the variables loads heavily on to only one of the factors, and the loadings on each factor exceed 0.5. The factors and associated variables are readily interpretable as operational risk in partnering (FACTOR 1), cost risk (FACTOR 2) and attitude risk (FACTOR 3)

Table 5 Varimax rotated matrix for partnering risk factors

Risk Factors	Code	FACTOR 1	FACTOR 2	FACTOR 3
Managers unwilling to relinquish control	R4	0.738		
Increasing dependence on partner	R2	0.707		
Partner reverts to adversarial approach	R10	0.682		
Suppliers add contingency to tenders	R11	0.680		
Limited competition leading to cartels	R3	0.646		
Partners become complacent	R1	0.645		
Cost of increased communication	R7		0.866	
Cost of new procedures/systems	R8		0.724	
Costs of understanding partner	R6		0.630	
Downsizing results in inadequate resources	R5			0.713
Pressure to perform	R9			-0.601

Operational risks relate to the process, cultural, structural and organisational requirements for the effective and successful implementation and adoption of partnering for construction. Some elements of the operational risk factors are considered and evaluated as follows:

Managers are unwilling to relinquish control

Overall, this risk is considered the most significant which supports the view held by Gattorna and Walters (1996). All categories, with the exception of consultants, are extremely concerned about management control. If senior management refuse to allow a project team to act in accordance with project goals then this will obviously impede a successful partnering relationship. Partnering tends to involve TQM principles including involvement in decision-making at all levels. If management refuse to support such a culture then benefits are likely to be reduced. Additionally, if low level managers feel that they have no control over decisions that affect them the feeling of ownership, which helps people to work together, is also diluted.

Partners become complacent

Complacency is regarded as a significant risk by the respondents which supports Lamming (1993). Contractors and clients are both highly concerned about this with consultants less so. Overall, this is regarded as the second most significant risk. This concern is understandable as partnering relations are costly to establish. For example, BAA has paid about £50,000 per annum on its Heathrow Airport/Paddington Station rail link project simply for contractors to meet regularly with each other and the client [Thatcher (1997)]. Similarly, Beefeater Restaurants provided contractors' personnel with training in interpersonal skills in addition to the customary health and safety and site specific training [Tulip (1997)]. Obviously, if an organisation has committed considerable resources to meetings, training etc. a partner becoming increasingly complacent could reduce the potential return on these resources.

Increasing dependence on partner

The respondents regard increasing dependence on their partner as a significant risk. This is not surprising as one way to cope with risk is risk sharing, in which case the occurrence of any single risk event is less likely to threaten the viability of the organisation. If an organisation is heavily dependent on another organisation, it may not survive the severing of the relationship. Consultants regard this as the greatest risk associated with partnering. This risk is also identified by Saunders (1994) and Ramsay (1996) as being important in partnering.

Partner may revert to adversarial approach

Although the presence of adversarial relationships in construction is one reason why partnering is currently being used in the industry, this adversarial culture is still regarded by the respondents as a risk factor to the success of partnering within the construction industry. While all categories believe that the risk of a partner reverting to an adversarial approach exists, clients ranked this as the second most significant risk. Considering that Watson (1997) regarded this risk as particularly significant it may indicate that the parties involved in partnering are still less trusting which may be due to their exploitative experience from traditional procurement methods.

Conclusion

Traditional adversarial procurement is increasingly being recognised as an inefficient procurement method. The short-term focus of this approach encourages poor performance from parties involved in construction projects due to the lack of incentives to perform and the tendency for exploitation by organisations involved in construction projects. The partnering procurement route is expected to address these shortcomings. However, this procurement route cannot eliminate all problems associated with the client-contractor or contractor-sub-contractor relationships. To be successful it requires fundamental changes in the participants' attitudes. Successful partnering will require trust, openness and a win/win approach to negotiation and problem solving.

The study shows an increasing trend in the use of partnering since the publication of the Latham report. The majority of the firms that participated in the study have been involved in partnering. In general, contractors and clients are supportive of partnering principles. Consultants are less enthusiastic due to fears of loss of control. It is expected that consultants will change this attitude with time considering that partnering looks set to play an increasing role in construction procurement.

Construction projects are inherently risky. This is due to many features, which are unique to the industry, including the prototypical nature of projects, the large number of firm involved which results in conflicts of interest and communication difficulties. Conflict is very common and therefore a significant risk faced by parties to a project. Although partnering represents a risk reduction technique, the use of partnering itself is not risk free. The risks must be analysed and strategies must be developed to mitigate them. This study shows that operational risks are most fundamental to the use of partnering for construction. It is the responsibility of the parties involved in the procurement route to develop strategies to respond to these risks very early in the process by avoiding the risks where practicable, reducing the probability of the risk events occurring or reducing the severity of the risks. By working together under an attitude of trust, open-book policy, a clear understanding of roles etc. these operational risks are less likely to arise and those that do should be responded to by all organisations thereby reducing the impact.

References

Al-Bahar, J and Crandall, K (1990) Systematic Risk Management Approach for Construction Projects. *Journal of Construction Engineering and Management,* Vol. 116, No. 3, September, pp. 533-545

Baker S T (1990) Partnering: Contracting for the future. *Cost Engineering*, Vol. 32, No. 4 April, pp. 7- 12.

Baxendale T and Greaves (1997) Competitive Partnering - A Link between Contractor and Sub-Contractor. in *'Procurement - A Key to Innovation'*, CIB Proceeding, Publication 203, pp. 21-28

Bennett, J., Ingram, I. and Jayes, S. (1996) Partnering for construction. Centre for Strategic Studies in Construction. Reading, UK.

Bova, A. J (1995) Managing Contractor Risk., *Risk Management*, January, pp. 45-51

Cox, A and Townsend, M (1996), Client Server. *Supply Management*, 23 May, pp. 30-31

Chadwick, T. and Rajagopal, S (1995) *Strategic Supply Management.* Butterworth-Heinemann Ltd., pp. 92-117

Gattorna, J.L. and Walters, D W (1996) *Managing The Supply Chain.* Macmillan Press Limited, pp. 189-203

Griffiths, F, (1992) *Alliance Partnership Sourcing - A Major Tool for Strategic Procurement.* Frank Griffiths Associates Limited

Harback H. F., Basham, D. L., and Buhts, R. E (1994) Partnering Paradigm. *Journal of Management in Engineering*, January/February, pp. 23-27

Kaiser 1974

Lamming, R (1993) *Beyond Partnership: Strategies for Innovation and Lean Supply.* Prentice Hall International (UK) Limited, pp. 168-175

Latham, M (1994) *Constructing the Team.* Department of the Environment, pp. 14-85

Lorraine (1994) Project Specific Partnering. *Engineering Construction and Architectural Management*, Volume 1, Number 1, September, pp. 5-16

MacBeth, D.K. and Ferguson, N (1994) *Partnership Sourcing.* Pitman Publishing, pp. 96-140

Mathews, J, Tyler, A and Thorpe, A (1996) Pre-Construction Project Partnering: Developing the Process. *Engineering Construction and Architectural Management*, Volume Number 1&2, March/June, pp. 117-131

McLellan, A (1995) Take Your Partners. *New Builder*, 12 May, pp. 16-19.

NEDO (1991) *Partnering: contracting without conflict.* London: HMSO

Norusis M. J. / SPSS Inc. (1992) *SPSS for Windows, Profession Statistics.* Release 5. SPSS Inc. Chicago.

Ramsay, J (1996) 'Partnership of Unequals. *Supply Management*, 28 March, pp. 31-33

Saunders, M (1994) *Strategic Purchasing & Supply Chain Management.* Pitman Publishing, pp. 215-239

Shash, A (1993) Factors Considered in Tendering Decisions by Top UK Contractors. *Construction Management and Economics*, 11, pp. 111-118

Smit, J (1995) Trusting the Partner. *New Civil Engineer*, 22 June, pp. 14-15

Thatcher, M (1997) A One-Track Approach. *Supply Management*, 13 November, pp.28-30

Tulip, S (1997) The Commitments. *Supply Management*, 13 November, pp. 24-26

Vidogah, W and Ndekugri, I (1998) Improving the Management of Claims on Construction Contracts: Consultant's Perspective. *Construction Management and Economics*, 16, pp. 363-372

Watson, K (1997) No Hiding Place. *Construction Manager*, February, pp. 12-14.

THE IMPLICATION OF PARTNERSHIP SUCCESS WITHIN THE UK CONSTRUCTION INDUSTRY SUPPLY CHAIN

AMANI HAMZA and RAMDANE DJEBARNI
Nottingham Trent University, Construction Procurement Research Unit, Nottingham, NG1 4BU. United Kingdom. amani.hamza@ntu.ac.uk

PETER HIBBERD
CRBE, University of Glamorgan, Pontypridd, Mid Glamorgan, CF37 1DL, United Kingdom. phibberd@glam.ac.uk

Abstract

The concept of partnering in construction seems to be the subject of the decade within the industry. Large construction companies in developed countries such as Japan and USA use partnership sourcing with suppliers, clients, and even competitors as a strategic choice to improve their effectiveness, innovativeness and competitive edge. But above all, these relationships can create harmony that can lead to increase returns for all parties.

The construction industry in UK tends to be fairly fragmented and adversarial although parties are increasingly trying to use alliances and partnering techniques to add value to their projects. While the material procurement services process is not the only factor known to contribute to the effectiveness of the construction industry, it has always been identified as an important causal factor. Nevertheless, Supply Chain Management has taken a low priority in the construction industry thinking and in particular its importance in determining business success.

Partnering is currently held by many as the way forward in construction. There is a belief that this form of procurement will produce 'win-win' results for clients, contractors, and suppliers. This paper will illustrate the implication of partnering success in the supply chain. It will outline partnership success in other industries, particularly in the manufacturing industry, and show what lessons are to be learnt from this success in order to be successfully implemented within the construction industry supply chain.

Key words: Construction industry, manufacturing industry, partnering, supply chain

Introduction

The phrase 'supply chain partnership sourcing' has become common in today's business language. While there are other forms of business relationships, it appears that, based upon academic and trade publications on the subject, supply chain partnerships are currently generating a great deal of interest (Ellram and Cooper, 1990; Miller and Treece, 1992).

During the last ten years, European and American companies have shown considerable interest in the concept of partnership sourcing, in contrast to the more traditional adversarial purchasing relationships (Lamming, 1993). Though this has been led by the automotive and electronics industries seeking an effective response to Japanese penetration of their home markets, its influence has spread into other sectors as well (Morris and Irmie, 1992). Certainly the United Kingdom has taken the idea with interest, at least at the level of government departments such as the Department of Trade & Industry and other industrial bodies such as the Confederation of British Industry and the Society of Motor Manufacturing and Trades (DTI, 1992).

Partnership success in the manufacturing industry

In industries other than the construction sector, much attention has been shown in the concept of supply chain management and its contribution to business performance. Benefits such as minimisation of waste, efficiency improvement, greater productivity and overall improved supply coordination all culminate in a more competitive business operation.

This approach has been put to test in the manufacturing industry, such as the automotive industry, where the adoption of the Japanese 'lean supply' concepts throughout the supply chain has proved its success in terms of improved competitiveness. The most notable example is that of Toyota Motors and its tier suppliers which was reported in 'The Machine that changed the world' by Womack (1990).

Many international surveys' results show that supply arrangements can provide significant business benefits to companies practising them (cf. Cox and Townsend, 1997). Research also shows that there is an increase in the proportion of the UK suppliers that feel customers can help them improve their performance in the face of new competition. UK suppliers with partnering relations have been able to contain cost increases and to protect their profit margins in a much more effective way than non-partnership suppliers (Sako, Lamming and Helper, 1994).

As a result, some researchers suggested that if partnering works for non-construction industries, then it should also work for the construction industry. However, Thompson (1997) argued that the supply chain characteristics of the construction and manufacturing industries (e.g automotive industry) are wholly different and thus blindly applying supply chain management principles to construction without any prior research is very risky and may require the procurement process to be re-engineered.

Partnering in the construction industry

Partnering appears to be a confused concept, meaning different things to different people. To some it means a single-sourced relationship, while to others it means effective project management (Larson, 1995). There are innumerable definitions of partnering, many of which seem to be derived from similar sources. One of the earliest definitions of partnering in construction appears to have been developed by the Construction Industry Institute's Partnering Task Force. Partnering has been defined as:

'A long-term commitment between two or more organisations for the purpose of achieving specific business objectives by maximising the effectiveness of each participants resources. The relationship is based on trust, dedication to common goals, and an understanding of each other's individual expectations and values.' (Construction Industry Institute, 1989).

Built on the thoughts and views of the Latham report, the Construction Industry Board (CIB) has developed the most recent defintion. The CIB, 1997 states that 'partnering is a structured management approach to facilitate team working across contractual boundaries. It should not be confused with other good project management practices, or with long-standing relationships, negotiated contracts, or preferred suppliers arrangements, all of which lack the structure and objective measures that must support a partnering relationship.' The inference here is that partnering can occur irrespective of the type of the supply relationship involved.

Types of partnering

Table 1: Different forms of partnering

		DIFFERENTIATING FEATURES		
Forms of Partnering	Sources	Relationship duration	Basis of partner selection	Condition for Use
Project	Bennett& Jayes Badden-Hellard CIB	One-off	Competition/ negotiation	All projects. Best for high value.
Strategic/Full	Bennett& Jayes NEDC CIB	Long-term	Competition/ negotiation	Where good business case, part of medium long term strategy
Post-award	ECI	One-off	Competition	Public projects, including series of small projects
Pre-selection arrangement	NEDC	One-off/ Long-term	Negotiation	Any project. Advanced selection of contractors.
coordination agreement	NEDC	One-off/ Long-term	Competition/ negotiation	any project. Agreement overlaid on standard contract.
Semi-project	Mathews, Tyler and Thorpe	One-off	Limited competition	All projects where scope of negotiation is limited

Source: Institute of Civil Engineering Surveyors.

The literature reveals that partnering is not a unified concept. It takes on a number of different forms, including:

- Project partnering
- Strategic/full partnering.
- Post-award project partnering.
- Pre-selection arrangements.

- Coordination arrangement
- Semi-project partnering.

The main differentiating features between these different types appear to relate to relationship duration, the basis of selection and the most appropriate conditions for application as presented in the table1.

Application of the supply chain management in the construction industry

Supply chain management is a new way of managing the supply chain. It has been defined as the process of activities which transforms raw materials to finished goods and services for use by the end consumer, irrespective of corporate boundaries (Thompson, 1997). This approach has been developed from the Japanese management practices. It has aims of building trust and cooperation, improving coordination, exchange market information, develop new products and streamlining material flow among all parties in the supply chain. The supply chain members in construction industry can be, clients, consultants, contractors, subcontractors or suppliers. Each member has his own role to do. Communication with different parties and team operation are essential to provide better performance. Kanaji and Wong (1998) suggest that there is an inherent conflict between many of the parties goals. The contractor, architiect, or engineer, consider maximising profit as their goal, while the owner has the goal of minimising costs. Consequently the application of partnering relationships among the different parties of the supply chain is suggested as a better means to coordinate the parties goals and build trust among them so that the whole supply chain can work as a single unit.

Problems of the supply chain in the construction industry

Disputes, claims, lawsuits, delays, cost overruns and so on are words which are all too common in today's construction industry. These issues have made the construction environment unbearable for both clients and contractors.

The construction industry is plagued by supply chain difficulties which are generally characterised by high levels of interdependence and uncertainty. The majority of the problems lie in completion dates, quality standards, cost and clients' satisfaction with the services of the construction industry. There are many aspects of designing a complex building, and choices or changes in one area: the structure, the services, and the finishes inevitably have consequences somewhere else. The whole process of designing and constructing a project is therefore characterised by discovering inter-dependencies, which makes for uncertainty.

The problem with the coordination of design, construction and supply is that it is ineffective in managing these interdependencies. Designers may not be fully aware of the all the complex interdependencies that exist between different aspects of the design of a building. They may also fail to understand the implications of design choices for construction methods, or for materials or components and so contractors have to find or improvise solutions on site. The result is extra work, with scope for conflict over who should undertake it, given that contractors have typically worked under fixed price contracts.

A related problem is that everyone is working on the basis of what the 'centre' specifies, rather than on the basis of what is actually needed by the person who is going to use their work. So the contractor building the structure of an office building may find themselves installing a cladding system specified by the architect which requires expensive facing to the structural members specified by the structural designer, while if the cladding supplier and contractor were together could have specified a much cheaper system to do the same thing. In reality, there is a lot of informal problem solving and adaptation of interfaces, but this happens inspite of the formal system, rather than because of it, and indeed may be held up by the need for formal approval (Holti, 1996).

Implication of partnership success

Until 1986, Laing Homes, a leading construction company, used a traditional adversarial approach in its buying. Analysis shows that supplier services were generally poor and product quality was patchy at best, particularly for timber products. Since that date the company has operated in co-operation with suppliers involving them in product design and performance improvement programs. Suppliers now have reasonable assurance of continued business and are given open access to their client forecast demand projections to enable them to improve forward planning. Although many of the derived benefits are difficult to quantify, Laing Homes' emphasis was on driving cost down. They illustrate their approach to improved quality and saving through partnership. With their suppliers, they jointly developed a specification which although achieved a higher unit price, met its needs precisely. Although the company paid more for the product it reduced wastage by nearly 20% and cut the rejection rate by a factor of ten and reduced overall costs substantially. To meet their client increased specification, the suppliers also formed ongoing relationship with exporters and suppliers in a Knock-on effect of improvements and mutual benefits (Ashmore, 1995).

Marks and Spencer, the Ministry of Defence and a host of other organisations which together spend £40 billion a year on construction have called on the industry to achieve 30 percent cost reduction by 2000 in return for establishing client policies to cut waste from the supply chain (Construction news, 1998)

Construction Clients' Forum commits the clients, who represent 80% of Britain's construction expenditure, to drive up standards for the 21st century. In return, it calls on the industry to collaborate to reduce poor standards of work and cut cost and time overruns.

It can be seen that a lot of work has been undertaken concerning the partnering of the supply chain in different industries. However, the recent movement to partnering offers an exciting alternative to the present system of doing business. Partnering is seen to provide the construction industry with a fundamentally different approach to teamwork. This revolutionary management process emphasises co-operation rather than confrontation and it calls for a simple philosophy of trust, respect and long term relationship. Partnering focuses on team building, conflict management, trust and mutual goal and objective development between contracting parties. A charter is developed between partners involved in the process, tools and techniques are identified to overrun the partnering relationship.

Partnering: a tool for improving performance

The central purpose of partnering is to increase productivity. Once this aim has been achieved, the greater productivity can be used to pursue other secondary goals. For example, productivity gains can be used to secure lower prices, increase profits, raise wages, increase quality, deliver better designs, make construction safer, make competition deadlines and provide every one involved with bigger profits.

It is acknowledged that the extension of partnering down throughout the supply chain in the UK construction industry is still in its fancy, but examples are beginning to emerge as cited above. All parties of the construction industry chain can benefit from partnering. The benefits would be greatest if partnering is applied throughout the supply chain, rather than simply between clients and main contractors. Partnering is a philosophy of teamwork and cooperation. Supply chain management supports partnering in the sense that it sees different parties as a supply chain, which needs integration to ensure best performance. In the supply chain, different parties perform different functions of production process. Hence the quality of the final product depends on the performance of each party. Without the application of the appropriate approaches, the effective and efficient performance of the supply chain cannot be achieved. This research is based upon these observations.

Current research

This paper provides the background of a research undertaken in the UK. The objective of this research to determine the use of partnering as a tool to enhance the performance of the supply chain of the construction industry and to answer the following questions:

1. What true partnering means and how it can benefit all parties concerned in the construction chain?
2. What are the appropriate basic tools and techniques used to manage the partnering relationship?
3. What are the behavioral characteristics (e.g. trust, co-operation), which enhance the success of the partnering relationship?
4. How to encourage the use of partnering on appropriate projects?
5. What are the managerial activities required to ensure the efficient performance of the partnering relationship and what are the tools required to support these activities?

Conclusion

The construction industry is plagued by many difficulties of the supply chain. The majority of the problems lie in the competition dates, quality standards and cost and the dissatisfaction with the services of the construction industry.
Partnering is a process of teamwork between various parties. Construction industry depends on working together in various parties in the supply chain in order to provide quality performance. Since partnering is a process of teamwork, it is seen to be one of the concepts of harmony of relations, which would positively affect the performance of the supply chain.

Partnering can be implemented as an approach to overcome these difficulties and enhance the performance of the industry. Consequently, there is a need for researchers to assess the current and future expectations of partnering satisfaction as perceived by business executives responsible for implementing partnering. More needs to be established as to the types of partnering suitable for the construction industry supply chain and the strategic performance to be adopted that will help the procurement process to undergo successful changes.

References

Ellram, L., and Cooper, M., (1990) Supply chain management, partnerships, and the shipper third party perspective. *International Journal of Logistics Management*, Vol. 1, No. 2, pp 1-10.

Miller, K., L., and Treece, J., (1992) The partners. *Business Week*, Vol. 10, pp 102-107.

Lamming, R., (1993) Beyond partnership: strategies or innovation and lean supply. London Prentice Hall International.

Morris, J., and Imire, I., (1992) Transforming buyer-supplier relationships. Macmillan.

DTI (1992) Supplier innovation: the role of strategic partnerships in the UK automotive component Sector.

Womack, J., P., Jones, D. T., and Roos, D., (1990) The Machine that Changed the World.

Cox, A., and Townsend, M., (1997) Latham as a half-way house: A relational competence approach to better practice in construction procurement. *Engineering, Construction and Architectural Management*, 4,2.

Sako, M, Lamming, R and Helper, S., (1994) Supplier relations in the UK car industry: good and bad news. *European Journal of supply management and purchasing*, Vol.1, No 4, pp 237-248.

Thompson, I., (1997) Is there one supply chain for construction? Centre for strategy and procurement management, university of Birmingham.

Larson, E., (1995) Project partnering: results of study of 280 construction Projects. *Journal of Management in Engineering*, March/April, pp20-27.

Construction Industry Institute (1989) Partnering: meeting the challenges of the future interim report of the task force on Partnering, University of Texas, Austin: CII.

Construction Industry Board (1997) partnering in the team. *A Report by working group 12 of the construction industry board.,London, Thomas Telford.*

Kanaji, G. K., and Wong, A., (1998) Quality culture in the construction industry. *Total quality management*, Vol. 9, No. 5.

Holti, R., (1996) Designing supply chain involvement for UK building. *Programme for organisational change and technological innovation.* Tavistock institute

Ashmore, C., (1995) Partnership sourcing approach-love match or shortgun wedding. *Engineering Management Journal*, August, pp148-152.

Construction news, (1998) industry pact promises to build up standards. *Supply management*, No. 26, pp10.

Robinson, G., (1998) Industry pact promises to build up standards. *Supply management*, No. 26, pp10.

PROJECT TEAM PERFORMANCE - MANAGING INDIVIDUAL GOALS, SHARED VALUES AND BOUNDARY ROLES.

M. C. JEFFERIES, S. E. CHEN and **J. D. MEAD**
Department of Building, The University of Newcastle, Callagham, NSW 2308 Australia.
bdmcj@cc.newcastle.edu.au

Abstract

The project team consists of participants from a variety of different organisations to form a temporary organisation in order to achieve the common objective of procuring the project. Each member of the team will have their own important objectives and self interests. The lack of concern for other participants' risks on the project are major contributors to team failings. Construction project performance is dependent on a team effort and good communication. A constructive team environment is one in which members are interactive and shared values provide the foundation for goal setting. This paper discusses these issues drawing on the results of a survey of 166 project participants on their perceptions on issues relevant to project team performance. Partnering identifies with the management of team building and emphasises an environment of trust, teamwork and co-operation. Partnering is an example of the need to balance individual goals, shared values and boundary roles.

Key Words : Goals, interaction, team performance, team roles, values.

Introduction

From initial inception and financing, to the design and then to the actual construction, the building process is largely a team effort. While teams and teamwork are important in the construction project process, it appears that they are commonly encumbered by problems. Teams in project based manufacturing industries such as construction face higher levels of uncertainty and consequently higher levels of problems encountered than in permanent organisations. The adversarial nature of business, and in particular the construction industry, can be considered an aim of each team member attempting to win at all costs. This creates a hostile and uncompromising team environment which is not often resolved but merely discarded at the end of the project [Loosemore (1995)]. As the construction industry is characterised by temporary teams it is vital to establish certain ground rules. In construction these 'rules' take the form of a building procurement method or the contract. The various procurement methods set the boundaries of conduct which the team members must abide and function. The behaviour and development of project teams consequently has been and remains very much a function of the procurement method adopted.

Procurement methods have undergone significant changes in recent years. It is apparent that between traditional systems and design and construct, there are a number of significant and direct, team organisational and operational implications. Design and construct has changed much of the mechanics of building procurement but not enough to eradicate the problems and biases held by stakeholders. The reality that there is still far from ideal teamwork operating as most publicly evident in the findings of the 1992 Royal Commission [Royal Commission into Productivity in the Building Industry in New South Wales, Australia (1992)]. The problems which project teams encounter must be faced within the framework or confides of the procurement system adopted. Problems within the project team are a complex function caused by dissimilar objectives, motivations and perspectives held by team members [Franks (1995)]. The magnitude of the negative effects that poor teams and teamwork can impact upon the project can be illustrated where management methods have been found to lower budgets by up to 10% and place projects ahead of schedule by up to 18% [Latham (1995a)]. These results have been achieved by what has become the most recognised team management method in the construction industry, termed 'partnering'. The success of management methods such as partnering is heavily reliant upon all stakeholders being open with each other as to their individual goals of the project. Understanding each parties motives and goals assists communication and formation of common project goals [CSPC (1993)].

Understanding individual goals in a team situation

The identification and understanding of team dynamics and decision making from the perceptions formed by the main project stakeholders as suggested by [Shapira et al. (1994)] is significant, yet appears to have been neglected in sufficient detail in published research to date. Traditional procurement has been identified as contributing to the fragmentation of the construction process and producing an adversarial climate for team members [Blake (1996)]. The development of alternative procurement methods such as design and construct, report to have increased the level of interdependence and teamwork between all team members, resulting in traditional adversaries required to work together [Blake (1996)].

[Hill (1995a)] supports the need for understanding the motives and agendas of individual team members and states "...the best way to prevent disputes from arising is to make sure that each party knows what the other party wants...". It is the participation of the stakeholders as a team from the very beginning of the construction cycle which [Mathur and McGeorge (1991)] see lacking as they identify the participating stakeholders in the construction process as being fragmented in their objectives, practices, methodologies and mechanisms. Further support is provided by [Latham (1994)] who in his report highlights the lack of trust and teamwork in the construction industry drawing particular attention to clients insisting on onerous contract conditions and consultants regarding clients not as the core of the process, but as a nuisance. Communication is essential to successful team forming, however the industry has been criticised for its poor communication practices, which consequently lead to reduced productivity and project performance [CSIRO (1996)].

The importance of clients to the construction industry cannot be over-emphasized, they have the right to expect that projects meet their needs and aspirations [Holt etc (1995)]. According to [Latham (1995b)] clients need a 'strong and influential voice'. Clients in general are becoming more sophisticated and experienced seeking more involvement in the building process [Nahapiet (1993)]. Client performance can not be solely attributed to whether or not the client is experienced as [Kometa etc (1994)] found that the success of the project with regard to the client, involved a number of closely interrelated issues, most importantly those concerning financial stability and project feasibility.

Consultants traditionally have been the leaders, however they have been relegated to a lower position on the team [Allen (1996)] mainly due to the evolution of design and construct procurement methods. This has created a significant amount of hostility as they attempt to retain their valued position in the team [Ndekugri and Turner (1994)] and [Gordon (1994)]. The client highly regards the effectiveness of consultants, and it remains the first group which is called upon when considering to commission a project. The [Building (1995)] survey found when asking over 400 clients who they first approach when considering the commissioning of a project that over 60% will approach a consultant first. In the growing environment of professional liability, consultants fear clients will hold them responsible for extra expenditure [Stewart (1994)]. This litigation prone industry makes for a very negative environment for the consultant to work within. The consequence is likely to result in reduced and heavily qualified communication being transferred between team members. The importance of avoiding litigation is evident with all parties and the search for a less combative situation has assisted the growth of design and construct procurement methods.

While the stakeholders work together towards the common goal of the project, their actions and motives do not always align, often resulting in potential conflict. The source of conflict can be crudely seen as a fight for control or power within the team. [Al-Sedairy (1994)] found that the most heated and frequent conflicts occurred between the contractor and the client, and the contractor and the consultant; where the most significant causes of conflict were found to be timing, project concept, costs and specifications. [Loosemore (1995)] realised the limitations of working in a project-based industry by highlighting the fact that participants are well aware of the limited life span of each job as it reflects the uncooperative, differentiated nature of the temporary organisation which is created to produce the project and contributes to the divisions and general lack of collective responsibility of the problems when they arise.

The team in construction - shared values

The advent of team failings can largely be attributed toward member's self interest. The lack of concern for other team members risk on the project is common [Loosemore (1995)]. [Gabriel (1991)] found that that "there are those who, for various reasons, are disruptive, i.e. they damage relationships and synergy for reasons including individual objectives that are different from the main objectives, a lack of understanding of priorities, or simply to be an 'anti project manager'." The NSW Public Works Department [CSPC (1993)] views coincide, attributing much of the problems team encounter are due to the individualistic nature of members within the team fostered by the adversarial nature of the industry. [Murphy (1996)] draws on the fact that even the gaming theories which believe zero sum or win-lose strategies are often successful if not necessary for survival. [Loosemore (1995)] found that the importance of an effective, structured and harmonious team seems very important. Whereas [Franks (1995)] goes further by stating that "...effective teamwork is essential to success."

Goal setting appears to be an underlying requirement for a team to work with any productivity. [Reijniers (1994)] found that a common cause of team failure was due to the non-goal directed preparation of teams. It seems apparent for a team made up of individuals bringing different motives together that common goals must be set before these members are able to work concurrently [Swierczek (1994)]. It is therefore advantageous that in the setting of common project goals that all members are involved to some degree as changes are likely to induce conflict [Williams and Lilley (1993)]. Increased understanding is the first step towards achieving better teams in the industry by highlighting the fact that the key to team development is to study the character of the individuals and then work at the relations of the group [Franks (1995)]. A degree of tolerance must be attained, in that there will nearly always be minor interpersonal problems within groups [Stronhmeier (1992)]. [Myers (1996)] emphasises Belbin's 'Apollo Syndrome' where teams of highly capable individuals can collectively perform badly.

[Ward etc (1991)] found that clients were strongly influenced by their previous experiences and consequently often brought to the team preconceptions based on how relationships with other participants worked in terms of both harmonies and conversely arguments and distrust. [Fellows and Langford (1993)] sugested that the construction industry will see Joint Ventures becoming increasingly important.

Management of the team - boundary roles

Team partners in Joint Ventures, at project level, more often than not have some degree of conflicting interests [Reijniers (1994)]. The management of conflict has been identified as a means for successful team forming, and to be achievable "...requires specific organisational structure so that the conflicts of interest can be managed...the structure should focus on the constraints and the common goal." A proactive approach to team building will lead to a greater chance in achieving success [Swierczek (1994)]. The construction project team is similar to a joint venture where the parties mostly come from different company, professional or cultural backgrounds. These 'cross-functional teams' [UCSD (1996)] are clearly advantaged by managed team forming to attain common goals. Team forming difficulties within the industry are partly due to the limited life span of the team, which consequently gives little incentive for the members to move too far from their organisational goals in favour of more common project goals [Loosemore (1995)].

Partnering is the term which this paper identifies with as the management of team building [CSPC (1993)]. According to [Hill (1995b)] partnering should be viewed as a concept to establish working relations among the parties through a mutually developed, formal strategy of commitment and communication where trust and teamwork prevents disputes, create a co-operative bond, and facilitate the completion of a successful project. Partnering was identified as one of the possible keys by Latham where studies have shown a wide range of projects that 10-30% cost savings were established [Latham (1995b)]. [Stevens

(1993)] saw partnering created an environment where trust and teamwork prevent disputes.

According to [Chandler (1994)], Australia has been slow to adopt partnering, but, some cynicism has been eliminated by a push by public sector clients to include partnering within contracts. A pilot study by [CSIRO (1996)] found that Australian contractors believe that partnering has improved communication between parties. The construction industry has been slow to adopt a managed approach towards team forming with [Ridout (1994)] highlighting through case study research in firms which fully adopt partnering, that, initially the contractors members believed that it was exposing the company to high levels of risk from being open. Designers were cautious, according to [Wright (1993)] as it opened up additional avenues for the other team members to criticise the designs.

The way in which each team members' organisation works also has significant implications to the effective co-operation of the team. [Williams and Lilley (1993)] identified a significant problem as being the differing procedures between the team members organisations followed. It can not be assumed that team members come from similar size firms, consequently, procedural requirements are often seen to frustrate. For example, larger firms tend to have a structured decision making hierarchy which will seem slow compared to the smaller, more responsive firm. Team forming problems are further enhanced by individualism as a common cause. According to [Tampoe and Thurloway (1993)] in a successfully managed team forming process, the individual within the team is recognised and the team will move towards fostering and development of project environment. This will encourage mutuality, belonging, rewards, banded power and creative autonomy is likely to result in improved project performance.

The survey

The aim of the survey was to provide an insight into the perceptions of project participants on the issues of firstly procurement and the team, secondly obstacles to team formation and finally team management. A structured questionnaire was administered to 166 building contractors, project managers, building owners and clients, architects, engineers and quantity surveyors in Sydney and Newcastle in Australia. Table 1 summarises the respondents by project team role.

Table 1. Respondents by project team role

Respondent group	Sub-group	Number
Contractors		31
Consultants	Architects	32
	Engineers	31
	Quantity Surveyors	18
Clients	Clients	27
	Client's Representatives	27

Procurement approach and team reliance

Respondents were asked about their preferences between the traditional and design and construct procurement approaches. Tables 2 summarises their responses by the different groups.

Table 2. Respondents' preference for procurement method

Respondents	Preferred method of project procurement	
	Traditional Contract	Design & Construct
Contractor	21%	79%
Consultant	64%	36%
Client	44%	56%

Building contractors showed a strong preference for the design and construct procurement approach in contrast with the consultants who generally still prefered the traditional approach.

The respondents were asked to rate how the two procurement approaches influenced team formation (ie team members working better together) on a 5-point Likert type scale where 1 was "Very Little" and 5 was "Very Significant". Table 3 summarises these results.

Table 3. Perceptions of the influence of procurement approach on team formation.

Respondents	Degree of influence on team formation									
	Traditional Contract					Design & Construct				
	1	2	3	4	5	1	2	3	4	5
Contractor	17%	40%	30%	13%	0%	3%	3%	14%	53%	27%
Consultant	8%	23%	43%	18%	8%	12%	14%	30%	35%	9%
Client	11%	39%	37%	13%	0%	7%	6%	20%	58%	9%

The design and construct approach was generally perceived to be more influential on team formation. Consultants as a group did not express this as strongly as the other groups.

Obstacles to team formation and success factors

Respondents were asked to assess and rank from a list of 43 factors the importance (seriousness and frequency) of these factors in hindering team formation. The factors were derived from an extensive review of the literature. Table 4 reports the top 10 factors ranked by each group of respondents.

Table 4. Perceived factors hindering team formation

Factors	Contractor ranking	Consultant ranking	Client ranking
Selection of a contractor who has tendered too low	1	3	4
Errors in documentation and drawings by consultants	2	4	2
Slow supply of documentation by consultants (drawings etc)	3	7	3
Members operating on tight profit margins	4	1	5
Slow resolution of disputes and conflicts	5	8	10
Unfair contract conditions/low re-gard for design issues	6	6	7
Unrealistic deadlines being set by team leaders	7	2	9
Team members held in low esteem	8	10	8
Communication shortcomings between members	9	5	1
Differing perceptions of project goals & quality requirements at project outset	10	9	6

All the respondent groups chose the same factors for the top 10 factors hindering team formation. Within these 10, the different groups of respondents perceived different rankings, probably related to their own roles and goals.

Respondents were also asked to rank 14 factors in terms of their perceived importance to successful teaming.

Table 5. Perceived factors important to successful teaming

Overall ranking	Factors	Contractor ranking	Consultant ranking	Client ranking
1	Ability of team to resolve conflicts quickly	3	2	1
2	Ability of members to negotiate and reach compromises	1	3	2
3	The project is feasible in terms of time and budget	2	1	8
4	The setting of clear and realistic milestones for the team	7	4	3
5	Confidence in other team members' abilities	5	5	9
6	Establishing mutually agreed goals early in the project process	6	6	6
7	Team organisational authority structure defined at start	9	9	5
8	Smooth operation of formal communication channels	8	7	7
9	Realistic profit margins for all team members	4	8	12
10	Trust exists between team members	10	10	10
11	Informal communications channels open	13	12	4
12	Understanding of other team members' objectives	12	12	11
13	Members are benefiting in proportion to their risk and effort	11	13	13
14	Similar levels of experience between team members	14	14	14

Factors to do with managing relationships, individual and shared goals featured strongly in the perceptions of all the respondent groups.

Team management

Respondents with personal experience with formal partnering arrangements (48) were asked to respond to three statements using a 5-point Likert type scale from "Strongly Disagree" (SD) to "Strongly Agree" (SA).The aim is to determine whether managed approaches to the team are viable and if they can improve team performance. Table 6 reports that responses given by respondents who have personal experience of partnering.

Table 6. Perceptions: "Partnering improved team formation"

Respondents	"Partnering improved team formation"				
	SD	D	N	A	SA
Contractors	0%	17%	25%	50%	8%
Consultants	21%	22%	21%	32%	5%
Clients	6%	20%	21%	53%	0%

Table 7. Perceptions: "Managed techniques can improve team formation"

Respondents	"Managed techniques can improve team formation"				
	SD	D	N	A	SA
Contractors	0%	8%	25%	58%	8%
Consultants	7%	7%	35%	42%	10%
Clients	18%	0%	11%	58%	13%

Table 8. Perceptions: "Team members were cynical towards partnering"

Respondents	"Team members were cynical towards partnering"				
	SD	D	N	A	SA
Contractors	0%	33%	8%	58%	0%
Consultants	0%	16%	20%	33%	31%
Clients	0%	6%	21%	38%	36%

Discussion

Procurement methods have been a strong theme throughout this research because they set the boundaries within which the team operates. While there are a great variety of forms of procurement currently available they can be typified by the two most prevalent, being traditional procurement systems and design and construct procurement systems [Ndekudrin and Turner (1995)]. Recent research [Fellows and Langford (1993)], [Ndekugrin and Turner (1994)], [Building Magazine (1995)]. suggests that alternate procurement methods, led by design and construct methods will gain popularity at the expense of the traditional approach in the future.

The favouring of design and construct by the contractor is quite understandable considering it generally elevates the contractor to a team leader position where the contractor has increased control over the direction of the team. Perhaps the reason why design and construct procurement methods are so popular with the contractor is because it eliminates the traditional consultant barrier of communication with the client. As the contractor usually carries the heaviest burden of risk and responsibility, it is critical that they be given the opportunity to participate at client level in order to monitor the risks. Consultants on the other hand report a preference for traditional procurement methods. The high dependency of consultant input with traditional systems is the most likely reason for their preference. Clients also showed a slight preference for design and construct procurement methods. The increase in sophistication and experience of clients generally means that they wish to become increasingly involved with the construction process. Design and construct type systems offer the client increased involvement and monitoring with the majority of risk transferred onto the contractor.

The contractor, consultant and client groups all reported that design and construct methods required a greater level of team work. This was particularly strongly expressed by the contractor group. When working in a traditional procurement environment the contractor must take a harder line with the other team members namely the consultants, than they would with design and construct. The harder line taken by the contractor in traditional procurement systems itself spells out the significance of procurement systems to teams. The traditional 'hard dollar' procurement drives contractors to bid on price alone resulting in minimal margins if at all.

Team hindering factors have been shown to arise from virtually the slightest weaknesses or problems encountered by the teams [Loosemore (1995)]. This study has focused on the problems identified by the review of the literature. The problems can be categorised into four clear areas concerning money, performance, communication and industry culture. Money has been reported as being a major hindering factor for project teams and is the element that affects the ability of members to perform. [Akintoye and Skitmore (1992)] noted that the construction industry is typically operating on tighter margins than other industries, consequently financial considerations are likely to be the origin of the many problems team face. [Latham (1995b)] pointed out the single-mindedness of clients towards cost minimisation as a significant cause that has led to a chain reaction creating a number of problems within the industry. The erosion of profit margins highlighted by [Ndekugri and Turner (1994)] is a function of an increasingly competitive environment and the increase in popularity of newer procurement systems. Although there have long been critics of the practice of selecting the lowest bid rather than an overall value for money basis selection policy [Ward etc (1991)] and [Latham (1995b)], it is pointed out by [Fellows and Langford (1989)] and [Latham (1995b)] that accountability issues make movement away from this policy of lowest bidder selection very difficult

Performance related issues are reported to be some of the top factors causing problems within the project team. The ability to perform is dependent upon each member performing to expectation. Errors and slow supply of documentation are highlighted as causing project performance difficulties. The fact that team members are highly interdependent of each others' services means that failure by one is likely to have a detrimental 'domino' effect amongst all team members. Onerous contract conditions that are seen to disproportionately shift the burden of risk between the stakeholders, as highlighted by [Latham (1994)] can be overcome by the use of design and construct methods of procurement due to the single point of responsibility which is commonly the contractor. Communication problems within the construction industry are well acknowledged by [CSIRO (1996)]. The survey respondents reported that good communication was an important ingredient of a successful team and conversely found that communication related problems apparently caused a large portion of team disharmony. Poor understanding or communication is also reported to lead to unrealistic time frames being set.

Both [Franks (1995)] and [Hill (1995a)] consider that the first step in addressing problems within teams is made through a better understanding between members of individual motivations, perspectives and problems. This is also the case as reported by the contractor and consultant as an important step to team improvement. While it is generally agreed that better teams will ensure a greater likelihood of eventual project success, there has been little effort directed towards managing or directing project teams on the basis of encouraging trust, teamwork and co-operation. A better understanding of how boundary roles between participants should be managed is essential.

Managed methods such as partnering can improve team dynamics although it is clear that a great deal of cynicism, particularly on the part of clients, exists. For the success of partnering this issue must be addressed before clients can begin to see the benefits in adopting such strategies with their projects.

Conclusions

The potential for the performance of project teams to improve will be enhanced as the building industry moves from the traditional procurement approaches towards alternate procurement methods such as design and construct, and adopt managed team formation approaches such as partnering. Some of the key issues in achieving this improvement are the management of the individual goals of the team members, developing strong shared values for the project and effectively managing the interfaces between team members and the external environment.

References

Allen, L. (1996), The builder is boss. *The Weekend Australian*, 29 June, pp. 6.

Al-Sedairy, S. (1994), Management of conflict; public sector construction in Saudi Arabia *International Journal of Project Management,* 12 (3), p. 143-151.

Akintoye, A. and Skitmore, M. (1992), Pricing approaches in the construction industry. *Industrial Marketing Management, 21*, pp. 311-318.

Blake, B. (1996), Tendering: The lumbering dinosaur. part 1. On Site, 5(6), pp.5-8.

Building (ed.) (1995), Customer services. *Building,* 28 July, pp.26-27.

Chandler, D. (1994), Partnering seen as the way to go. *The Chartered Builder*, May, p.22.

Construction Policy Steering Committee (CPSC). (1993), Capital Project Procurement Manual: Partnering. NSW Public Works Department, Sydney.

CSIRO, (1996), Collaboration in communication - leading research points to improvement areas. Building Innovation and Construction Technology, (9), pp.25-26.

Fellows, R. and Langford, D. (1989), Marketing and the construction client. CIOB, UK.

Franks, J. (1995), Building construction teams. *Construction Management*, November, 1(2), pp.16-18.

Gabriel, E. (1991), Teamwork. Int'l Journal of Project Management, 9(4), pp.195-198.

Gordon, C. (1994), Choosing the appropriate construction contracting method. *Journal of Construction Engineering and Management,* 120(1), pp.196-210

Hill, R. (1995a), Overview of dispute resolution:
URL: *http://www.batnet.com/oikoumene/arbover. html*

Hill,R.(1995b), An introduction to partnering for telecommunications.
URL: *http://www.batnet.com/oikoumene/arbpart.html*

Holt, G. Olomolaiye, P. and Harris, F. (1995), A review of contractor selection practice in the UK construction industry. *Building and Environment*, 30(4), pp.553-248.

Johns, T. (1995), Managing the behaviour of people working in teams-Applying the project management method. *International Journal of Project Management,* 13(1), pp.33-38.

Kometa, A., Olomolaiye, P. and Harris, F. (1994), Attributes of UK construction clients influencing project consultants' performance. *Construction Management and Economics*, 12, p.433-443.

Latham, M. (1994), Constructing The Team. HMSO, London.

Latham, M. (1995a), Reviewing Lathams review. *Building*, 2 June, pp.5.

Latham, M. (1995b), Team talk; at half time. *Building*, 28 July, pp.28-30.

Loosemore, M. (1995), Reactive management: Communication and behavioural issues dealing with the occurrence of client risks. *Construction Management and Economics*, 13, pp. 65-80.

Mathur, K. and McGeorge, D. (1991), Towards the achievement of total quality in the building process. *Australian Institute of Building Papers*, 4, p.109-120.

Murphy, P. (1996), Game theory models for organisational/public conflict. *Canadian Journal of Communication*, August, URL: http://edie.cprost.sfu.ca/cjc/cjc-info.html

Myers, S. (1996), Team technology - the Apollo syndrome.

Nahapiet, H. (1993), Future shape of a people business. *Chartered Builder*, May 1993, 5.

National Committee on Rationalised Building (NCRB), (1985), Glossary of Australian building terms (3rd ed.). *Sydney Building Information Centre*, Sydney.

Ndekugri, N. and Turner, A. (1994), Building procurement by design and build approach. *Journal of Construction Engineering and Management*, 120(2), pp.243-256.

Reijniers, J. (1994), Organisation of public-private partnership projects - The timely prevention of pitfalls. *International Journal of Project Management*, 12(3), pp. 137-142.

Ridout, G. (1994), Consort pitch. Building, 5 Aug, pp. 30-33.

Shapira, A., Laufer, A. and Shenhar, A. (1994), Anatomy of decision making in project planning teams. International Journal of Project Management, 12(2), p.172-182.

Royal Commission into Productivity in the Building Industry in NSW, (1992), *Final report of the Royal Commission into productivity in the building industry*, Sydney.

Shapira, A. Laufer, A. and Shenhar, A. (1994), Anatomy in decision making in project planning teams. *International Journal of Project Management*, 12(3), pp. 172-182.

Stevens, D. (1993), Partnering and value management. *The Building Economist*, Sep, pp.5-7.

Strohmeier, S. (1992), Development of interpersonal skills for senior project managers. *International Journal of Project Management*, 10(1), pp. 45-48.

Swierczek, F. (1994), Culture and conflict in joint ventures in Asia. *International Journal of Project Management,* 12(1), pp. 39-47.

Tampoe, M. and Thurloway, L. (1993), Project management: The use and abuse of techniques and teams (a motivation and environment study). *International Journal of Project Management,* 11(4), pp. 245-250.

UCSD - Human Resources Department, (1996), Managing team performance. University of California, San Diego,
  URL: http://www-hr.ucsd.edu/online/staffedu/guide/index.html

Ward, S., Curtis, B. and Chapman, C. (1991), Objectives and performances in construction projects. *Construction Management and Economics*, 9, pp. 343-353.

Williams, R. and Lilley, M. (1993), Partner selection for joint venture agreements. *International Journal of Project Management*, 11(4), pp. 233-237.

Wright, G. (1993), Partnering pays off. *Building Design and Construction*, Apr, pp.36-39.

2. Harmonious Relations

NETWORKS IN THE CONSTRUCTION INDUSTRY: EMERGING REGULATION PROCEDURES IN THE PRODUCTION PROCESS

MARTINE BENHAIM,
Ecole de Management, Lyon, France benhaim@em-lyon.com

DAVID BIRCHALL
Henley Management College, United Kingdom davidbi@henleymc.ac.uk

Abstract

This paper highlights interfirm relationships within the French and UK Construction Industries, focusing on the latest organisational form to emerge: partnering.

A comparative overview of the Industry reveals similarities in the development of the two national industries with variations related to the players' influence and structural roles within their respective construction processes. The Industry's fragmentation, generated by the main contractors' merchanting role and their positions of power, and by changing market conditions over the past 20 years, has created imbalances in production conditions. These dysfunctions can, in part, be attributed to the disappearance of informal regulation procedures, based on traditional codes of practice, linking contributors to a construction project. An emerging organisational form - partnering - provides the basis for an overall reassessment of the relationships between clients, designers, engineers, main contractors and subcontractors.

Three case studies are used to explore examples of partnering implementation and, in the case of the French study, the reasons for its failure. This qualitative research emphasises the importance of renewing the regulation procedures between the contributors to the construction process.

Finally, evolutionary theories provide some useful concepts to help understand the UK and French pioneering construction companies' approaches to network creation and management.

Keywords: Co-operation, interfirm relationships, networks in the construction industry, partnering

Introduction

In France and in the UK, a number of initiatives have taken place, aimed at reducing the drawbacks of the Industry's fragmentation. Within these initiatives, some main contractors chose to re-examine their position in the construction chain and to test other forms of relationships with their clients and subcontractors. Their endeavours created a distinct trend in the Industry that was defined as "partnering" in the UK or "partenariat" (partnership or network) in France.

Three different cases of partnering implementation have been studied (M. Benhaim, 1997) through longitudinal research methods in the UK and historical research methodologies in France. The aim of this paper is to show that, in the light of the human ecology/collective-action theory, the UK experiments in partnering are providing the basis of an alternative to the traditional and conflictual merchanting system, particularly through the creation of new, formal, regulation procedures between contributors to the construction project, whereas in France the situation does not permit such a conclusion. We will, therefore, examine the major institutional differences between the French and the UK Construction Industries in order to contextualise the experiments on partnering collated in both countries A theoretical interpretation of the case studies will then be presented, followed by the limitations of the present study and orientations for future research.

Vicious circles and imbalances in the Construction Industry

The entire construction process is regulated by mechanisms combining contractual and informal rules, which bear the characteristics of national cultures and an historical balance of power between players and their institutional representatives.

Differences in roles and contracts between France and the UK

The principal functions (ie conception, construction and control) are often performed in both countries by different organisations or consultancies. It would also appear that, as one of the rare comparative surveys between France and the UK (Winch and Campagnac, 1995) pointed out, the role of architects is different in the two countries. UK architects expand their working area to include detailed design, a role carried out in France by the main contractor. As a consequence, the specifications produced by UK architects are far more detailed than they are in France, leaving the main contractor with a narrower margin to manoeuvre, after completion of the conception phase, when his input is unlikely to be taken into account.

The pre-eminence of the designer in the UK is compounded by the contractual system which - in the traditional contract, still the most accepted of contractual forms in the UK - establishes him as the leader of the construction team. In contrast, the most frequent contract in France, "contrat en entreprise générale", allows the main contractor to have direct contact with the client, and assigns the major responsibility to him, together with total control over the subcontractors. As a result, the French main contractors have been able to make technical recommendations to their clients at the detailed stage of the design (APD: Avant Projet Détaillé), allowing them a wider range of initiatives ("variantes" often evaluated by the independent "Bureau de contrôle" in the building sector) as well as permitting more efficient optimisation of the production process.

Now in the UK, new forms of contracts, such as Design and Build, are becoming increasingly popular, as they allow the main contractor to assume responsibility for the whole project and to organise conception and construction choices in a co-ordinated manner. In that respect, D & B contracts are going further than the French "entreprise générale" contract, where the main contractor is not responsible for the entire design stage.

The regulation mechanisms

In both countries, autonomous players are introduced into the conception/production process at different phases, bringing with them highly differentiated views of the project. However, none has enough power to integrate the entire process or to prioritise its different aspects (Gobin, 1993). According to Brousseau and Rallet (1993), centrally monitored information would delay the whole decision process on a site and the use of contracts can never anticipate every conceivable event in the production process, nor avoid potential suboptimisation. Other soft, process-based, co-ordination methods, such as a code of practice, the value of a contractor's and individual site worker's reputation, and the informal, verbal nature of relationships in the Construction Industry, exist between players and will accommodate a relatively fluid adjustment between them in non-contractual circumstances (GRIF, 1992, Brousseau and Rallet, 1993). The emergence, in the 70s in the UK and in the 80s in France, of tougher forms of contractual management, where general contractors shift financial and technical risks onto subcontractors, has weakened these integrative tools, whose disappearance may, in turn, explain some of the current conflicts in the Construction Industry.

The merchanting system

UK main contractors have adopted a merchanting system based, according to Ball (1988), and Campinos-Dubernet (1988), on financial optimisation logic. Contractors' fixed costs are kept at the lowest possible level, with extensive recourse to plant-hire and subcontracting of the workforce. The main contractors' search for financial flexibility has an important impact on production, leading to the Industry's fragmentation (a consequence of vertical disintegration and extensive use of independent subcontractors); increasingly complex contracts (contractual management of the subcontractors); and an ongoing negotiation between players, sometimes at the client's expense (ie through delays, quality or final price), or at the subcontractors' expense (by renegotiating estimates after winning a contract, claims during completion of the project, delayed payments, etc.).

The overall result of the main contractors' wide recourse to market forces in their business appears to have negative consequences. Profitability, on average, is weak in the Construction Industry, with detrimental cycles between bleak and boom periods, encouraging widespread short-termism. High costs and low productivity are endemic, as is the clients' dissatisfaction with the quality and price of the final product. One of the reasons for this inflection in results is the fact that the main contractors - through the merchanting system - are dragging the flexible construction process so far that the process-based regulation mechanisms, ie the professional codes of practice and informal relations on site, which enable the contracts to be applied in operational situations, have been damaged. As time passes, the UK Industry sees its future at risk (Latham, 1994). In France, where the merchanting system has been introduced in a milder way - as seen before, the French main contractors have been given more contractual licence to organise the production process for which they are responsible and liable (garantie décennale) - the main contractors have analysed the crisis which occurred between 1992-1997 in terms of it being a classic downturn in an economic cycle that will eliminate the weakest companies and reduce the Industry's overcapacity.

Partnering: Experiments in France and the UK

Partnering experiments in France:

Early partnering experiments were carried out, in isolation from the rest of the French Construction Industry, by one main contractor, Campenon Bernard Construction (CBC) - a building subsidiary of the former Générale des Eaux, now Vivendi - from 1982/83 to 1988/89. In 1985, CBC's turnover was FFr 6 billion with losses of approximately FFr 500 million, due to the problems incurred on a Berlin project. At that time, its parent company was in the process of restructuring its construction subsidiaries, which include SGE and CBC.

The case study presented here is thus the product of historical research based, for the most part, on evaluations written at that time (Weiller, 1987) and afterwards (Campagnac, Bobroff and Caro, 1990, Roch and Colas, 1993, Colas, Mathieu and Roch, 1994), and on retrospective interviews carried out during 1995.

The partnering experiments have been undertaken mainly in CBC's subsidiary in the Paris region, under "Entreprise générale" contracts, when possible in a "marché négocié" situation (without tendering and when dealing with private clients). At the conception and pricing stage, CBC, supported by its important internal design and engineering team, was able to significantly influence the design and timetable, requesting some of its subcontractors to react and improve on the project's technical options. CBC provided its subcontractors with managerial tools to help them to assess their costs more accurately. At the construction stage, CBC programmed and formalised their procedures by using CAD to limit changes made later on in the project. Subcontractors were bound to CBC by a contractual system which they could not control, but to which they could contribute with some productivity improvements on site, after they had completed two or three projects together. Specifically, CBC organised its relationships with its partnering subcontractors on the following principles:

1. **Selection**: Three preferred subcontractors are chosen among technical trades, for their specific assets and knowledge. No tender is implemented or only a restricted one, if the preferred contractors submit proposals that cannot be accommodated within the main contractor's cost calculations.

2. **Co-ordination**: Subcontractors are asked to improve the design at the operational stage (APD: Avant-Projet Détaillé). They are part of the construction team from the project's outset and sign an operational contract ("Bon pour exécution") in order to set out their respective roles and contributions.

3. **Partnering management**: The main contractor meets all his preferred subcontractors twice a year in order to establish and retain personal and social relationships. A survey is carried out regularly in order to measure the subcontractors' level of satisfaction.

This partnering agreement, worked out by CBC as a means of differentiating itself (through better quality and lower production costs), suffered from a number of dysfunctions related to stable and closed networks. Opportunistic attitudes arose among the preferred subcontractors, who formed a cartel in order to impose their pricing on the main contractor when the economic climate improved, and actually overheated between 1988 and 1990, making it easier to risk jeopardising their relationship with CBC. Simultaneously, CBC suffered from a "lock-in" effect, as the remaining subcontractors in the market were unwilling to submit a tender that would provide the Company with a benchmark to control their existing preferred subcontractors. Furthermore, no evaluation tools were available to compare the subcontractors' performance from project to project and to share the benefits of improved productivity.

As a result of these difficulties, CBC abandoned their stable network solution, based on preferred subcontractors and returned in 1988/89 to a looser relationship which was revised for each project. Limited tenders on a broader subcontractor base reasserted the main contractor's power over the situation and suppressed the opportunistic and cartel-type behaviours of the subcontractors.

Experiments in the UK

The publication of the Latham report (1994) crystallized a number of thoughts within the Industry, regarding the contracting forms and procurement systems applied in the UK, seen as the cause of a number of dysfunctions and excessive costs in the British Construction Industry. As a result, some partnering trials, regarded as a possible remedy to these problems, were carried out (Barlow 1996), particularly within two main contractors, Kyle Stewart (1995 turnover: £154.2m), a building contracting subsidiary of the Dutch construction company HBG, on a 12,000 m^2 office building project in Birmingham (No 5 Brindleyplace) and the civil engineering branch of Balfour Beatty (1995 turnover: £1,701m), part of the UK group BICC, on a road enlargement project (the M25, near London). The two experiments have been tracked through longitudinal field research (Pettigrew, 1995), over a period of 16-18 months, from conception to completion of the projects, with interviews at critical periods such as the conception/design period, the choice of subcontractors in the initial months of the construction period, during implementation of the partnering policy and, over the last few months of the construction period, to obtain an overall evaluation of the partnering process. These interviews targeted the functional and operational directors and managers directly involved in the partnering experiments and were held with them on an ongoing basis.

Compared with the early French experiments of partnering in the 80s, UK firms adopted a more careful and pragmatic approach in the 90s in implementing project-based partnering in Design and Build contracts. As in France, subcontractors were encouraged, during their tender phase, to submit new ideas at the design phase in order to improve the "buildability" of the project (thereby increasing the level of quality at a lower production cost) whilst the monitoring of the partnering processes involved the same teamwork and social events, designed to collate the principal comments and criticisms from the players during the construction phase. However, the UK firms differed from the French contractor, mainly in their management of an unstable network. Neither of them used preferred subcontractors, nor did they allocate resources to extensive programming, tying the subcontractors to the main contractor's objectives and organisation. On the other hand, both of them designed new procedures in order to manage their relationships with their partners; Balfour Beatty resorting to ex ante risk management and Kyle Stewart to ex post evaluation methods and ongoing productivity improvement.

Balfour Beatty's site team on the M25 project implemented a risk management framework with a priviledged subcontractor (a piling company which was part of the same parent group, BICC), designed to reduce information asymmetry, and therefore to lessen the risk of opportunistic behaviours between specialists and main contractors on construction projects. In this scheme, the risk-sharing agreement was established on the basis of in-depth, formal negotiations which could be extended to other subcontractors familiar with their cost structure. The method involved compiling a list of the project's potential risks in accordance with information available at the tender stage and assessing each risk; its exact nature; its possible costs and effects on the subcontractor and the main contractor; the means by which it could be reduced and controlled by changing some operational routines or decisions, and the costs of adapting the work processes in order to reduce these risks. The costs and benefits obtained by adopting this procedure were shared among Balfour Beatty and the piling company.

Kyle Stewart's ex-post evaluation plans

According to Kyle Stewart, long term partnering - seen as a commitment between specific trade contractors and the general contractor throughout different projects - provides a response to the transaction costs still occurring in project partnering, as long as pricing procedures and technical evaluations of the subcontractors' performance on former projects exist. With the abolition of tenders, pricing must be monitored through indirect methods, such as retaining data on market conditions (through other non-partnering projects implemented by the main contractor), and kept under control by means of limited competition established between the few preferred subcontractors selected by trade. The level of technical and organisational innovation must also be controlled, to prevent the partnering agreement from restricting any collective effort towards improving processes and techniques.

Procedures are therefore discussed by the main contractor and some voluntary subcontractors, not with a view to evaluating each contractor's performance (with the difficulty already recognised by the French in evaluating the rather informal operational modes of subcontractors), but in order to find areas for improvement within the subcontractors' and the main contractor's internal organisation and in their relationships, which could have an impact on overall costs.

A collective-action interpretation of partnering

The Construction Industry's fragmentation and its consequences in terms of interfirm conflictual relationships have long been targeted by experts as one of the major difficulties for contractors. Partnering - seen at first as a set of techniques promoting "openness, trust, cooperation, unanimity of decisions, sharing of benefits and a fair allocation of risks" (Loraine, in MPA, 1995) on site - can therefore be primarily regarded as an integrative tool between the contributors to a construction project. Nevertheless, its experimentation, both in France and in the UK, shows that partnering can provide an alternative to the merchanting system as a structured base for interfirm relationships within the Construction Industry, under certain conditions. Among the different theoretical frameworks that can be used to interpret the Industry's situation in the two countries, the voluntarist sociological approach of the collective-action theory (Emery and Trist, 1965, Trist, 1983, Astley, 1984, etc.) provides fruitful concepts and analytical tools, incorporating the pioneering firms' autonomous strategic decisions within the context of a broader change in the Industry's culture.

The collective-action theory emphasises that collective survival is achieved through voluntaristic human action which occurs at a collective level. This action results in the creation of a regulated and controlled social environment that mediates the effect of the natural environment (ie the negotiated order of Emery and Trist). In this view, theorists supplant the determinism of an exogenous "natural" environment (supported by population ecologists) by the idea that "organizational environments are primarily socially constructed", and "resources are not simply fortuitously discovered and exploited, they are produced" (Astley, 1984, p.533). They advocate that "the meaningful environments are outputs of organizing, not inputs to it" (Weick, 1979, cited in Astley and Van de Ven, 1983, p.257). In particular, a social innovation, such as network formation, is viewed as a processual, empirical phenomenon, evolving through three principal stages: (1) *enactment* : firms in an industry can modify their environmental situation through collective reinterpretation, ie *domain formation;* (2) *selection*: the successful *experiments* undertaken by pioneering companies will provide their innovators with an improved level of performance; and (3) *retention*: these experiments, currently being diffused through imitation by a growing number of firms, will create the next stage within the Industry as well as influence the shape of the firms' environment.

1. Enactment

As previously mentioned, institutional differences may explain, in part, the higher degree of conflict existing between clients and contractors in the UK, mentioned by many authors (Ball, 1988, Barlow, 1996, Bresnen, 1996), compared with France, where the problem appears to be less acute, albeit present in the last decade. The merchanting model seems to have exhausted most of its possibilities for incremental adaptation, as evidenced by the clients' criticisms as well as the emergence of a collective survival theme, now recurring in the UK. This suggests that UK firms are searching for another systemic framework, whereas French firms, who adopted the merchanting system more cautiously are, in general, still interpreting their positions within its mainframe.

New forms of relationships, currently being sought by UK firms, are under collective discussion on the initiative and with the aid of an Industry referent body (The Construction Industry Board (CIB) established by Sir Michael Latham).

2. Selection

Given the present turbulence in the Industry's environment, this framework also assumes that networks (ie partnering, linking both main and subcontractors) will be increasingly used as a major form of interfirm relationship and regarded as a voluntaristic response offered by UK construction firms to environmental change. In the UK, contractual and regulation procedures on site are currently being reviewed. D&B contracts establish the central position of the main contractor and partnering appears to respond to the need for renewed regulation procedures between the contributors on site. It acknowledges the trade contractors' technical roles, involves them in the conception phase and substitutes more formal risk-assessment and ongoing improvement procedures to the former professional codes of practice and informal, verbal communication on site.

3. Retention

The UK's features appear to closely match those cited by Trist (1983) and by the human ecology/collective-action theorists. In contrast, the French situation can still be interpreted - but as a counter-example - under this framework, as it presents different institutional and historical characteristics. In particular, the referent body has not yet emerged in the French Construction Industry, where overall conditions are not viewed so seriously. Isolated French experiments can be interpreted as being mutations that were not successful enough to be applied to the whole Industry. French innovators experimented with collaborative interfirm relationships as another form of project management, within an Industry context of conflictual traditions.

Retention mechanisms are not based on a single appraisal of the experiments' results, but on their potential to provide an alternative systemic framework for interfirm relationships in the Industry. Innovation is accepted as an alternative if, as a pre-condition, the previous patterns of behaviour have been challenged and the Industry recognises that they are no longer appropriate to the current situation. The collapse of CBC's implementation after five years is a further indication that any innovation which influences a collective issue (interfirm relationships) cannot be accepted without collective reflection. In other words, subcontractors' opportunism was permitted in France because partnering was not regarded as a collective cause by the Industry.

Research limitations and future direction

Due to its exploratory characteristics, this research has not provided data that is extensive enough to draw wide-ranging conclusions and normative recommendations for the Construction Industry's management. In particular, the scarcity of experiments on partnering, in both France and the UK, may well have magnified biased or unusual situations that might be explained by the positioning of the firms that implemented them at specific times in the construction business cycle. This risk, inherent in the qualitative approach chosen in this work, must be reiterated as well as highlighting the cautious interpretation that a comparative bi-national research design merits.

Further and more wide-ranging experiments are still required to test and expand the meaning of networks within the Construction Industry. Nevertheless, it is interesting to note that former experiences in the US and in Australia (Loraine, 1994, Baden Hellard, 1995) display similar traits to the ones recognised in France a decade ago and in the UK today. In this respect, this research reveals a persistent feature that may be of use to companies willing to initiate future partnering experiments or to deepen existing ones.

Future research programmes could focus on at least two areas. Initial research could consist of following up the current UK partnering experiments in order to ascertain whether relational quasi-rents can emerge in transient networks, such as project partnering. A further research area could be a study of the lock-in effects of long-term partnering and its potential solutions. This subject is of particular interest to the Construction Industry, as most of the anticipated benefits of a networking organisation are a consequence of recurring - if not stable - relationships supported by trust, intangible and long-term investment and collective working routines.

References

Astley, G.W. (1984), "Towards an appreciation of Collective Strategy", *Academy of Management Review*, Vol 9, No 3, p 526-535,

Akintoye A.(1994), "Design and build: a survey of construction contractors' views", *Construction Management and Economics*, 12, p 155-163,

Akintoye A. and FITZGERALD E. (1995), "Design and build: a survey of architects' views", *Engineering, Construction and Architectural Management*, 1, p 27-44,

Baden Hellard, R. (1995), *Project Partnering, Principle and Practice*, Thomas Telford, London,

Ball, M. (1988), *Rebuilding Construction*, Routledge, London,

Barlow, J. (1996), "Towards positive partnering. Cultural and managerial challenges for the UK Construction Industry", paper presented at the 3rd International Workshop on Multi-Organisational Partnerships, University of Strathclyde, 5-7 September 1996,

Benhaim, M. (1997) "Interfirm relationships within the Construction Industry: Towards the emergence of networks? A comparative study between France and the UK", DBA Thesis, *Brunel University*,

Bennett J. and Ferry D.(1990), "Specialist contractors: A review of issues raised by their new role in building", Construction Management and Economics, 8, p 259-283,

Bresnen, M. (1996), "Cultural change in the construction industry: developing the client's management role to improve project performance", in *Partnering in Construction*, Proceedings of a one-day conference, University of Salford, 13/05/96,

Broussaeu, E. & Rallet, A. (1993), *Développement des systèmes télématiques et évolution des relations interentreprises dans la construction*, Plan Construction et

Architecture, Ministère de l'Equipement, du Logement, des Transports et de l'Espace,

Campagnac E., Bobroff J., Caro C.(1990), *Approches de la Productivité et Méthodes d'organisation dans les grandes entreprises de la construction*, Plan Construction et Architecture, Ministère de l'Equipement, du Logement des Transports et de la Mer, Paris,

Campagnac, E. (1991), "Flexibilité, informatisation et changements organisationnels" in *Europe & Chantiers. Le BTP en Europe, Structures industrielles et Marchés du travail,* Proceedings of a conference held on 28/29 Sept. 1988, Plan Construction et Architecture, Ministère de l'Equipement, du Logement des Transports et de la Mer, Paris, p 201-210,

Campinos-Dubernet, M. (1991), "Diversité des formes nationales de gestion de la variabilité des processus du bâtiment" in *Europe & Chantiers. Le BTP en Europe, Structures industrielles et Marchés du travail* , Proceedings of a conference held on 28/29 Sept. 1988, Plan Construction et Architecture, Ministère de l'Equipement, du Logement des Transports et de la Mer, Paris, p 149-175.

Colas R., Mathieu C. and Roch C.A. (1994), "La gestion de projet dans le process CBC" in *Gestion de Projet et gestion de production dans le bâtiment*, Plan Construction et Architecture, Paris,

Eccles, R. G. (1981), "The Quasifirm in the Construction Industry", *Journal of Economic Behavior and Organization* n°2 : 335-357,

Emery, F.E. and Trist, E.L. (1965), "The Causal Texture of Organizational Environments", *Human Relations,* Vol 36: 269-284,

Gobin, C. (1993) "Le Cycle conception-construction-maintenance: la démarche pro-active, une méthodologie reproductible à d'autres opérations", in BOBROFF, J. (ed), *La gestion de projet dans la construction*, Presses de l'Ecole Nationale des Ponts et Chaussées, Paris,

Gray C. and Flanagan R.(1989), *The Changing Role of Specialist and Trade Contractors,* The Chartered Institute of Building, London,

G.R.I.F. (Groupe de Recherche-Innovation-Formation) (1992), *Qualité et partenariat dans le Bâtiment,* Plan Construction et Architecture, Ministère de l'Equipement, du Logement, des Transports et de l'Espace, Paris,

Guilhon B. and Fuguet J.L.(1992), *Dynamique des performances et organisation des relations inter-entreprises dans le bâtiment*, Plan Construction et Architecture, Ministère de l'Equipement, du Logement, des Transports et de l'Espace, Paris,

Hillebrand P.M., Cannon J. (1989), *The Management of Construction Firms,* Macmillan Press, London,

Hillebrand, P.M., Cannon, J. & Lansley, P. (1995), *The Construction Company in and out of Recession,* Macmillan Press, London,

Latham, M. (1994), *Constructing the team,* HMSO, London,

Loraine, R. (1994) "Project specific partnering", *Engineering, Construction and Architectural Management,* Vol 1, 1, p 5-16,

Major Projects Association (1995): "Partnering: Competition vs Cooperation", Proceedings of Seminar 67 held at Templeton College, 4-5 December 1995,

Pettigrew, A.M. (1995), "Longitudinal Field Research on Change", in HUBER, G. and VAN DE VEN, A. (eds) *Longitudinal Field Research Methods, Studying Processes of Organizational Change*, Sage, London,

Roch, C.A.and Colas, R. (1993), "Remontée amont et nouvelles formes de coopération, le process CBC" in Bobroff (ed.) *La gestion de projet dans la construction*, Presses de l'Ecole Nationale des Ponts et Chaussées, Paris.

Trist, E.L. (1983), "Referent Organizations and the Development of Interorganization Domains", *Human Relations,* Vol 36:269-284,

Weiller, D. (1987), *Filières d'acteurs et partenariat, vers de nouvelles formes de sous-traitance dans le bâtiment?*, Centre Scientifique et technique du bâtiment, Plan Construction et Architecture, Paris.

Winch, G. and Campagnac, E. (1995), "The Organisation of building projects: an Anglo-French comparison", *Construction Management and Economics*, 13, p 3-14.

HARMONIOUS CONTRACTS AND HIGHER PROFITS: DILEMMA OR OPPORTUNITY?

W.TIJHUIS
Department of Construction Technology and Construction Process, Faculty of Technology and Management, University of Twente, P.O.Box 217, NL 7500 AE Enschede, The Netherlands. w.tijhuis@sms.utwente.nl

Abstract

In several actual situations in the industry there is an increasing effort to improve profit, which is becoming common in the construction-industry. The represesenting (buzz?)words like "business-process re-engineering", "holistic approach", "lean production" etc., are both parts of well known strategies of action in these situations. Remarkable however is, that for example or maybe especially in construction industry these developments seem to lead to an increase of contractual arrangements: parties try to reduce their risks as much as possible, wishing to lead to a reduction of costs and in that way to possible higher profits. But these attitudes often do not work that way: a lot of clients feel embarassed about the delivered projects, several contractors get involved in conflicts with sub-contractors and suppliers, etc. So in the end nearly very few of them got the result he or she targeted.

This paper focusses on aspects about the contractual arrangements and the "matching"of attitudes between parties: Do they really go for the same goal, and especially: are they forced together into relationships? Influences of regulations and directives are described, coupled with specific procurement-aspects to the relationship of European tendering of projects. Situations are analyzed by describing a case-study about a European tender procedure, leading to a forced relationship between client and contractor. It seemed to be then that working within a scope of mutual trust has to be one of the most important points for starting, processing and finishing projects succesfully. But governmental directives, leading to an extreme of open competition and forced relationships seem to be not really succesful. A forced openness in the choice of parties and onerons contractual arrangements should in general be the "last and least" way for keeping the parties involved going in the same direction, reaching their win-win goal in close harmony and mutual profit. Trust must be generated through a free choice, leading to pursuing harmony and profit together as a team.

Keywords: Contracts, directives, harmony, profits, relationships, tendering.

Introduction

Creating an optimized construction process often focusses on the purely technical aspects. Especially at project level, improving of technical details and solutions often seems to be the way construction professionals look to this item. This attitude seems to be quite obvious, leading to the general impression that there is quite a lack of really proactive ways of working in the construction industry: parties don't (re)act as long as there are no problems. Thus, an attitude of looking in advance at aspects of improvement of existing action and/or ways of working at project level, is quite rare. The action (or in fact *re*action) which takes place on construction improvement on project level, looks as if it is working "from critical incident to critical incident".

The focus on *critical incidents* is in research practice described by Schein, who uses it as a way of observing people into organizations and/or situations [Schein (1985)]. In that way it is interesting to look how to avoid such critical situations between parties in construction process: thus how to create an atmosphere of harmony instead of conflicts. Introducing win-win goals plays an important role in this development, although parties must then have the same scope.

Relationships in construction projects

General

In construction industry there are several relationships to be distinguished. An important reason for this is the fact that it is a very *fragmented* part of industry: several small and medium-sized firms play the main role in this area. Certain trends in the construction industry are even established, for instance, main contractors discontinued their employees, and prefer to subcontract work to them as several small firms (with one or more employees) per project [Scholman (1998)]. Thus, the larger contractors are transferring themselves instead of a (large) contractor more and more towards a sort of *quasi-firm* or a so called *virtual-organization* [Eccles (1981); Van Tongeren and Doree (1997)], whose capacity in fact only exists on temporary project level. As a result, the number of quite big and strong players in the field is quite low. This can seen in the several branches in construction industry, e.g.:

*   *contractors;*
*   *sub-contractors;*
*   *architects;*
*   *consultants and engineers;*

In addition to that it seems that the (financial) barrier for entering the market as a small firm is quite low: "You only need a hammer or a pencil to start a carpentry-firm or a designer-office", although in the European Union there are still special directives for assuring a certain knowledge and experience for craftsman, engineers, etc. [EC (1985)]. However, competition from companies from "free and wild" non-member states has increased significantly, due to the very open policy of the European Union on the other hand.

The above mentioned groups or branches often do have correlation with each other, for example, many of larger contractors own specialized engineering-consultant offices [ENR (1997)]. For several architects and consultants there can also be distinguished quite a lot of combined functions (as e.g. archineering firms, etc.). Besides that the larger conglomerates of contractors, general contractors or management contractors were often established by the merging or take-over of smaller firms.

In a comparative way one can also see the growth towards merging or collaboration in the following group or branch:

- *suppliers.*

This branch consists of large players, quite often belonging to large conglomerates. While the barrier of entering the industrial manufacturing market is quite high in terms of cost and investment, it is difficult to start individual enterprises within this branch; therefore resulting in only a small percentage of fragmentation.

The manufacturers of building materials become increasingly nationally and internationally spread, for example, by following their existing clients abroad [Tijhuis (1998b)]. This "piggy-back" approach can be successful when having strategically important products, but when the distance for transportation becomes too large, it isn't interesting at all; especially when the local suppliers can deliver quite cheap due to currency exchange rates, cost of labour, etc. Thus, when looking to the global construction industry as still a local oriented branch, it is quite necessary to act locally and have a local subsidiary or partnership. *"Knowing someone in the market is more important than having market-knowledge"* seems to be true in such circumstances [Tijhuis and Maas (1996)].

Aspects of relationships

For establishing a relationship, there are in general at least two parties necessary. This seems quite logical, but still can be quite problematic, especially when looking to a relationship as a means of exchanging information (e.g. feelings, papers, money, etc.) between parties. The following aspects of relationships can be distinguished [Bos (1994)]:

- *Contents*   : *What information is being exchanged?*
- *Structure*   : *How is information being exchanged?*
- *Atmosphere*   : *With what mutual feelings is information being exchanged?*

The above mentioned aspects do interfere with each other in the process of establishing relationships.

Types of relationships

The above mentioned interference between the three aspects indicates that relationships and their creation is a very dynamic process. Therefore relationships can change into different types, with different meaning. Thus, although the above mentioned aspects represent the essentials of relationships, an extra aspect should be added:

- *Goal*   : *What should be the purpose of the exchange of information?*

This last aspect often seems of great importance especially as a way of creating teams. For example, Luck and Newcombe mentioned that several *teams* are in fact *coalitions* [Luck and Newcombe, 1996]. It also plays an important role in the field of types of relationships. Several types can be distinguished for the construction industry. This is related to the way in which the construction process is being organized like e.g. *traditional, building-team* and *turnkey*, etc. [Tijhuis, Maas and Dowidar (1995)]. Some examples of the types of relationship with regard to the "goal" of a relationship are:

(1)   Strategic (often with reference to long-term);
(2)   Temporary (often with reference to short-term);

These aspects are quite widely known and recognised, when looking into the field of *strategy* in construction industry. And not only in industry, but also into construction research, as described by Shaffer [Shaffer (1987)].

There is on the other hand, the temporary, short-term relationship between parties-a very common way of working, also in relationship to the development of the above mentioned quasi-firms and, as Mohsini mentioned, the temporary multi-organizations, with different scopes and viewpoints [Mohsini (1984)].

In the scope of this paper however, I would rather put the emphasis on an increasing important point of view, seeming to become the basis of many relationships in the (European) construction industry:

(3)   Forced (obliged, e.g. by regulations);
(4)   Unforced or free (not obliged).

Especially in the field of European tender-procedures for public market, the directing "forces" for creating certain relationships are quite strong: There is an obligation for certain tender-procedures, with public selection rules, according to which the public client can be forced to choose not only his most likely partner, but just a partner who may be quite unacquainted with him; and the forced "marriage" often ends up in trouble. Although these practices are quite evident nowadays from the European scope of view, they seem to be in conflict with other trends in the private markets: several developments like BOT, DBM supply-chains and lean-production, co-makership, etc. [Walker and Smith (1995); Womack et al (1990)].

Harmony and profit within relationships

Although some of the above mentioned developments seem to be like modern "buzzwords", there can also be seen a realization of an important goal by using those developments: reduction of construction costs especially by introducing trust and alliancing or partnering models, not only with contractors and suppliers, but also with the principal client [DACE (1997); Tijhuis (1998a)]. However, the aspect of "being forced" or "unforced" within relationships plays an important role in it.

Similar to the situation within forced marriages, the use of forced relationships in the construction industry often ends up in trouble. For example, conflicts and unsatisfied or embarrassed clients when contractors calculate a lot of extra costs, delay of construction processes, etc. The Latham-report on *trust* and *money* describes several of these aspects [Latham (1993)].

A tool for trying to reduce risks **and** uncertainty for such unsatisfying situations still stays the contract and additional insurances [IMSA (1981); Boyd (1998)].

However, in general the "thickness" of the contract increases when the trust between parties decreases, which generally leads to a vicious circle (and not often in opposite directions!; contracts therefore still seem to stay quite thick).

But with reference to the Latham-report, one can translate *trust* into *harmony* within relationships, and *money* into *profits* on projects. However, it seems they cannot be achieved simultaneously by the use of a contract: It always incorporates a certain amount of distrust, while each party generally seeks for their highest profit, which can cause a loss for the other party.

Therefore, focussing on harmony (trust) *or* profit (money) instead of harmony (trust) *and* profit (money) is often the attitude, practiced by the parties involved even more, when the parties are being forced together into a relationship for the project. These attitudes hardly give any opportunity for reaching the goals that both parties seek: They simply do not match with each other.

The following case-study will highlight some aspects of these practices nowadays, which will be used for the conclusions and recommendations at the end of this paper, leading to a proposed attitude and opportunity for reaching harmony *and* profit in parts of construction industry.

A case-study on tendering and forced relationships

Background

As it has been pointed out above, the use of forced relationships in the construction industry is not a very succesfull way of reaching a win-win situation: harmony and profit are separated in such situations. However, when looking at the trend for opening markets by stimulating competence between parties, the result is often a forced relationship especially between European public clients and its delivering or servicing parties (like contractors, architects, suppliers, etc.). The use of European directives for tendering of public works and services is being forced by the European Commission [EC (1993)]. These obligation cannot be avoided by the national governmental laws or whatever, but should be obeyed. That these directives can cause much discussion and argument is shown by the number cases nowadays, which are being put forward to the Euopean Court, although this "development" seems to "fit" into an international European view on a general increase of putting conflicts in construction industry to court: people seem to reach the claiming-stage quickly, which should be avoided when possible [Turner (1989)].

The project

A European semi-governmental client had to build a new office for its employees. Although this client usually had a lot of smaller construction works carried out by contractors with good experience, this project was too large for contracting to the most likely party; at least, with reference to European directives on tendering procedures. While the contract-amount was larger than 5 million ECU (European Currency Unit; nowadays in 1998, 1 ECU = ca. US$ 1.15), the client was forced to tender for it according those European regulations.

Normally, when clients do have a lot of "European-sized"projects, they can handle them by using a pre-selection procedure for work-packages over a long period. So, by officially pre-selecting their parties before the projects, they can create a list of pre-qualified contractors who can tender for the project. This list is being known and registered by the European offices. But this was not the case with this client, while they did not have that large projects in general: they only had to tender European-wide for this project, so having such a list did not seem quite necesseray or usefull, at least up till then.

The client

In the two-stage procedure used, the client felt quite concerned by being forced to tender European wide, they were quite worried that they would be forced to choose a contractor which they did not trust at all. So, while becoming aware of this aspect, they started to define as precisely as possible the qualifications which the selected contractor had to meet, due to European references to such procedures, and this resulted in a long list of aspects, introducing a lack of trust. This led in fact to a situation in which harmonious aspects did not get any consideration but only the reduction of trust by using a lot of contractual selection-aspects resulted.

The tender procedure

After publishing the tender announcement, there were about ten interested contractors for the project. The selection from this group in the first stage to reduce the number of contractors to at least five, is normally a difficult process: how to compare the several offered profiles of parties? But, while using an accurate formulated selection list with measurable criteria, it was possible to select this group in an obvious and clear way. However, the creation of this list with appropriate criteria takes quite a lot of time, while it can give quite a large risk when not doing it properly. So, the extra effort being undertaken in advance of the procedure was not only a manner for reducing risk for failure into these European procedures, but it was also quite costly, while in much cases specialized parties are being hired to advice the client on these items.

In the second stage of the procedure, the party with the most economic offer was selected out of the group of five contractors. But in fact it was not the party which the client would have chosen, had he been completely free in making his choice.

The contract

Although the party was selected in a clear way, the relationship was still forced: There wasn't a real *team!* The result of having a party who was not the client's favorite was lack of trust. It also posed difficulties for the contractor realizing some profit satisfactory to the project: The client was not really happy, and therefore created a formal way of working, resulting in a very thick contract document.

Results

Doing projects within forced relationships does not create real teams. This project was an example of such a situation. This meant a continuous risk for conflicts in the project, fed by a feeling of distrust. With reference to the increasing product liability for delivering parties, in favour of the client, this increased the risk for failure costs for the contractor. These negative aspects nowadays are being improved by using other means, for example partnering, creating trust and harmony. These way of working should result in an improvement of construction processes, leading to a reduction of costs for both parties: thus, creating real profitability into projects for all parties involved in the project, as also the Dutch Association of Cost Engineers mentioned in their survey [DACE (1997). But still European directives seem to be certain barriers for (semi)governmental clients for using the opportunities of like partnering.

Conclusions and recommendations

Using modern concepts like partnering and alliancing, gives opportunities to clients to choose their most suitable partners. Not only should they be just the cheapest one, but also they should trust them to be cheap *and* qualified for realizing their projects. In such a way, the quality of parties entering into construction process really becomes a means for creating trust and harmony in construction, leading to the lowering construction costs. So, harmony and profit for both parties can be a way of working in such situations, also becoming real teams.

However, when looking to the aspects of forced relationships like it is in the case of certain (European) tender procedures, there seem to be quite a lot of barriers for introducing trust and harmony at all, with the focus only on introducing competition. This results in the clients in several cases choosing the aspect of "most economic offer"instead of "lowest price", while it gives them more opportunities to choose other aspects than only price. They want to have as much as possible selection aspects, which can reduce the risk of being confronted with the choice of an unlikely party. But it still does not give them the opportunity for the free choice, which they would prefer when possible. It certainly does not give the guarantee for lowering total construction costs at all for both parties: clients add a lot of distrust leading to (indirect) costs, while contractors bear a lot of risk as their costs of failure.

The freedom for choosing parties themselves, leading to harmonious contracts and serious offers, should create an atmosphere of total improvement in construction. Together, this will lead to increasing profits for all parties involved. Whenever a party tries not to be serious when using this attitude and only tries to go only for the profit and not for the harmony, he profiles himself as one, not worth trusting, loses his good name and goodwill in the market. Creating a good name and goodwill costs a lot more effort and money than may be gained while losing it! This aspect should keep the parties serious and qualified, for reaching the same goals together, i.e., *Harmony and profit.*

References

Bos J. (1994) *Communicatie in teamverbanden*; paper; lecture in course: 'Projectorganisatie in de bouw'; vakcode 7T580; 22nd December, 1994; Eindhoven University of Technology; Eindhoven.

Boyd F. (1998) *To insure or not to insure*; article in journal: 'Management Services'; March 1998.

DACE (1997) *Betere kansen door lagere kosten – Van Conflictmodel naar Alliantiemodel*; Bureau NAP & DACE; Dutch Association of Cost Engineers; October 1997; Leidschendam.

EC (1985) *European Directive Nr. EC/85/384*; 10th June, 1985; Publication Office of the European Commission; Luxembourg.

EC (1993) *European Directives Nr. EC/93/36 – EC/93/37 – EC/93/38*; 14th June, 1993; Publication Office of the European Commission; Luxembourg.

Eccles R.G. (1981) *The quasi-firm in the construction-industry*; article in: 'Journal of Economic Behavior and Organisation'; Vol.2, pp.335-357.

ENR (1997) *International Construction Sourcebook*; supplement to journal: 'Engineering News Record – ENR'; 22nd December, 1997; Mc.Graw-Hill Companies; USA.

IMSA (1981) *Construction Risk Identification & Prevention Techniques – CRIPT'; Internet – International Management Systems Association (IMSA)*; Proceedings of conference: 'State-of-the-art-methodology for controlling cost overruns on major construction projects – A limited participation seminar'; October 1 and 2, 1981; Massachusetts Institute of Technology; Cambridge; Massachusetts.

Latham M. (1993) *Trust and Money*; Interim Report of the Joint Government / Industry Review of Procurement and Contractual Arrangements in the U.K. Construction Industry; HMSO; London.

Luck R.A.C. and Newcombe R. (1996) *The case for the integration of the project participants' activities within a construction project environment*; Reading University; proceedings of CIB W65 and W92 conference: 'The Organization and Management of Construction'; Volume Two: 'Managing the Construction Project and Managing Risk'; pp.458-470; Ed.: D.A.Langford and A.Retik; September 1996; University of Strathclyde; Glasgow.

Mohsini R.A. (1984) *Building Procurement Process: A Study of Temporary Multi-Organizations*; Ph.D.-thesis; unpublished; Faculte d' Amenagement; Universite de Montreal; Montreal.

Schein E.H. (1985) *Organizational Culture and Leadership: A Dynamic View*; San Francisco; Jossey-Bass.

Scholman H.S.A. (1998) *Weinig zzp-ers groeien door naar bedrijf met personeel*; article in journal: 'Bouw/Werk – De bouw in feiten, cijfers en analyses'; pp.2-4; Vol.23, Nr.2; June 1998; Economisch Instituut voor de Bouwnijverheid; Amsterdam.

Shaffer L.R. (1987) *Strategic Overview of W65 "Organization and Management of Construction"*; paper; proceedings of CIB-W65 conference: 'Managing Construction Worldwide'; Volume three: 'Construction Management and Organisation in Perspective'; pp.27-28; Ed.: P.R. Lansley and P.A.Harlow; E.&F.N.Spon; London, New York.

Turner D.F. (1989) *Building Contract disputes – Their Avoidance and Resolution*; Longman Scientific & Technical; Essex; UK.

Tijhuis W., G.J.Maas and A.Dowidar (1995*) Keuze-instrument voor opdrachtgevers – Traditioneel, bouwteam of Turnkey?*; article in: 'Bouw'; journal; Vol.50, Nr.5; May 1995; pp.37; Misset Uitgeverij; Doetinchem.

Tijhuis W. and G.J.Maas (1997) *Project-development in Germany: Example of a Way of Working*; Eindhoven University of Technology, Eindhoven – The Netherlands; paper on CIB-W92 Conference: 'North meets South – Developing Ideas'; 14-17 January; pp.592-601; Ed.: R.Taylor; University of Natal, Durban.

Tijhuis W. (1998a) *Procurement and sustainability in construction industry: Tendering for durable relationships with process and projects*; paper; University of Twente, Department of Construction Technology and Construction Process, Enschede, The Netherlands; proceedings of CIB World-Building-Congress 1998: 'Construction and the Environment'; 7-12 June 1998; Symposium C: 'Legal and Procurement-Practices - Right for the Environment'; 5 Volumes; pp.1635-1643; Gavle, Sweden.

Tijhuis W. (1998b) *Connecting Marketing to Construction Process: The Impact of Internationalization*; University of Twente, Enschede and WT/Consult BV, Rijssen – The Netherlands; paper on conference: 'The 1st International Construction Marketing Conference – Opportunities and Strategies in a Global Marketplace'proceedings; 26-27 August; pp.141-148; Ed.: Chr.N.Preece; University of Leeds, Leeds.

Van Tongeren H. and A. Doree (1997) *Quasi-firms for real innovations*; University of Twente - Construction Technology and Construction Process; Enschede; CIB-W92 proceedings: 'Procurement: A key to innovation'; Publication 203; pp.761-770 and 852; Eds.: Davidson C.H. and T.A.Meguid; IF Research Corporation, Universite de Montreal; 20-23 May 1997; Montreal, Canada.

Walker C. and A.Smith (1995) *Privatized Infrastructure: The BOT Approach*; Thomas Telford Ltd.; London.

Womack J.P., D.T.Jones and D.Roos (1990) The Machine that changed the world – The story of lean production – How Japan's secret weapon in the global auto wars will revolutionize western industry; Harper Perennial; New York.

PROFITING FROM HARMONY - IN CONSTRUCTION PROCUREMENT AND MANAGEMENT

MOHAN M. KUMARASWAMY

Dept. of Civil Engineering, The University of Hong Kong, Pokfulam, Hong Kong
mohan@hkucc.hku.hk

Abstract

While task specialisation benefited many industries following the 'industrial revolution', the construction industry moved further towards even greater role fragmentation and adversarial relationships in a system of supposed checks and balances. It is now increasingly asserted that this evolution went too far - by (creating and) polarising industry (and project) groupings, inhibiting innovations, stifling teamwork; and contributing significantly to unacceptable cost and time over-runs, large contractual claims and prolonged disputes.

This is turn led to dispute minimisation initiatives such as in 'partnering'. This paper traces the foregoing developments in construction project management, with examples from Hong Kong. A survey of factors influencing building project performance in Hong Kong also confirms the importance of mutual confidence between teams, in addition to appropriate procurement paths. Untapped strengths are observed - in not mobilising the potential for multi-faceted individual development, harmonious relationships and allegiance to common goals. It is concluded that construction procurement and operational systems should be holistically re-engineered to profit from harmony; in order to properly reap the potential benefits of recent specific managerial initiatives such as those targeting 'lean construction'.

Keywords: Holistic reengineering, lean construction, operational system, procurement, profit from harmony.

The search for new paradigms to boost productivity

Turning points in the development of general management thought / theories were apparent responses to both techno-social transformations and the growing demands for enhanced productivity. Continued calls for increased productivity in the construction industry have recently reached a crescendo, not surprisingly soon after studies in the U.K. (Latham, 1994) and Australia (Sidwell, 1997) for example, confirmed the potential for achieving cost and time savings of the order of about 30%.

That there was such waste was of course becoming increasingly apparent with the 'inertia' against innovation, as well as the adversarial attitudes engendered by the marked industry divisions between clients/ consultants and contractors. At times, specialisation and the hoped for checks and balances perhaps went too far in one direction - with overkill on contractors who were presumed to be unscrupulous; but often not far enough in another direction - with little if any checks on designer/ supervisor/ contract administrators who were presumed to be professionally above-board and meticulous in their multiple duties (that even involved conflicts of interest at times, for example when 'ruling' on contractors' claims where they themselves may have been at fault, say, during design, documentation or transmission).

Such noted shortcomings have led to experimentation with a proliferation of alternative procurement paradigms ranging from various 'Design and Build' type 'management'· type and 'PFI/ BOT' (Private Finance Initiative/ Build-Operate-Transfer) - type protocols. While this aspect will be expanded upon in the next section, the following sections will deal at greater length with two parallel needs - to improve operational procedures, as well as the human resource base itself.

In terms of the former, it may be noted at this point that some recent initiatives in general management/ industry have indeed been drawn upon to different degrees in attempted adaptations to the construction industry. For example while 'Value Engineering' and 'Total Quality Management' have been introduced in varying doses, 'business process re-engineering', concurrent engineering' and 'lean production' are being advocated in construction-specific adaptations termed 'construction process re-engineering', 'concurrent construction' and 'lean construction' respectively. For example, Alarcon [ed (1997)] draws attention to the claims of dramatic performance improvements in automobile manufacturing with lean production techniques using 'half the manufacturing space, half the human effort, half the product development time, half the investment in tools'. Such quantum leaps in productivity are reportedly achieved by (a) differentiating value-adding activities (termed 'conversions') from non-value adding activities (termed 'flows' - such as inspection waiting and moving); and (b) by then reducing the latter (the 'flows') and increasing the efficiency of the former (the value-adding 'conversions'). The adaptations, experiments, achievements and potential for further gains in construction itself are explored in Alarcon (ed.); 1997. Proponents of 'lean production' itself have gone to the extent of promoting it as a paradigm shift in production philosophy from mass production, just as the assembly line aspects of mass production heralded a paradigm shift from previously craft-based production.

However, a true paradigm shift in construction, may necessitate a complete and radical rethink of procurement, as well as operational processes - more reminiscent of the 're-engineering' philosophy. Still the parent philosophy of 'business process re-engineering' itself has been said to be deficient in its applications, for example in that 'two thirds of re-engineering projects would fail'; and the need has thus also been recognised to go 'beyond re-engineering' and to emphasis the 'people' side of the process [(Hammer, 1994) and (Finerty, 1993)]. The following sections thus examine areas that may be profitably re-engineered, also incorporating relevant aspects of human re-engineering.

Re-engineering procurement systems for ehnanced harmony

Traditional procurement systems separated the design and construction functions, also introducing ambiguities/ conflicts in the management role (e.g. as cited in paragraph 2 of the previous section). Such systems apparently evolved to provide intensive controls for the existing types of projects and the industry composition and characteristics at those times. However, inherent flaws have emerged, for example in the ambiguous/ dual role of design and supervision/ management consultants - that inhibit progress in present scenarios. This has triggered the experimentation with alternative procurement protocols (eg: 'design and build', construction management, BOT); as well as the consequential initiatives to provide tools to aid the selection of the 'most appropriate' procurement system for a given project scenario (Kumaraswamy, 1997a). This in turn prompted a study of both 'procurement' and 'non-procurement' related factors that could affect performance levels in different building project scenarios (Dissanayaka and Kumaraswamy, 1998) - so that these relationships could be incorporated in building procurement system selection tools.

Data derived from 32 project data-sets and 46 questionnaires responses (in the first phase) and more detailed data from 30 projects (in the second phase) was analysed using a series of methods, including multiple linear regression and neural networks (Dissanayaka, 1998). The results indicate the importance (influence on performance levels) of both:

1. procurement related variables such as 'payment modality', 'change order/ variation' 'risk retained by client for quantity variations' and 'construction complexity due to sub-contracting'; as well as
2. non-procurement related variables, such as 'client type', 'client confidence in the construction team', 'project team motivation and goal orientation' and 'construction complexity related to new technology.

The study outcomes thus pointed to the twin needs:

1. to re-engineer procurement systems appropriately (i.e., to assemble each system suitably to match particular project conditions and priorities, from the range of possible procurement options in aspects such as work packaging, contract type and contractor selection sub-systems); and
2. to re-engineer/ revamp the human resource and organisational sub-systems that could themselves enhance performance levels - through increased confidence, improved teamwork, better motivation and common goal orientation.

Promoting internal harmony through revamped human resource management

Human resource management in the construction industry has rarely attracted the attention it deserves, perhaps: (a) because of perceptions of the transient nature of temporary teams and the increased tendencies to sub-contract (both work elements and the 'human issues'); and (b) because of the inherent difficulties in 'making sense' of the multitude of variables that could cloud human-related issues. Unfortunately, opportunities to profit from a mobilisation of human factor strengths are also frittered away in such delegations of duties and relegation of responsibilities (through sub-contracting).

Exceptions have emerged, indicating an appreciation of the value of a motivated and committed core team for example:

1. the continued deployment of 2,500 full time employees and 4,000 directly employed skilled workers by a leading contractor in Hong Kong (Tam, 1998);
2. the appointment of a 'Culture Director' by a leading British Contractor; followed by the formulation of their 'learning liberator' programmes intended to release and share a wide range of skills and knowledge, while aligning team attitudes (French, 1997); and
3. the formulation of a relevant set of 'Common Learning Outcomes' by the U.K. Construction Industry Board (1997) - for integration into all construction-related degree programmes - which have also been endorsed by all major construction-related professional bodies: these outcomes have been identified as required outputs in three groups - 'communication', 'group dynamics' and 'professional awareness'.

Unfortunately, the qualities related to the foregoing three groups were rarely evaluated during personnel selection in the past, which often focused on paper qualifications and specific (if not 'the number of years of') experience. However, the importance of such additional dimensions of human potential is being increasingly recognised. For example:

1. 'emotional intelligence' (and corresponding EQ) is increasingly considered, alongside the traditional manifestations of logical/ mathematical and verbal/ linguistic skills as evaluated in IQ ('Intelligence' Quotient) tests (James, 1997);
2. the 'multiple intelligence' framework - first presented by Howard Gardner of Harvard University in 1983 - is being increasingly used in classrooms to develop the other component of intelligence (Chapman and Freeman, 1996) - to activate and test the hitherto 'less fashionable' components of 'intelligence': i.e., music/ rhythmic intelligence, visual/ spatial intelligence, bodily/ kinesthenic intelligence, naturalist intelligence, intrapersonal intelligence and interpersonal intelligence (apart from logical/ mathematical intelligence and verbal/ linguistic intelligence); and
3. the need to harmonise 'left brain' and 'right brain' processing, is being increasingly recognised, along with the importance of empowering 'lateral thinking' to generate alternatives to traditional approaches. The 'people issues associated with knowledge' were recently highlighted by the Australian Human Resources Institute in their focus on the 'growth of intellectual capital', 'raising group intelligence' and 'knowledge management' (Shaffran, 1997).

Along similar lines, Kumaraswamy (1997b) proposed fresh approaches to improving industry organisational and project performance levels through integrated human resource development. This involved educational, training, re-training, continuing professional development (CPD) and 'on the job' components. It was stressed that such components need to focus more on the individual 'attitudes' (and organisational 'cultures') that need to be instilled, where for example 'achieving the right results' should supersede merely 'performing a task'.

This while individual performance levels would be enhanced by upgrading knowledge, skills and attitudes; the latter is crucial in enhancing the 'internal harmony' needed to synergise team outputs; as well as to promote 'external' harmony across project team interfaces. Such factors have also been confirmed as important from the study of the various variables affecting performance, as indicated in the previous section.

Promoting external harmony

Apart from the facilitation of harmonious interfaces through more appropriate procurement and human resource development, as examined in the previous two sections, the information technology explosion has empowered managers to maintain multiple multi-media links with even remotely located teams. A recent CIB W82 report (CIB, 1997) examines projections of the desired future organisation of the building process - for example, in terms of markets, information technology, management philosophies and integrated markets.

The fostering and maintenance of good relationships with clients is receiving increasing attention, specially with the realisation that repeat orders, or even longer term 'partnering' or strategic alliances, are usually more profitable than one-off jobs which are won through keen competition.

A critical area to be addressed in promoting short and long term harmony is in the conflicts, claims and disputes that arise in many construction projects. Kumaraswamy (1997c) illustrated the relationship between these aspects, as well as the advantages of (a) promoting constructive conflict at the front-end of a project, for example to generate better design solutions at the outset, while reducing destructive conflicts that may usually develop later; (b) minimising avoidable claims, and (c) controlling disputes that arise from the unavoidable or unavoided claims.

The disillusion with cumbersome litigation processes had led to the almost mandatory arbitration provisions in construction contracts, which had in turn given way to less complex alternative dispute resolution (ADR) techniques such as mediation, conciliation, adjudication and mini-trials. However, even these are 'after the event' dispute resolution mechanisms which are activated after relationships have already soured. More proactive and preventive dispute minimisation methods are now being promoted. For example (a) in 'Dispute Review Boards' where panels comprising a group of independent experts are set up in advance to monitor the project and help anticipate and avoid disputes; while (b) in 'Partnering' the contracting parties themselves set up a secondary (non-contractual but formalised) system of working relationships geared towards achieving common objectives, solving problems at the lowest possible level and thereby minimising disputes - whether contractual or otherwise.

A case study of partnering in Hong Kong by Dissanayaka and Kumaraswamy (1997) indicated the usefulness of this approach, while exposing some current shortcomings that need to be addressed. For example the co-operative teamwork 'within' organisation needs upgrading (as also discussed in the previous section) in order to extract full benefits from partnering 'between' organisations. A specific area to be addressed was the incorporation of the sub-contractors into the partnering process, given the dominance of the sub-contracting element in most projects. This study illustrated the need for the upgrading both internal and external co-operative relationships, and adds weight to the general argument in this paper for a holistic re-engineering of the overall system.

Concluding observations - on profiting from more harmonious sytems

In drawing together the threads traced in the foregoing sections, the need for integrated solutions is increasingly evident. For example, dispute minimisation through a 'partnering' sub-system (while an improvement on 'after the event' dispute resolution) will only 'plaster over the cracks' rather than provide the structural re-adjustments called for - in more appropriate project-specific procurement systems. Ad hoc quality management systems may churn out impressive documentation, but may not always achieve net benefits, unless integrated with 'dispute minimisation', 'safety management' and 'productivity enhancement' operational systems.

Human factors merit more attention in releasing the vast untapped potential of human resources through the development of individual knowledge, skills and attitudes, together with co-operative intra-team and inter-team relationships. Although institutional and technological system improvements are not discussed in detail in this paper, these must also be targeted and integrated into the procurement, human resource and operational system improvements - in an overall re-engineering exercise as illustrated in Figure 1:

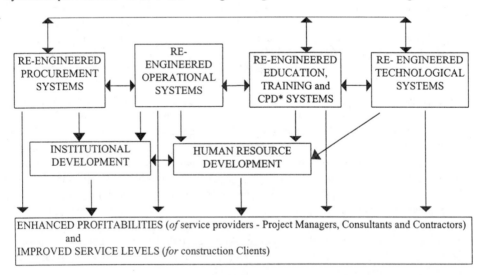

* CPD - Continuing Professional Development
Figure 1. A basic Model of the necessary integration of principal approaches needed to increase profitabilities through harmony

It is also evident that however well-designed the systems, instilling the right attitudes is as important as installing the right systems. Generating harmonious relationships though cooperative team cultures is therefore an essential (although often neglected) aspect in the pursuit of profitability.

References

Alarcon, L. [ed.] (1997) *Lean Construction*. A.A. Balkema, Rotterdam, Netherlands.
   Chapman, C. and Freeman, L. (1996) *Multiple Intelligences Centers and Projects*. Skylight, Arlington Heights, Illinois, U.S.A.

CIB (1997) Future Organisation of the Building Process, CIB Publication No. 172, March 1997, CIB W82, International Council for Building Research Studies and Documentation, Netherlands.

Construction Industry Board (1997) Common Learning Outcomes to be introduced for all Construction Professionals. *News Release*, CIB, U.K.

Dissanayaka, S.M. (1998) *Comparing Procurement and Non-Procurement related Contributors to Project Performance.* unpublished M.Phil. thesis, The University of Hong Kong, Pokfulam, Hong Kong SAR.

Dissanayaka S.M. and Kumaraswamy, M.M. (1997) Partnering begins at home. *First International Conference on Construction Industry Development, Singapore*, Dec.'97, Vol. 1, pp. 340-346.

Dissanayaka S.M. and Kumaraswamy, M.M. (1998) Identifying Factors affecting Project Performance: through Multiple Regression Analysis and Artificial Neural Network approaches. *'Engineer',* Institution of Engineers Sri Lanka, Vol. XXVII, No. 1, pp. 23-27.

Finerty, T. (1993) (part of) What the Experts Say. *Management Today*, Aug. 93, 20.

French, S. (1997) 'Comment'. *Construction Manager*, Chartered Institute of Building, U.K., 3.

Hammer, M. (1994) *Beyond Reengineering.* Harper Business, New York U.S.A.

James, J. (1997) *Thinking in the Future Tense.* Simon & Schuster, New York, U.S.A.

Kumaraswamy, M.M. (1997a) AI Aids for creative Construction Procurement. *International Conference on Computational Intelligence and Multimedia Applications (ICCIMA 97)*, Goldcoast, Australia, Feb. 1997, Proceedings, pp. 96-101.

Kumaraswamy, M.M. (1997b) Improving Construction Industry Performance through integrated Training Programmes, *ASCE Journal of Professional Issues in Engineering*, Vol. 123, No. 3, July '97, pp. 93-97.

Kumaraswamy, M.M. (1997c) Conflicts, Claims and Disputes in Construction. *Engineering, Construction and Architectural Management Journal*, Blackwell Science, 4, 2, 95 - 111.

Latham, M. (1994) *Constructing the Team.* HMSO, U.K.

Shaffran, C. (1997) Marketing Knowledge. *Sunday Morning Post*, Nov. 23, 1997, Hong Kong.

Sidwell, A.C. (1997) Effective procurement of Capital Projects in Australia. *Conference on Construction Industry Development, Singapore,* Dec. 97, Vol. 1, 38-62.

Tam, A. (1998) Gammon in Hong Kong - 40 years of excellence. *Asia Engineer*, Hong Kong Institution of Engineers/ Henderson & Associates, pp. 10-11.

HARMONIOUS AND SYMBIOTIC RELATIONSHIPS TOWARDS PROJECT SUCCESS: MYTH OR REALITY?

P. D. RWELAMILA

Department of Construction Economics and Management, Private Bag Rondebosch 7700, Rondebosch, Cape Town, South Africa. pdr@centlivres.uct.ac.za

Abstract

Lack of harmonious and symbiotic relationships between project stakeholders in the Southern Africa Development Community (SADC) public building sector has been surrounded by controversy and strongly held opinions. The work reported in this paper attempts to indicate some salient issues affecting the relationships between project stakeholders. The Botswana public building sector is used as a case study. The paper addresses two propositions. *First that lack of harmonious relationships between project stakeholders is primarily due to an inappropriate project organisational structure. Secondly, that the traditional construction procurement system (TCPS) provides a poor relationship management system.* Information is obtained on the research areas through questionnaires to construction firm executives, contracts managers, site managers, trade foremen and skilled trades-persons on the dominant procurement system used.The primary conclusion to be drawn is that the building procurement system purported to be in use in the SADC public building sector differs significantly from that recommended in the theory, resulting to poor relationships between project stakeholders. This is primarily due to the use of inappropriate building procurement systems. In general the TCPS in the SADC public building sector is used as a 'default system'. This situation has led to a situation where project management is a 'fire fighting' activity, where harmony between project stakeholders is out of reach. Salient steps are proposed with a proviso that the SADC public building sector should establish appropriate methods of selecting building procurement systems as a prerequisite in formulating appropriate project organisational structures which will bring the spirit of real co-operation between project stakeholders towards project success.

Keywords: Botswana, construction procurement system, harmony, public building Sector.

Introduction

Project management is about controlling the demands of the project, and making choices, within constraints [Stewart (1982)]. The client's objectives (project parameters) are *cost, time, quality and utility* [(Rwelamila and Savile (1994)]. Cost generally refers to first cost and in this context, choice means the allocation of budgets and contingencies. Time is better described as timing; choice means the adjustment of timing, timescales and float(s). Quality is the level of specification, and choice therefore means the adjustment of the specification, i.e. to basic, medium or high. Utility includes such things as running costs, maintenance issues, buildability, and flexibility for alterations or other uses. Choice, in this context, usually means making value judgements between higher initial costs and longer-term savings.

For the sake of completeness, it is important to recognise that projects do not occur in isolation, and that consequently they must be considered in relation to the prevailing environment. According to Rwelamila and Savile (1994), this background is specific to the country, year, location, project type, industry etc. There is little choice for the project manager (PM); any changes that occur will probably be due to outside agencies, and will cause a risk of serious disruption to the project. These changes may be in regulations, union policy, markets, technological innovation etc. Thus an appropriate procurement system is necessary for the PM in order to balance the project parameters and allocate risk appropriately, hence the foundation for harmonious and symbiotic relationships towards project success.

This paper endeavours to identify one of the primary reasons, which has contributed to poor public building sector general performance, in terms of time, cost, quality and utility. It will provide evidence which suggest that incorrect choice and use of procurement systems has contributed to creating tensions between project stakeholders and this has consequently contributed to poor project performance in the SADC public building sector.

Defining terms

- **Construction procurement system**

The principal argument of this paper rests on one fundamental aspect of the building process that requires early and particular attention if project success is to be achieved. This involves the selection of the most appropriate organisation for the design and construction of the project – herein referred to as the construction procurement system (CPS).

Reporting on an examination of past research and literature, Masterman (1992) refers to phrases such as *'building procurement method', 'procurement form'* and *procurement path'* which have been used by various authorities when referring to this concept.

According to Franks (1984), CPS is *'the amalgam of activities undertaken by a client to obtain a building'*. The term *construction procurement system (CPS)* has therefore been adopted and used throughout this paper. This term is generally used in this paper to describe:

The organisational structure adopted by the client for the management of the design and construction of a building project.

- **Harmony**

The Oxford illustrated dictionary defines harmony as *'a combination or arrangement of parts to form consistent and orderly whole', 'agreement', 'congruity'* or *'agreeable effect of apt arrangement'*. The first definition will be used throughout the paper, but with a qualification Since the study referred to in this paper is in Southern Africa, *'harmony'* in this paper, *will* be used in the African context as a metaphor that describes the significance of group solidarity, on survival issues. The word *'ubuntu'*, *'obuntu' or 'utu'* will be used interchangeably with 'harmony'.

Looking at the 'spirit of African transformation management', Mbigi and Maree (1995) argue, that for African organisations and firms, the challenges of social and political innovation far exceed the technical challenges. They thus suggest that it is important to harness the social experience and innovation of the African people and align them with successful management techniques from the West and the East.

- **Symbiosis (adjective: symbiotic)**

The Oxford illustrated dictionary refers to *'symbiosis'* as a noun, and is defined as an *'association of two different organisms living attached to one another (used especially of associations advantageous to both organisms)'* Using the ubuntu concept, and focusing on a construction project, *'symbiosis'* is one of the four cardinal principles which make-up the values of an African. Mbigi and Maree (1995) refer to this principle as *'the principle of interdependence'* – the belief that collective co-operation of all stakeholders in any project can only be achieved by acknowledging interdependence.

- **Combination of the terms**

The integration of the above terms is achieved by using two equations below. The equations are focussing on the public building sector's general performance, in terms of the usual criteria of **time, cost, quality and utility**. The starting point in looking at performance is to accept the fact that the correct choice and use of the most appropriate CPS is one of the principal requirements for an efficient project management system. But selecting an appropriate CPS needs to be supported by good management, management that is responsible to both internal and external clients. In Africa, the appropriate way of dealing with the internal clients is to apply the principles of 'ubuntu', 'obuntu' or 'utu' in addition to using an appropriate CPS.

Selection of an appropriate CPS $\Rightarrow$ balancing (Q, C, T and U) -----------Equation A1

Where:

CPS	=	Construction Procurement System
$\Rightarrow$	=	Will lead to
Q	=	Quality
C	=	Cost
U	=	Utility

Balancing (Q, C, T and U) as in Equation I + applying 'ubuntu' principles = +P
---- Equation A2

Where:

+P   =   Project Success

equation A1:

Equation A1 is based on the argument that once a client is satisfied about real need and feasibility of the building project within overall budgetary constraints, the instinctive reaction is to retain a consultant to help in the assessment of risk towards devising an appropriate CPS. The client should decide how much risk to accept. No construction project is risk free. Risk can be managed, minimised, shared, transferred or accepted. It can not be ignored. The client who wishes to accept little or no risk should take different CPSs from the client who places on detailed, hands-on control.

The basic decision on the appropriate CPS to take should precede the preparation of the outline (project) brief, since it necessarily affects who shall assist with the design brief as well. That choice of an appropriate CPS must be determined by the nature of the building project and the clients' wishes over acceptance of risk. Such decisions are very difficult. There are a number of publications [for example Masterman (1992), Hughes (1990), and Hughes (1992)], which give, detailed account of different CPSs, their risk distribution potential and merits and demerits of the same.

Once an appropriate CPS has been determined, the first stage of managing the project parameters (Q, C, T and U) has been reached. You now have a system with its associated project management structure, which will be partly, able to manage the whole spectrum of dynamics associated, will balancing the project parameters. *Why partly?* The second equation gives the answer to this question.

equation A2:

For the selected CPS to lead to project success (+P), it is important to identify and understand the project client. In fact there are many clients for a building project, e.g. a funding financial institution [external client (EC)], venture capitalist (EC), developer (EC), purchaser on completion (EC), the tenant (EC), and *in our case, a public client (EC) -using tax payers money*, contractor [internal client (IC)], and subcontractors (IC) too, have different expectations of the project Although profit may be a common desire, no one except the end user cares what is built. If the PM does not take it upon him/herself to assess the needs of all the building clients, especially the internal clients (ICs) who ranges from highly skilled to site cleaners, it is unlikely that anyone else will, and project success (+P) may not be achieved.

This raises many hard questions. Who are we building for, apart from the external client who sets the brief? How can the PM take account of other clients' (internal clients') demands? With who is the PM's building contract, and does this create a conflict of interests? The idea of the universal client (EC + IC) [for example in Rwelamila and Savile (1994)] is not new, and nor is it impossible that we should be able to satisfy the universal client's demand Focusing on the location of Botswana, there are two important facts, which need to be considered First, the majority of internal clients are Africans, secondly these ICs have, as part of their culture, a social survival strategy which if challenged could make them feel threatened, and thus rebel against the project management through different ways to the detriment of the project.

According to Mbigi and Maree (1995), there are four cardinal principles, which are derived from the values of any African community. These are given an acronym **MIST** (**M**orality, **I**nterdependence, **S**pirit of man and **T**otality) which are essentially based on 'ubuntu', 'obuntu' or 'utu' – literally translated as 'a person is a person'. This person (an African) is therefore entitled to unconditional respect and dignity.

The project manager should understand the meaning of these four principles in order to manage the ICs. In other words, if the selected appropriate CPS is going to lead to project success, the ICs must feel part and parcel of the whole building project management system. Harmony, and symbiotic relationships will depend on how the PM and his or her subordinates uses **MIST** to manage the ICs.

What is in MIST for the PM?

The principle of morality: the belief that no organisation can attain its highest potential without touching its moral base. This is best expressed through a passionate living of the code of trust. The project and its management structure will lose credibility and effectiveness if there are indications of corruption in the project management system. If there is an indication to suggest that some members of the management team are involved in corrupt activities, e.g. diverting some project materials to the black market, favouring some labourers in various ways, outside the agreed principles etc The PM must declare his or her commitment to fair practices.

The principle of interdependence: the belief that the task of achieving success in the project requires the collective co-operation of all stakeholders in the enterprising community. This can only be achieved by acknowledging interdependence. Every worker from the PM to the unskilled operative cleaning the site yard is part of the project success formula, and any move to belittle any operative on the basis of qualification or position on the project management hierarchy, will bring negative results to the project. This is the superiority of the Japanese firms because of their focus on stakeholder unity.

The principle of the spirit of man: the belief that the spirit of man recognises that man is the creator and benefactor of all wealth creation. Every person is entitled to unconditional respect and dignity. It does not matter if you are an engineer or a tea maker on site, you are all human beings first and your humanity deserves respect. The belief goes further to suggest that man is the purpose of all organisations and they must work in harmony with him or her in the spirit of service and harmony. In other words, a project is present to serve man. When it fails to do this, it ceases to exist. If the PM treats ICs as just cogs in the machine, he or she creates a situation where the operatives work in order to just make money for their livelihood and very little commitment to see the project successful.

The principle of totality: the belief that the task of making profit in a project is highly complex and involves the attention and continuous improvement of everything in the project, by every IC of the project. A project management system is made up of a number of variables. Therefore a successful project requires a number of improvements by everyone in the project. The journey towards a successful project must start with little improvements, in the manner in which every IC does his or her job and the PM relates to every IC, in terms of the improvements in universal standards which include the project parameters and two additional 'ubuntu' standards, which are *relationships and quantity* – hence the principle of totality rests on six universal standards of ***relationships, quantity, quality, cost, timing and utility***. This is the essence of 'ubuntu', 'obuntu' or 'utu' – collective participation of every IC through freedom of enterprise is a precondition to the creation of harmonious and symbiotic relationships in a construction project. These six 'ubuntu' principles can simultaneously affect the project management issues of co-ordination, communication, competence, competitiveness and compassion.

Applying 'ubuntu' in Equation A2 means applying MIST in managing the ICs. Any conscious or unconscious decision to neglect MIST principles will lead to poor project performance, because the human effort which is responsible for using the CPS to balance the project parameters will be frustrated and unable to be committed to the task in hand.

Project management in the Botswana public building sector

The dominant CPS in the Botswana public building sector is the conventional or traditional construction procurement system. It follows a strictly sequential path of four phases: preparation, design, preparing and obtaining tenders and construction.

Briefly, the building project organisation structure of this system is characterised by the appointment of a principal adviser, an architect (the PM), who leads the design team (the team may include structural, building services engineers, a landscape architect etc.) which is assembled at his or her recommendation. The building project is designed and detailed up to a point where the various elements of the design can be taken-off and worked up into a bill of quantities or a schedule of rates. At this stage the building contractor is invited to bid for the construction work and, if successful, is expected to start on site within a few days (quite often) with very little knowledge or understanding of the building he or she is to construct and probably having made no acquitance with the client for whom the building is to be produced. This project organisation structure is normally formulated parallel with a standard project contract between the client and the building contractor.The contract defines what is to be built the roles of the various parties concerned and the terms of the bargain between them. In so doing, it provides the framework of parameters balancing system. It specifies the client requirements, it stipulates the measures to be taken to assure compliance and it states the remedies available to each party in the event of default.

The Botswana public building sector purports to use this project organisation structure, and hence the framework of a project parameters balancing system in most public building projects, despite the basically transient and unique nature of various building projects. This is contrary to increasingly strong suggestion that, organisational structures should be tailored to meet particular project needs [Barrie and Paulson (1978); Walker (1984); Hughes (1990); Rwelamila and Hall (1994)]. As will be demonstrated later, this underlines the nature of project performance problems in the Botswana public building sector.

The current practice of selecting procurement systems: a survey

This section is devoted to reporting the results of an empirical survey to construction firm executives, site managers, trade managers and skilled operatives, which was carried out in 1992. Discussion of these results and their implications and relationship to the theory contained in Equations A1 and A2 will follow after this section.

For lack of space and brevity statistical analysis of the results will be kept very simple, but retaining the logic of the results in relation with the established theme of this paper.

In general, the percentages associated with opinion option are given, but where these results are misleading, this is explicitly stated. Hence, unless otherwise stated, where the term 'respondents' is used, it refers to item respondents as opposed to the overall response rate to the survey.

- **A questionnaire survey**

The number of respondents who replied by the due date, stipulated in the covering letter accompanying the questionnaires, was very good. A response rate of approximately 50% for each of the five groups canvassed was achieved. This is depicted in Table 1 below.

Table 1.   Response rates

	CF (N=369)		CM (N=123)		SM (N=492)		TF (N=615)		ST (N=2460)	
	No.	%	No.	%	No.	%	No.	%	No.	%
Description Total actual sample	60	100	60	100	60	100	60	100	60	100
Total Responses	30	50	35	58	28	47	32	53	25	42

CF = Construction firm executives, CM = Contracts managers, SM = Site Managers
TF = Trade foreman, ST = Skilled tradesman

As reported above, the dominant CPS in the Botswana public building sector is the TCPS. This system is discussed in detail elsewhere [for example, Franks (1984) and Rwelamila (1996)]. Because so little exists to confirm the general use of a TCPS, beside the extensive use of TCPS standard contract forms and traditional project management structures, it was necessary to examine the circumstances in which it is used in practice in the Botswana public building sector. The circumstances (statements) where the TCPS is likely to be successful as advocated by Franks (1984) and Hughes (1992) as shown in Table 2, were tabulated, and construction firm executives, contracts managers, site managers, trade foremen and skilled tradesmen, were asked to indicate how true those statements were applicable to most of their public building projects undertaken within the last 6 years. The results are shown in Table 3.

Table 2. Characteristics of the TCPS

A. The client commissions, and takes responsibility for the design of the works
B. The design is complete at the time of selecting the contractor
C. Prime cost sums, including nominated subcontractors do not form the major proportion of the contract sum
D. The architect appointed by the client is adequately experienced to cope with the co-ordination of the design team, to lead the design effort, and to co-ordinate the interface between design and fabrication
E. The client uses the quantity surveyor to plan and control the finance of the project, in conjunction with the architect
F. The client requires the contractor selection process to be based upon the contractor's estimate of price and for the contractor to bear the risk of costs exceeding price
G. The client reserves the right, via nomination, to select the subcontractors for certain parts of the work (but see C above)
H. An acceptable negotiated project contract form is used in order to ensure a fair and familiar distribution of risk
I. The client does not know what else to do, and the consultants do not raise the choice of procurement method as an issue

Source: Franks (1984) and Hughes (1992

Table 3. Opinions regarding the characteristics of projects (mean %)

Project Characteristics	True	More true than false	Difficult to say	More false than true	False
A	93%	-	6%	-	-
B	-	-	5%	5%	90%
C	88%	-	12%	-	-
D	-	-	40%	5%	55%
E	90%	-	10%	-	-
F	85%	-	5%	-	-
G	90%	-	10%	-	-
H	-	-	-	20%	80%
I	60%	-	20%	-	20%

- These percentages on the opinions of construction executives, contract managers, site managers and skilled tradesmen are simple averages.
- It is important to report that telephonic interviews to 10 of each of the five groups of the respondents were carried out (as a pilot study), to get an indication of the dominant CPS and to test the respondents understanding of the TCPS.
- A 100% (50 respondents) response rate was received.
- The response indicated that all the respondents were using the TCPS in all their projects. Based on the researcher's experience in Botswana (the researcher worked in Botswana for five years, and was generally conversant with procurement practices). This prompted the researcher to change the questionnaire focus, from direct reference to TCPS to identification of TCPS by its major characteristics as indicated in Table 2.

From Table 3, it is clear that, construction firm executives, contracts managers, site managers, trade foremen and skilled tradesmen are in agreement that in most of their public building projects, the following TCPS characteristics apply:

- the client commissions, and takes responsibility for design of the works (A);
- prime cost sums, including nominated subcontractors do not form the major proportion of the contract sum (C) ;
- the client uses the quantity surveyor to plan and control the finance of the project, in conjunction with the architect (E);
- the client requires the contractor selection process to be based upon the contractor's estimate of price and for the contractor to bear the risk of costs exceeding price (F);
- the client reserves the right, via nomination, to select the subcontractor for certain parts of the work , without violating the requirements of characteristic (G); and
- the client does not know what else to do, and the consultants do not raise the choice of procurement method as an issue (I).

While six TCPS characteristics apply in the majority of public building projects in Botswana, three TCPS principal characteristics do not apply:

- the design is most instances not complete at the time of selecting the contractor (B);
- the architect appointed by the client is inadequately experienced to cope with the co-ordination of the design team, to lead the design effort, and to co-ordinate the interface between design and fabrication (D); and
- a project contract form used, is not normally negotiated in order to ensure a fair and familiar distribution of risk (H).

A brief analysis and synthesis of findings

In this section a brief discussion of the significant results reported in the above section is made. Some findings are expounded by reference to the questionnaire results as a means of explanation and provision of further supporting evidence.

Each respondent group showed a fairly high general level of agreement with the others on the TCPS characteristics, which are general to most public building projects. These are characteristics A, C, E, F, G and I. The views expressed of course represent the opinion of those responsible for the management, supervision, and inspection of construction work, and it was considered that a more complete picture would evolve if the views of those physically producing the work were obtained and compared with those established from the general perception as described in the section on 'Project management in the Botswana public building sector'.

The majority of respondents, were in agreement that, for the majority of their projects it is false to conclude that "*the design is normally complete at the time of selecting the contractor*" *(B)*. This suggests that the majority of public building project designs in Botswana is incomplete at the time of selecting the contractor. Furthermore, this suggests that the selection of a contractor is based upon a provisional price, though referred to in the contract as 'a firm price'. Under this situation, the client can not describe with certainty what it is that the contractor is being invited to construct (from the documents available).

On the aspect of the architect being adequately experienced to cope with the co-ordination of the design team, to lead the design effort and co-ordinate the interface between design and fabrication, the majority of respondents were in agreement that this is not the case or it is difficult to say. A conclusion could be made from these results that, although the architect is contractually in-charge (client representative – and PM), the majority of contractors are not confident with the architect's role and his or her ability as a project manager. Why are they not confident? The architect under the TCPS is required to supervise and co-ordinate the project teams. This is a very serious situation, which need to be addressed, if project parameters are to be balanced.

Summary of findings and their implications

One of the unique features of the TCPS is that it follows a strictly sequential path, where each of these four phases as indicated in the early sections of the paper, can be viewed as separate entities and carried out, to a certain extent, in isolation of the others. The four phases are now used to summarise the findings and their implications to the project, with respect to the characteristics of the TCPS.

Summary of findings

- **Preparation**

This is the inception phase of the building project when the client establishes his or her needs in principle, but not in detail, appoints the a PM (architect) and selects and appoints his or her design team. This phase is covered under the TCPS characteristic A, which is supported by the majority of respondents as one of the characteristics of public building projects which seems to be adhered to.

- **Design**

This phase sees the appointment of the design team who develop the project through a series of sub-phases: briefing, feasibility, outline design, scheme design and detailed design with the scheme's configuration and features becoming firmer at each sub-phase. According to Masterman (1992), progress during this phase should be carefully controlled and not unreasonably forced. Hastily prepared design details can lead to major misunderstandings and disputes during the construction stage, which may result in details and cost penalties. There is no evidence from the questionnaire results to suggest that 'designs are hastily prepared', but there is strong evidence to suggest that 'designs are normally incomplete' (TCPS characteristic B) – are they incomplete because clients are anxious to see work commence on site or because there is very little understanding of the TCPS?

• **Preparing and obtaining tenders**

Tender documentation on the TCPS normally consists of drawings, specification(s) and bills of quantities, with the letter document being prepared by a quantity surveyor (TCPS characteristic E, which seem to apply in Botswana) on the basis of measurements 'taken off' from the designer's drawings in order to provide each tenderer with a common base from which to price the bid. For the TCPS to operate successfully, and to minimise the financial risk to the client, it is imperative that the design is fully developed before the bills of quantities are prepared and tenders invited. If this is not done, as clearly is the case in Botswana (TCPS characteristic B not being adhered to), excessive variations and disruptions of works are likely to occur.

Whilst selection of the contractor by limited competitive tendering should offer the assurance of achieving the lowest price for the project, the Botswana case seem to be the opposite, the respondents comments that TCPS characteristic B does not apply in their projects suggest that designer's drawings are rarely in sufficient detail to enable a bills of quantities to be prepared with any accuracy and the art of evaluating from the drawings the exact amount of work required must be a tall order to achieve.

• **Construction**

When using the TCPS, an adequate period should be allowed for the contractor to plan the project thoroughly and organise the required resources. Undue haste in making a physical start on site, which seems to characterise the Botswana situation, may result in managerial and technical errors being made by both the design team and the contractor, which could lead to lengthening rather than a reduction of the construction period.

What are the implications of these findings?

• The characteristics of the majority of public building projects in Botswana do not conform to the principal characteristics of the TCPS (characteristics B, D and H). This strongly suggests that the TCPS is basically used as **a default system**. In other words, the TCPS management structure and its respective contract arrangements are used merely because the clients and project consultants do not seem to consider the issue of selecting an appropriate procurement system.

• The extensive use of default TCPSs in the Botswana public building sector suggest that in the majority of projects, PMs do not have appropriate management structures to balance project parameters. This further suggests that the gap between an appropriate CPS and an inappropriate one (default CPS) remains a hidden challenge to the PM in trying to balance the project parameters.

• The gap between an appropriate CPS and a default CPS leads us to speculate with a strong argument, that any PM's efforts to balance project parameters under this condition will automatically lead to more pressures, and the possibility of balancing project parameters are very remote indeed.

The use of a default TCPS changes Equation A1 to:

A default TCPS ⇓ balance (Q, C, T and U) --------------------------Equation A3

Where:

⇓ = is unable to

A default TCPS is unable to balance the project parameters, and the PM's work is reduced to 'fire fighting'. Because the project organisation structure falls short of what is required, the PM's ability to keep track of the IC's dynamics is lost. As a last resort, the PM keeps his or her focus on the ECs, and uses the ICs, as cogs in the machine and conflicts become a common phenomenon between ICs (operatives) and management.

- The inability to balance parameters due the use of a default TCPS, changes Equation A2 to:

Unable to balance (Q, C, T and U) as in Equation A3 (−) 'ubuntu' =   −P
                                                                    ----------Equation A4

Where:

(−) 'ubuntu' =   MIST principles are not applied (all the forces of harmony are lost)
− P   =   Project failure

The fact that the PM does not have an appropriate CPS to balance project parameters, which leads him or her to 'fire fighting', suggests that he or she does not have time at all to deal with the human side of the ICs. The 'ubuntu' principles do not mean anything to the PM, and the majority of ICs interprets this as follows:

In the eyes of the ICs, the project has no moral base. Since the PM does not show any interest with the majority of ICs, the credibility of the PM is lost and the forces of effective strategic implementation in the project are lost − there is zero principle of morality.

The PM's inability to pull every project stakeholder together leads to a situation where the majority of ICs do not feel as a collective towards project success. The project collective becomes a myth and hence interdependence is lost in the eyes of the ICs, hence zero principle of interdependence.

The PM's attitude of trying all aspects of manoeuvring to turn the default CPS into a workable system − an impossible task, makes the PM and his or her subordinates to become aggressive. This leads to a situation where work and work only becomes a measure of every ICs contribution. The project becomes important than those who are working in the project, hence the ICs are no longer the purpose of the project. The stakeholders become servants of the project; hence the principle of the spirit of man is lost.

The PM's attitude towards ICs when trying to deal with the odds of a default CPS, becomes that of divide and manage − coming closer to those he or she feels are putting more effort to turn the project around, and those he or she feels that are not pulling themselves together become project eye sores. The essence of 'ubuntu', which is collective participation of every project stakeholder through freedom of enterprise becomes nothing in the project success equation. In short the principle of totality or system thinking becomes an empty shell.

Conclusions and recommendations

• **conclusions**

This study has shown that in the majority of public building projects, no project characteristics are identified, hence inappropriate CPSs are used. This leads to a situation where the ICs are neglected and feel unimportant. This leads to a situation where all their 'ubuntu' principals are lost and their commitment to the project lost – the harmony foundation get lost and hence the relationships between project stakeholders remain in the peripheral of the project domain.

The extraordinarily use of a TCPS as a 'default system' in most public building projects in the Botswana public building sector shows that the sector needs a paradigm shift in its choice of CPSs.

Under this culture of default CPSs, there is no chance for harmonious and symbiotic relationships to be part of the Botswana public building sector; project success will remain a dream until such time when the practice of selecting appropriate CPSs will become a norm.

There are strong indications to suggest that the Botswana public building sector procurement practices are primarily based on the legacy of her colonial past. While the British have significantly moved from the TCPS, towards embracing other CPS, Botswana like other SADC countries have remained predominantly using the TCPS. There has been very little change in procurement practices in Botswana since colonial days. There are strong indications to suggest that the same argument is relevant to the other nine of the 14 SADC countries.

• **Recommendations**

In order to establish the basic principles of formulating appropriate CPSs, the Botswana public building sector and other nine SADC countries should establish appropriate methods of selecting appropriate CPS. There are a number of ways in which this can be achieved; these are discussed in detail by various authors listed in the reference section of this paper. As a starting point a literature survey should be carried out as an audit of available methods of selecting appropriate CPS. This should be followed by intensive validation and checking exercises in order to establish methods, which are appropriate to the Botswana or SADC public building sector.

It is recommended that work be done in establishing a standard flexible contract document, which could be used for any selected CPS. This means that it will be possible to have a contract document, which could go along the above recommendation. For any selected CPS based on the actual tasks peculiar to the project, the contract document will be adjusted to deal with respective tasks.

References

Barrie, D S and Paulson, B C (1978) *Professional Construction Management*, McGraw Hill, New York.

Book Club Associates (1981) *The Oxford Illustrated Dictionary*, Coulson, J, Carr, C T, Hutchinson, L and Eagle, D Edition, Oxford University Press.

Franks, J (1984) Building Procurement Systems – a Guide to Building Project Management, *The Chartered Institute of Building*, Ascot.

Hughes, W (1990) Designing Flexible Procurement Systems, *Proceedings of CIB W92: Procurement Systems Symposium,* Zagreb, Yugoslavia.

Hughes, W P (1992) An Analysis of Traditional General Contracting, *Construction paper series No. 12, The Chartered Institute of Building*, Ascot.

Masterman, J W E (1992) An introduction to building procurement systems, *E & FN Spon,* London.

Mbigi, L and Maree, J (1995) Ubuntu – the Spirit of African Transformation Management, Knowledge Resources, Randburg.

Rwelamila, P D (1996) Quality Management in the public building construction process, *Unpublished PhD. Thesis*, Department of Construction Economics and Management, University of Cape Town, South Africa.

Rwelamila, P D and Hall, K A (1994) An inadequate traditional procurement systems? Where do we go from here? *Proceedings of CIB W92: Procurement Systems Symposium*, Hong Kong, pp. 107-114.

Rwelamila, P D and Savile, P W (1994) Hybrid value engineering: the challenge of construction project management in the 1990s, *International Journal of Project Management, Vol. 12, No. 3*, pp. 157-164.

Stewart, R (1982) A model for understanding managerial jobs and behaviour. *Academy Management Review, Vol. 7*, pp. 7-14.

Walker, A (1984) *Construction Project Management*, Granada, London.

----

* The financial assistance of the Centre for Science Development (HSRC – South Africa) towards the collection of data for this research is acknowledged.

SEARCHING FOR TRUST IN THE UK CONSTRUCTION INDUSTRY: AN INTERIM VIEW

GRAHAM. WOOD, Department of Business Studies, University of Salford, Salford, M5 4WT, G.Wood@business.salford.ac.uk

PETER. McDERMOTT
Research Centre for the Built and Human Environment, University of Salford, Salford, M7 9NU, P.McDermott@surveying.salford.ac.uk

Abstract

The pursuit of harmony and profit in the UK construction industry has seen firms eschewing traditional competitive behaviour in favour of some forms of co-operative behaviour. Partnering, as advocated, is the institutional form that co-operative behaviour is taking. The relationship between the adoption of Partnering and the creation of harmonious relationships is not yet clear. The Partnering literature does refer frequently to trust as being the foundation of partnering. However, there is no clear conceptual view of what is meant by trust. This paper reports on a pilot study of trust in the UK construction industry: interviews were conducted with five senior managers from differing kinds of firms in the industry. These managers were found to look beyond narrow self-interest, to a concern with establishing relationships which earn a profit, yes, but which produce higher returns for all if their adversarial nature can be changed. There is a clear desire to move beyond narrow self-interest to a philosophy of partnering and co-operation to gain those higher returns by lowering transactions cost and reducing conflict. However, these respondents were as concerned that the processes of the relationship were fair, open and honest as with outcomes, that is, they were as concerned about procedural justice as about distributive justice. From this they are able to offer a clear agenda for further progress in developing co-operative partnering. The pilot study has produced a clear agenda for a larger and more comprehensive study within the UK construction industry.

Keywords: Commitment, competence, motives, partnering, trust.

Introduction

This paper reports on a pilot study of trust in the UK construction industry. What picture emerges from these interviews? These managers look beyond narrow self-interest, to a concern with establishing relationships which earn a profit, yes, but which produce higher returns for all if their adversarial nature can be changed. There is a clear desire to move beyond narrow self-interest to a philosophy of partnering and co-operation to gain those higher returns by lowering transactions cost and reducing conflict. However, these respondents were as concerned that the processes of the relationship were fair, open and honest as with outcomes, that is, they were as concerned about procedural justice as about distributive justice. To produce these benefits trust has to develop between the co-operating partners (Fukuyama 1995, Brenkert 1998). These respondents have a clear concept of trust, even if they have difficulty articulating it. From this they are able to offer a clear agenda for further progress in developing co-operative partnering. The findings reported in this paper offer confirmation of previous results from studies of other industries, for example, Swan et al (1985), Anderson & Narus (1991), Moorman et al (1992 and 1993), Ganesan (1994), Kumar (1996) and Clark & Payne (1997). The pilot study has produced a clear agenda for a larger and more comprehensive study within the UK construction industry.

Procurement and contractual relations in the UK construction industry

There have been countless governmental, institutional and academic publications over the last forty years that have reported on the performance of the UK construction industry. Comparisons of the cost and time performances of UK projects with those in other developed countries were almost invariably unfavourable. The adversarial nature of contractual relationships was a dominant theme in many of the reports.

This concern with the performance and productivity of the UK construction industry led to Latham being charged (Latham 1993) with investigating the procurement and contractual relations within the industry and the structure of the industry itself. Reporting on the prevailing state of the industry, Latham commented: "....disputes and conflicts have taken their toll on moral and team spirit. Defensive attitudes are commonplace...." (Latham 1993:5)

The launch of his final report (Latham 1994) was seen by many as a watershed for the industry. Latham (1997) claimed that changes introduced since 1994 have amounted to a cultural change in the construction industry, but McDermott et al (1997) argue that the extent to which the adversarial culture of construction has changed in the wake of Latham is still open to debate. What is agreed is that a major thrust of the Latham Review (Latham 1994) has been the attempt to re-build trust in the construction industry, and this is the focus of the current research activities reported in this paper. The attempt to re-build trust has been both through the advocating of partnering at project level and through encouraging the re-structuring and realigning of the existing client, contractor, sub-contractor, supplier and consultant institutions.

Trust

Traditionally organisations have looked to the institutions of property rights, contracts and commercial law as a mechanism for sealing-in the benefits from relationships. However, Williamson (1975) observed that because of bounded rationality, and, the cost of negotiating, writing and implementing a contract, no totally comprehensive contract is possible. If confidence between partners is to develop they need to build trust in each other. Fukuyama (1995) argues that transaction costs can be lowered by social capital and trust. Although Fukuyama was applying this to nations, Latham in titling his interim report "Trust and Money" (Latham 1993) was signalling the importance he attached to the lack of trust in the UK construction industry.

Trust is a multidimensional (Sako 1992, Ganesan 1994, McAllister 1995), multifaceted social phenomenon (Fukuyama 1995, Misztal 1996), which is regarded by some as an attitude (Luhmann 1979, Flores & Solomon 1998), by others as a personality trait (Wolfe 1976) and by yet others as a vital social lubricant (Gambetta 1988, Fukuyama 1995). In spite of the large literature on the subject, Gambetta (1988) still saw trust as an elusive concept, and ten years later Misztal (1996) noted the continuing conceptual confusion that surrounded this social phenomenon. It is not within the scope of this paper to review the whole of this literature, but there are good reviews in Mittal (1996) and Misztal (1996). Rather, we wish to focus on those particular sections germane to our interest in partnering approaches to construction procurement.

We define trust as the willingness to rely upon the actions of others, to be dependent upon them, and thus be vulnerable to their actions. Trust always involves an element of risk, that a partner will abuse the trust placed in them. Where there is no vulnerability, there is no need for trust. Trust is built up over a series of interpersonal encounters (Moorman et al 1993), in which the parties establish reciprocal obligations (Nooteboom 1992). We are mainly interested in trust as it affects the willingness to co-operate. Gambetta (1988) and his contributors see trust as a precondition of co-operation because partners need some assurance that the other parties will not defect.

Ganesan (1994) and McAllister (1995) identified two components, or dimensions of trust, whereas others (Sako 1992, Mittal 1996) have argued for three dimensions of trust. For our purposes it is sufficient to note the emphasis in trust on competence (behaviour), motives (feelings) and commitment (beliefs). Issues of competence are established by market reputation, but goodwill or benevolence can only be verified in a context of mutual expectations and social bonds, that is, co-operative relationships.

An extensive list of qualities of trust, or more accurately descriptors of trust could be generated from the literature, although there tends to be considerable overlap between the various lists. Kumar (1996) listed dependability, honesty, interdependence, openness and fairness: Clark and Payne (1997) offered integrity, competence, consistent/fairness, openness, respect shown: Morgan and Hunt's (1994) lists included, consistency, competence, honesty, benevolence and fairness. All these are marked by a pronounced ethical slant, and Wolfe (1989) saw the ambiguities of modern societies as creating opportunities to "take moral shortcuts" and that "not everyone resists them." (p.219). He continued that we are not social because we are moral, but are moral because we live together. Morality matters because we have reputations to protect and co-operative tasks to carry out. Brenkert, in noting the growing importance of trust within and between business organisations,observed that: "Trust is said not only to reduce transaction costs, make possible the sharing of sensitive information, permit joint projects of various kinds, but also to provide a basis for expanded moral relations in business" (Brenkert 1998: 195).

Although time and experience are the two most commonly cited antecedents of trust (Anderson & Narus 1991, Ganesan 1994, Morgan & Hunt 1994), reputation is the most significant (Gambetta 1988, Nooteboom 1992, and Ganesan 1994). Without a reputation for trustworthiness possible partners are unlikely to enter into the first steps of a partnership. Reputations are expectations others hold of your likely behaviour in a partnering relationship and need the accumulation of positive experiences to develop. Concrete demonstrations of reliability and competence, honesty and fairness, and, helpfulness and benevolence confirm that reputation once established. A partner with a 'good' reputation is more likely to be trusted.

Methodology

We conducted semi-structured interviews focused around questions about the respondent's conception of trust, how they tried to get others to trust them and how they learnt to trust others. Five interviews have been conducted to date. In themselves these interviews provide an interesting, informed and current perspective on the nature of relationships in this industry, the prospects for the partnering philosophy and the extent to which trust is already present in relationships.

The five respondents have been drawn from different professional and organisational groups. There is a Director of a major construction company (R1); an industrial and commercial designer running his own practice (R2); a project manger working for a large commercial property developer (R3); and, a construction contracts consultant (R4); an architect/town planner with a mixture of public and private sector experience. Their length of time in the industry varies from 10 to over 30 years giving them considerable accumulated experience of how their industry operates.

Results

Characteristics of trust

Respondents were asked to identify those characteristics of trust that they sought in others, and tried to communicate to those by whom they wished to be trusted. The following five major characteristics all found in previous studies were identified.

1. *Competence/ credibility/ reliability* -Competence is here interpreted to mean that they offer or gain satisfaction from the relationship, or that value added is created. If people are not competent then we can assume that these respondents do not wish to form a relationship with them.

   We have to show we have the technical ability and a proven track record. This is essential, the number one priority (R4)

   Reliability is the key factor in my willingness to trust (R3)

2. *Promise keeping* - Two partners who cannot rely on each other to keep promises are unlikely to be able to develop trust in each other, and thus go on to develop the style of relationship which is central to partnering as a philosophy.

   You should never promise what you cannot deliver - do not mislead with false promises (R1).

   For these respondents failure to keep to what was promised is a breach of trust or betrayal.

3. *Confidence* - If the first two characteristics are present then it leads to confidence. Having confidence in the other leads to trust because there is confidence in their

competency and reliability, that they will keep their promises, that they are offering value added, give satisfaction, and can do the job.

I need to bring my client into my confidence and increase their confidence in me (R1)

Confidence is built in the early meetings through a step-by-step process, within which creating a favourable initial impression is crucial if the relationship is to get started.

4. *Communication* is the medium through which the respondents' confidence is built. They emphasised the need to keep lines of communication open.

Honesty and truthfulness are the number one requirement (R2)

Where the other is revealed as deceitful or lying is one of the few instances where trust is immediately and irrevocably withdrawn. Failure to share information openly leads to distrust.

5. *Reciprocity* is an important characteristic of trusting relationships: the partners work to their mutual advantage and seek to ensure that the relationship produces benefits for both. The benefits do not have to be equal, but they do have to be equitable.

People have to be fair and reasonable (R2)

We have to show mutual respect and tolerance for each other (R4)

These are business relationships even though some of them develop a social dimension, so the desire for mutuality is about fair exchange not altruism.

Reflection on these five characteristics suggests that to establish any relationship, a minimal degree of trust has to be offered, *a step of faith* (R3). From then on, trust deepens only where the co-operation of the other gives evidence of their competence, their ability to keep promises, to communicate openly and honestly, sharing information to produce mutually beneficial outcomes. We have a virtuous cycle where trust and openness leads to increased confidence to be more open and trustful.
The five characteristics identified by the respondents from the construction industry mirror the findings from other industries reported in the literature.

Emerging Issues

Through this pilot study we have identified five practices that form a research agenda for the larger study and these will be outlined in the rest of this paper.

1. *'Sacrificing' Behaviour* is the name we have given to actions where one actor is making some sacrifice now in the expectation of future returns. Thus the analogy is not with ritual or altruistic sacrifice, but with the 'pawn sacrifice', or 'gambit' of chess. It is a considered option, which involves the risk that it may not prove beneficial in the future, and, may involve both short-term and the long-term costs.

I am willing to open the books to reveal costs, margins and required profit where this might help establish a future profitable relationship (R4)

And:

I am happy to work with no tightly defined contract, just a guaranteed maximum price (R1)

Instead the offer is to respect the need for both parties to make a reasonable profit and for any unforseen cost increases to be negotiated to achieve an equitable sharing of the burden. All respondents show a willingness to engage in this behaviour, and accept the risk they might be taken advantage of. Engaging in trusting actions or responses, or sacrificial behaviour is used to demonstrate how open and honest you are, that you are prepared to place confidence in the other, that you will keep your promises.

Sacrificing behaviour does lead to the development of long-term reputation and hence trust (R2)

We need to identify the incidence of such behaviour, its justifications and the extent to which it produces an appropriate response. In Construction, where the industry itself has identified the lack of trust as a major problem, this is an important area for future investigation.

2. *Problem solving* - Most construction projects, even the smallest, are a series of problems to be solved if the building is ever to be erected. To the respondents this is a key test of whether a trusting relationship can be forged. Through problem solving others reveal their openness and honesty, willingness to share information and develop a reciprocal relationship. Respondents looked for a joint approach to problem solving that produces joint satisfaction and benefit.

The important point is to solve the problem with everybody's agreement and thus move the ground for everybody (R2)
Respondents readily cited instances where difficulties or obstacles, were used by others to ease the tightness of a tender offer, or increase their margins, but they also cited examples where mutual problem solving had produced solutions relatively easily and with a beneficial impact on the overall project.

Client and Contractor act as one in a positive atmosphere, whereby contractual issues take second place to a problem solving 'can do' which produces for the Client a project which meets his ambitions in terms of time, budget and quality. (R1)

3. *Reputation* is the key to being trusted and our willingness to trust. Gambetta (1988) reported that a favourable reputation influences the decision to trust, because building a reputation takes time and resources, so people can be trusted because they do not want to lose this valuable asset. Our respondents agreed in wanting to build a reputation for fairness and honesty, for a willingness to work for their partner's advantage as well as their own, and for a willingness to protect the interests of their partners.

The reputation of my business is very important to me. The reputation has been gained because this practice is able, honest and truthful (R2)

What is not clear, is on what evidence this favourable view of reputation is based. It makes sense, a priori, but reputations are intangible and fragile, easily damaged, yet many firms continue to work profitably even after damage to their reputation. Reputation is not the only factor, or even the most important factor, in deciding whom to do business with. Hall (1992) produced evidence of the value of reputation as a strategic asset but it remains an ill-defined concept. Reputation is important to these respondents but how reputations are formed, or indeed damaged, is far from clear, although it probably arises from a complex of actions, motives, intentions, attitudes and values. In an industry such as Construction, where the overall reputation of many of the companies is not very positive, this may be an even more complex issue.

4. *Interaction between business and social relationships* - Two respondents identified social situations as providing opportunities to assess potential clients' trustworthiness and demonstrate their own desire for an honest and open relationship. Social situations may provide a non-threatening situation in which to explore:-

...shared values and similar attitudes, to develop a common perspective on the proposed project (R1)

This respondent talked about the need to:-

meet people, shake their hand and look into their eyes (R1)

When one respondent reported that:

Any relationship involves the exchange of tokens. Tokens can be gifts, but they could be intangibles such as acceptance, status, and friendship (R4)
We raise the question how widespread is the use of gifts or corporate hospitality to cement business relationships? This raises serious ethical concerns.

5. *Long-term relationships* - There is an implied desire for long-term relationships, though it has rarely been made explicit in the interviews conducted to date.

Establishing a relationship is the priority rather than gaining business. Relationships lead to business and a relationship is essential though the objective remains to gain business (R4)

One illustration of this is that they all said that existing partners are given more than one chance if they abuse the trust placed in them. However trust is not easily earned and:-

trust needs to be constantly earned, it is not granted forever (R3)

Trust leads to a long-term orientation to a relationship (Ganesan 1994, Morgan & Hunt 1994), but we felt we detected some reluctance among our respondents to place themselves at too great a degree of dependence upon their partners. Perhaps this industry has too much mistrust for its members to be other than cautious in developing relationships with others. Or perhaps they are as sensitive as Husted (1998) to the dangers of a relationship descending into cronyism and favouritism.

Conclusions

No conclusions can be confidently drawn from such a limited pilot study, but the managers reported on in this paper are moral managers though they might reject that label. In their daily business practice they engage with values, principles and moral precepts as they seek to establish relationships, built on trust not contract, with those with whom they do business, though they would see this as being pragmatic. Working this way is better. Better in the sense that the returns, both short- and long-term are higher.

We can also note that the findings presented in this paper confirm earlier work by such as Ganesan (1994) and Kumar (1997) on trust and Bennett & Jayes (1995, 1998) on partnering in construction. Bennett & Jayes (1998) from their case studies have identified three stages of partnering in the UK construction industry. Those who have reached the third stage of partnering are beginning to adopt the behaviours which our respondents described as desirable. Further the five emerging issues identified find echoes in a report published recently on a consultation organised for the construction industry (Higginson 1998) and a series of academic papers on trust published this year (see Brenkert, Flores & Solomon and Husted 1998).

This pilot study does offer a research agenda that requires larger scale and more representative studies of the professional or trade groups, or different business organisations that populate this industry. An alternative is to focus on a particular client's or contractor's supply chains. Future research needs to be directed towards the development of new models of partnering for the construction industry allowing the promise of the Latham Report to be delivered (Green & McDermott 1996). It should also provide:

- Confirmation that the five characteristics identified in this paper contribute to most individuals' concept of trust, and that these five characteristics are used to establish the trustworthiness of potential partners and demonstrate their own trustworthiness.

- Confirmation that there are degrees or levels of trust (see Brenkert, Flores & Solomon and Husted 1998).

Establish the evidence to confirm or refute that the five emerging issues in this paper are the key to developing relationships based on trust and a partnering philosophy as the anecdotal evidence in Higginson (1998) suggests.

References

Anderson, J.C. & J.A. Narus (1991) 'Partnering as a focused market strategy', California Management Review, Spring: 95-113.

Bennett, J. & Jayes, S. (1995) Trusting the Team, Centre for Strategic studies in Construction, University of Reading.

Bennett J. & Jayes S. (1998) The Seven Pillars of Partnering, London, Thomas Telford.

Brenkert, G. (1998) Trust, Business and Business Ethics: An Introduction, Business Ethics Quarterly, vol. 8, no. 2: 195-203.

Clark, M. C. & Payne, R. L. (1997) The Nature and Structure of Workers' Trust in Management, Journal of Organisational Behaviour, vol.

Flores, F. & Solomon, R. C. (1998) Creating Trust, Business Ethics Quarterly, vol. 8, no. 2: 205-232.

Fukuyama, F (1995) Trust: The Social Virtues and the Creation of Prosperity, Harmondsworth, Penguin Books.

Gambetta, D (ed) (1988) Trust: making and breaking cooperative relations, Oxford, Basil Blackwell.

Ganesan, S (1994) 'determinants of long-term orientation in buyer-seller relationships', Journal of Marketing, vol. 58 (April): 1-19.

Green, C & P. McDermott (1996) An upside down and inside out approach to partnering research, Proceedings of the 12th Annual ARCOM Conference, York.

Higginson. R. (1998) Establishing Trust in the Construction Industry, Cambridge, Ridley Hall Foundation.

Husted, B. W. (1998) The Ethical Limits of Trust in Business Relations, Business Ethics Quarterly, vol. 8, no. 2: 233-250.

Kumar, N. (1996) 'The power of trust in manufacturer-retailer relationships', Harvard Business Review, vol. 74, no. 6: 92-106.

Latham, M. (1993) Trust and Money, Interim report of the Joint Government/Industry Review of Procurement and Contractual arrangements in the United Kingdom Construction Industry.

Latham, M (1994) Constructing the Team, Final Report of the joint Government/Industry Review of Procurement and Contractual Arrangements in the United Kingdom Construction Industry, HMSO.

Latham, M (1997) Procurement - the present and future trends in Procurement the Way Forward, Proceedings of CIB W92 Procure Systems, Montral, Canada: 61-74.

Luhmann, M (1979) Trust and Power, Chichester, Wiley.

McAllister, D.J. (1995) 'Affect and cognition based trust as a foundation for interpersonal cooperation in organisations', Academy of Management Review, vol. 38, no. 1: 24-59.

McDermott, P., S. Rowlinson & D. Jaggar (1997) Foreword to CIB W92 Proceedings: Procurement: a key to Innovation, (eds) C. Davidson & Tarek Meguid, Montreal, May.

Misztal, Barbara A. (1996) Trust in Modern Societies, Cambridge, The Polity Press.

Mittal, Banwari (1966) 'Trust and relationship quality: A conceptual excursion', in Parvatiyar, Atul and Jagdish N. Sheth (eds) Contemporary Knowledge of Relationship Marketing: Research Conference Proceedings, Emory University.

Moorman, Christine, Gerald Zaltman & Rohit Despandé (1992) 'Relationships between providers and users of market research: The dynamics of trust within and between organisations', Journal of Marketing Research, vol. 29, August: 314-329.

Moorman, C., Deshandé, R., and Zaltman, G (1993) 'Factors affecting trust in market research relationships', Journal of Marketing, vol. 57 (January): 81-101.

Morgan, Robert M. & Shelby D. Hunt (1994) 'The commitment-trust theory of relationship marketing', Journal of Marketing, vol. 58, July: 20-38.

Bart Nooteboom (1992) 'Marketing, Reciprocity and Ethics', Business Ethics: A European Review, vol. 1, no. 2: 110-116.

Sako, M (1992) Prices, Quality and Trust: Interfirm Relations in Britain and Japan, Cambridge, Cambridge University Press.

Swan, J. E., Trawick, I.F & Silva D.W. (1985) 'How Industrial Salespeople Gain Customer Trust', Industrial Marketing Management, vol. 14, 203-211.

Williamson, Oliver E. (1975) Markets and Hierarchies: Analysis and Antitrust Implications, New York, The Free Press.

Wolfe, A. (1989) Whose Keepers? Social Science and Moral Obligation, Berkeley, University of California Press.

Wolfe, R. N. (1976) Trust, anomia, and the locus of control: alienation of US College students in 1964, 1969, 1974, Journal of Social Psychology, 100: 151-172.

TRUST AS A HUMAN FACTOR IN MANAGEMENT IN GENERAL AND IN CONSTRUCTION

H.L. LAU
Division of Building Science and Technology, City University of Hong Kong
83 Tat Chee Avenue, Kowloon, Hong Kong. bsellenl@cityu.edu.hk

Abstract

The word trust has been invariably included in literature relating to partnering. The question then is whether trust will induce good working relationship or business relationship when establishing long-term relationship is the goal.

In a Chinese community, one may consider 'quanxi' as relationship, but the meaning of 'quanxi' is twofold. One is related to personal networking that people mostly hear when doing business in China. One refers to relationship that exists among people, organizations and business activities. Relationship involves two parties. Relationship is built on trust. There is not much discussion about trust in construction so far except in conflict resolution, subcontracting and partnering. This paper attempts to understand trust in a comprehensive way so as to interpret trust in management in general and in construction. The purpose is to find out how it relates to construction and to explore the possibility of carrying out research in this area.

Key words: Construction, management, relationship, research, trust

Introduction

'Trust' has been the word constantly mentioned in management literature in the late 80's and the 90's. It has been examined closely with relationship in the marketing field for buyer and seller relationship or salesman and customer relationship. It has been reviewed when business ethics becomes one of the hot topics in the business world. When partnering and strategic alliance issues are discussed, the significance of trust has been invariably brought up.

Trust as defined in Oxford Dictionary is belief or willingness to believe that one can rely on the goodness, strength, ability etc. of somebody or something. I do not use the definition in Webster Dictionary which America writers use because the definition there appears to have more moral content, as 'trust' is regarded as the belief or confidence in the honesty, integrity, reliability and justice of another person or thing. Trust is too vague if regarded as a feeling although it has been used in sociology and psychology. It should have something to base upon, particularly when used in business and management.

The components of trust

Cummings and Bromile (Kramer & Tyler 1996) state that trust has three components: the affective, cognitive and behavioural components. The affective component is personal and individual whereas the cognitive and behavioural components are relevant to what is being addressed in this paper. Both Chatfield (1997) and Whitney (1993) describe trust factors that can only be measured in a qualitative way and contains more of the affective and cognitive content.

Whitney (1993) further describes mistrust and finds roots of mistrust and proposes to be used as a diagnostic framework for most contexts relating to trust. He confirms in his research findings that when the following sources appear, there exists mistrust.

- The misalignment of measurements and rewards
- Incompetence or the presumption of incompetence
- Lack of appreciation for a system
- Untrustworthy information
- Integrity failure

He also suggests identifying mistrust in organization through locating the symptoms of recurring problems, but one should not mistake the problem itself as symptom. The above sources are related to human management, communication and value system.

Trust is developmental. There are three stages where it can progress to a stage where understanding and appreciation is acquired: the identification based trust, which is the highest form of trust (Kramer & Tyler, 1996). The three stages are:

Calculus based trust———➤ Knowledge based trust———➤ Identification based trust
(Conforming)   (Predicting)   (Understanding)

Handy (1995) explains the need to move beyond the fear of losing efficiency and the desire to improve checks and controls. He mentions seven rules of trust and three of them appear significant in enhancing relationship and reducing uncertainty. They are:

Flexibility - Trust demands learning and openness to change
Cohesiveness - Trust needs bonding: the goal of a small unit must gel with a larger group
Leadership - Trust equates leader

Trust must be earned (Whitney, 1993). It is not something that can just come along automatically, and organizational trust will not be automatically realized by the announcement of a trust statement. The importance of trust is now realized in enduring change and sustainable improvement in business activities. Business is built on trust. If trust can be built upon, then it can either be inside a firm or external to a firm. The bridge to it is relationship. Kenneth Arrow (1974), the Nobel laureate, has said that trust is the lubricant of society. It is not only the lubricant that helps to get things done; it is also the glue that holds the organization.

Powell (1992) in his paper on 'Conflict in the Context of Education in Building Ethics' states that 'Trust pertains to relationship, the future, confidence, morality and mutuality. Trust only comes about gradually and painstakingly. It may be perceived as a climate.' Trust has a future, but it needs to be developed progressively and it can turn into distrust if it is not monitored carefully. Trust contains mutual benefits, but it can be eroded without warning.

The levels of trust

Two levels of trust have been mentioned in the literature, high level and low level. Mechanized production requires little trust whereas automated production requires high trust. When risk and uncertainty is high, high level of trust is required such that the organization can be of a less hierarchy structure and be more flexible in responding to change. When the environment is more stable and organization is more in control, then a low level of trust is required.

Needless to say, trust can be a word used among individuals like that of faith, belief, and confidence. When it relates to management in general and in construction, trust exists in different contextual situations and in organizations, such as in a conflict situation or in a negotiation process, and in the implementation of strategy. They are now discussed under three parts:

Part I	Trust in management
Part II	Trust in contextual situations
Part III	Trust in construction

Part I: Trust in management

Trust in people management

Trust has been a word used in organization in relation to people management. Trust implies a moral content in itself which the management writers claim necessary for the implementation of strategy. It is people-centered, usually regarded as the job of the human resource managers. But the reality in many organizations is that human resource managers are mainly concerned with aligning a suitable person to a job and having him/her properly rewarded, and are seldom involved in ensuring that company goals are accomplished.

Trust (Parsons and Smelser 1956) is seen as representing a 'diffuse loyalty to the organization' which prompts the individual to 'accept, as the occasion demands, responsibilities beyond any specific contracted functions'.

The implication of high trust is that one would be confident and psychologically secure and that s/he is more relaxed, less suspicious and defensive towards the organization s/he is entrusted to (Kao & Ng/Westwood 1992). The Japanese system of lifetime employment is one example of high trust and the Chinese family business is another example.

Trust in organization

"A system is a network of interdependent components that work together to accomplish the aim of the system. A business is a complex system. All of the components are interdependent and must work together to produce services that accomplish the aim of the system. Optimization for accomplishment of the aim of the system requires cooperation between the components of the system.

Trust is mandatory for optimization of a system. Without trust, there cannot be cooperation between people, teams, departments, and divisions. Without trust, each component will protect its own immediate interests to its own long-term detriment, and to the detriment of the entire system. Transformation is required. This means adoption of and integration of new principles." W.E. Deming 1993

Information technology has open up abundant information that can be immediately available, making knowledge transferable, creating a workplace in which trust cannot only survive but also thrive. It is a people-centered work environment. Organizations are seen as dynamic systems that will naturally have disturbances, which create tension and conflict. This conflict is not only tolerated, but also seen as a sign of growth (Chatfield 1997).

When **commitment** is laden with moral significance, there is a web of mutual obligations and rights that are reciprocal -- but often subtle, understood and implied -- existing between the parties concerned and linking or binding them together (Kao & Ng/Westwood 1992). Organization commitment is often attributed to an element of **trust,** which the individual develops towards his/her working organization (Fox 1974).

The dynamic and logic behind a person's attachment to his work organization are not always calculative and rational. Both the individual and the work organization may feel mutually identified and committed to each other. But in Hong Kong the scenario may not be the same, as construction professionals demonstrate little commitment to their organization and is indicative of a rational-economic being in the community (Rowlinson and Root 1996). This can be the reason why professionals are more prone to attachment to projects rather than organizations. They are motivated by the successful performance in projects rather than the contribution to the successful performance of their firms. Could this be because professionals are unlike normal employee of an organization? Or because professionals have different codes that refrain them from sharing organization goal? This is yet to be answered. Attachment and commitment must then be viewed separately. The person can be attached to a firm but s/he is committed to the job. In Japanese organization, a person can both be attached and committed to his/her organization. So, commitment can appear in two forms: attached to task and attached to organization.

R. Likert's (1961) model of the organization (democratic/participative work groups) is a good example of linking social psychology with organizations. He viewed organizations and work groups with a managerial **human relations** perspective. He considered that work groups were an important source of the individuals' need satisfaction and that, if managers created work groups which fulfilled this need by developing 'supportive relationship', the work groups would also be more productive. Whether need satisfaction will increase productivity is still arbitrary. As for the relation factor, it is totally unknown.

The participative management system is however regarded to be successful because it associates with the motivation of the employee: *develop positive attitudes towards other members and commitment to the organization; involve group participation in setting goal, improving methods and appraising progress; establish a climate of trust and confidence and a sense of responsibility for the organization goal.* Some writers believe that participative system refers only to techniques of communication and motivation. When human relation is considered, it refers to motivation, training and job satisfaction. On a different level, it further relates to leadership, management style, organizational behaviour, decision-making, culture, job attitudes etc.

Trust in group

Collective cultures are typified by their distinction between in-group and out-group (Jackson 1993). In-group members are seen as having more commitment to the organization as inter-personal trust is difficult to generate towards people in an organization who do not consider each other to be a member of the same in-group. The psychological security in in-group relationship is formed by the implication that the member's interest is being protected and yet provides a problem-solving attitude to deal with problems associated with the organization. This further reinforces the concept of trust being in place when one is considered as a member of the in-group. Chinese family business is an example of in-group.

Views and opinions will be heard if expressed by a member of the in-group. Such in-group is built where members share the same goal towards the organization. In a project organization, if a shared goal is created, then an in-group is formed. When an in-group is formed, there exists trust among people. When an in-group relationship exists in a project team, the team members will work in harmony. They are less defensive and less confrontational, even though they come from different organizations. But then it is not their organizations that they are protecting; it is rather their professional integrity that they are maintaining. The relationship is therefore project-based. The Hong Kong view is that loyalty lies with the project (Rowlinson and Root 1996). It displays a problem-solving attitude differing from that of the western culture when such attitude is more job-oriented. The problem-solving attitude demonstrated in Asian culture is more people oriented. This is reinforced by Rowlinson's findings that professional decision-making style in Hong Kong is a more consensus based approach, and consensus is based on groupthink. Also, the tendency of having informal arrangements in project management in Asian culture is another form of working relationship and is also people-oriented.

If trust is an essential element of successful project performance, then how do we build trust within the group and outside the group? In other words, what will be the input if trust is the expected output? What are the basic components or criteria required to be satisfied as setting the first platform for trust? What has to be done to induce high trust and good working relationship such that decisions and propositions are unquestioned and accepted?

Trust in strategic management

To have strategy successfully implemented, trust is required from the top executive down to the line supervisors. This accords with Ansoff's (1987) theory on strategic action plan where planners are also implementers. The planners and implementers here may not refer to a single group of people. Due to the division of labour in organizational structure, this can be described as forming a link between planners and implementers and therefore they are 'one' – through a shared goal. This is where the beautiful word 'harmony' can come about. In Chinese it means mixed in peace.

The action plan is not in itself constant. It is developing and changing. The focus is whether the goal formulated by strategic decision-making will bring one to the desirable outcome. Therefore parallel planning and implementation are desirable whereas the management style should bear co-existence of profit-making and entrepreneurship; and the organizational culture must be defined and redefined as changes occur.

Part II: Trust in contextual situations

Trust in working relationship

It would then be fair to say that successful project management is dependent on the balance of the formal structure and the informal structure. The formal structure is the framework of working relationship, rigidly defined in agreement, in contract, or on paper record. It pushes things together. The informal structure is verbal, cannot be traced and yet exists in the word of month. The only traceable element is trust, but yet it is difficult to describe it in any physical form. It pulls things together and gives rise to an inbound for providing 'energy' for the performance of the work. It smoothes the process.

A complex project is exposed to more uncertainty than a simple project. This high exposure of uncertainty and risk can only be resolved in a flexible manner. Such flexible attitude cannot be controlled by formal structure, but rather by informal structure where good working relationship exists. So good working relationship coupled with adequate knowledge and capability is an essential element to the successful performance of a complex project. The opening of the airport is a good example: an in-group relationship has not been built at the top while exposing to a problem of high complexity. It is also fair to say that good working relationship helps to eliminate uncertainty because members are more willing to express their views and opinions before any real crisis is at stake. As Ansoff puts it: there is always a signal before a crisis arrives.

So, the higher the uncertainty or risk, the more a cohesive working relationship is required. This allows solving problems in an efficient way, which is usually in an informal manner. A collective working spirit is therefore much in demand. The Chinese saying, 'When two persons thinks alike, they are able to accumulate gold', is none the less true to a certain extent.

What kind of relationship can trust give rise to? Is this relationship sustainable? Can it be changed and under what circumstances is it changed? Is it on single route or does it appear in multitude? Can it be taken as a behaviour that can be affected, managed and controlled?

Trust in communication

In the **communication** process, a lack of trust inhibits horizontal communication (Chua/Westwood, 1992). Trust is an important dimension in interpersonal communication in business and organization relationships. High trust is necessary for all business relationships and transactions in Chinese business practice (Baron, 1983). Baron observes that business relationships begin with a transaction that involves little risk. When one party satisfactorily fulfils his/her part of the transaction, the relationship progresses to another level involving higher risk. The progression of taking greater risk will continue until it reaches a level of trust where decisions and propositions are unquestioned and accepted. Understanding one another can create shared vision or shared goal. Therefore listening is necessary in the communication process in establishing an atmosphere of trust.

Trust in negotiation

In **negotiation**, the difference degree of trust is identified as a variable of an essential precondition for successful negotiation (Fisher, 1981). A precondition of successful cross-cultural negotiation is the ability of each party to enter into the cultural space of the other party. French negotiators may come to the table mistrusting the other party until they can establish an element of trust while American negotiator may come fully trusting the other unless led to believe that the other person is untrustworthy (T. Jackson 1993). Japanese tends to have a tolerance of ambiguity and rely on mutual trust (shinyo) for the implementation of agreed settlements which are seldom in written form, not until they are faced with internationalization of business. This is a way of avoiding making offensive statement (Yuen/Westwood, 1992). Many managers in Asian countries negotiate in a subtle and indirect manner to avoid confrontation. This is also true for the Chinese. The Chinese negotiation process is in order of preference, compromising, avoiding, accommodating, collaborating, and competing (Westwood, 1992) with competing being the last resort. Their behaviour displays a low-key style at the expectation –constructing stage like other Asian managers. In fact, when they accept negotiation, it has already displayed an element of trust, just that they would buffer their bargaining positions with wide margins for a gradual compromising process to retain 'face' (Yuen/Westwood, 1992). In Chinese way, 'face' goes with respect.

Trust in conflicting situation

Learning to trust others first of all we must accept diversity (Whitney, 1993) in people. Accepting differences in others like race, gender, age, experience, and culture means that we can trust them. Therefore, trust starts with acceptance. Accepting others can only happen if we understand them. The key word is communication. It is not easy to tell whether trust leads to communication or communication leads to trust, like the chicken and egg problem. Therefore communication needs to be redefined here as conflict takes place because of communication.

In a conflict situation, accepting diversity in people will help to resolve the problem. In business, the win/lose situation is being emphasized too much. It has adopted the concept of warfare. With such mentality, there is little room for agreement or consensus to be made.

Working through the conflict may increase trust. When it gets out of hand it will destroy trust. When conflict surfaces, it must be managed in the interest of trust, otherwise it will reach a level capable of eroding trust. Whether trust and conflict can exist at the same time in the same situation is still a question to answer.

Part III: Trust in construction

Trust in partnering

The National Economic Development Council's publication defines 'Partnering' as 'the **relationship** based upon **trust**, dedication to **common goals** and an understanding of each other's individual expectations and values. Expected benefits include increase opportunity for innovation, and the continuous improvement of quality products and services.'

The words trust, commitment and relationship, frequently appear in the literature relating to partnering in construction. Partnering agreement has been well developed in the U.S. on government projects. It has been client-led for the majority of the projects. The tangible benefits of partnering have been viewed from the reduction of number of disputes, reduction in value of claims, saving in time, saving in cost and improvement in quality. The non-tangible benefits are working in harmony and maintaining good working relationship. In Hong Kong, partnering is contractor-led. It is possible that such arrangement is included to build relationship, or even target for establishing long-term relationship.

When partnering agreement is contractor-led, then such agreement can be regarded as a 'marketing tool'. It is then possible to test 'trust' in the formation of relationship and hence relationship exchange. Of course, this is yet to be proved.

Trust in sub-contracting relations

The construction industry is working with a sub-contacting system. It is necessary for sub-contractors to maintain good working relationship with main contractors such that a long-term business relationship can be formed. This has to be built on trust that one party will perform to the other party expectation, while the other party is more willing to enter into sub-contract agreement with him/her. It demonstrates cooperative behaviour, willingness to negotiate, and constant communication including formal and informal functions which all contribute to relationship building,

Trust in research

The research on trust found so far is more of qualitative test designed with questions to examine the existence of trust factor (Chatfield, 1997 & Whitney, 1993) in organization or between individual and the organization. However, there exists research with empirical studies on relationship in marketing where trust is taken as a critical factor for relationship exchange (Simpson and Mayo, 1997).

Similar research on relationship exchange is possible if applied to construction to test relationship exchange between main contractors and sub-contractor or suppliers. There is limitation in applying it to relationship exchange between clients and contractors due to the prevailing competitive tendering system when price is still the dominant criteria in selecting contractors. This is because the transactional exchange is only money exchange (Fontenot & Wilson, 1997). If construction transaction is on negotiation basis or a simple buyer-seller relationship, then the test will appear viable. It may be however possible to view client-contractor relationship in a long-term perspective to find out whether repeated transactions constitute relationship building in construction activities.

Conclusion

Since 'trust' emerge in business, marketing, partnering, sub-contracting, it must have some substance that we should not ignore. As human factor has been emphasized in the management field, America management writers and researchers are establishing 'trust' as a new principle and an underlying theory to explain organizational behaviour. Trust is therefore an area for further investigation. But trust cannot stand alone, it goes together with relations. It causes cooperation that can be considered as desirable in many situations.

References

Ansoff, I.H. (1987) Corporate Strategy, Penguin Books

Arrow K. J. (1974) The Limits of Organization, Norton, New York

Chatfield, C.A. (1997) The Trust Factor: the Art of Doing Business in the 21st Century, Sunstone Press USA

Baron R.A. (1983) Behaviour in Organizations: Understanding and Managing the Human Side of Work, Allyn & Bacon

Convey, S.R. (1989) The Seven Habits of Highly Effective People, Simon and Schuster N.Y.

Fisher R. & Ury W. (1981) Getting to Yes, Business Books, London

Fontenot, R.J. & Wilson, E.J. (1997) Relational Exchange: A Review of Selected Models for a Prediction Matrix of Relationship Activities, *Journal of Business Research* 39, pp. 5-12.

Fox A. (1974) Beyond Contract: Work, Power and Trust Relations, Faber & Faber London

Handy C. (1995) Trust in the Virtual Organization, *Harvard Business Review May/June 1995*

Hollway, W. (1991) Work Psychology and Organizational Behaviour, Sage Publications

Jackson, T. (1993) Organizational Behaviour in International Management, Butterworth Heinemann

Kramer R.M. & Tyler T.R. (1996) Trust in Organizations: Frontiers of Theory and Research, Sage Publications

Likert R. (1961) New Patterns of Management, McGraw Hill, New York

Parson T. & Smelser K.P. (1956) Economy and Society, London Rouledge and Kagan Paul

Powell, M. (1992) Conflict in the Context of Education in Building Ethics, *Finn, P. & Gammon, R. (ed.), Construction Conflict Management and Resolution* E & FN Spon pp. 388-405

Rowlinson, S.M. & Root, D. (1996) The Impact of Culture on Project Management, Unpublished paper

Sewell, R. (1996) The 12 Pillars of Business Success. Kogan Page, London

Simpson, J.T. & Mayo, D.T. (1997) Relationship Management: A Call for Fewer Influence Attempts? *Journal of Business Research* 39, pp. 209-218.

Westwood, R.I. (1992) Organization Behaviour: Southeast Asian Perspectives, Longman.

Whitney, J.O. (1993) The Trust Factor: Liberating Profits and Restoring Corporate Vitality, McGraw-Hill

PROJECT PERFORMANCE ENHANCEMENT-IMPROVING RELATIONS WITH COMMUNITY STAKEHOLDERS

KRISEN MOODLEY

Construction Management Group, School of Civil Engineering, University of Leeds, United. Kingdom. K.Moodley@Leeds.ac.uk

Abstract

The level of world-wide development continues unabated and the construction industry plays a key role in the provision of the necessary infrastructure. In developed economies the impact of the pace and scale of development is scrutinised in more detail. The construction industry and its activities have to operate within the social and political norms of the time. The industry in its widest context has an impact on every part of society and consequently, has a large group of stakeholders - those groups or individuals that are influenced by or influence the industry.

A key group of stakeholders comes from the communities in which construction projects occur. Their "stake" can vary from being initiators, to approvers, to users, to objectors, etc. Local communities are becoming more powerful as a socio-political force. They have the ability to lobby, protest, take legal action, and support construction. There is also growing sophistication in the tools and techniques used to exercise influence. The construction industry has to respond to the scrutiny it is now facing and re-evaluate the position it adopts in an increasingly stakeholder society.

The aim of this paper is to examine the issues facing construction in responding to the needs of local communities. It elucidates how the procurement of construction projects can be influenced by local community actions and what strategies can be adopted to facilitate the smooth progress of construction projects. This paper explores how greater harmony can be achieved between communities and the construction industry that should improve elements within the procurement of projects. A more inclusive and non-confrontational approach to procurement of projects is the desired aim.

Keywords: Community, construction projects, harmony, procurement, stakeholders

Introduction

The construction industry in its widest definition has a major impact on society. It uses vast quantities of resources and produces physical products that are visible all around us. The contribution construction makes to development is immense but the world is changing. There is no longer the complete and unrestricted acceptance of new development. Increased concern about the impact of new development on our daily lives, society, the environment, etc means that we are starting to adopt new perspectives on activities and actions that influence our lives. Society is taking a greater interest in their "stake".

Stakeholding

The idea of stakeholders is quite old, starting in the 1960's through the work of Ansoff, Rhenman, Ackoff and their collaborators (Freeman, 1997). It is connected with the tradition that see business as an integral part of society rather than an entity that is separate and purely economic. Identifying and analysing stakeholders were a simple way to acknowledge the existence of multiple constituencies in a company. The classic definition of a stakeholder was provided by Freeman (1984)

"A stakeholder in an organisation is (by definition) any group or individual who can affect or is affected by achievement of the organisations' objectives"

The main insight of stakeholder identification was that strategic managers might pay some attention to those groups that were important to the success of the corporation. As we have moved through the last few decades more management thinkers began to advocate that stakeholders have some sense of participation in the affairs of the firm. The emergence of consumer advisory panels, quality circles, environmental advisory panels, etc are all designed to get the firm more in touch with key relationships it faces in the future.

During the late 80's and early 90s we also saw the emergence of a strong movement concerned with business ethics. Most of the business ethics movement has been to respond to the excesses caused by financial scandal, oil spills, corruption, and damage to the environment. Questions were being asked about the very purpose of the organisation. Universal prescriptions such as, firms were primarily in business to increase shareholder wealth were under fire. Simply there are many ways to manage successful companies. Despite differences between these companies they will all involve the intense interaction between employees, management and non-management alike, with critical stakeholders. The concept of stakeholder management is being recognised as an essential part of strategic planning.

Community

The concept of society has been attacked a great deal over the last two decades in favour of the cult of the individual. During the same period the growth of monetarism and the decline of Keynsian economic philosophies increased the emphasis on the market. The late 90s have started a shift back towards the importance of collective values. Moreover there has been the realisation that power can be derived from groupings of people acting together in a cohesive manner.

The concept of community has been an integral part of the human fabric. The idea that socially and politically cohesive groups of people or communities are present have been usually associated with residential clusters. Proximity to other people is not the sole function of establishing the idea of community. Groups of people who feel they belong can create a sense of community. The very nature of community suggests that they are as diverse. Inner city neighbourhoods will be as diverse in their attitudes to suburban middle class areas as are rural villages. There will be differences in economic, social and political attitudes. The value systems that each group adopts will be determined by the particular circumstance. The use of the term community can also be attributed to collections of people with similar interest such as the financial community or academic community. The power they wield will also be determined by their political influence and the degree of cohesion.

Community is becoming a more influential group of stakeholders to large corporate organisations outside the construction industry. Consumer related industries have realised the need to be more community orientated. Consultative panels have been formed to provide information on new products, process and directions. These focus groups provide useful comments to firms on their activities and help ensure successful implementation of schemes and products. The case of Johnson and Tylenol is well known, however the attention the company paid to its communities created goodwill and trust that allowed then to reintroduce the brand successfully under the most difficult of circumstances after the poisoning of eight people. Customers, clients and employees live in communities and their well being is inextricably tied into their communities.

Community and construction

The construction industry has often faced hostility over its activity. The positive aspects of construction activity are rarely seen but the negative accentuated. The 'Blots on the landscape' gain the greatest publicity. In the UK, a major battleground between the industry and community is on roads and motorway development schemes and the extent to which these projects degrade the environment. The arena has now moved towards the protection of 'greenfield' sites for housebuilding and new development. Any scheme that represents a threat to the environment is regarded as a legitimate case for opposition. This confrontation pits the industry against a variety of interest groups and communities. These confrontations are not isolated and extend to other types of construction and development projects. The coalitions that oppose construction have also gained additional support from parts of the public that feel their quality of life is threatened or compromised. A new more affluent and influential group of people are joining elements of the anti-construction lobby in certain cases. These groups have a great deal of resources at their disposal and the ability to frustrate projects through more sophisticated means than demonstrations.

The industry also suffers a poor image by being seen to be dirty, noisy and disruptive. It also suffers the effects of 'short termism', here today but gone tomorrow without concern about the aftermath. With these credentials construction activities are often not welcomed by the communities they find themselves yet the industry is inextricable linked to the communities within which they operate through the planning processes, labour, resources, production processes and the final finished products. As communities become more sophisticated they acquire the means to disrupt, delay or ultimately stop projects from achieving completion or conversely help expedite them. It would therefore be logical acquire the support of local communities to ensure projects are completed satisfactorily. To function more effectively the industry needs to consider different strategies when interacting with communities.

Clients with a social remit require greater social input from their contractors and consultants. Local authorities are at the forefront of this drive. They are looking for planning gain before projects are allowed to proceed. Regeneration projects rather than being prescriptive have moved towards closer co-operation with the communities they are involved in. The considerate contractor's scheme which has been introduced in the UK is largely a result of a local authority initiative. The City of York have set up a group called Local Agenda 21 to develop green ideals with practical measures to ensure quality of life and environmental protection. It uses the Rio Protocol as the starting point for its activity. This body is will influence thinking on areas such as transport, housing, pollution, leisure education, etc. Its remit will have an influence on all future construction work and development in the city. Such initiatives are not isolated and are part of a growing trend to create a sustainable development in sustainable communities.

Large retailers particularly the supermarket chains are concerned about the impact their construction projects have on their local communities. They are in locations for a long term and do not wish that their projects create any disharmony or negative publicity within the communities they operate in. They ultimately recognise that community provides the environment in which the business can grow. A very conscious effort is being made by the large supermarkets to ensure that they do become involved with projects that portray a negative image. The pressure from clients will grow as they attempt to differentiate themselves from each other. A socially conscious client is likely to demand greater social input from contractors that work for them. The greater use of framework agreements between certain clients and contractors leads to the implication that the client has greater leverage to shape how the construction work proceeds.

The relationship between construction and community is clear. Community led responses are becoming more sophisticated and vociferous. The move towards recognising communities as important stakeholders is developing with clients who view communities as a key constituency. Those organisations that are reliant on consumers or have a social remit realise that to alienate communities are bad for business. They are therefore adopting strategies that suit their businesses.

Community influence on projects and business

Success in construction is normally judged in economic and technical terms. The management of construction projects revolves around the ability to meet cost, time and performance targets. Meeting these targets are not easy and as greater pressure from clients to improve performance grows, it will become even more difficult. The Rethinking Construction Report (1988) suggests that British construction can reduce capital costs by 10% per annum, construction time by 10%, predictability of cost by 20%, and defects and accidents by 20%. These targets should make construction more productive and efficient but would also create an environment where there is little scope for contingency for unforeseen events. Construction projects are now very much a target in the current environment in the UK if they are seen not to meet environmental and sustainable ideals.

Community influence appears very early in the life cycle of the project during the feasibility and approval stages. All projects require approval at either a local or national level. The process of approval often requires consultation in a public arena. These activities take time and can have a knock on effect on the project success. The proposed terminal five facility at London Heathrow airport is in the third year of its public enquiry. The cost of the enquiry is expected to be approximately £100 million. The scheme has eleven local authorities and a variety of local, national and international interest groups opposing the scheme. It is expected that as a result of this process there will be considerable difficulty in ensuring the original cost targets are met should the project be approved. The inflationary impact of delays has required a rethink on the budgets. This high profile project has met a great deal of local resistance ensuring that a comprehensive enquiry is undertaken. This case demonstrates how projects can be frustrated is there is opposition on any major scale.

The process of approval has failed in many cases to meet the desires of local communities and interest groups. The approval process is very formalised and is not seen in many cases to give an adequate voice to dissent. Failure to get a project stopped does not end at public enquiry stage, as attempts through the litigation to hinder projects are another form of action. It is a costly way to resolve problems but with the support of special interest groups or local authorities, local communities can form powerful alliances in their actions. The influence of political power can also be exercised and its use has a far greater impact at local level where relatively small numbers of people can have an impact. A well organised local community with a determined agenda can make their voices heard at the highest possible political levels.

Once the project has completed its feasibility and design stages it moves into the execution phase of the project. Timing, budgets and technical targets have been finalised and the project has to be executed efficiently. In the previous phases of the project, the project champions and the approving authorities may have been the targets of community action but during the execution the construction firm is very much at the centre of all attention. Once the execution starts then the contractor assumes responsibility. Most of the anti-construction activity focuses on direct action involving activities like picketing, occupation of the site and to a certain extent sabotage. Common tactics include occupation of trees, tunnelling and setting up protest camp. To protect the sites or remove occupiers costs both time and money. Events such as the openings of schemes are also targeted for protest. The main thrust of these actions is to generate as much negative publicity as possible for the project.

There is another level of activity that is aimed at a corporate level. This is where the parent firm faces community action at a corporate level. Ownership of shares in the firm allows access to annual general meetings where it possible to protest a companies actions. Many institutional shareholders are wary of investing in companies that are seen to be acting contrary to wishes of their shareholders or public perception. The public perception of the construction firm may also be influenced by any negative publicity. The improvement of information transfer now makes it possible to run protest campaigns on the internet that can disseminate information to a wide audience. This is now a very effective way of spreading information and galvanising opinion. Companies need to consider the impact this can have on the long term future of the firm.

As a constituency, community cannot be ignored. The growth in awareness and changing social values is probably best illustrated by Greenpeace that has a membership of four and a half million, and offices in thirty countries. The issue of environment was seen to be on the fringe of consciousness but is now very central to most political parties in Europe. Successful coalitions between well organised interest groups and communities have launched a number of successful campaigns to hinder projects. The most successful of these has been forcing the British government to rethink its road building policy. The question is how do we integrate community stakeholding into the way we operate.

Strategies for community interaction

The development of a community strategy is concerned with the orientation of the company to a key external environment influences. It is also a reflection of the norms and value that the firm adopts. An emerging theme is that world class companies develop within the communities they operate. Corporate involvement in the community is not only geographically and projects specific but requires a long term outlook that requires a philosophy that is contrary to the short term approach of construction projects. There has to be a conscious decision at a strategic level that community engagement is a worthwhile activity. The level of corporate commitment will determine whether the firm is not hypocritical in its behaviour.

Once the strategic approval is given the strategic development has to take place. The involvement of construction is not only geographic but also projectised which would suggest that the strategy needs to operate at two levels, a project specific level and an industry/environmental level.

The Industry/Environmental Strategy - this strategy is aimed at issues that influence the firm irrespective of the project or location. Community policy aims to support activities that are of strategic interest to the firm, as well as its norms and values. The other influential factors are general social, economic, political and environmental issues. This part of the emerging strategy is concerned with the long term approach needed to make community policy meaningful. This part of the strategy is more closely allied to activities that may take place in more consumer orientated industries where engagement is widespread and general.

This part of the strategy is easily implemented because it relates more directly to the corporate ethos of the firm. The values a firm adopts will provide the springboard to community related activity. The translation of the values of the firm throughout its activity was identified as one of the features of "excellent" companies. A company committed to education and training of its workforce would be able to translate these activities into a community policy. A literacy initiative, skills development programme, etc may follow as a community strategy. The community policy meets the needs of corporate goals and actively engages an important part of the community. Kanter (1995) identifies many similar features in her examination of what constitutes world class.

Project Strategy- this part of the strategy is aimed at local and project concerns. Ideally local policies fit into the industry wide strategy but sufficient flexibility has to be built in to allow implementation at a project level to be meaningful. Local economics, local values, local politics and local infrastructure play an important part in community strategy at a project level.

The project level strategy in construction is important due to the contribution project success makes to the overall success of the firm. Projects are the production mechanisms of firms. Community interaction is also greatest at the project level as the firm is operating at the heart of the community. The purpose of developing a strategy is to take cognisance of many of the issues that communities feel need to be addressed when construction activity enters their environment. These are project and community specific and may vary. Issues such as pollution, noise, disruption, safety, and privacy are often raised as standard problems that communities encounter but may vary in intensity with each new project. Standard procedures conceived in advance of execution of projects will ensure the smoother implementation. This type of forward planning can only be achieved if there is organisational commitment to the improving community interaction.

Relationship building at a project level can also ensure that disruptions are kept to a minimum. This could take place at a school, religious group or other community group. Dissemination of information about the project helps improve understanding about the project. Good relations and a sense of community ownership of the project reduces activities such as theft and vandalism and provides a support structure for the project. Relationship building is often misguided in that only those that support the project are engaged. Wherever possible the different community stakeholders should be engaged to build relations. Good relations help to facilitate projects.

Customer service/community service orientation.

A tried and tested approach to community interaction is to develop a customer service orientation for each project. A customer service approach is used successfully in other industries. There has been a reluctance by many parts of the industry to adopt this type of approach. One of the excuses is who is the customer? It is easy to become caught up in semantics while there are negative perceptions about the industry. There is very little information on how customers perceive the industry. Communication and information are the key to engagement.

Information about the project should be made available to interested parties. The site information boards should clearly state the nature of the project and the duration the project is likely to take. This gives the community an indication why their lives are going to be disturbed, what benefit will they get from it and how long is the project going to last. This information could also be made available at central locations such as shopping centres. Local residents likely to come into the contact with the project should also be informed were possible. As part of developing a service orientation ethos there should be complaints or information hotline for the project. If the cost of having a dedicated line is too expensive then a corporate level decision about the importance of a such a service should be investigated. While some in the construction industry may balk at the cost, it is imperative to remember that most customer orientated industries have a dedicated information and complaints service. Setting up such a service is no good if there is no follow up to complaints. It is also the sign of a mature and confident company that a contact person is provided to allowed for on the project.

The development of a community orientation should look to minimise the impact construction has on the local environments. The site should attempt to reduce any detrimental impact on the immediate environment such as pollution and noise. Paying attention to these will also reduce the number of complaints to the firm and to the nearest local authority. The hoardings on the site should be kept in a tidy condition, free from graffiti and unsightly posters. It is also useful to remember that not everyone wants to see large parts of a messy site. If you are going to provide viewing the use of viewing portholes or windows are a better idea. Promises must be kept as a demonstration of good faith. When the company promises not to remove a tree it should do so or if a playground is going to be created it should be done. Nothing sours community relations like bad faith.

Although the terminal 5 facility mentioned earlier does not have planning permission, the client project managers for the scheme are planning ahead. They have outlined stringent controls for environmental and pollution standards. A code of practice to prevent disruption of local communities has been devised. Actions include a specified transport manager to ensure materials are delivered on specified routes to avoid traffic congestion, and restrictions on the movement of deliveries during peak traffic periods. Work will be have to be conducted within set times and twenty-four hour working period limited to tunnelling work. The client intends setting up a 24 hour complaints line with replies guaranteed within a day. Contractors will also be required to meet local residents quarterly to discuss any problems. Such schemes may not be common place but it does show that some clients see the importance of good community relations. It is probably not surprising that the client for this project is involved on a day to day basis with thousands of travellers and probably realises the importance of good customer service.

Disruption to communities also occurs outside urban or built up areas. On road projects the dreaded cone or diversion has almost become a feature of the landscape. The use of the local media to inform motorists of diversions, delays or other distractions is also part of the development of customer orientation. Information should be provided on a continuous basis. Getting information to media sources helps reduce the impact on the community. The less the disruption of local life the more successful the project is.

Community interaction - the positives

Positive community policies can improve a company's market position both at a corporate and community level. The immediate impact of a community policy is to the profile and image of the organisation. Involvement in community policy increases the profile and recognition of the company. Recognition creates a 'brand' identity associated with the particular activity. The image of the company is enhanced through the work it does. At a local level the image of the firm is directly associated with the projects they undertake. The manner in which it operates at this level will have either a positive or negative influence on the firm. The identity is still improved irrespective of the message that is generated.

Involvement brings the company into a wider network of interaction. The power of community politics particularly in the planning and approval process should not be underestimated. Local people power can have a much greater influence on the direction a project takes. The development of technology means that access to information and the transmission of information is easier. A developed social policy should provide an additional resource to test ideas and concepts before entering the formal system. The concept of consultation groups on the environment indicates that it possible to create community based groups to provide opinion on ideas, issues and problems facing construction companies.

Competitive advantage can be gained where clients require construction projects to provide more than the facilities. Firms with a community policy are better able to market themselves to such clients. Community involvement indicates that firms are better able to operate in partnership arrangements with communities. These firms enter into partnership arrangements with a clearer vision of what is achievable under such arrangements and with a more realistic approach to the expenditure involved in these processes. This is evident in new trends in regeneration projects. This competitive edge is further enhanced by the firm's improved image and access to existing community networks.

The general public and ethical investors are looking increasingly at the activity and behaviour of firms. People prefer avoid companies that are regarded as not socially responsible. The growth in ethical investment funds also indicates that ethical investors are looking more closely at the activities that firms are involved in. Involvement in community projects increases the ethical dimension and social responsibility profile of the firm. As prices, quality and service differences narrow, firms will have to differentiate themselves. In the marketplace, the perception of the public to a firm is critical to its future viability. It is perhaps interesting that the UK supermarkets are taking the biggest lead in organic produce that damage the industry less and in sourcing their products under "fair trade schemes". As an industry sector they are probably closer to the consumer opinion. Large scale supermarkets are also particularly concerned about the image that is created when they construct new facilities and are particularly concerned about minimising disruption to their communities.

The advantages of developing community based project interaction are evident. The success for individual companies is dependent on their commitment to such schemes.

Conclusion

Community engagement mirrors the environmental movement in many ways. When environmental concern and issues began to emerge it was treated as a fringe activity. Rather than engaging the concerns it was ignored until the momentum for change was such that it was no longer possible to control the agenda. The power of communities has not had a significant opportunity to express itself. Over time as development continues quality of life, environmental and sustainable issues will become more evident. Communities will disrupt and delay project and possibly force policy rethinking with greater vigour. This negative ability will in future take on more significance as the drive to reduce costs and improve delivery of projects continues. The main aim should be to turn the negative forces to a more positive approach where communities will act as facilitators and champions of projects. It is time that aspects of the industry became more consumer/community service orientated. The pressure to implement these ideas will come from clients that have their business interests tied to community and consumer interests. Government and consumer based industries are at the forefront of the movement for change. Community stakeholders are likely to become the next big movement for change in the way construction operates.

References

Freeman, R.E . (1984) Strategic Management: A Stakeholder approach, Marschfield, M.A: Pitman

Freeman, R.E. & Liedtka, J (1997) Stakeholder Capitalism and the Value Chain. European *Journal of Management*. Vol 15, No 3. Elsevier Science

Kanter R. M .(1995) World Class -Thriving Locally in a Global Economy. Simon and Schuster. New York. Pp 29-33, 174-197

Rethinking Construction (1998) The Construction Task Force. Department of the Environment, Transport and the Regions. HMSO

HARMONY AND PROFIT LESSONS FROM HISTORY: THE CASE OF SOLOMON'S TEMPLE

STEPHEN O. OGUNLANA

School of Civil Engineering, Asian institute of Technology, PO Box 4, Klong Luang, Pathum Thani 12120, Thailand. ogunlana@ait.ac.th

Abstract

The construction industry is in search of new ideas capable of ensuring harmony and profitability in the management of projects. New systems and methods are being developed within the industry and others are being adapted from other industries. In procurement, there is general agreement that adversarial relationships are not beneficial and, as such, they should be avoided. Consequently, design and build and other partnering systems are in vogue. In this paper, it is argued that the industry can learn lessons from history. The Jerusalem temple built by King Solomon is a classic example of a project executed to the satisfaction of all the parties involved. Being a project executed in an atmosphere of trust, good organization, recognition of expertise and good communication, it provides valuable lessons that may help in improving harmony on today's projects and for ensuring that the owner and other parties 'profit' from projects.

Keywords: Trust, relationships, culture, recognition of expertise, good communication

Introduction

Hegel once proposed a depressing hypothesis thus: "What history and experience teach is this - that people and governments have never learnt anything from history, or acted on principles deduced from it". Construction professionals need not run foul of Hegel's hypothesis. It can be argued that the hypothesis was proposed to encourage people to learn from experience. The experience in construction is rather rich. From the beginning of civilization, many impressive structures have been built for habitation, worship, to commemorate civic occasions, etc. It is heartening to note that most of the seven wonders of the ancient world were either directly civil engineering structures or structures needing civil construction to stand: the hanging gardens of Babylon, the mausoleum at Halicarnassus, the Colossus of Rhodes, the lighthouse at Alexandria, the pyramids at Giza, the statue of Zeus in Olympia, the Temple of Artemis in Ephesus. Moreover, the only structure reported to be visible from space, the Great Wall of China, is also a civil engineering project.

The temple built by King Solomon has been chosen for this paper for two principal reasons: (1) the details about its construction are available; and (2) it is a rich example of the good qualities we are advocating in the procurement group. The rest of this paper concentrates on the construction of the temple and the procurement lessons that can be learnt from it.

The building of the temple

The accounts of the construction of the temple in Jerusalem are taken from I Kings 5:1 – 9:9, I Chronicles 28:1-29:9 and II Chronicles 2:1-7:22. In order to prevent distracting the reader from the major issues, constant references to different sections in the passages will be avoided. Therefore, interested readers may refer to the relevant accounts in the Holy Bible.

Project initiation

King David first had the idea to build a temple. He finally settled as King of Israel after many battles which he won because "the Lord was with him". After establishing himself, he built a palace in Jerusalem. It was while relaxing with his friends in the palace that he observed that there was no accommodation for the ark of the covenant of the Lord while he, a servant of the Lord, was "living in a palace of cedar". However, David was instructed, by the Lord through Prophet Nathan, not to build a temple for the Lord. God promised to set up a dynasty in David's name. His immediate successor and son, Solomon, was chosen to build the temple.

King David, however, proceeded to make elaborate preparations (See Appendix I) for building the temple because he reasoned that his son was too young to do the work unaided. He purchased the land, organised people to prepare building materials and exhorted the people to cooperate with his son in building the temple "for God". In addition, he prepared a whole tribe, the Levites, to supervise the construction, and staff the building when completed. King David also prepared the blueprints for construction and gave detailed specifications for the furnishings including the objects to be used for temple worship.

Project briefing

The biblical accounts record that King David prepared the blueprints with "the hand of God upon him". Every detail was included in the blueprint King David handed down to King Solomon. Thus, it was impossible to misrepresent the intentions of the owner. Unlike the temple project, today's projects struggle to cope with inadequate briefing which often results in numerous change orders (Ogunlana and Promkuntong, (1996).

Solomon's organization

Seventy thousand carriers and eighty thousand stone cutters served under 3,600 foremen. Solomon requested (ordered) cedar logs from a friendly nation, Tyre, whose king had earlier made provisions for building a palace in Israel (King David's palace). Since the temple to be built was expected to be a "great temple", he requested for

"a man skilled to work in gold and silver, bronze and iron, and in purple, crimson and blue yarn, and experienced in the art of engraving, to work in Judah and Jerusalem with my skilled craftsmen, whom my father David provided."

In addition he requested for cedar, pine and algum logs from Lebanon, for:

"I know that your men are skilled in cutting timber there. My men will work with yours to provide me with plenty of lumber, because the temple I build must be large and magnificent. I will give your servants, the woodsmen who cut the timber, twenty thousand cors (about 4.400 kilolitres) of ground wheat, twenty thousand cors of barley, twenty baths (440 kilolitres) of wine and twenty thousand baths (440 kilolitres) of olive oil."

The use of an international construction manager

The King of Tyre sent a skilled craftsman that would be acceptable to the people of Israel, Huram-Abi, whose mother was an Israelite from the tribe of Dan. He was trained to work in gold and silver, bronze and iron, stone and wood, and with purple and blue and crimson yarn and fine linen. He was also experienced in all kinds of engraving and could execute any design given to him. The king also proposed that the logs be delivered on rafts from Lebanon to a city where Solomon's men could easily transport it. He was satisfied with the payment offered by King Solomon.

The temple building

The temple was built on a location selected by King David, on mount Moriah, which used to be the threshing floor of Araunah, a Jebusite. Going by today's standards, the building itself was not very big (Figure 1). It measured 27 meters in length, nine meters in width and was 13.5 metres high. The portico extended the width by another nine metres. Solomon made narrow clerestory windows in the temple. The design was essentially patterned after the tabernacle that had been built earlier by Moses (Numbers Chp. 7).

"Against the walls of the main hall and inner sanctuary he built a structure around the building, in which there were side rooms. The lowest floor was 2.3 metres wide, the middle floor was 2.7 metres and the third floor 3.2 metres. He made offset ledges around the outside of the temple so that nothing would be inserted into the temple walls". (I Kings 6:5,6).

Judging by the standards of those days, there was a great element of innovation in construction.

"In building the temple, only blocks dressed at the quarry were used, and no hammer, chisel or any other iron tool was heard at the temple site while it was being built". (I Kings 6:7)

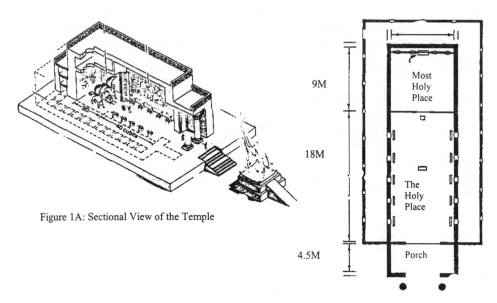

Figure 1A: Sectional View of the Temple

9M

18M

4.5M

Most
Holy
Place

The
Holy
Place

Porch

Figure 1B: The Plan of the Temple

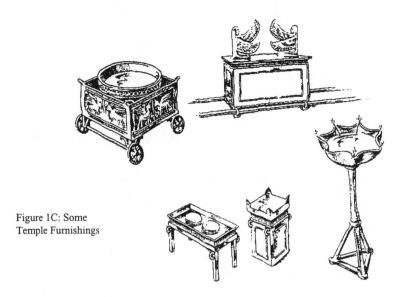

Figure 1C: Some
Temple Furnishings

Figure 1: Sectional View, Plan and Some Temple Furnishings

This must have been a great achievement considering the technology available to the builders and the transportation facilities available in those days.

The temple building was roofed with beams and cedar planks. Solomon lined the interior walls of the temple with cedar boards, panelling them from the floor of the temple to the ceiling, and covered the floor of the temple with planks of pine. The interior panelling was also overlaid with fine gold decorations.

The interior of the temple was divided into two main sections: (1) the inner sanctuary, referred to as the Most holy Place, and (2) the outer sanctuary, the Holy Place. The inner sanctuary was a square, nine meters in length, leaving the outer sanctuary to be another square 18 metres in length. The ark of covenant was housed in the Most Holy Place under two cherubims each with two wings. The wings of the two cherubims extended from wall to wall and were joined at the center. Constructional details for the temple is available in I Kings Chapter 6 while the temple's furnishings are described in I Chronicles Chapters 3 and 4.

The construction of the temple lasted approximately seven years; from the fourth year of King Solomon's reign to the eleventh year. Judging from the size of the structure and using today's standards, it may be deemed that this was not a great achievement. However, it should be remembered that the technology available to Solomon's workers was rudimentary. In addition, it was perhaps not the actual framing of the temple that required much time to complete. The furnishings and decorations were very intricate, requiring close attention by skilled craftsmen (Figure 1C and Appendix B).

The result

When the temple was completed, Solomon held an extravagant dedication ceremony. The highlight of the ceremony was described thus:

"The temple of the Lord was filled with a cloud, and the priests could not perform their service because of the cloud, for the glory of the Lord filled the temple" (2 Chronicles 5:13,14).

This was a demonstration that God was pleased with the end product. Moreover, when Solomon finished praying:

"Fire came down from heaven and consumed the burnt offerings and the sacrifices.." (I Chronicles 7:1).

God promised to answer prayers offered in the temple. Thus, Solomon and all the workers were rewarded for their effort.

The lessons to be learnt

Several lessons can be learnt by today's project management from the building of the temple. The lessons include: building on relationships; trust; proper organization; recognition for expertise; and generous remuneration.

Building on relationships

Solomon's basis for approaching the King of Tyre can be summed up in these words, "you have been friendly with my father". Thus, the initial approach was made on the basis of existing relations between the two nations. The participation in the project by the people of Tyre was a means of building or strengthening relations. Partnering has become an increasingly appealing word in procurement circles in the last few years. Records show that when 'partners' work together on a project, harmonious relations are possible and both parties benefit (profit) a lot as the mutual understanding that we need each other leads to considerations of not just the interest of one party but what is good for both.

Organization

When involved in strenuous and dangerous work like felling timber, it is not unlikely that emotions could easily fray. This may result in fights between the men from the two countries. By making the Sidonians fell the timber and floating it to Joppa from where the Israelites could collect it, quarreling, and the consequent negative impact on relations, was avoided.

However, the craftsmen from Tyre were able to work with the Israelites in casting all the objects needed for temple worship. This was, perhaps, an activity that required more intelligence and skill than physical effort. Since they were also working in Israel and only a few highly skilled and highly valued professionals sent by the express command of the King, it was unlikely that any major squabbles would develop.

Recognition for expertise

The pioneers in the field of management were quick to value individual expertise. Taylor's principles of scientific management separated between the role of the manager and that of the worker. Henri Fayol's fourteen principles of administration put much emphasis on division of labour. Since then the principle has always been that people should be engaged to do what they are best at. Solomon approached the King of Tyre for assistance because "we have no-one so skilled in felling timber as the Sidonians". Solomon's men readily accommodated the involvement of the highly skilled Sidonians and all was well.

Experience in Thailand shows that the introduction of construction management has been welcomed by other professionals (Ogunlana and Malgren, 1996). Architects, systems engineers and subcontractors cherish the coordinating role of professional construction managers. However, contractors (main contractors) have been reluctant to welcome professionals who have eroded their powers which in the past have been used to oppress subcontractors. The introduction of new procurement arrangements usually result in shifts in power balance (Newcombe, 1994). Unless we all are willing to accommodate the shift, the new forms are unlikely to produce the best results.

Trust

Unless one is familiar with history, the role of trust in the construction of the temple may be undervalued. In those days, it was the habit of kings to go to war whenever there was less work to be done in the farms. Kings used also to send spies to study the territories of those they intended to raid. Since alliances have never been permanent in any society, Solomon could have been opening his country to potential invaders by inviting the workers from a foreign country to work on what was, undoubtedly, the most important national monument. However, the high level of trust between the two kings is based on a long history of cooperation by their two countries and it produced a good result.

Generous remuneration and good treatment of workers

King Solomon's offer of payment was very generous. The result was that King Hiram was "greatly pleased". The book of the law in Israel specifically instructs every citizen not to "muzzle the ox that treads the wheat" (Holy Bible: Deuteronomy 25:4) and elsewhere in the Bible it is stated that "a worker deserves his wage" (Holy Bible: Luke 10:7). Therefore, it was important that the supplier of an essential service be well remunerated for the service. Whether or not procurement has a role to play in the treatment of workers is subject to debate. However, inadequacy of pay is a hygiene factor capable of preventing employees from performing at the normal level (Herzberg, 1959).

Experiences with procurement all over the world have shown that the use of the low bidder on construction projects is being critically re-examined. This is because the low bidder may often engage in activities that are harmful to harmonious relations when his/her profits are eroded through the foolishness of 'winning at minimum price'. Professionals need to be paid adequately for their services. Only when this is done are they likely to give their optimum to the client. Thus, the client and the contractor have much to 'profit' from giving generous remuneration for services.

Thirty thousand labourers were conscripted from Israel to work alongside the Sidonians. The management was very mindful of the welfare of the conscripts. They were sent off to Lebanon in shifts of ten thousand, so that they spent one month in Lebanon and two months at home. Thus, they never had to spend more than one month away from their families. Their conditions were definitely better than those of today's migrant workers who are often poorly paid and may have to spend up to three years from home (see Wells, 1996).

Cultural sensitivity

King Hiram's choice of a skilled craftsman to send to Israel showed great respect for culture. The earlier biblical records show that the Israelites were specifically instructed to separate themselves from foreigners who were not allowed into the assembly of worship (Deuteronomy 23). Whereas it was acceptable to conscript people of other races living among the Israelites to serve as labourers on a national project, the use of a foreign project manager might have met with some form of opposition. The man chosen for the task, Huram-Abi, had 'dual citizenship', which was pointed out at the onset through him being described as the son of a woman from the tribe of Dan (one of the twelve tribes in Israel). This again was an important factor in smoothening relations between the two countries.

Experience in Bangkok show that things work well when projects are staffed with customer satisfaction as a major goal (Siddiqui, 1996). As project management is becoming increasingly globalised, professionals need to be trained in managing in a cross-cultural context and especially to respect the culture of the environment where projects are located otherwise, the goal of ensuring harmony will be difficult to achieve.

Good communication

By having the intentions of the client made clear at the onset, the temple builders were able to complete the project with no record of rework. If today's owners are careful in preparing good briefing documents and are equally careful in communicating their needs to professionals, frequent change orders can be avoided.

Conclusion

The temple in Jerusalem, constructed by King Solomon, was a project of national significance. The temple though small by today's standards, required expertise that was best met by combining indigenous knowledge with foreign know-how. Its construction provides a case from which valuable lessons of harmony and profit can be learnt. The important lessons include:

- partnering builds on long term relationships;
- careful work organisation facilitates working in harmony;
- recognition for expertise yields profits;
- generous remuneration is mutually beneficial to the worker and the employer;
- trust is an important factor in maintaining relations;
- respect for the local culture enhances harmony in relations; and
- good communication prevents rework.

These are valuable lessons that have potentials for improving project performance.

References

Herzberg, F., Mausner, B. & Synderman, B. (1959) *The Motivation to Work*, 2^{n}edn.Wiley, New York.

Holy Bible (1991) New International Version, Hodder and Stroughton, London.

Newcombe, R (1994) Procurement paths – a power paradigm. *Proceedings of CIB W92 Symposium: East Meets West,* Hong Kong, pp. 243-250.

Ogunlana S.O. and Malmgren C. (1996) Experience with professional construction management in Bangkok, Thailand, *Proceedings of CIB W92 Symposium: North Meets South*, Durban South Africa, pp. 483-491.

Ogunlana, S.O. and Pomkuntong, K. (1996) Construction delays in a fast growing economy: comparing Thailand with other economies. International Journal of Project Management, Vol. 14, No. 1, pp. 37-45.

Siddiqui, Z.H. (1996) Factors and procedures used in matching project managers to construction projects. *AIT Thesis No. ST96-34.*

Wells, J. (1996) Labour migration and international construction. *Habitat International*, Vol. 20, No. 2, June, pp 295- 306.

Appendix A: Temple building materials prepared by King David

National preparations for the temple

1. Dressed stones
2. Nails for doors and fittings
3. Bronze for the fittings
4. Cedar logs
5. 3,450 metric tonnes of gold
6. 34,550 metric tonnes of silver
7. Iron - too great to be weighed.

King David's personal contributions

1. 100 tons of gold
2. 240 tons of silver

Contributions by King David's officials

1. 170 tons of gold
2. 345 tons of silver
3. 610 tons of bronze
4. 3,450 tons of iron
5. Unspecified amount of precious stones.

Appendix B: Temple Furnishings

Those made by Huram-Abi

1. A bronze altar
2. Sea of cast metal with 2 rows of bulls
3. Twelve bulls as stand for the Sea
4. 10 basins for washing
5. 10 gold lamp stands
6. 10 tables
7. Pots and shovels and sprinkling bowls
8. Two pillars with bowl shaped capitals
9. 400 pomegranates
10. Stands and basins
11. The golden altar
12. The tables for the bread of presence
13. Gold floral works
14. Golden doors for the two sanctuaries.

CUSTOMER ORIENTATION IN THE DELIVERY OF PUBLIC WORKS – THE IMPACT ON BUILDING PROCUREMENT IN HONG KONG

BRIAN WILKINS
Department of Building and Construction, City University of Hong Kong, Tat Chee Avenue, Kowloon, Hong Kong. E-mail: bcbrianw@cityu.edu.hk

Abstract

The paper introduces two trends. One relates to changing attitudes and expectations of the public regarding public services provision, and the other is the introduction of a range of alternative procurement possibilities into the mainstream of construction activity worldwide. The paper then analyses the extent to which the first is influencing the second, within the context of Hong Kong. The analysis focuses on two major public sector services, health care and housing. It shows how public agencies, in response to consumer orientated demand, are requiring construction industry firms to provide a unified and cooperative approach in the supply of their services for the production of public capital works. Significant changes are therefore occurring, but the paper concludes that a fundamental shift in attitudes and cooperative practices within the construction industry is needed, to support the contractual systems providing integrated alternatives in construction procurement.

Keywords: Customer orientation, procurement, public works.

Introduction

Public accountability has been a major element in the funding, design and delivery of public works, manifested most clearly in the need to demonstrate that value for money has been obtained. This in turn has required the client to be reasonably certain of project costs at each stage through the use of fully measured estimates based on comprehensive drawings and specification. Lengthy approvals procedures and a very conservative approach to project procurement [Lynn and Jay (1984)] has led to public works being generally portrayed as centrally controlled, bureaucratic and inefficient. In addition the procurement of some public works, notably, for example, health care, is characterised by a "multi-headed client problem" – a proliferation of end-users, enjoying high levels of authority [Smith and Wilkins (1996)]. This scenario has often, ironically, produced cost increases and programme overruns. The additional need for competitive bidding to ensure perfect competition has led to long lead in times to construction without necessarily guaranteeing completed designs, or even adequate briefing material [Wilkins and Smith (1996)]. This has in turn tended to generate a proliferation of changes and claims from an industry which has managed to institutionalise conflict, particularly in situations where public accountability dictates that the lowest bid will be selected. Periods of extremely tight competition and low volumes of work exacerbate these problems, and much reliance is placed on contractual arrangements which reinforce and formalise confrontation rather than cooperation and partnership.

Two trends relevant to the discussion in this paper have emerged in recent years in many societies, and certainly in Hong Kong. One relates to changing attitudes and expectations of the public, the taxpayers, based on increasing affluence and demographic changes. The other is the introduction and implementation of an increasing range of alternative procurement possibilities into the mainstream of construction activity. These two trends are summarised in turn in the two paragraphs which follow.

The first trend relates to the mechanisms for the delivery of public works, which are slowly beginning to change, and in some sectors are undergoing significant development. This is particularly so for health care and housing, the two public works sectors selected for discussion in this paper. Greater public expectations in respect of both these traditional public works sectors are causing a shift away from the notion of passive populations receiving publicly provided social services, to a more consumer driven approach. This shift is given impetus by the ever increasing costs of technological advances in housing and, particularly, healthcare. These costs are beginning to exceed society's willingness and ability to pay. There is therefore a general worldwide decentralisation and privatisation of some public services for reasons of political ideology, as part of public spending cuts, or in the belief, for example, that everyone prefers to own their own home, or to seek quality "patient-centred" healthcare [Wilkins (1997)]. Diversification in the funding of publicly procured services away from the public purse and onto the consumer and commercial providers, would suggest a strong market driven model based on quality, price and reliable and speedy delivery of the facilities in demand. The taxpayer is becoming transformed into the "customer".

The second trend is in the extent to which construction industries have adopted new methods over the last twenty years, arising from, on the one hand, recession in some economies (e.g. in Britain in the eighties and into the nineties), and on the other, phenomenal growth in culturally diverse regions of the world (notably the Middle East and South East Asia). These societal and economic influences have demanded change in order to compete and survive, and have also promoted the cross fertilisation of radical ideas about roles and structures in project delivery systems. This sea change can be seen in the research and literature about procurement published over the last twenty years or so. The beginning of this significant two decades is marked by an early publication by Cocks (1977), which describes how not to procure a building. Later work began to focus very specifically on issues relating to procurement and procurement selection [e.g. Skitmore and Marsden (1988), Bennett and Grice (1990), Lam and Chan (1994)]. In particular a number of authors were able to demonstrate that design and build procurement methods can result in a reduction of overall project duration [e.g. Pain and Bennett (1988), Griffith (1989), Ndekugri and Turner (1994)]. The most recent work, significantly, is concentrating more and more on the cooperative and integrated aspects of procurement systems. Sheath *et al* (1994) refer to the extent to which a procurement system can facilitate the management of all the resources within a project. Dulaimi and Dalziel (1994) take this idea further and discuss how the overall efficiency of a project is dependent on the coordinated efforts of the client and individuals working together in groups. Bhuta and Karkhanis (1995) refer to the control of risks to benefit all members of the project team. Ashworth (1996), while focussing on the client, acknowledges the need to weigh all interests. Finally much of the recent literature relating to "partnering" concepts in construction suggests that shared attitudes, goals and expectations are important components of project success. These notions are far removed from the acrimony which characterised, and continues to characterise to some extent, many major public works contracts.

The paper suggests that these two trends are now being brought together. Analyses of publicly funded health care and public housing in Hong Kong are presented to illustrate how the first trend is influencing the second. The intention is to demonstrate how evolving customer driven "public sector" environments, with consequent demands for high levels of quality, time and cost performance, are requiring a significant degree of cooperation and partnership from a fragmented and disputatious industry.

Sources of data

The arguments and discussion presented in this paper are drawn together from the findings of various investigations relating to project delivery systems carried out in the Department of Building and Construction at the City University of Hong Kong by the author, colleagues and postgraduate students over the last five years. Funded fieldwork has taken place in Hong Kong, Britain, China and the United States. Much of the information related to public housing in Hong Kong arises from a MSc Dissertation prepared by one of the author's students [Au (1998)]. The data relating to healthcare has been developed on the basis of research funded by the Hong Kong Hospital Authority and a City University Strategic Research Grant [Smith and Wilkins (1995)]. Methodologies have included questionnaire surveys, project and company case studies, structured and semi-structured interviews, and the examination of the data collected using both quantitative and qualitative methods. The work is ongoing and several papers have been published, some of which are cited in this paper.

The next two sections use the data arising from these studies to show how the policies and practices of two "customer focused" public authorities in Hong Kong are fostering structural changes in the construction industry and its operations.

The case of publicly funded healthcare

Procurement and accountability

The research referred to in the previous section included in-depth analysis of a number of case studies of hospitals, recently completed or under construction, in Hong Kong, Britain, and the United States. Statistical information relating to the time and cost performance of several projects was also available from the NHS Estates in Britain and the Architectural Services Department in Hong Kong. The objective was to assess various procurement strategies in terms of time and cost performance.

In Hong Kong, public accountability was of paramount importance in the case studies investigated, and this gave rise to lengthy and complex approval procedures producing preconstruction periods as long as seven years. The traditional form of procurement, involving complete drawings, fully measured bills of quantities, competitive bidding and awards to the lowest bid, was the norm in all case studies except two. Typically, large numbers of end users were involved in approvals and thus able to influence and impede the progress of design and construction. This led to extended briefing and design stages and claims for extra time and payments during construction. Substantial client changes continuing into the post contract stage of projects, resulted in, for example, the reissue of all drawings, the addition of entire buildings to a contract, and as many as three thousand variations.

More recent projects incorporating variations on the traditional procurement approach have shown improvements in project performance. These variations have generally sought to introduce measures that would reduce damaging claims to everybody's advantage, or unite and integrate the parties for the overall benefit of the project. An example of the first was the formal incorporation of a dispute resolution adviser within the project team, a measure which is thought to have contributed to the project being completed within budget and only seven months late. Examples of attempts to induce cooperation through integrating the various firms and organisations include the introduction of overall project management, and experiments with variations on the design and build route. The project management case study was a project, which was going very badly wrong, resulting in major delays in the preconstruction stage. Appointment of project management external to the government bureaucracy is thought to have remedied the situation and achieved a completed project only two months late and US$20million under budget. In the last 4-5 years, design and build methods have been increasingly employed with some success in achieving higher levels of time and programme performance, without sacrificing quality, or compromising public accountability. Measures used to ensure public accountability and achieve high levels of cooperative effort have included strategic project management, novation of client's design services to the contractor, and multidisciplinary value management, alongside the design and build contractual arrangement. In the design and build cases, public expectations, impinging on political sensibilities, have forced health care authorities towards innovation so that facilities come on stream quickly. Speed in procurement methodology, within defined budgets and in line with the principles of patient centred care, has become a key criterion in the selection of procurement methodolgy. The growing assumption now is that the more firms and organisations are able to combine efforts in the procurement process, the more likely is success to be achieved for all parties.

The end-users

Measures have also been introduced to formalise end user input from the earliest stage, and to take account of the likelihood of change occurring throughout the various stages of the project. This is particularly important in some procurement methods, such as design and build, where a clear brief or statement of requirements is essential. Some of these measures have been extensively discussed elsewhere by Wilkins (1997) and will only be briefly referred to here. The major initiative was to introduce multidisciplinary interactive decision making at the early inception stage of projects, prior to the inclusion of the planned project within the Government's capital works programme. The system applies to all major public works projects, including hospitals, and is particularly useful where there are large numbers of end users. The principle in the case of hospitals, involves the appointment of hospital planners, cost consultants, and any other relevant expertise, to work with the client and end users in the assessment of feasibility, the formulation of a schedule of accommodation, the determination of a budget and programme, and the selection of a procurement method. A concept design is usually produced together with a sixteen page report and estimate, which then proceeds quickly through defined approval routes in the government bureaucracy where it is "signed off " at each stage. Estimates incorporate the likelihood of client change through the assessment and costing of this uncertainty using risk analysis. The document produced is known as a Preliminary Project Feasibility Study (PPFS). Its approval confirms the project's inclusion within the capital works programme at the stated budget, which contains an average risk allowance in respect of areas of uncertainty, the major area in a hospital project being client change.

Once in the capital works programme, the next priority is to select a team appropriate to the procurement method selected. This involves price bidding by interdisciplinary teams and the production of a Technical Proposal within the context of the approved budget and programme. Selection of a successful team is based partly on the Technical Proposal and partly on the financial package offered. A committee formed of end users and other client representatives score the submissions against a predetermined scale, and the lowest bid is not necessarily selected. Heavy weighting may be assigned, for example, to the team's approach to achieving cost effectiveness or to the quality and experience of the team and its individual members. Alternatively, the identification of key issues and constraints, and the innovative ideas proposed in response may count more heavily.

An approach to design and build

Changes in procurement methodology affecting hospitals have been facilitated by the creation of a Hospital Authority which was established in December 1990 and assumed responsibility for managing all public hospitals and related institutions one year later in December 1991. The Authority contains a permanent capital works department with a strong project management ethos. The Authority was set up to deal with the problems associated with the procurement of public hospitals and one of its first acts was to introduce a Patients' Charter dedicated to the promotion of patient centred health care. Its first major project was based on a modified design and build procurement route selected to achieve very rapid delivery of the completed project to the highest standards of patient centred care. On this project, the Hospital Authority acted as agent of the Government to plan and procure the project. The procurement method eventually chosen to fast-track the project was officially defined as a "design, development and build system" in which the contractor was expected to start work on site prior to completion of construction documents. Various accountability measures were incorporated. These included strategic in-house project management, limited cost adjustment on the basis of prescribed criteria (there were no Bills of Quantities), substantial pre-design by client's consultants, and then novation of the consultants to the contractor, who was responsible for design development. The exclusion of Bills of Quantities represented a major departure from traditional Hong Kong practice

Within the health care sector, construction industry firms are being forced into cooperative and integrated situations. It is not clear to what extent changes evident in the last few years in this particular sector represent a fundamental shift in attitudes and practices, or merely a superficial and limited response to a particular market. The next section considers the work of a much older public authority and describes how a culture of quality through cooperation does indeed seem to be evolving.

The case of public housing

Approaches to public housing procurement

The modern Hong Kong Housing Authority has been in existence much longer than the Hospital Authority, having been established in April 1973. It was reorganised in 1988 as an autonomous body with independent finance to implement the Government's Long Term Housing Strategy. The Housing Department is the Authority's executive arm containing Works Groups, which are responsible for the planning, designing and construction monitoring of low cost public rental housing schemes and a home ownership scheme. The Housing Authority considers itself to be "customer-focused". High standards are expected by the community in terms of choice, size, quality and overall estate environment. The Government wishes to provide public rental housing for all those who need it. Home ownership is however strongly desired. It is intended to raise the number of new flats built by the Housing Authority for sale under home ownership and private sector participation schemes from the 57% figure of 1995/96 to 70% within ten years. Overall, an average of 56,000 public housing units are to be built annually.

Speed of construction is crucial in achieving these output requirements, and a radical approach to procurement is achieved at a strategic level. This includes industrialisation of many aspects of production requiring a restructuring and realignment of productive capacity within the industry, as well as the acquisition and application of new technologies and management methods. With regard to home ownership, the private sector participation scheme (PSPS) was introduced as long ago as 1978 to enhance productivity and quality, effectively through a single point of administrative control - a design, manage and construct developer. Under the scheme, private developers design, finance, build and manage the flats. The Housing Authority determines buyers and guarantees to take up any unsold flats at an agreed price. Quality is vital too if expectations regarding standards and home purchase are to be met. As part of its strategy in achieving quality, the Housing Authority requires all contractors permitted to bid for Housing Authority projects to have achieved ISO 9002 certification by the Hong Kong Quality Assurance Agency.

Attitudes to procurement

The study by Au (1998) referred to earlier in this paper included an attitudinal survey of a cross section of people in the construction field who have been involved in the implementation of public housing programmes. The sample of 100 was selected randomly from internal project teams of the Hong Kong Housing Authority, housing project teams within Chartered Quantity Surveying Consultant firms, and building contractors. The methodology and results of the survey merit a separate paper and this is currently in preparation. Some of the main points of the study will be drawn out and summarised here to support the discussion in this particular paper. The questionnaire used in the survey was divided into two linked parts. Part I questions relate to commonly identified factors in procurement and contract selection. Respondents were asked to rate these factors on a five point scale. Part II contained reinforcing questions relating to perceived advantages of alternative procurement methods in relation to various factors. Respondents were asked to rank their preferences of the considerations used in the choice of procurement system. Procurement systems used in the assessment were traditional lump sum, design and build, design and manage, construction management, and management contracting. The data was analysed, average weighted scores computed, and factors ranked. An assessment was then made of the importance of various factors and the perceived performance of various procurement methods.

All respondents agreed that good performance in relation to time, price and quality is an essential consideration in the selection of a procurement method but that only design and build and design and manage performed well in all three areas. The traditional approach was thought to be deficient in time performance owing primarily to long preconstruction periods. Construction management and management contracting performed less well owing to the very low probability of achieving price certainty, and the likely incidence of contractual claims. All respondents, including contractors, preferred a procurement system which provided for selection of a contractor on the basis of quality (rather than lowest bid), and quality assurance in workmanship was considered by all respondent groups to be more important than in any other aspect of the construction process. All respondents thought that a procurement system which shared risks equally was superior to those which did not. The traditional system fared well here, while all the others were considered to be poor performers. Buildability was considered to be an important factor in the selection of a procurement system and all respondents thought that this should be achieved through the involvement of the client's designers, and the contractor, working together. Design and build and design and manage methods were ranked highest here, with the traditional method being rated as very poor. Analysis could continue, but what is most significant for the purposes of this paper is the remarkable unanimity recorded by all respondent groups: client, contractor and consultant, regarding fundamental issues affecting procurement selection, and the performance characteristics of the various procurement systems themselves. Respondents appeared to place considerable value on cooperation, integrated procurement methods, and the achievement of quality.

Conclusion

Hong Kong is still very much an autonomous, self-contained society comprising a large population concentrated into a limited geographical area, and responsible for the conduct of its own affairs. The actions of the Government of the SAR and its agencies therefore have a profound and multiplying effect on society and the economy as a whole. Conversely, Governments in the territory before and after the transfer of sovereignty have tended to be very sensitive and responsive to local needs and aspirations. Hong Kong is very much a consumer-orientated society. The Government, through its two large public authorities, is responding to "customer" need and consequently significantly influencing the construction industry in all its aspects. The Housing Authority has been developing and perfecting its methodologies for rapidly providing ever-increasing quantities of housing to a diversifying market for twenty years. Producing quality housing is as much a part of its objectives as meeting the demands of an ever increasing population. The Authority is Hong Kong's biggest client and its approach to procuring housing is encouraging the growth of a cooperative and quality culture within the construction industry. This is seen in the opinions and attitudes displayed in Au's survey, and this bodes very well for other clients, public and private. The Hospital Authority has benefited from these trends and developments in the ease with which it has been able to introduce, very quickly, innovative procurement methods which fast track projects, through integrated groups and teams. Interviews with Hospital Authority staff display a ready willingness to explore new approaches, including fast-tracking, value management, and partnering.

Some philosophical resistance to these developments has come from the architectural profession, which has enjoyed a traditional position of authority and independence in Hong Kong. New methodologies have tended to be viewed as a threat to this position. In practice, though, economic imperatives prevail. Architects too are recognising the structural changes which are taking place, and are beginning to adapt to new modes of working, at least on an individual basis, whatever the codes of conduct might say. The profession as a whole, however, is at least displaying a willingness to promote research and debate relating to alternative procurement methods, although debate has initially been limited to rather outmoded assumptions about the inevitability of declining quality.

The ingredients of success in public services, as far as customer demand is concerned, appear to be quality, diversity and "ownership". The public agencies serving the customer believe that these ingredients are most easily supplied when design of capital works, their construction and sometimes part or all of their funding, are devolved away from the public bureaucracy and onto a single, unified point of contractual and administrative control. Costs can then ultimately be transferred to the customer in return for the services, medical or housing, provided in a consumer orientated market environment.

The contractual divide is however still there with the potential for acrimony and dispute, whatever the formal procurement route adopted. To complete the construction industry's transformation into an industry which is capable of providing unified services to affluent and discerning customers in the public sector, something of a paradigm shift is needed in cooperation and attitudes. According to Masters (1998), this goal is attainable. The construction of the UK Inland Revenue's new office building in Manchester, England, on time and within budget despite major changes to the building's footprint, he describes as a testament to the success of a "culture of close liaison and cooperation between all parties to the contract". It represents, he asserts, a "new approach" to design and build. Evidence from the Hospital Authority and Au's study relating to public housing, would suggest that a similar goal is achievable elsewhere.

References

Ashworth, A (1996) *Contractual Procedures in the Construction Industry*, 3rd Edition, Longman.

Au, W F (1998) *Evaluation of the Procurement Methods Employed in the Works Group of the Hong Kong Housing Department*, unpublished MSc. Dissertation, Department of Building and Construction, City University of Hong Kong.

Bennett, J and Grice, A (1990) Procurement Systems for Building. *Quantity Surveying Techniques, New Directions,* Brandon, P (ed.), BSP Professional Books, Oxford.

Bhuta, C and Karkhanis, S (1995) Risk Control in Construction using Contractual Strategy. *Asia Pacific Building and Construction Management Journal*, Vol. 1, pp. 71-101.

Cocks, B Sir (1977) *Mid-Victorian Masterpiece: the story of an institution unable to put its own house in order*, Hutchinson, London.

Dulaimi, M F and Dalziel, R (1994) The Effects of Procurement Method on the Level of Management Synergy in Construction Projects. *Proceedings East Meets West CIB W92 Procurement Systems Symposium*, Hong Kong, December 4-7, pp. 53-58.

Griffith, A (1989) Design-build procurement and buildability. *Technical Information Service paper no. 112*, Chartered Institute of Building , U.K.

Lam, P T I and Chan, A P C (1994) Construction Management as a Procurement Mehod – a New Direction for Asian Contractors. *Proceedings East Meets West CIB W92 Procurement Systems Symposium*, Hong Kong, December 4-7, pp. 159-167.

Lynn, J and Jay, A (1984*) The Complete Yes Minister.* Guild Publishing, London.

Masters, J (1998) Taxing problems overcome. *Construction Manager,* Vol. 4, No. 5, June, pp. 10-11.

Ndekugri, I and Turner, A (1994) Building procurement by design and build approach. *Journal of Construction Engineering and Management*, Vol. 120, No. 2, ASCE, pp. 243-256.

Pain, J and Bennett, J (1988) JCT with contractor's design form of contract: a case study. *Construction Management and Economics*, Vol. 6, United Kingdom, pp. 307-337.

Sheath, D M, Jagger, D, and Hibberd, P. (1994) Construction procurement Criteria : A Multi-national Study of the Major Influencing Factors. *Proceedings East Meets West CIB W92 Procurement Systems Symposium*, Hong Kong, December 4-7, pp. 361-367.

Skitmore, R M and Marsden, D E (1988) Which procurement system? Towards a universal procurement selection technique. *Construction Management and Economics,* Vol. 6, United Kingdom, pp. 71-89.

Smith, A J and Wilkins, B (1995) An investigation into improved methods of procurement for major publicly funded building projects in Hong Kong with special reference to health care buildings. Research Report, Department of Building and Construction, City University of Hong Kong.

Smith, A J and Wilkins, B (1996) Team Relationships and Related Critical Factors in the Successful Procurement of health Care Facilities. *Journal of Construction Procurement,* Vol. 2, No. 1, pp. 30-40.

Wilkins, B (1997) An Integrated Approach to the Formulation of Design Briefs for Publicly Procured Health Care Facilities. *Australian Institute of Building Papers,* Vol. 8, pp. 159-169

Wilkins, B and Smith, A J (1996) The Management of Project Briefing: The Case of Hospitals. *Australian Institute of Building Papers*, Vol. 7, pp. 87-95.

THE EFFECT OF PRIVATE ASPIRATIONS IN CONSTRUCTION ALLIANCES

A. B. NGOWI
Department of Building and Quantity Surveying,University of the Witwatersrand
Private Bag 3, WITS 2050, South Africa, Ngowiab@noka.ub.bw

Abstract

The primary purpose of forming an alliance in the construction industry is to pool together the resources of the participating partners in order to form a team that has a competitive advantage. Each partner in an alliance has its own competence and market share that do not necessarily fall under the alliance as common resources. Therefore, although the competitive advantage aimed at when forming an alliance is for common profits, each partner has a possibility of using it (the competitive advantage) for private profits (i.e., activities that do not fall under the alliance)

Using a case study from Botswana, this paper argues that a construction alliance strives as long as the profits created by common activities are substantially higher than the ones that can be created by private activities. Once one of the partners in the alliance can create the competitive advantage in question on its own, it will opt out of the alliance through such mechanisms as withdrawing some of its key contributions to the alliance.

Keywords: Common profits and private aspirations, competitive advantage, construction alliance, private profits,

Introduction

The development of a highly sophisticated and economically accessible global transportation system has opened up local construction markets to the access of outside specialists in what were previously inaccessible areas. Additionally, widespread and efficient telecommunications systems have produced a network of information that creates virtually equal access to information anywhere on the globe. These developments have brought about increased competitiveness within the construction industry.

The barriers that separated economic and vertical market sectors and the firms that operated within them are quickly falling. Competition can arise unexpectedly from anywhere, which means that firms can no longer be overconfident about their market shares and their competitive positions. With markets and their players constantly changing, the possibility of establishing a sustainable competitive advantage no longer exists.

Traditionally, construction firms have attempted with some success to face the challenges brought by competition by being self-sufficient through vertical integration within the enterprise. In the intensified competitive environment, this strategy can no longer work.

Firms must look to a variety of other strategies to assemble the services required to successfully acquire new projects and new revenues. Combining the strengths of firms that provide complementary services through alliance formation is one way to provide both harmony and profits to the participating firms. However, alliances are characterized by common activities and private activities (activities that do not fall within the alliance) which create varying degrees for competition and cooperation between the firms in the alliance. The success of an alliance is influenced by a firm's "relative scope", which captures the initial conditions likely to influence the competitive and cooperative dynamics, and for each firm is the ratio of the scope of the alliance to the total set of markets in which the firm is active (Gulati, 1998).

Against this background, this paper reports on a study carried out in Botswana to determine the influence of private activities on the performance of construction alliances.

Importance of alliances in the construction industry

A study conducted by the World Bank in 1984 showed that nearly 80% of all formal construction projects in developing countries were accomplished by foreign firms (Simkoko, 1989). To the governments of these countries, this is not an acceptable situation because it would seem that they depend entirely on foreigners to develop their countries. To provide sufficient opportunities for small and medium sized domestic construction firms to participate and learn during the nation building process, most developing countries see alliances as one of the best instruments to achieve this (Sornarajah, 1992).

Moreover, since construction industry cannot use the conventional forms of market penetration such as agents, distributors and licencees, alliances in the form of joint venture is the most popular form of entry because of the perceived benefits it brings to the host country through transfer of technology, job creation and capital inflow (Sornarajah, 1992). Based on this fact, most governments in developing countries, particularly South East Asian countries, require a minimum percentage of local participation in most of business ventures operating in their territories. Besides the pressure exerted by governments to pair up foreign and local firms, firms may on their own volition decide to work together on a short term or long term arrangements. This form of cooperation is known as strategic alliance and is defined as voluntary arrangements between firms involving exchange, sharing or codevelopment of products, technologies or services (Gulati, 1998). They can occur as a result of a wide range of motives and goals, take a variety of forms and occur across vertical and horizontal boundaries. Similarly, the Construction Industry Institute (CII), defines alliance as "a long-term association with a non-affiliated organization, used to further the common interests of the members" (Badger et al, 1993). Kogut (1988) highlighted three main motivations for formation of alliances: transaction costs resulting from small numbers bargaining, strategic behaviour that leads firms to try to enhance their competitive positioning or market power, and a quest for organizational knowledge or learning that results when one or both partners want to acquire some critical knowledge from the other or one partner wants to maintain its capability while seeking another firm's knowledge. Some of industry-level factors linked with alliance formation include the extent of competition, the stage of development of the market, and demand and competitive uncertainty (Harrigan, 1988; Eisenhardt and Scheonhoven, 1996). If an alliance functions successfully, the partners find themselves working in harmony besides getting higher profits arising from combining their competencies.

Private and common benefits

Private benefits and common benefits are two qualitatively different kinds of benefits available to participants in an alliance. Private benefits are those that a firm can earn unilaterally (from activities in markets not governed by the alliance) by picking up skills from its partner and applying them to its own operations in areas unrelated to the alliance activities. Common benefits are those that accrue to each partner in an alliance from the collective application of the learning that both firms go through as a consequence of being part of the alliance; these are obtained from operation in areas of the firm that are related to the alliance (Khanna et al; 1998).

Intuitively, the ratio of private to common benefits for a particular firm will be higher when it has more opportunity to apply what it learns to its businesses outside of the scope of the alliance (and thus earn private benefits), than opportunity to apply what it learns to business within the scope of the alliance (and thus earn common benefits). This intuition crucially impacts the behaviour of a firm within the alliance because the different incentives to invest in the alliance are a result of the competitive aspects of what is simultaneously a cooperative and a competitive enterprise. The cooperative aspect arises from the fact that each firm needs access to the other firm's know-how and that the firms can collectively use their knowledge to produce something that is beneficial to them all (common benefits). The competitive aspect is a consequence of each firm's attempt to also use its partners' know-how for private gains, and of the possibility that significantly greater benefits might accrue to the firm that finishes learning from its partner(s) before the later can do the same. While the cooperative aspect of alliances creates harmony in the market, the competitive aspect tends to create an adversarial environment.

Each firm in an alliance operates in a set of markets each element of which can be described by its products/services and geographic characteristics. The scope of the alliance refers to a need that both partner firms have agreed to target (e.g. the introduction of a new construction service), typically corresponding to some subset of markets in which the firms are themselves involved. The larger the overlap between alliance scope and firm scope, the higher is the common benefits and the lower are the private benefits. Khanna et al (1998) introduced the notion of "relative scope" of a firm in an alliance, to refer to the ratio of the scope of the alliance to the total set of markets in which the firm is active. Thus, the relative scope, which is a ratio that lies between 0 and 1, is a measure that is particular to a given firm in a given alliance. Different firms in the same alliance, and the same firm in different alliances, would have different relative scope values. Correspondingly, a greater ratio implies more opportunity for a firm to apply skills acquired in the course of the alliance to markets not involved in the alliance, and therefore, less harmony in the market.

Factors other than relative scope affect the magnitude of private and common benefits as well, and thereby their ratio for a particular firm. Given a set of markets outside the scope of a particular alliance, a firm's ability to earn private benefits by applying what it has learned to those markets is affected by the extent to which these markets are related to those within the scope of the alliance and the extent to which the firm has skills to accomplish the transfer to learning (Cohen and Levinthal, 1990). The type of knowledge transferred need not be restricted to R&D issues. It could relate to understanding a particular customer base, understanding marketing in a new country, or learning the use of new production techniques.

It is important to note that it is the ratio of a particular firm's private to common benefits that affects its decision to stay in or quit the alliance, as the firm in question compares its already existing private benefits to its potentially attainable common benefits in trying to decide whether to continue its involvement in the alliance. In contrast, the ratio of one firm's private benefits to the private benefits of its partner, are not relevant to the individual firm's decision to continue in an alliance (Khanna et al, 1998).

Performance of alliances

Numerous studies have reported dramatically high failure rates of alliances. In an in-depth study of 59 alliances, Bleeke and Ernst (1991) reported that, in about half the cases, at least one of the partners felt that the alliance had been a failure. Other studies have reported failure rates as high as 80 percent, usually leading to the dissolution of the alliance or acquisition by one of the partners (Harrigan, 1988; Geringer and Herbert, 1991). Many of these problems can be traced to the cultural differences that exist at both the national and organizational level, i.e., country of origin of partners as developed or developing (Datta, 1988). Cultural differences can often lead to a breakdown of communication, create mistrust and sometimes result in eventual termination of an alliance (Peterson and Shimada, 1978). Other factors which may contribute to problems in alliances are the presence of current ties, partner asymmetry, age dependence or the duration of the alliance, characteristics of the alliance itself such as autonomy and flexibility and importantly, the competitive overlap between the partners (Beamish 1985; Kogut, 1989).

On the other hand, termination of an alliance is not necessarily a failure as some successful alliances terminate because they are predestined to do so by the parent firms at the very outset. In other instances, an alliance may simply be a transitional arrangement that the parents plan to terminate when their objectives are met or when they have valuable new information that makes viable an acquisition or divestiture of that business (Kogut, 1991; Bleeke and Ernst, 1991; Balakrishnan and Koza, 1993).

The following section reports on a study carried out in Botswana to determine the influence of private activities of some known construction alliances on their performance.

A Study in Botswana

Background

Botswana is a sparsely populated country with a total area of 582,000 square kilometres and a population of 1.4 million people (National Census, 1991). The country has experienced rapid economic growth since the time of independence in 1966, and development of the infrastructure has been one of the country's priorities. Construction activities in general constitute an average of 7.5 percent of the Gross Domestic Product (GDP) and is estimated at Pula 2 billion (1 Pula = 0.21 US$) in 1998. The estimated employment in the formal sector in 1995 was 234500 of which 22600 or 9.6 percent was in the construction industry (Annual Economic Report, 1996). Informal employment in the construction industry may be as high as formal employment.

Unlike most countries in South East Asia, Botswana does not have a restriction against entry of foreign firms to the local construction market. Botswana provides free access to foreign firms as part of her free market policy. As result of this, alliances in the construction industry are formed for motives other than government pressure.

Objectives of the study

The main objective of the study was to determine the influence of private benefits on the performance of construction alliances.

Methodology

The projects that have been executed by alliances of firms between 1980 to-date were identified from the records of the major employers in the Botswana construction Industry. Semi-structured interviews based on the questions stated below were held with the chief executives (or their representatives) of the firms in the alliances in their offices. Where the chief executives could not be appointed for face to face meetings, the interviews were conducted by means of telephone. Each interview lasted for half an hour, and where necessary additional explanations were given for some of the questions. In one case the respondent volunteered to send a written reply by fax.

Construction alliances in Botswana

From the records kept at the Ministries of Works, Transport and Communication (MWTC); Local Government, Lands and Housing (MLGLH); and Finance and Development Planning (MFDP) reliable information was obtained on five alliances. For the sake of confidentiality, which was the condition for obtaining information throughout the study, the alliances are named A, B, C, D, and E while the firms in the alliances are named a_1, a_2; b_1, b_2; c_1, c_2, d_1, d_2 and e_1, e_2 respectively. The origins of the firms in the alliances, the number and approximate value of the projects they have executed and the present status of the alliances are shown in table 1.

Table 1: Construction alliances operating in Botswana between 1980 and 1998

Alliance	Origins of the Firms	Projects Executed	Appr. Value Projects (Million Pula**)	Status of the Alliance
A	a_1. Botswana a_2 - Sweden	3	6.3	Terminated
B	b_1 - Botswana b_2 - RSA*	2	9.8	Terminated
C	c_1 - Kuwait c_2 - U.K.	1	100	Terminated
D	d_1 - Kuwait d_2 - RSA	1	179	Sustained
E	e_1 - Botswana e_2 - RSA	2	29	Sustained

*RSA -   Republic of South Africa
**1 Pula (P) = 0.21 US$

Although the main focus of the study was the activities of the firms in the alliance after their formation, it was necessary to gather information about the choice of partners and the motivations for forming the alliances. This information was important in assessing the behaviour of the firms in the alliance. The following section is the summary of the responses elicited from the chief executives of the firms on the interview questions. The questions appear in past tense for the alliances that have been terminated, but they were amended accordingly when directed to the existing alliances.

Question 1: What type of alliance was your association?

Responses to question 1 established that, except for firms a_2, c_1, c_2 and d_1, the respondents were not aware of the different forms of alliances. However, their responses indicated that the form of alliances which they intended to form were joint ventures. The chief executives of firms c_1 and c_2 clearly stated that their alliance was a single project partnership which was then terminated on completion of the project.

Question 2: What motivated your firm to form an alliance?

Responses to question 2 established that the motivations of the firms whose origins are Botswana in all alliances were acquisition of key technical and management knowledge to execute large projects. The firms in alliances C and D were motivated by their enhanced combined capability in handling complex projects, ranging from infrastructure to multi-storey buildings. Perceived increased efficiency was another factor that motivated the firms in alliance D to form the venture. According to the chief executive of firm d_1, it is much cheaper and quicker to get skilled personnel from RSA because of the distance and familiarity with Botswana environment than getting them from Kuwait.

Question 3: What motivated your firm to form an alliance?

Generally, choice of partner was based on the record and type of operation carried out by the prospective firm. All responses to question 3 indicated that the firms sought out ties with partners who had strategic interdependence with them and who could help them manage such interdependencies. While firms a_2, b_2, and c_2 had superior technical and management capabilities to those of their partners, the later were familiar with the labour market at the artisan level and were conversant with the local environment. Responses from the firms in alliance C, further indicated that the firms wanted to introduce a novel method of producing low-cost in-situ concrete housing units by casting them in a single mould. Due to uncertainty in the acceptability of the method in the market, none of the firms was willing to pursue it alone. This finding concurs with the observation by Kogut (1991) that many joint ventures occur as options to expand in the future and are interim mechanisms by which firms both buffer and explore uncertainty.

Question 4: What type of activities did your firm do that did not fall under the alliance?

The responses to question 4 established that all activities performed under the alliance (common activities) were similar to the activities of the firms (private activities) in the alliance. General differences between the common activities and the private activities were the geographical locations of the projects and/or their complexities. While for firms a_2, b_2, c_2, d_1, the private activities were carried out outside Botswana, the activities of firms a_1, b_1, c_1, and part of the activities of firms c_1, d_2 and e_2 were carried out in Botswana.

It was also established that common activities in firms a_2, b_2, c_1, and c_2 formed less than 25 percent of the firms' activities, while for firms d_1, d_2, and c_2, this percentage is between 25 and 50, and for firms a_1, b_1 and e_1, it was more than 50. This finding indicates that the majority of the firms that originate from Botswana in the alliances got their revenues from common activities.

Question 5: What was the structure of your alliance?

The responses to question 5 determined that the partners in all alliances created new entities, at the project level, in which they shared equity. Key personnel for the projects were in principle drawn from both partners' staff. However, it was established that there were asymmetries in alliances A, B and E because most of the key technical and management staff were drawn from one of the partners, in this case, firms a_2, b_2, and e_2 respectively. The respondents from the firms that contributed more resources to the alliances, further stated that they had feelings that their partners were either free-riding by limiting their contributions to the alliances or were simply behaving opportunistically. They were of the opinion that, being the main contributors, they had the right to make key decisions about the alliance and direct their operations. On the other hand, respondents from firms a_1, b_1, and e_1 were of the opinion that, being citizens of Botswana where the projects were been conducted, they had the right to transfer of technology in key project areas, and therefore, their partners should have taken this into consideration. In addition, they thought that the alliances were run just like the partners' firms and not according to what was agreed. Responses from the firms in alliances C and D indicated that all partners in the alliances contributed to the established entities as per agreements.

Question 6: How were the activities of your firm that did not fall under the alliance run?

The responses to question 6 established that, while firms a_1, b_1, and e_1, contributed very limited resources to their respective alliances they used their key personnel to run private activities. An indication that the alliance was used as a mechanism for transfer of technology is that firms b_1 and e_1 had a policy of attaching a key personnel with the alliance for periods not exceeding four months, after which a transfer was made and replacement was attached to the alliance. The idea behind this, according to the respondents from these firms was to ensure that the key personnel in these firms were exposed to the operations of the alliances, which were deemed superior to those of their firms. Firms e_1, and d_1 had private activities in Botswana, but once they assigned personnel to a project, they did not remove them before the completion of the project.

Question 7: How loyal were you employees to the alliance?

As responses to question 6 indicated, private activities had influences on the loyalties of the partners to the alliance. The responses to question 7 generally established that the individuals involved in the alliances were torn between the loyalties to the ventures themselves and to the parent organizations from which they originally came. As it was established by responses to question 5, while individuals from firms a_2, b_2 and e_2 tended to run the alliances in the same way their parent organizations were run, individuals from firms a_1, b_1 and c_1 concentrated on learning the skills of their partners for use in their parent organizations.

Question 8: What were the reasons for terminating/sustaining the alliance?

The original information about the alliances established that, of the five alliances, three had terminated and two are still in operation. Responses from firms c_1 and c_2 to question 8 established that alliance C was formed as a single project partnership and this was terminated on completing the project. The main reason given for the termination of alliances A and B is that the construction environment changed, leading to the partners altering their needs and orientations which affected the partnership. However, more fundamental reasons given by firms a_1 and b_1 was that after working within the alliances for more than four years, they had learned the skills of their partners and so they did not see any economic sense in sustaining the alliance as their private activities gave them more income than the common activities. The experience and knowledge learned from the alliances enabled them to raise their grades in the classification of construction companies, and being citizen companies, they enjoy 2.5% price preference on all public financed projects. This factor contributed substantially in their decision to terminate the alliances. Similarly, firms a_2 and b_2 became familiar with the labour market and the local environment. So the original needs of forming the alliances had been fulfilled. Responses from firms d_1, d_2, e_1 and e_2 to question 8 established that the reasons for forming their alliances were still valid and although they were not running perfectly, there were still some benefits to be gained by sustaining them.

Conclusions

This paper attempted to highlight the importance of alliances in the construction industry and the influence of private aspirations on their performances.

The review showed that the development of efficient transportation and communication systems have made construction markets all over the world accessible to specialists from different parts of the world. Competitive advantages that were originally enjoyed by certain firms as a result of their presence within the market or in their vicinity can no longer be sustained. As a result, firms need to form alliances so as to pool their resources together and compete favourably.

The study in Botswana established that such reasons as acquisition of know-how and sharing of risks in uncertain market sectors lie behind the formation of alliances. It also established that in the alliances, firms aim at learning the skills of their partners and apply them to their private activities. The study also showed that, although the general objective of forming alliances, i.e., pooling together of resources for competitive advantage was also valid in Botswana, once incomes from private activities are higher than those from common activities, incentives to sustain the alliances run out and often lead to their termination.

References

Annual Economic Report (1996). Government Printer, Gaborone.

Badger, W et al. (1993) Alliance in International Construction. *Source Document 89, Construction Industry Institute*, Austin, Texas.

Balakrishnan, S and Koza, M P (1993) Information, asymmetry adverse selection and joint ventures: Theory and evidence. *Journal of Economic Behaviour and Organization*, 20, pp. 99-117

Beamish, P (1985) The Characteristics of Joint Ventures in developed and developing countries. *Columbia Journal of World Business*, 20, pp. 13-19.

Bleeke, J and Ernst, D (1991) The way to win in cross-border alliances. *Harvard Business Review*, 69(6), pp. 127-135.

Cohen, W and Levinthal, D (1990) Absorptive Capacity: A new perspective on learning and innovation. *Administrative Science Quarterly*, 35, pp. 128-152.

Datta, D K (1988) International Joint Ventures - A framework for Analysis. *Journal of General Management*, Vol. 14 (2), pp. 78-91.

Eisenhardt, K and Schoonhoven, C B (1996) Resource-based view of strategic alliance formation: Strategic and social effects in entrepreneurial firms. *Organization Science*, 7(2), pp. 136-150.

Geringer, M A and Herbert, L (1991) Measuring Performance of International Joint Ventures. *Journal of International Business Studies*, 22, pp. 249-264.

Gulati, R (1998) Alliances and Networks. *Strategic Management Journal* Vol, 19, pp 293 - 317.

Harrigan, K R (1988) Joint Ventures and Competitive Strategy. *Strategic Management Journal* 9 (2), pp 141-158.

Khanna, T; Gulati, R and Nohria, N (1998) The Dyanamics of Learning Alliances: Competition, Cooperation, and Relative Scope. *Strategic Management Journal*, Vol. 19, pp.193-210

Kogut, B (1989) The stability of joint ventures: Reciprocity and competitive rivalry. *Journal of Industrial Economics*, 38, pp. 183 - 198.

Kogut, B (1991) Joint ventures and the option to expand and acquire. *Management Science*, 37 (10), pp. 19-33

National Census (1991) Central Statistics Organisation.

Peterson, R B and Shimada, J Y (1978) Sources of Management Problems in Japanese - American Joint Ventures. *Academy of Management Review*, Vol. 3, pp. 796-804.

Simkoko, E E (1989) Analysis of Factors Impacting Technology Transfer in Construction Projects - Case Studies from Developing Countires, Swedish Council of Building Research, Sweden.

Somarajah, M (1992) Law of International Joint Ventures, Longman, Singapore.

3. Contract and Contractual Relationships

A STRUCTURED METHODOLOGY FOR IMPROVING THE OWNER-CONTRACTOR RELATIONSHIP IN CONSTRUCTION PROJECTS

ALFREDO F. SERPELL

Department of Construction Engineering and Management, Pontificia Universidad Católica de Chile, Vicuña Mackenna 4860, Santiago, Chile. aserpell@ing.puc.cl

Abstract

Given the complexity of the relationship between owners and contractors in construction projects, research work has been carried out to identify mechanisms to facilitate the understanding between these parts. Based on the Quality Function Deployment (QFD) approach, an exploratory study on the form of rationalising this relationship has been conducted and its preliminary results are presented in this paper. Within this research effort, a model that puts emphasis in the factors that, depending on the type of contract, can improve the performance of the project has been developed. This model helps to structure a set of information through a matrix scheme that allows the weighting of each relevant characteristic of the commercial relationship between the client and the contractor with the purpose of prioritising between them. The scheme uses a cross-impact approach and provides a sequential methodology that conditions the achievement of the owner's objectives to the actions of the contractor and to his own key contract management elements that influence the contractor's behaviour. It is proposed here that this methodology could constitute a useful decision tool for the owner when developing and executing a construction project.

Keywords: Contract management, construction projects, decision tool, owner – contractor relationships

Introduction

The relationship between owners and contractors in construction projects has always been a matter of concern for practitioners and researchers in the area of construction management. This paper presents a preliminary research effort being carried out with the purpose of identifying mechanisms to facilitate and improve the understanding between the participants involved in this relationship. This objective has been approached from the viewpoint of the client-supplier relationship used in quality management schemes, exploring the possibility of applying some of the quality tools to the analysis of the owner – contractor relationship.

Thus, using an approach that emulates the Quality Function Deployment (QFD) scheme, an attempt has been developed to structure a set of information through a matrix display. This set of information would contain the most relevant characteristics of the owner – contractor relationship and provide a sequential methodology that would condition the achievement of the owner's goals to the contractor actions and to its own contracting management actions. This methodology could become a decision tool for the owner when developing and executing a construction project. The next sections present a description of this methodology.

Background

The QFD is a planning tool that is used to help a supplier to focus on the needs of its customers when setting the design and its specifications [Munro-Faure and Munro-Faure (1992)]. This model is very flexible and allows an organisation to identify the characteristics of a product or service and prioritise the most important ones for the client. This model was first developed in Japan in the seventies and its application in the western world has widely expanded in the last years. The method presents a great potential of application to the design and conceptualisation of houses, being one of the first applications accomplished in the construction sector [Shiino and Nishihara (1990); Serpell and Wagner (1997)].

The most known characteristic of the QFD is the matrix display of what is denominated the house of quality. This matrix help to show the relationships between the customer requirements and the supply voice which is represented by specialists on product or service characteristics. Figure 1 shows a simplified example of this matrix.

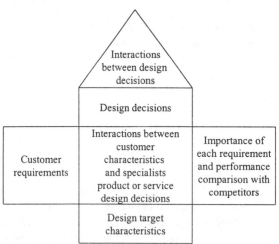

Figure1 The matrix display of the house of quality

As shown in the matrix, several elements are included in the model. First, the customer requirements that are obtained using an appropriate tool to acquire their opinion in front of a product, service or problem like, for example, surveys, interviews and others. These requirements can be weighted according to the importance given to them by customers. Second, the design characteristics or decisions are those defined by specialists regarding the product or service analysed.

The two sets of information are then related by a matrix of interactions. Through these interactions, it is possible to use the customer requirements to weight the design decisions and better focus them to the customer needs. This is the most important feature of the QFD and can be repeated along four stages:

1. Customer requirements are related to design decisions,
2. Design decisions are related to product or service specifications,
3. Product or service specifications can be related to process definition, and
4. Process definition can be related to production or service requirements.

At the end of these stages, the customer requirements have been carried through to the production or service requirements allowing an organisation to adequately address its production processes and requirements to the customer needs and improving the potential of success of its products or services.

The owner – contractor relationship

The working relationship between an owner and its contractors is one of the most important determinants of project success. A positive working relationship can help overcome the problems that inevitably arise on every project [Wakefield (1989)]. However, it is not uncommon that the owner and the contractors lack the necessary communication and information to better understand what are the expectations that each one has of the other. This is a main factor in making the relationship between these parties a poor one.

The owner – contractor relationship should and can be modelled as a customer – supplier relationship. In this relationship, it is important to understand what are the main factors that impact it. Some of these factors are:

- Type of contract
- Relative capabilities of owner and contractors
- Co-operative attitude
- Personalities of key managers
- Scope of the work
- Third-party involvement
- Contractual liabilities
- Senior management support
- Planning abilities
- Other factors

Most of these factors are established on a formal level, in the contract and its documents. However, the relationship between an owner and its contractors presents also an informal level. On the formal level, the contract defines the scope of the work, the commercial terms and conditions and the understanding achieved by the owner and the contractor at the time the contract is signed. It reflects the favourable expectations of each party at the beginning of a construction project. Also, the contract assigns the responsibility of each party over hypothetical circumstances that will happen during the execution of the project.

On the informal level, the owner – contractor relationship operates within a context that is influenced by many factors. These factors are related to the competence of each party to manage the project as follows: project management ability, construction expertise, planning ability, supportive attitude, organisational ability, timing of conformity and others. Generally, owner and contractor have different opinions about the ability of the other party to perform on these different levels of competence.

Then, both levels should be incorporated into the agreement established between the owner and the contractor to reduce the potential of conflict that usually exists in this relationship. The concept of partnering is an approach that attempts to handle these both levels, in particular, the project specific partnering scheme as discussed by Loraine (1994). As stated by this author, the attitudes of both clients and contractors need to be fundamentally changed. Clients need the skills and confidence to analyse bids more completely, and to genuinely seek a fair sharing of risk, while contractors should be more open and freely share information.

It is proposed in this work that some elements in the analysis of the owner – contractor relationship are still not well considered. First, the expectations of each party and how to communicating them to the other. Second, how can the owner incorporate these expectations in its contracting approach to motivate a better contractor performance. The methodology described in the next section addresses these elements.

An structured methodology for analysing the owner – contractor relationship

The proposed methodology is based on the idea that it is necessary to analyse the expectations of owner on the relationship with the contractor and to incorporate them into the relationship using a cross impact approach. This approach also uses some of the properties of the Quality Function Deployment methodology. The sequence of analysis is shown in Figure 2.

Figure 2 Sequence of owner – contractor relationship analysis

The first stage corresponds to the identification of the owner's contracting goals for a project and the weighting of them to obtain a prioritisation of the goals. Using the QFD approach and having the owner's goals as inputs, it is possible to identify the most important actions that the contractor can include in its performance to satisfy the owner. These actions are also weighted to determine a ranking of their impact on the achievement of the owner's expectations. Finally, the identified contractor's actions are used as inputs to the third stage, and now it is the owner who has to identify which actions can he take to incentive the contractor to act in a way of maximising his own expectations.

An example of the application of this methodology is shown in the next sections. For the example, a lump-sum type of contract has been used and the information was generated through interviews to an experienced owner's project manager (OPM).

Identification and weighting of owner's contracting goals

The first step of the methodology was to identify the owner's contracting goals. According to the interview, four parameters were considered important by the OPM: cost, time, quality and functionality. These parameters can be obtained using several techniques, like the interviews, team – work techniques or the Delphi technique. In this way it is possible to include several opinions from the most relevant project stakeholders.

Later, it was necessary to obtain the relative importance of each of the goals compared to each other. This step produces the required weights for each goal. To do this, a pair-wise comparison technique was applied as shown in Figure 3 [Krishnan et al. (1993)]. This technique is based on constructing a matrix with the same factors in rows and columns. All the elements in the rows (i elements) are compared to the elements in the columns (j elements). The basic question used for the evaluation is: between element i and element j, which is the most important and by how much? A typical scale used is from 1 to 7.

		Normal Values				Normalised Values					
		Cost	Quality	Schedule	Functionality	Cost	Quality	Schedule	Functionality	Total	Weigths
Owner's Goals	Cost	1.0	4.0	4.0	7.0	0.61	0.39	0.75	0.37	2.11	0.53
	Quality	0.25	1.0	0.2	5.0	0.15	0.10	0.04	0.55	0.55	0.14
	Time	0.25	5.0	1.0	6.0	0.15	0.49	0.19	0.32	1.14	0.28
	Functionality	0.14	0.2	0.17	1.00	0.09	0.02	0.03	0.05	0.19	0.05
		1.64	10.2	5.37	19.0	1.00	1.00	1.00	1.00	4.00	1.00

Figure 3 Pair-wise comparison matrix

As a general convention, if the i element is more important than the j element, then a number between 1 and 7 is assigned. If the j element is more important than the i element, the number assigned is the inverse of the assigned in the first case. After all the cells are filled, the matrix is normalised by dividing each number in the cells by the sum of the corresponding column. Finally, the normalised cells for each row are summed up and the total is normalised again on base 1 to obtain the weight of each goal. These weights are using in the next step.

Identification and weighting of contractor's performance priorities

The identification of the contractor's priorities corresponds to what is named "the voice of the supply" in the QFD scheme. To define these priorities, a working team formed by contractors' specialists should be created to analyse the owner's goals and to propose performance priorities that will facilitate a contractor performance that will highly satisfy the owner's requirements. At this stage of research, this process was simplified by using bibliographical sources of information and the interview of an experienced construction manager. A small set of priorities was generated in this way as shown in Figure 4 for the purpose of illustration only.

As can be observed in the matrix of Figure 4, the contracting goals of the owner are maintained but the weights of each goal is now included. To weighting up the contractor's priorities as a function of each of the owner's goals, an evaluation process is carried out by the contractor's specialists. This process consists of evaluating the relationships between the goals and the performance priorities using an appropriate scale of influence. As explained before, one of the main features of QFD is the fact that utilises the demand for pondering design aspects that the supply itself has defined. In this case, the scale used for pondering was as follows:

- No influence or relationship: 0
- Low influence or relationship: 1
- Medium influence or relationship: 3
- High influence or relationship: 5

		Weights	Contractor's performance priorities						
			Good management capacity	Top management participation	Application of constructability	Good planning system	People well qualified	Effective control system	Effective quality management
Owner's Goals	Cost	0.53	5.0	5.0	3.0	5.0	3.0	5.0	3.0
	Quality	0.14	3.0	5.0	3.0	5.0	5.0	3.0	5.0
	Time	0.28	5.0	3.0	5.0	5.0	5.0	3.0	5.0
	Functionality	0.05	1.0	1.0	5.0	1.0	1.0	1.0	3.0
Weights			4.5	4.2	3.7	4.8	3.7	4.0	3.8	28.7
Normalised weighs			0.16	0.15	0.13	0.17	0.13	0.14	0.13	1.0

Figure 4 Identification and weighting of contractor's priorities

The selection of a not continuous scale is to reinforce the difference of evaluation to achieve a better discrimination at the end of calculations. After filling all the cells with the influence numbers, it is possible to combine the evaluations for each performance priority by multiplying each number of the corresponding column by the weight associated to each goal. In this way, for each performance priority a weight is obtained. Later these weights are normalised on base one as shown in the last row of the matrix.

Until this point, the application of the QFD scheme has been straight forward, using its capability of relating the supply priorities to the demand goals. Through this approach, a contractor would be able to analyse and identify which of his performance factors are more important to provide total satisfaction of its clients' needs. In this way the contractor can obtain a clear understanding of the key performance functions and characteristics that shape its service following the same process that QFD provides for product design.

Identification and weighting of owner's contracting strategies

An additional development of the research effort was to make use of the information obtained so forth as an input for the owner when deciding about his own contracting strategies. It is clear that the owner strategies can have a strong impact over the contractor's performance but it is not always evident which ones are the most influencing on the project performance.

Then, a cross impact matrix was created with the purpose of identifying the most important strategies for influencing the contractor's performance applying the QFD approach in an inverse way, i.e. pondering the demand voice (owner) by the supply voice (contractor). In this way the owner can identify strategies that will help produce the best performance of the contractor, by including those factors that will incentive the contractor to take care of his performance priorities that would maximise the potential satisfaction of the client. These performance priorities were identified in the previous stage.

The inverted matrix obtained in this process is shown in Figure 5. As seen in the matrix, now the answers that are looked for by the owner are to the question: what can we do to incentive the contractor to give importance to the performance priorities that, according to the previous analysis, would maximise our satisfaction potential? Using the inverted matrix is possible to answer this question for each of the most relevant performance priorities applying the same kind of process carried out in the previous section.

Following this last matrix, the owner can then devise the best strategies for contracting that will take care of many of the risk associated to the contractor's performance. This analysis can be expanded with other matrices to obtain more detail for each of the strategies. For example a matrix can be created between owner's strategies and the required actions to achieve the objectives of each strategy. Later, another matrix can be developed to transform the actions into contract clauses and specific activities during the bidding stage and after contract signing. Figure 6 displays this additional sequence.

This approach takes advantage of the QFD methodology but instead of using it for the design of a product, here the last result would be a specific plan of contracting activities that the owner should develop to increase the contractor's performance. These activities also include the design of suitable contract documentation to support the owner's goals. This set of activities and the plan are in fact, elements of a plan to manage the risk associated to the contractor behaviour and capabilities during the execution of the project. A summary of the general methodology and its results is shown in Figure 7.

The model shown indicates that the owner's action contracting plan will act over different areas or functions associated with the contracting process. These impacts will produce the necessary requirements or drivers to influence the contractor's performance with the

		Weights	Good contractor selection process	Early integration of contractors	Better scope definition	Clear statement of expectations	Good project management capacity	Inclusion of quality management system as contractual requirement	Use of contractual incentives	Effective control system	
			Owner's contracting strategies								
Contractor's performance priorities	Good management capacity	0.16	5.0	3.0	1.0	3.0	3.0	3.0	5.0	1.0	
	Top management participation	0.15	5.0	5.0	0.0	5.0	3.0	3.0	5.0	0.0	
	Application of constructability	0.13	3.0	5.0	5.0	1.0	1.0	3.0	3.0	0.0	
	Good planning system	0.17	5.0	1.0	3.0	5.0	5.0	3.0	5.0	5.0	
	People well qualified	0.13	5.0	1.0	0.0	5.0	1.0	5.0	5.0	3.0	
	Effective control system	0.14	3.0	3.0	3.0	5.0	5.0	5.0	3.0	5.0	
	Effective quality management	0.13	5.0	3.0	5.0	5.0	5.0	5.0	5.0	3.0	
Weights			4.5	3.0	2.4	4.2	3.4	3.8	4.5	2.5	28.3
Normalised weights			0.16	0.11	0.08	0.15	0.12	0.13	0.16	0.09	1.0

Figure 5 Defining owner's contracting strategies

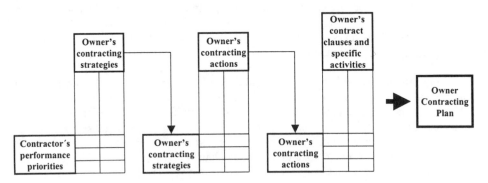

Figure 6 Sequence of steps for further analysis

purpose of increasing the likelihood of achieving the owner's contracting goals.

This approach has been already applied informally by a mining company when it decided to incorporate quality assurance management systems in its contracting scheme. The project team of this company performed a thorough analysis of the risk associated to this decision and how to introduce a group of actions destined to achieve the owner's goals by

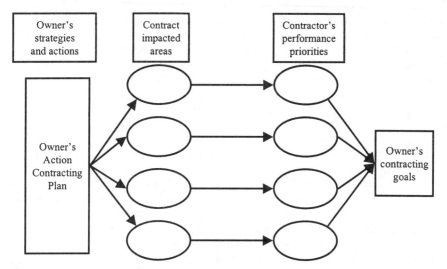

Figure 7 Summary of proposed methodology

acting over the contractors to maximise the probability of success of new contracts. For example, some of the actions included in the owner's strategy were [Morales (1998)]:

- Special pre – bidding meetings with contractors' representatives to communicate them the owner's expectations.
- Training activities for contractors' teams with the purpose of assuring the necessary capabilities to work under the new contracting approach, paid by the owner.
- Designing of contracting documents to make them consistent with the new approach and including different types of incentives to motivate contractors to make the required effort to work with quality assurance systems.
- Sharing of quality consultants costs with contractors.

Conclusions

This paper has presented the preliminary developments of a research project on the owner – contractor relationship that is exploring a QFD based methodology for its improvement. It is important to note that all the information included in the examples above is not complete and is used for illustration purposes only. This methodology can be utilised as a decision tool by the owner when planning and deciding about the most effective contracting strategy for the procurement of construction services.

Although the research is still in its preliminary stages, some validation of the approach has been carried out through the analysis of the contracting processes followed by two owners who, though informally, applied a number of steps that are very similar to the ones included in this work. In both cases, the owners are very effective in their contracting approach and perform very detailed analysis of the risks associated to contractors' performance that can strongly impact their own goals. Then, they have been quite successful in incorporating suitable strategies to manage these risks.

The QFD approach, mainly used for the design of products and services offers a very high potential for other kind of applications like the one presented in this paper. The structured analysis offered by this tool seems very suitable for creating a decision tool for contracting management. More research effort should be carried out to confirm this appreciation and to fully exploit its features.

References

Munro-Faure, L and Munro-Faure, M (1992) Implementing Total Quality Management, Pitman Publishing, London, U.K., chapter 7.

Shiino, J and Nishihara, R (1990) Quality Deployment in the Construction Industry, in Quality Function Deployment: Integrating Customer Requirements Into Product Design, edited by Yoji Akao, cap. 10, Productivity Press, USA.

Serpell, A and Wagner, R (1997) Application of Quality Function Deployment (QFD) to the determination of the design characteristics of building apartments, in Lean Construction, edited by L Alarcón, Balkema, Rotterdam.

Wakefield, W (1989) Owner/Contractor Relationship, in Project Management, edited by R Kimmons and J Loweree, Marcel Dekker, Inc., New York, p. 61-65.

Loraine, R (1994) Project specific partnering, Engineering, Construction and Architectural Management, Volume 1, Number 1, p. 5-16.

Krishnan M., Houshmand A., McDonough J. F. (1993) Applying TQM in the College Environment: Using QFD to Design a Curriculum, College of Engineering, University of Cincinatti.

Morales, H (1998) Experiences of quality assurance system implementation in construction projects, Presentation to the Diploma on Quality Assurance and Management in Construction Projects (in Spanish), Department of Construction Engineering and Management, Catholic University of Chile, 12 August.

CONSIDERATE CONTRACTING – ALTRUISM OR COMPETITIVE ADVANTAGE ?

S. BARTHORPE

School of the Built Environment, University of Glamorgan. Pontypridd. CF38 1AA. United Kingdom. sbarthor@glam.ac.uk

Abstract

The construction industry suffers generally from a poor public image and one of the main reasons for this is the poor relationships that exist between construction sites and their neighbours. The intention of this paper is to consider the social responsibility of developers and contractors towards the general public and to discuss several innovative initiatives devised to minimise the disruptive nature of contracting. The reasons for creating a more harmonious relationship with the public are not necessarily entirely altruistic since improved profitability and performance are likely to result from this neglected aspect of marketing.

Keywords: Considerate contracting, marketing, public relations, social responsibility,

Introduction

The impact of the construction industry on the environment at all stages of operation is enormous. Construction activities include; mineral extraction, pre-fabrication, construction, maintenance and demolition. These activities are often carried out without due consideration given to the public, which contributes to their negative perception of the construction industry.

There is evidence to suggest that the construction industry's interpretation of public relations is limited to its product marketing orientation, ignoring the more holistic application, which includes social responsibility and ethics. The industry's commercial interests and operational practices would appear mutually exclusive and serves to illustrate that the negative public perception is justified.

Public relations should be considered a strategic management tool which organisations could use to establish and improve their reputation. Organisations should take a proactive 'stakeholder' approach to maintain good public relations with the community. Initiatives that tackle social problems such as crime prevention will be mutually beneficial so that effort expended in maintaining good public relations need not be considered by organisations as a financial burden.

Innovative considerate contracting schemes emanating from London Metropolitan authorities have provided the mechanism and impetus for improving the public face of construction. These schemes are discussed in this paper and practitioners are encouraged to implement them in order to maintain a harmonious relationship with the public.

Public relations and marketing

The UK's Institute of Public Relations (IPR) define public relations as;

> *The deliberate, planned and sustained effort to establish and maintain goodwill and mutual understanding between an organisation and its publics.*

Publics would include; employees, investors, customers, governments, neighbours, communities, pressure groups, supply-chain companies and even competitors. Public relations embraces public affairs, corporate affairs, community affairs, community relations and corporate communications. Positive publicity is dependent upon a good relationship with the media and some organisations have specific individuals or departments that manage their public relations. External pressure groups and lobbying organisations are becoming more demanding and some companies are beginning to demonstrate their social responsibility on a global basis.

It is presumed that the marketing departments of contractors expend considerable resources in their efforts to maintain market credibility and achieve corporate image enhancement. The same contractors however appear to ignore the detrimental effect that their often, inconsiderate site based production activities have upon the public.

The broadening procurement options available and the fragmentation of the construction industry into specialist services has created the dilemma of whether construction is a product or service industry. Preece and Male (1997) contend that the intangibility of the product in construction has meant that many practitioners have found it difficult to apply the concepts of marketing with that of the contractual service. Male and Stocks (1991), whilst applying the concepts of services marketing, assert that the construction contractor promotes an image based on its reputation to complete on time, to budget and to the desired quality.

The apparent dichotomy between the corporate manifestations of its marketing and those of its operational activities indicate the limited application and even understanding that contractors have of marketing and public relations. Hillebrandt and Cannon (1990) suggest that orientation to the marketing concept in construction has been slow and that it is perhaps the least well developed of the management activities. A lack of understanding of what marketing is, and how it can be applied has, according to several writers, led to it being developed in an *ad hoc* fashion (Bell, 1981 and Newcombe *et al*, 1990). Latham (1994) recommends contractors to understand that marketing needs to facilitate deeper client satisfaction and mutually beneficial business relationships that promote 'win-win' outcomes. Partnering, Latham suggests is the most suitable 'vehicle' to promote this. The observations made by the Construction Industry Board Working Group 7 in their report 'Constructing A Better Image' (1997) highlight many contractors myopic vision of marketing :

> *Traditionally, contractors have seen marketing in the context of winning new business, but this selling role is only one aspect of marketing. To deliver the deeper client satisfaction, brighter image and better rewards for the industry marketing needs to be understood as being primarily concerned*

with the establishment of enduring and mutually profitable relationships between a firm and its customers.

Public relations according to Harrison (1995) has four main stages :

- Finding out where you are.
- Formulating a (corporate) public relations strategy.
- Implementing the programme that follows from it.
- Monitoring and reviewing the results.

Effective communication is essential to successful public relations. A communications audit is a useful tool to help organisations find out where they are in this respect. The results and feedback may indicate that an overhaul of the organisation's corporate identity is required as well as a critical review of its corporate advertising campaign, investor relations programme and community liaison. Internally, employees need to be fully aware of the corporate mission, goals and objectives. Good relations with the community are a pre-requisite for a successful organisation. Every organisation should identify and define its own community and consider its responsibilities to society at large. With construction companies this may apply at international, national, regional and local levels and at both office and project levels.

Construction projects are often sited in isolated locations and often prone to criminal attention. The reciprocal benefits to the contractor of maintaining a good relationship with the community evidence themselves in greater co-operation and a more sympathetic and tolerant attitude to those operations causing noise, dust and inconvenience.

Clutterbuck, *et al* (1992) suggest that, "what is good for society is good for business". Initiatives that tackle social problems such as crime prevention should be fostered. Mutual benefits include increased profitability for the contractor due to the reduction of thefts and delays caused by vandalism and improved quality of life for the community. If a good relationship exists between the contractor and their neighbouring community the contractor will effectively enjoy free public 'policing' of the construction works.

Organisations representing levels one and two indicate where companies operate within basic business parameters, observe legislation and discharge their responsibilities to society. The level three type of company displays a different outlook. It realises the important correlation between maintaining an environmentally friendly, community sensitive approach and its resultant commercial success. Peach (1987) argues that they are not a luxury add-on to help a company's conscience. They should guide the company into a new way of thinking about all aspects of its activities, a phenomenon he calls the 'stakeholder approach. Sir John Egan's government sponsored Construction Task Force report, *Rethinking construction* exhorts a radical change in the way contractors operate. "We are not inviting UK construction to look at what it does already and do it better, We are proposing a radical change in the way we build" (Egan 1998). Those companies with the right culture are likely to be the ones that survive.

Harrison (1995) suggests that public relations is a strategic management tool which an organisation can use to establish and improve its reputation among key groups.

Companies need not conduct their business in the community entirely out of altruistic or philanthropic motives, it makes sound business sense to foster good relations with the local community. Peach (1987) compares the impact of an organisation on its communities to that of the ripple effect created by a stone dropped into a pond. The three levels of impact are indicated in Figure 1 below.

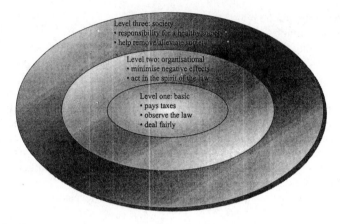

Level three: society
• responsibility for a healthy society
• help remove/alleviate societal ills

Level two: organisational
• minimise negative effects
• act in the spirit of the law

Level one: basic
• pays taxes
• observe the law
• deal fairly

Figure 1 :Impact of a business on its environment : ripples in a pond
Source: Adapted from Peach (1987)

Environmental factors

Environmental factors that create a negative public image of the construction industry include the hazardous working conditions and often disruptive, inconsiderate working practices that impact so significantly on the general public. The construction industry's poor record of waste control and its resource-exploitative nature, understandably cause ecological concern.

Although nation wide statutory control legislation is in place, it is the construction industry's erratic compliance with it that often causes concern to the general public. The relatively weak and inconsistent monitoring and control by many local authorities and environmental agencies in the UK have exacerbated the frustration and inconvenience experienced by the general public. Strict observance and innovative control and reward systems are however in operation in certain metropolitan areas of the UK and these provide much greater protection to the public. These initiatives have enhanced the reputation of those contractors that voluntarily comply with these tighter control procedures, resulting in higher safety standards on site and improved marketing opportunities.

Local government initiatives, particularly those emanating from metropolitan boroughs in London have made major contributions to the image enhancement of the construction industry. Projects operating under the 'Considerate Contractor Scheme' or the 'Considerate Builders Scheme' codes have significantly improved the general standards of public safety and cleanliness. The Considerate Contractor Scheme (CCS) and Considerate Builders Scheme (CBS) operate voluntary Codes of Good Practice but infringements of considerate conduct are likely to incur serious financial penalties.

According to Steadman (1987), one contractor was presented with a £80,000 ($135,000) bill for remedial work necessary to repair damaged paths, roads and drains surrounding a site. Although the schemes are voluntary, contractors are made accountable for their actions and transgressors can ultimately suffer the ignominy of being suspended from the scheme and by inference, be labelled inconsiderate (Smit 1996). A 'fail' mark from the inspectors can lose a company subsequent jobs and has proved a very effective deterrent according to the schemes' administrators. Terry Hines, the Corporation of London assistant city engineer is credited with inventing the Considerate Contractor Scheme in 1987 and explains its origins to Smit (1996):

> *The Scheme recognises the need to work with the public, sets standards and improves the way the industry works. In monetary terms, contractors get nothing, but the award (of Considerate Contractor of the Year'), means an awful lot in terms of recognition. Instead of recognising the directors of a company, it recognises the workers.*

The awards are increasingly seen as an indicator of contractor management skills which have enormous marketing potential. Disciplinary action is taken by contractors operating in Westminster for any of their workers that sexually harass women by using lewd behaviour, bad language or wolf-whistling. The City of Westminster's Considerate Builders Scheme officers operate a complaints 'hotline' in their *'Don't be a pest'* initiative. According to Prior (1996) several construction workers have already been disciplined and one worker has been dismissed.

The Considerate Contractor Scheme and Considerate Builders Schemes have arguably made the greatest contribution of any initiative to the improvement of construction operations in London. These improvements have led to a greatly enhanced reputation and image among the contractor's clients and public. The standards set now, in some of the most heavily trafficked and populated areas imaginable, can according to the CCS lay claim to be as good as anything in the world. Such has been the success of the CCS and CBS that other local authorities in the UK now operate similar schemes. According to Farrelly (1998), interest has also been shown from organisations abroad, notably from Malta, Australia, Italy and the United States of America.

The CCS and CBS Codes of Good Practice are in addition to the contractor's legal obligations and are based upon the compliance of the basic 'housekeeping' criteria of; care, consideration, co-operation, cleanliness, safety, and accountability. 'By offering coveted Awards based on the condition of the interface between construction sites and the public highway, the innovative Considerate Contracting Scheme induces a spirit of pride and excellence in the workforce and supervisors', according to Hines (1998). An overview of the schemes is shown in figure 2.

In English law, contractors owe a reasonable duty of care to minimise inconvenience to the public. Responsible contractors operate considerately by restricting hours of noisy work and liaising closely with the public. Tolman (1997) advises contractors to take

The Corporation of London 'Considerate Contractors Scheme': Code of Good Practice (launched 1987)

Care: All work will be carried out safely and in such a way that it will not inconvenience pedestrians or other road users. Special care will be taken to make sure that pedestrians with sight, hearing or mobility difficulties are not inconvenienced or endangered and that access is maintained for those in wheelchairs or pushing prams.

Clean : Footways and carriageways affected by works are to be kept in a tidy and safe condition. Hoardings, scaffolds, warning lights and other features are to be kept clean and neat.

Considerate : As far as possible, works are to be carried out in such a way that noise and dust are kept to a minimum and at times which will minimise the effect on city workers, residents and visitors.

Co-operative : The main contractor is to ensure that sub-contractors, suppliers and others working on, or near the site, maintain the standards of the Code of Practice.

The City of Westminster Council 'Considerate Builders Scheme': Code of Good Practice (launched 1989)

Considerate : All work is to be carried out eith due consideration for residents, workers, pedestrians, visitors, neighbouring occupiers, businesses and other road users, and at times and in a manner that will minimise disturbance. Special attention is to be shown to the needs of those who have difficulties with sight, hearing or mobility, and those in wheelchairs or pushing prams and pushchairs.

Quiet : Noise from works, machinery, workers, radios, music, vehicles and all other sources is to be kept to a minimum. There are to be no works that are audible at the site boundary outside permitted hours of work, unless prior agreement has been reached with the Westminster Council.

Clean : Footways and carriageways adjacent to the site, as well as all visible aspects of the site, such as hoardings, scaffolding and warning lights, are to be kept clean and in good order.Dust and smoke are to be kept to a minimum. Mud and spillage is to be cleaned off pavements and roads immediately.

Tidy : Pride in the condition and appearance of the site and the adjoining highway is to be shown in every way, including the tidiness of temporary structures, materials and machinery, and the constant removal of litter and rubbish.

Safe : Works and vehicle movements are to be carried out with utmost care for safety of passers-by as well as for workers. All plant and machinery is to be maintained in safe working order and the safety of structures is to be checked frequently.

Responsible : The main contractor is to ensure that all employees, agents, sub-contractors, suppliers, drivers and others working on or near the site maintain all aspects of the Code of Good Practice.

Accountable : A Contact Board is to be displayed outside the site, giving names and telephone numbers of staff who can be contacted promptly and take immediate action, in response to issues raised by residents, businesses and others.

The Construction Industry Board 'Considerate Constructors Scheme': Code of Considerate Practice (launched 1997)

Considerate : Be considerate to the needs of all those affected by the construction process and of its impact on the environment. Special attention to be given to the needs of those with sight, hearing or mobility difficulties.

Environmentally Aware : Be environmentally aware in the selection and use of resources. Pay particular attention to pollution avoidance and waste management. Use local resources wherever possible and keep to a minimum at all times noise form construction site activity.

Clean : Keep the site clean and in good order and ensure that the surrounding area is kept free from mud, spillage and any unnecessary construction debris.

A Good Neighbour : Be a good neighbour by undertaking full and regular consultation with neighbours regarding site activity from pre-start to final handover. Provide site information and viewing facilities where practical.

Respectful : Promote respectable and safe standards of behaviour and dress. Lewd or derogatory behaviour should not be tolerated under threat of the strongest possible disciplinary action.

Safe : Be safe. All construction operations and vehicle movements to be carried out with care for the safety of passers-by, neighbours and site personnel.

Responsible and Accountable : Be responsible for the understanding and implementation of the Considerate Constructor Scheme by all site-related personnel and be accountable to the public by providing site contact details. Take immediate action to ensure compliance to the Scheme's code of good practice and develop good local relations.

Figure 2 : An outline comparison of the main considerate contracting schemes necessary precautions to avoid prosecution under the law of nuisance because successful nuisance prosecutions usually result in court-imposed restrictions which are not 'relevant events' for extending time under most standard form contracts.

The major British supermarket companies each open a new store approximately every two – three weeks in the UK. This construction experience combined with their exacting retailing standards of customer-driven service ideally qualify them as 'enlightened clients', who are able to initiate innovation and change in the construction process. Tesco have played a significant role in effecting change and are anxious to get full value for their £500 ($840) million annual construction bill. Glackin (1997) cites Tesco's construction manager, Alison Hughes:

> *We see all our stores during the construction process as marketing possibilities. Potential customers pass by our sites every day and we want to show them a good image of Tesco that will complement our retail operation once the store is opened.*

Tesco's former property services director Mike Raycraft states that; 'We needed to establish continuity. We wanted contractors to become an extension of Tesco and to understand us culturally'. (Macneil 1997).

An innovative agreement has been made between the Rocco Forte Hotel Group (RFH), the Cardiff Bay Development Corporation, Laing Construction and the residents of Cardiff Bay, South Wales, UK. The unique 36 point Residents' Charter is designed to promote harmony with the residents during construction, while offering quality job opportunities to local people afterwards. (RFH Residents Charter 1997). Strict rules are encompassed in the Residents' Charter to minimise disruption to the residents and cash penalties are imposed if certain regulations are breached by the contractor. An overview of the Residents' Charter is provided below in figure 3.

Residents' Charter overview:
- Designed to ensure harmony with residents during construction.
- Training and job opportunities to local people after construction.
- Strict rules minimising disruption to residents.
- Cash penalties for breach of certain regulations.
- Construction traffic diverted from local homes.
- Fortnightly cleaning of nearby resident's windows.
- Construction vehicles washed before leaving site.
- Air quality monitoring.
- CCTV monitoring of activities and fines imposed for breaches.
- Regular meetings with community representatives.
- Special promotions offering resident's access to hotel facilities.
- Fines imposed will be used to benefit local community projects.

Figure 3. Overview of Cardiff Bay Residents' Charter
Source: RF Hotels Ltd.

RFH intend the Charter to demonstrate that the company is serious about working in partnership with neighbours, not only during construction but also as a member of the local community once they are open for business.

In accordance with one of the 'proposals for change' recommended in the Building Towards 2001 report (1990) the Construction Industry Board developed the very successful Considerate Contractors Scheme and Considerate Builders Scheme into a nationwide 'Considerate Constructors Scheme'(CCS). The CCS underwent a 'pilot' stage period and was officially launched nationally in July 1997. The CCS is administered on behalf of the CIB by the Construction Confederation and operates a voluntary Code of Practice which seeks to:

- Minimise any disturbance or negative impact (in terms of noise, dust and inconvenience) sometimes caused by construction sites to the immediate neighbourhood.
- Eradicate offensive behaviour and language from construction sites.
- Recognise and reward the constructor's commitment to raise standards of site management, safety and environmental awareness beyond statutory duties.

The CCS is a national initiative to improve the image of the construction industry through the better management and presentation of its sites. It aims to raise the standards of construction above statutory requirements. Contractors must adhere to a seven point Code of Considerate Practice which is indicated in figure 2. The benefits that contractors receive who participate in the scheme according to the CCS are, that it improves :

- their business reputation.
- perceived professionalism.
- relations with the local community.
- future tendering opportunities.
- the opportunity to receive national recognition for awards of excellence.

Nick Raynsford, the UK Minister for Construction states in the CCS promotional literature:

> *The establishment of the Considerate Constructors Scheme provides a real opportunity for the industry to improve its image. It has the full support of every sector of construction and is backed by the Government, who recognise the importance of the industry demonstrating consideration for both the public and the environment.*

> *I want the construction industry to be increasingly seen as a force for good – environmentally as well as economically. The Considerate Constructors Scheme is one way in which the industry can use its professionalism and technical expertise to work for the benefit of all.*

Conclusions

Environmental factors influencing a negative public image concern the generally hazardous, inconsiderate and disruptive working practices used by contractors. Although sufficient statutory legislation exists to control these environmentally-unfriendly activities, the monitoring and implementation of these controls appear to be limited to a minority of local authorities in the UK. Isolated examples of good practice exist where contractors and communities co-operate and these should be considered as benchmark projects, worthy of wider implementation.

Whether the construction product is bespoke or not, the construction industry should consider themselves as a 'service industry', concentrating more on the *process* and utilising their considerable abilities for resolving technical problems in other important areas, particularly in public relations and social responsibility. For those contractors willing to create harmonious relationships with their neighbours, the potential for capitalising on this for strategic competitive advantage with ever more demanding, image-conscious clients, is significant. Being socially responsible by applying considerate contracting techniques is therefore not altruism but sound business sense.

References:

Bell, R. (1981) Marketing and the Larger Construction Firm. *Occasional Paper No. 22 Chartered Institute of Building.* Ascot. UK

Clutterbuck, D. Dearlove, D. & Snow, D (1962) Actions Speak Louder : a management guide to corporate social responsibility. Kogan Page / Kingfisher. London.

Considerate Builder Scheme (1989) CBS Documentation, Westminster City Council. London.

Considerate Contractor Scheme (1987) CCS Documentation, Corporation of London.

Considerate Constructor Scheme (1997) CCS Documentation, Construction Industry Board, London.

Construction Industry Board (1997) CIB Working Group 7 report 'Constructing a better image'. CIB, London.

Egan, Sir J. (1998) Report of the Construction Task Force, 'Rethinking construction'. HMSO. London.

Farrelly, W. (1998) Interview with W. Farrelly, City of Westminster Council Considerate Builders and Roadworks Manager.

Glackin, M. (1997) 'Brand-new outfit' in Building, September 19 1997 pp 60-61. ABC Business Press. London.

Harrison, S. (1995) Public Relations : An Introduction. International Thomson Business Press. London.

Hilebrandt, P.M. & Cannon, J. (1990) The Modern Construction Firm. Macmillan.Basingstoke, Hants. pp 63-69.

Hines, T. (1998) Economic Opportunities in Sustainable Transportation, in a paper presented to an International Conference : 'Moving the Economy'. Toronto, Canada July 1998

Institute of Public Relations (IPR) Public Relations defined in the 'Mexican Statement' World Assembly of PR Associations. IPR. London.

Latham, M. Sir (1994) Constructing the Team. Final Report of the Government/Industry Review of Procurement and contractual arrangements in the UK Construction Industry, HMSO, London.

Macneil, J. (1997) 'Big changes in store at Tesco' in Building, December 5 1997 pp 14-15. ABC Business Press. London.

Male, S. & Stocks, R. (1991) Competitive Advantage in Construction : A Synthesis in Competitive Advantage in Construction. Male, S. & Stocks, R. (ed.) 1991. Butterworth-Heinemann. Oxford.

National Contractors Group (1990) Report of the NCG, Building Britain 2001. Centre for Strategic Studies in Construction. University of Reading.

Newcombe, R. Langford, D. & Fellows, R. (1990) Construction Management 2 : Management Systems. Mitchell. London.

Peach, L. (1987) Corporate Responsibility in Hart, N. (ed.) Effective Corporate Relations. McGraw-Hill. Maidenhead.

Preece, C. & Male, S. (1997) Promotional literature for competitive advantage in UK construction firms.Construction Management and Economics (1997) 15, 58-69.

Prior, G. (1996) 'Westminster cracks down on sexism' in Construction News, September 5 1996. EMAP Business Publication. London.

Rocco Forte Hotel (1997) Resident's Charter with Cardiff Bay residents. Cardiff. UK.

Smit, J. (1996) 'Compassion Fruit' in Procurement, July 1996 pp 14-16. ABC Business Press. London.

Steadman, R. (1987)'Pride amid the debris' in Building, September 11 1987 pp 30-31. ABC Business Press. London.

Tolman, S. (1997) 'Nuisance and the builder' in Building, June 27 1997 pp 29. ABC Business Press. London.

WINNING ALLIANCES FOR LARGE SCALE CONSTRUCTION PROJECTS ON THE WORLD MARKET

DRAGANA MITROVIC
School of Civil Engineering, University of Leeds, Leeds LS2 9JT, United Kingdom.
d.mitrovic@leeds.ac.uk

Abstract

The paper will discuss how changes in the nature of demand and changes in the business practice of the major construction industry clients, triggered by the increased global competition, are affecting the supply side of the market. It will also highlight how these changes shape the role of the construction industry on the global market and what opportunities they provide. In an attempt to answer if the harmonious relationships are becoming the pre-request for the success, the paper will examine the rationale behind harmonious relationship at the project level, and from a global construction company point of view.

Keywords: Core competencies, globalisation, privatisation, sourcing alliances, strategic partnering

Introduction

Globalisation of everything from industry to the lifestyle is almost certainly the most transformative global force since the inception of the industrial revolution. While technology is making globalisation feasible, liberalisation is allowing it to happen [Wolf (1997)]. Trade liberalisation, spread of financial deregulation, expanding capital markets, and the emergence of huge new markets, coupled with technological change and the communication revolution have created an increasingly integrated and institutionally harmonised world economy (Fig 1).

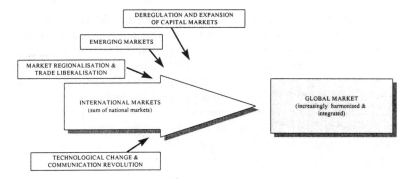

Figure 1 Transition of the International Market into the Global Market

Globalisation is opening up great scope for innovation and enterprise on a world-wide scale, free of the old constraints of distance, time and even borders. Multinational corporations are expanding in a greater number than ever into new markets. Whilst in some industries, there is a continuous search for low cost labour, the driving force behind most international investments appears to be the size and growth of the regional market itself. The expanding trade areas and growing economies are creating a huge demand for large scale construction projects, particularly new infrastructure, providing plenty of opportunities for the global construction industry. At the same time, increased competition generated by bigger, more liberal markets has profoundly influenced the business practice of the major construction industry clients, resulting in the new trends in the procurement of large scale projects. This paper will discuss how these changes are transforming the supply side of the market and what opportunities they provide for the future.

Research methodology

The paper builds on the initial findings of the research conducted for the ESPIRIT 20876 - eLSEwise project, addressing competitiveness of the large scale engineering industry. The study of the reconfiguration of roles and relationships at the market place, presented in this paper, is based on information gathered from:

a number of European national workshops hosted by the eLSEwise project partners. For each workshop's case study, one or two large scale projects were chosen and used as a catalyst to draw on the wider experience of all participants.
interviews with professionals involved in large scale projects at the global market
literature survey conducted in order to investigate the trends at the global construction market over 1993-97 period, comprising a review of newspaper articles, articles in business magazines and construction professional journals, and journal papers.

Changes in the construction clients business practice

Corporate clients

The tough mature markets of Europe, Japan and USA and increasing global competition have profoundly changed the way in which big corporations operate. They are coming to realise that they can no longer do everything well enough to sustain competitiveness and they are concentrating on the core operations, which are their reason for being in business. This is followed by outsourcing of non-core activities through the establishment of a whole range of sourcing alliances. While such clients usually keep control of the overall capital spending programme within their own companies, at the projects level more and more processes are outsourced. The clients have taken a view that the contractors are more efficient in managing a project process and project supply chain than the clients themselves, regardless of how knowledgeable they are.

Downsizing of clients' organisations and increasing pressure to get the product out into the market is resulting in increased trend towards design and build, and turn-key lump sum projects and also, in the last few years, towards total life-cycle service. In the case of the later, a contractor or a consortia is delivering project processes that go beyond the delivery of a facility into long term facility maintenance. Some clients, which are using the opportunity of the deregulated infrastructure market, are extending this co-operation to multiple projects and towards joint offering of BOT (Build, Operate and Transfer) projects at the global market [Tulacz (1997), Rubin (1997)].

Public clients

New economic reality is also transforming the public sector. Global competition and shrinking resource bases are pressuring governments to re-evaluate their roles. Once encompassing both policy setting and implementation, governments now are breaking away from the traditional roles focusing on policy setting alone and developing arrangements with the private sector to implement policies. Governments want private sector to provide services not just assets. They are turning to private sector to join in the construction, ownership and operation of national infrastructure assets. Two key development directions are: privatisation of the industry sectors (such as power, water, telecommunication, etc.) resulting in the new private infrastructure clients, and provision of public projects by the private sector.

The key driver behind the governments intent, to place developments, risk and financing of infrastructure projects within the private sector, apart from restrained government budgets, is recognition that some project processes are more efficiently done by the private sector. There are indications that current procedures for public work are among main sources of inefficiency. They involve long and complex planning procedures and sequential working methods resulting in project process disconnects and complex and costly contract administration [Mitrovic (1998)]. The World Bank has pointed out that cost overruns and time delays are common in purely public sector provision, leading to cumulative cost increase that can easily cancel out any interest rate advantage enjoyed by government [International Construction (1997)].

Changes in the nature of demand

Both trends, among major private and public clients, point in the same direction. Focusing on the core competencies, core processes and core skills enables clients to harness their resources to continually improve what they do best. There is readiness to abandon strict control, and ultimately the ownership, of the project processes, in exchange for a more efficient process.

The construction industry is increasingly asked to deliver holistic solutions that will satisfy clients, as well as end users and/or consumers needs, forcing project teams to take more responsibility for whole product and project life cycle from business development and into operation phase. To be able to fulfil these new requirements, the project supply chain, which for the large projects is usually lead by a consortium assembled around a contractor, will need much wider competencies and new types of relationships. Figure 2 presents the changes in the nature of demand and how they determine new competencies. Most of them will be best underpinned through long term partnering arrangements which allow continuous building of knowledge and facilitate seamless project interfaces.

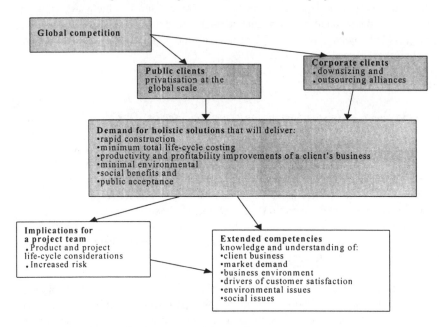

Figure 2 The changes in the nature of demand

The changing role of the construction industry in delivering large scale projects

Changes in the nature of demand are continually shaping and transforming the supply side of the large scale projects market. In this process of transformation, the following key evolution stages of the contractor's role can be recognised as [Mitrovic, (1997)]:

Delivering value to a client

A contractor is responding on a client request to take part in delivering a product required to satisfy client's business needs. Project process is client business process with which the contractor interacts. Contractor business process is order fulfilment. The focus is on product, and integration of a project process into client's business. Each project stage is triggered by a client and the ultimate project management responsibility stays with a client, who in the case of a large volume repetitive client is likely to have strong in-house project management team. A client is the party that brings knowledge about the business into project process, and a client's business needs are communicated through client's brief. He will choose key participants in each project stage, and within their scope of work they may manage their own supply chains. The 'sourcing' is done at operational level. Frequently, informal long term relationships can be traced among main project players.

The new trend among repetitive clients is the increased importance which they attach to the formation of long term partnership relations within the project supply chain. By offering continuity of repeat order business a client seeks to 'partner' with consultants, contractors and suppliers and in this way draw upon a wide skill base for a specialist knowledge input whenever needed in the project process. The co-operation, extended beyond a single project, places sourcing at the tactical level. These partnering arrangements have demonstrated significant product and construction project performance improvements [Mitrovic (1997a)]. The key advantage of this approach is that the close integration of a client and supply chain, brings better understanding of a client business into project process. The key disadvantages stem from the fact that by keeping control of a process within himself, the client may act as a barrier to better process integration.

Co-producing value with a client

A client is outsourcing a part of, or the entire project process to a contractor and/or a consortium, like in design and build, and turnkey projects. Also a contractor may take responsibility for the long term maintenance and consequently co-responsibility for operational performance, and he may be involved in provision of project funding. Project process is still client business process, but he does not keep control of the entire process himself and part of the project process which is outsourced becomes also a contractor's and/or a consortia's business process. More responsibility for project process is shifted towards contractor, but it also provides more opportunity to introduce improvements into the process. Depending on a particular arrangement, sourcing can be done on both, the tactical and strategic level. The contractor role has changed from delivering value to a client to co-producing value with a client. In the case of the total life cycle service, the focus is on improved operational performance, and risks and rewards are shared.

This approach needs high understanding of a client's business, particularly concerning total life cycle service. This understanding is best achieved through the long term strategic partnering alliances and integration of the total value system. The partners shift from a fragmented view of their respective contribution to a larger perspective that encompasses the entire project process. This enables them to align their processes toward the ultimate goal - the end user or the consumer at the market.

Delivering value to the consumers

A contractor is responding to consumer/end user demand and becomes involved into the business operation. A contractor, or more likely a consortium, is responding to business opportunity. It takes responsibility for the entire project process including provision of finance and responsibility for operation. A project extends into business operation which can range from property development business, to power generation business, to transport infrastructure provision business etc. Currently, two most frequent forms of such business projects are BOT and BOO (Build, Operate and Own) projects. While the BOT is a form of the concessionaire projects and may involve public client who will take accountability for a project, the BOO project is a pure private business venture. Consequently, delivery of a facility as part of this business operation, serves consortia/contractor business needs. Project process becomes consortia/contractor business process aiming to deliver value directly to the consumers. Traditional construction consortia is transforming into business consortia in order to mobilise creation of value in new forms and by new players.

Table 1 Transformation of the contractor's role

	Delivering value to a client	Co-producing value with a client	Delivering value to the consumers
scope	product delivery	product and service delivery	business function delivery
horizon	construction project	product life-cycle	business project life-cycle
risk allocation	client	shared	consortia
competence	understanding client's need	understanding client's business	understanding socio-economic need
dominant culture	bidding culture	co-operation	partnering
performance criteria	construction product performance	and operational performance	business and social performance
alliances	operational / tactical	tactical / strategic	strategic
benefits of partnering	cost savings	improved performance	competitive advantage
extent of partnering	within construction supply chain	includes client/operator/lender	includes stakeholders

The relationships as a source of competitive advantage

The notorious competitive pressure on prices, as a consequence of the construction industry bidding procedures and usually pre-determined product, has always pushed contractors to base their competitive strategies on differentiation. Which, in the context of the construction industry should, still, deliver the lowest bid price, or be able to prove that it is delivering better value for money (i.e. total cost of ownership). Under these constraints, it has always been important to try to enlarge the competitive scope of the company [Porter (1985)]. In the global construction industry, the notion was reflected in contractor's attempt to have wide sectors and geographic coverage, and in many examples of vertical and horizontal integration. In this light, can also be seen the transition 'from delivering a product to delivering a business', and Figure 3 shows how contractor's competitive scope is changing in this way. Design, Procurement and Construction are considered core construction industry functions. Each level, in the competitive scope pyramid, provides certain opportunities for competitive advantage, that are described below, and it is obvious that as the scope is widening the opportunities are increasing.

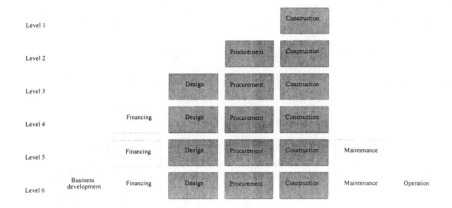

Figure 3 Contractor's competitive scope change

Level 1 & 2 - 'all competitors offer the same product' through competitive bidding. At level 1, the main part of the supply chain is predetermined by a client. In search for competitive advantage a contractor can look into alternative construction methods, optimising his own construction capabilities and product requirements, and integrating subcontractors interfaces. This will be reflected in better site mobilisation and organisation of works. At level 2, the choice of the supply chain is left to a contractor, although some key suppliers may be nominated by a client. In addition to the opportunities described for level 1, contractors may benefit from a good knowledge of global and local markets and linkages developed with supply chain.

Level 3 - 'all competitors offer similar product that has to serve the same business function', through competitive bidding. Suppliers of main equipment can be nominated. Addition of design service provides real opportunity to build quality and functionality into a product and also to introduce innovation. In this way product differentiation becomes possible. But design also provides opportunity for process differentiation, that spreads improvements into procurement and construction. A contractor has the opportunity not only to consider constructability, but to design for company specific construction capabilities and company specific supply chain strengths. The concept extends to consortium members.

Level 4 - 'all competitors offer a project that has to serve the same business function', through competition. Involvement in financing through participation in funding or by supporting funding arrangements is a step away from core construction industry functions that can be critical in job acquisition. Also, the ability to provide favourable funding arrangements can offset some pressure from the construction cost.

Level 5 - 'all competitors offer to support the same business function', through competition. Inclusion of long term maintenance services into offer, transform short to medium term client-contractor relationships into long term ones, where focus is shifting from construction performance to operation performance. Provision of total life-cycle service puts all upstream services into the function of operation and consequently, by changing focus from investment cost to total cost of ownership, gives contractors opportunity to compete on the ability to support client's value creation process.

Level 6 - 'all competitors offer to perform the same business function' in a competitive environment. New opportunities for competitiveness will be found in the ability to identify market demand and society needs, the ability to evaluate business opportunity, particularly in terms of possibilities it provides for synergy, and the ability to integrate different economic actors across industry borders into viable global offering.

The review of opportunities for building competitive advantage demonstrates that they can be broadly grouped under two headings: knowledge and relationships. It seams that these are the only two elements that really matter for a sustainable competitive performance.

Table 2 Opportunities for competitive advantage

Competitive scope	Knowledge	Relationships
level 1	construction management expertise	relations with subcontractors
level 2	as 1 + logistic expertise	as 1 + linkages with suppliers
level 3	as 2 + technical, process management, risk management, and R&D expertise, and understanding of client business function	as 2 + linkages with consortia partners
level 4	as 3 + financial expertise	as 3 + linkages with financial institutions
level 5	as 4 + operational management expertise	as 4 + linkages with client/ operator/ end users
level 6	as 5 + business development expertise and understanding of customers broad society needs	as 5 + relations with stakeholders and wider society representatives, and political authorities

At the point where competitive scope widens beyond the core construction industry functions: design, procurement and construction, a contractor's competitiveness begins to depend more and more on the quality of relationships which complement the contractors own knowledge. In this, strategic alliances are becoming crucial for the achievement of breakthrough strategies that completely redesign delivery processes and value chains. The value perspective of the strategic partners shifts from a primarily price driven one to one of true competitive advantage, in which quality, speed, innovation and other factors are considered along with price. In addition, as the project boundaries are extending, complexity of project interfaces is increasing. Ultimately, project success will depend on how well these external and internal interfaces are managed, and how well processes are integrated across physical and organisational boundaries. Long term partnering arrangements and collaborative approach extended towards wider society will pave way for seamless project interfaces.

Building winning alliances

The analysis of the case studies has stressed that winning alliances are made of partners with proven expertise, common experience, compatible business culture and objectives, preferably through long term partnership approach [Mitrovic (1997)]. Although, many companies tend to have, quite stabile, core supply chain, which they use across projects, they will assembly it again for each project, instead to forge long term strategic outsourcing alliances. It is fair to say that for a long time, contractors have just passed a client attitude, downstream, into supply chain. As more and more of the project process is transferred from clients to the contractors, they have more freedom to develop relationships with project supply chain (Fig. 5) and ultimately to reconfigure the whole value system in the way which best support value creation process.

Both, global and local component are equally important for contractors competing for the large scale projects as these projects bring international resources to be transformed at a local scene and output of the process has to serve local business needs and/or local market demand. Good global market coverage enables a contractor to establish links with potential clients, partners and suppliers and to procure globally. Lump-sum and BOT contracts mean search at cheaper places to buy material and equipment and to use lower-cost personnel outside the home country. Modern transportation and communication made it possible. Ability to exploit these opportunities better then others and build global networks gives a contractor a competitive edge.

Well established local links provide a contractor with understanding of socio-economic environment within which it has to operate, understanding of local needs which a large scale project has to satisfy and understanding of local business environment a contractor has to interact with. Alliances with local firms may also help market penetration. Once, the market is entered, increasingly, international contractors tend to become part of local scene - a corporate citizen. A sound understanding of local culture and knowledge of local conditions are important success factors at the changing market place where a contractor has to respond directly to consumers demand.

As the contractor competitive scope is enlarging alliances go beyond the industry boundaries and they may develop from well established links with potential clients, lenders, and political institutions. Under conditions when securing finance is becoming decisive factor for the viability of many large scale projects access to financial markets and well established links with financial institutions provide competitive advantage for a company. Close links with a client preferably developed through long term partnership relations, gives a contractor better chance to acquire knowledge about a client business and to understand and respond better to clients needs, within dynamic business environment. The world market survey indicates that in some industry sectors, like manufacturing, process industry, off-shore, close and well established long term relations provides contractors with significant competitive advantage [Mitrovic, (1997)].

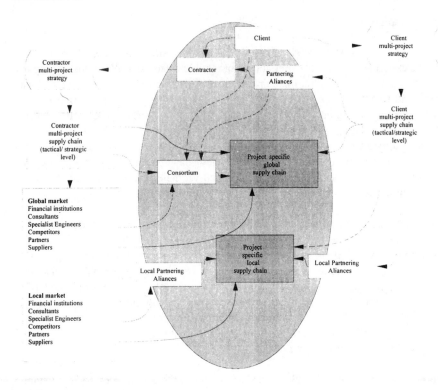

Figure 5 Project supply chain

Most public works projects are not profitable enough and involve high risks on a private only basis if customers are to be guaranteed low cost of services. On the other hand, a purely private-funded project would be tempted to maximise revenue, to the extent that it may fail to deliver socio-economic benefits. Public-private partnership seems to provide the optimum solution for these potentially conflicting objectives. Such partnership should go beyond risk and finance sharing, aiming to understand drivers of local customers' satisfaction in an attempt to jointly produce and deliver customer oriented value. By its nature and significance, the effects of a large scale project will spread far beyond the parties directly involved within the project increasing the importance of public acceptance. Therefore the concept of customers' satisfaction needs to be extended towards wider society addressing issues such as environmental, safety and health considerations.

Conclusion

The construction of networks of strategic alliances enables contractors to achieve maximum gains in value creation process. Networks of organisation working together to meet customers' needs will generate competitive advantage by enabling companies to focus simultaneously on their core and their customer. At the same time these companies will create competitive advantage by drawing upon the best resources at every step in the value chain. The following principles are proving to be important in building a networked enterprise: the relationships should be forged around a combined vision with opportunities for synergy, and it is important to assess potential alliances in terms of strategic value-added and customer value-added as well as economic value-added. At the same time, due to the socio-economic significance of the large scale construction projects, the awareness of the interdependence of the success factors is paving the way to extend partnership spirit towards all stakeholders.

References

Wolf, M (1997) Perspective: The heart of the new world economy. *Financial Times*, October 1, 1997, pp.

Tulacz, G J, Krizan W G and Tanner, V (1997) The top owners, *Engineering News-Record*, November 21, 1997, pp. 30-32

Rubin, D (1997) Need for goods spurs good market. *Engineering News-Record* , December 22, 1997, pp. 65

Mitrovic, D and Mangels B (1998) German national workshop, *Espirit 20876-Elsewise, Deliverable D107*, April 1998

Public-private partnership, *International Construction*, February 1997, pp 12

Mitrovic, D Male, S and Brohn, C (1998) The current LSE industry, *Espirit 20876-Elsewise, Deliverable D105*, September 1997

Mitrovic, D and Davis, T (1997a) UK national workshop, *Espirit 20876-Elsewise, Deliverable D101*, January 1997

Porter, M (1985) Competitive Advantage, *The Free Press,* 1985

HARMONISATION BETWEEN THE SMALL AND LARGE FIRM: SMALL FIRMS ADDING VALUE BUT AT WHAT COST?

CHRISTOPHER MILLER, TREFOR WILLIAMS and **LYN DAUNTON**
School of the Built Environment, University of Glamorgan, Trefforest, Pontypridd. CF37 1DL. Wales. CJMiller@glam.ac.uk, Twilliams@glam.ac.uk, Ldaunton@glam.ac.uk

Abstract:

Small business theory recognises that there are significant differences between the large and small firm in terms of structure, management style and organisational governance. These differences are especially indicative of the construction industry in Industrial South Wales (ISW). Despite this diversity of form, it is important to acknowledge that interdependencies exist between large and small construction firms. In fact, it could be contended that large construction firms have encouraged the manifestation of small subcontracting entities within the industry through the process of retrenchment and disinvestment.

Market conditions however, necessitate that small subcontracting firms are constantly in competition with each other and that increasing demands for low cost and flexibility by large contracting organisations effectively reduce the profit accrued by the small firm. This paper investigates this relationship and asks whether it is possible for harmonisation to exist between the subcontractor and the contractor that leads to mutual profit and added value for the client.

Keywords: Harmonisation, Industrial South Wales, subcontracting, technologies and Processes.

Background

The Small and Medium sized Construction Enterprise (SMCE) and the industry as a whole has been criticised by many with regard to its adversarial nature, the take up of new technologies and processes and issues associated with organisational management (Simon Report 1944, Emmerson Report 1962, Banwell Report 1964, Building Britain 2001 1988, Latham Report 1994, Small business Research and Consultancy Report 1996, South Glamorgan TEC Report 1996). Historically, large construction firms have taken remedial action to negate the effects of declining profits and economic recession. One strategic option available to the large firm is to retrench back into core business areas and organisations (Wimpey and Tarmac offer perfect examples of this process). Retrenchment and disinvestment strategies enable larger diversified firms to disregard peripheral business activity and instead concentrate on areas in which they enjoy distinct competencies or superior competitive advantage. Three aspects of this process are cost reduction, asset reduction and revenue generation. These processes, together with a need for greater flexibility, have led to the manifestation of the small subcontracting firm and the Specialist Trade Contractor (STC).

Retrenchment strategies and the emergence of the small subcontracting firm suggest that there is an implicit interdependent relationship between the small and large firm. Interdependence implies that both parties require mutual cooperation to ensure the success of a project and secure customer satisfaction. Harmonisation however, can often prove to be difficult. The very fact that small and large firms are fundamentally different leads to as many problems as it does opportunities for both parties. The Cambridge Study (1992) quoted in Storey, (1994) and Stokes, (1995) suggests that Small and Medium sized Enterprises (SMEs) are not adept at marketing, financial control, or management, and often possess little motivation in terms of accelerating the firm's innovative capacity. These factors can be construed as contrary to the atypical large organisation. The paper will also attempt to evaluate the relationship in the context of a specific region, namely Industrial South Wales (ISW). It is recognised that cultural and regional differences are significant in terms of sustainability of a given locality (Perrow 1970, Malmberg et al 1996). This paper sets out to investigate these interdependencies and ascertain how the adoption of new processes and technologies can benefit both the small subcontracting firm, the large contractor and add value to the process of construction within ISW.

The significant differences between small, medium and large enterprises (SMEs)

There is a wide range of disparities between large and small construction firms. SMEs differ from large enterprises in terms of size, structure, management style, and organisational governance. The Bolton Report (1971) described a small business as having a small share of its market or alternatively a large share of a very small market. The firm must be managed in a personalised way, and not through the medium of a formalised management structure by the owners or part owners and should be independent in that it does not form part of a larger organisation. Wynarczyk et al (1993) provided an alternative economic definition to that put forward by the Bolton Report (1971). The theory put forward by Wynarczyk et al (1993) builds upon the observation put forward by Penrose (1959, 1995). The emphasis was that small and large firms are fundamentally different from each other in terms of the administrative structure and it is insufficient to treat these distinctive forms as a homogenous grouping for the purposes of comparative analysis. The work undertaken by Wynarczyk et al (1993) argues that there are three central differences between large and small companies: uncertainty, innovation and evolution. The uncertainty within the market ensures that small firms are generally price-takers, ensuring that SMEs offer products or services to the market, at the market price. SMEs in general have a limited customer and product base. An important dimension of uncertainty relates to a much greater mix of objectives for the SME when compared to that of large firms.

The second area, which signifies the differences between large and small firms, is that of innovation and the take-up of new technologies and processes. Stokes (1995) suggests that, in terms of innovation, the management structure of small firms offer a strategic advantage over that of large organisations. The entrepreneurial nature of the small firm enables it to take risks, whereas large firms are essentially risk averse It is the management of complexity that is becoming increasingly important to new and established SMCEs. To this end, it is contended that SMCEs require technologies and processes that can increase the quality of the specialism offered to the market. For harmony to exist it is contested that the differences between large and small organisations in respect of the take-up of new technologies and processes are not fully appreciated by the construction industry.

The third area of difference between large and small firms is the likelihood of evolution and change within the smaller firm. The issue of firms changing in terms of management or structure has been discussed above using the work of E. Penrose (1959, 1995). Management theorists however, explain change within Small and Medium sized Enterprises as being multiple-stage. Thus as firms grow, many changes occur which influence the style and role of management, and the structure of the organisation. Many models exist that purport to assess the growth stages of small and medium sized firms. The problem with models is that they assume all firms try to go through many stages or fail in the process. That is to say models assume that small and medium firms can be compared to large organisations. However it is argued that firm development is unique and that the accumulation process is specific to each firm. As small firms grow there are various changes within the organisation that often create problems for the management. Many writers within the field of entrepreneurship indicate that small firms often stay the same size with no wish to grow due to the necessary management of complexity

Saad and Jones (1998) emphasise the need for specialists to play a more important role within the construction process, as they account for as much as 80% of contract expenditure. It is worthy of note that the recommendations of the report offer excellent solutions as to the strategic fit between contractors and smaller firms but fail to account for the significant differences between the parties. It is recognised and accepted that the more professional established organisations can and will attain a significant short-term competitive advantage by adopting the recommendations of such reports but it is argued that many established SMCEs will fail to attach significance to the findings. It is the owner/manager and the management skills possessed that are a key criterion for the success of the SME (Jennings and Beaver 1997, Jennings and Beaver 1995, Westhead 1995). It is argued that within small construction firms' ineffective management has a causal effect on firm performance.

Table 1, below, offers an adaptation of the work of Ghobadian and Gallear (1996) that attempted to indicate some of the major differences between large and small engineering firms. This table is by no means exhaustive in nature and it is argued should be tested within the SMCE sector to assess its relevance. It is contended that the differences between large and smaller firms within the construction arena need to be understood

From the table it is evident that key differences exist between large and small firms. The network structure of large firms is often wide whereas within small firms networks are often limited to the locality. Limited network activity can often have an adverse affect upon firm performance, as the environmental scanning necessary to increase the knowledge base (and decision-making ability) is limited. The culture of small firms is important in assisting firm performance. Small firms are essentially task/result oriented and flexible, which it is argued actually assists the process of construction. The limited strategic vision within small firms has for many years been a contentious issue within management thinking. Small firms often adopt unstructured and informal strategies that can often be the cause of failure. It is contended that the differences between large and small companies are so great that the traditional subcontracting arrangements currently in situ within the construction industry unless properly managed will not offer many SMCEs

Table 1: Differences between Large and Small Firms

Large Organisations	Small and Medium Organisations
Hierarchical with several layers of management	Flat with few layers of management
Clear and extensive functional division of activities. High degree of specialisation	Division of activities unclear and limited. Low degree of specialisation
Strong departmental/functional mind set	Absence of departmental/functional mindset. Corporate mindset
Governed by formal rules and procedures	Not governed by formality
High degrees of standardisation and formalisation	Low degrees of standardisation and formalisation
Mostly bureaucratic	Mostly organic
Extended decision making chain	Short decision making chain
Top management distanced	Top management close to delivery
Wide span of activities	Narrow span of activities
Multi-sited and often multi-national	Single sited
Cultural diversity	Unified culture
System dominated	People dominated
Cultural apathy	Fluid culture
Rigid organisation and flows	Flexible
Many interest groups	Very few interest groups
Evidence of fact-based decision-making more prevalent	"Gut Feeling" limited strategy
Dominated by professionals	Pioneers, entrepreneurs
Management style: Participative, paternal, etc	Range of styles, directive, paternal, autocratic
Extensive external contacts	Limited network of contacts
Normally slow to respond to environmental changes	Rapid response to changes within the local environment
Low incidence of innovation	High incidence within certain industries
Formal evaluation, control and reporting procedures	Informal evaluation, control and reporting procedures
Control oriented	Result, task oriented
Rigid corporate culture dominating the firm	Owner manager ethos focus and outlook

within regions the opportunity to form long term partnerships. The flexible subcontracting arrangement reduces costs for large firms but it could be argued that the SMCE is no longer the "slave to the master" but an independent small or medium sized business that faces the same difficulties, opportunities and threats of any other business from any other sector. Operating within a turbulent environment causes great difficulties for any smaller firm. The opportunity exists however, for large firms to work in harmony with small firms in an attempt to minimise the level of uncertainty the environment offers. The very nature of the industry ensures that demand is derived and transient in nature. Regional specialists can provide cost effective solutions and provide local knowledge to assist the construction process.

Many practices are in place within the large players of the construction industry that purport to offer the industry as a whole a competitive advantage over that of competitors. Many reports offer advice and the way forward for SMCEs and STCs. O'Farrell and Hitchin (1988) amongst others agree that small firms cannot and should not be treated as "little big firms". The research will assess if the key differences that are in place within the plethora of empirical data within the SME sector are true of the SMCE within ISW. To attempt to implement large company processes and technologies within SMCEs could offer a short term competitive advantage. It is argued however that if the needs and requirements of the SMCE or STCs are satisfied, it is possible to negate the current adversarial nature of the current contractual arrangements (Saad and Jones 1998). It is further contended that best practices within subcontracting can only be successful if a "win-win" situation exists that offers the smaller firm the opportunity to satisfy its needs and aspirations. Many practices are in place within the large players of the construction industry that purport to offer the industry as a whole a competitive advantage over that of competitors.

Harmonisation between large and small firms: The need for process and technological integration?

If we accept that the construction market for many small companies is constrained geographically, and that economies of scale to undertake large projects are not in existence, then it could be argued that some degree of knowledge must exist pertaining to the opportunities, that particular market offers (Malmberg et al 1996). This is further supported within the same paper that utilises the work of Hoover (1948), and Lloyd Dicken (1977), to indicate that firms benefit from the development of general labour markets and specialised skills; enhanced interaction between local suppliers and customers; shared infrastructure and other localised externalities. These elements are requirements within the construction process. The work of Barkham (1992) of 303 small firms further supports, this by suggesting that the continued development of lagging regions, for example Industrial South Wales, will ensure entrepreneurial success. Cultural and regional settings of organisations are important in terms of performance within the market place (Perrow 1970). It is therefore contended that for success within regions to be attained it is necessary to understand the differences between organisations and discount the current homogeneous grouping of sectors.

The small business theory in existence recognises that small and medium sized firms would benefit from increased awareness in management skills, financial planning and control, marketing and possess the necessary motivation to be entrepreneurial in nature. The plethora of reports in regard to the construction industry concludes that the industry must adopt new technologies and processes. The work of Hall (1991, 1995) within the SMCE sector indicates that sub contracting poses severe problems, especially if SMCEs (who are mainly, but not always sub-contracting) perceive adding value of no benefit. This is essentially due to the fact that as the client and contractor are so distant from the sub contractors the SMCEs can attain no benefit by the provision of quality work. Hall also places great emphasis on the small construction firm's (defined as having less than 100 employees) ability to manage finance effectively, and that the firm must possess a skilled workforce. It is interesting that from these studies that the education of the owner may be positively correlated with the survival of the firm. The work of Saad and Jones

(1998) identifies that many of the problems faced by the industry can be minimised through the co-operation between large and smaller firms. It is argued however that for harmony to exist the smaller firm would require benefits to accrue not to just the large contractor, but to the individual specialist firm. The strategy of the small firm is dependent on the strategy of the contractor, that is, the interdependency of the parties ensures the project is completed. Questions remain as to whether harmony can exist if the environment remains hostile and is one of little trust. The industry continues to be cost led which poses problems for specialists. If "Dutch auctions" continue to be the focus of specialist selection the requirements of the small firm and contractor will not be satisfied. Harmony can exist if the power held by large organisation works toward minimising the problems faced by specialist smaller firms. To understand the SMCE within regions would offer a richer insight than that offered by much of the current research.

The work of O'Farrell and Hitchins 1988 offers the argument that many craftsmen-entrepreneurs wish to continue to exercise their own trade skills, and may be reluctant or unable to become more heavily involved in the administration, management and paperwork required by technological complexity. It could be argued that future requirements could lead to a more fragmented industry due to the increased pressure being put upon smaller firms in terms of complexity in whatever form. This point is confirmed by Cockrill and Scott (1997), who point out that the growth of new fragmented specialisms and self-employment have encouraged the expansion of SMEs which further cements the variety of working relationships. Woo et al (1989) identified that craftspeople and independent entrepreneurs had less experience in marketing and sales. These are areas that current theoretical thinking consider to be imperative if the UK is to succeed within a global market. Many theorists within the small business sector have identified areas that can offer competitive advantage and arguably, the mechanisms that will allow smaller firms to work in harmony with larger organisations. It has been recognised within the small business arena that the motivation of the owner manager is the key criteria to the attainment of competitive advantage within a firm's environment (Gartner 1988). The literature within this area suggests that the personalities of owner/managers in terms of values and goals are indeed indistinguishable from the goals of their businesses (Kotey and Meredith 1997, Jennings and Beaver 1997, O 'Farrell and Hitchins 1988, Bamberger 1983). Research conducted by Meredith and Kotey (1997) of 224 small businesses indicated that personal values, business strategies and firm performance were empirically related. This research is important if the SMCEs ability to adopt new technologies is related to the values and goals of the owner/manager (Lefebvre, Mason and Lefebvre 1997). In a study by Wynarczyk et al (1993) of 150 small businesses, the management and marketing abilities of owner/managers has been shown to have a positive effect upon the financial success of SMEs. Within the SMCE the marketing efforts of owner managers are usually ad hoc and often the only effort made from within the firm (Carter and Dunne 1992).

Much research conducted within the small and medium sized sectors has indicated that the variances between large and small firms are so significant that the failing and indeed the success of small firms can be attributed to these differences. It is argued that for SMCEs to adopt, and manage new and diverse forms of technological complexity it is necessary to understand the heterogeneity of the sector. It is further contended that the sectoral differences in existence between regions are also important in terms of the way in which smaller firms manage the complexity of the environment.

The small and medium construction enterprise within Industrial South Wales: A research need?

Industry in Wales is predominantly dominated by the manufacturing and public sector. The public sector and manufacturing contribute a disproportionate percentage of the regions Gross Domestic Product (GDP) when compared with the national average. Conversely, the contribution of the construction industry is 1% below that of the GDP of the UK at 5% (Pathway to Prosperity 1998). Despite this trend, evidence suggests that contracting within the industry is increasing. In fact, between 1970 and 1994 there has been a 175% increase in the regional distribution of contracting firms within Wales (Harvey and Ashworth 1998). This would seem to suggest that the retrenchment strategies of large organisations have presented opportunities for the emergence of small subcontracting firms in Wales. Recent figures suggest that 85% of construction firms within the region employ less than 50 people and 71% less than 10 (Nomis 97).

The construction industry within the ISW region has enjoyed significant activity, but showed a greater degree of fluctuation than the rest of the United Kingdom. In 1995 the ISW region had 2600 SMCEs operating within the sector. The overall change in numbers of VAT registered SMCEs within the region has decreased by 16.1% between 1991-1995, resulting in a 14.8% decrease in the number of persons employed (Nomis 1997). This again is below the industry norm which is projected to increase by 3% (Harvey and Ashworth 1998) and emphasises the difficulties faced by the small sub-contracting firm in a changing environment.

Of significant importance to the ISW region is the availability of appropriately skilled people to enter the construction industry. The structure of the industry has changed from being dominated by contractors to being a highly flexible system that utilises subcontractors for a large percentage of the work. Such methods have enabled organisations to pull through recessionary periods, but have failed to retain the skills and knowledge required to take the industry into the next century (South Glamorgan TEC 1996). The industry is now experiencing the aftermath of retrenchment strategies that arguably can sustain competitive advantage only in the short-run. This should be of concern to policy makers and large contractors, if the local skills and knowledge inherent within SMCEs are acknowledged as being significant and a necessary input into the industries supply chain. What should be of concern is that the ISW region can offer the required labour, skills and knowledge, but lacks the support mechanism and arguably the motivation to succeed, and often survive, within this complex environment.

Within the Industrial South Wales (ISW) the plethora of SMCEs are firms that fit into the categorisation of much of the small business theory in existence. It is this theoretical grounding that will be the main focus for the research to be carried out. For SMCEs to be successful within their environment and embrace the required technological complexity it is contended that the SMCEs within the region must adopt new technologies and processes in harmony with the larger contractors (Hughes Gray and Murdoch 1997). It is further contended that the research arena within the construction industry should focus upon the sectoral differences of regions and the heterogeneity of smaller firms within these regions, however they are labelled. Richardson (1964) (in O'Farrell and Hitchins 1988) in a survey of managers, found that the availability of suitable managers was the major problem in terms of expansion. It is an objective of the research to determine the perceptions of owner managers in terms of the criteria necessary to adopt new processes and technologies. It is contended that the power and motivations of the owner manager is a barrier that can have an adverse and alternatively positive, effect on the success or failure of an SMCE. Moreover, the unequal power relationships that exist within the industry may impinge upon the profitability of the smaller firm that may not possess the skills and knowledge to adopt new technologies and processes at the rate required by the industry.

In conclusion, the ISW region provides a suitable arena in which to evaluate the interdependency of small and large firms within the construction industry. Furthermore, it offers an opportunity to investigate the consequences of disharmony between the parties involved in the construction process. It can be hypothesised that this harmony will have an impact upon the profitability of the small and large firm and ultimately the value offered to the client.

A research methodology for Industrial South Wales.

It is evident from the arguments put forward that for research to be a true picture of a SMCE, it is necessary to conduct research that will capture the motivations and aspirations of owner/managers. To achieve this aim it is deemed appropriate to use a case study approach to capture the culture and life style of SMCEs within the ISW region. The investigation will attempt to capture the idiosyncrasies that current theory purports to be in existence within regions. Very little grounded data has been captured within the small business or entrepreneurship arena by using an ethnographic approach (Curran and Barrows 1987).

The evaluation will initially take the form of a study to ascertain whether literature can be applied to the SMCE within the ISW region. The survey will offer a comparison between large and small firms, paying particular attention to issues such as the adoption of technologies and processes, firm governance, strategy and management and perceptions pertaining to the importance of harmonisation. The survey will aim to evaluate whether the region can be considered typical to the UK as a whole, or whether certain peculiarities need to taken into consideration when analysing the data set. From the study, a conceptual framework will be constructed for further testing within the ISW region. The conceptual framework will focus upon the processes inherent within the small sub-contracting firm.

A pilot study will be undertaken to test the validity of the conceptual framework focussing upon particular processes and their affect upon harmonisation, profitability and value added. Subsequently the conceptual framework will be amended to ensure that further research can confidently be used as a representative template of the processes that are evident within the SMCE. A multiple case study approach will be undertaken to test the robustness of the conceptual framework through the medium of replication. Multiple case designs offer distinct advantages and disadvantages over that of the single case study approach. The findings of multiple studies are often more thorough and considered to be more vigorous (Yin 1994).

A triangulation approach will be adopted to improve the validity of the data (Silverman 1985). This approach will assist the research in identifying previous important happenings within the sample that may be relevant to the strategies previously adopted as put forward by Denzin (1971) (quoted in Stockport and Kakabadse 1992). The method of triangulation to be adopted will be participant observation, interviewing and the study of non-sensitive company documentation (Stockport and Kakabadse 1992). The data will be classed according to the perceived category, whether Observational Notes (ON), Methodological Notes (MN), or Theoretical Notes (TN). The logic behind organising data in this manner is to assist in the facilitation of its analysis. It is understood that this method will allow the researcher opportunities to write-up methodological tactics as they occur, further validating the research tool.

References

Bamberger, I. (1983). Value systems, strategies and performance of small and medium sized firms. International Small Business Journal. Vol. 1, Issue 4.

Barkham, R. (1992) Regional Variations in Entrepreneurship: Some Evidence From The UK. Entrepreneurship and Regional Development. Vol.4, No.3.

Bolton, JE. (1971) Small Firms Report of Inquiry on Small Firms. HMSO London

Burrows, R (1991) The discourse of the enterprise culture and the restructuring of Britain. In Curran, J and Blackburn, R. (Eds.) Paths of Enterprise: The Future of Small Business, Routledge.

Carter, S. Dunne, A. (1992) Pre-tendering in the construction sector: A comparison of small and large companies. International Small Business Journal. Vol.11, Issue 2.

Central Council for Works and Buildings. (1944) The Placing and Management of Building Contracts, HMSO, London.

Chapman, T. (1983) The process and problems of achieving organisational access. Graduate Management research, Vol. 1, no.2, pp. 21-25.

Cockrill, A. Scott, P. (1997) Training, Skills Provision and Multi- skilling in the Construction Industry: a Welsh German Comparison. Regional Industrial Research Report No. 28

Committee on the placing and management of building contracts. (1964) Report of the Committee on the Placing and Management of Building Contracts. HMSO, London

Curran, J. Barrows, R. (1987) Ethnographic approaches to the study of the small business owner, in K. O' Neil, R. Bhambri, T. Faulkner and T. Cannon (eds.) Small Business Development. Some Current Issues, Avebury, Aldershot, pp3-24.

Emmerson, Sir H. (1962) Study of Problems before the Construction Industries. HMSO, London

Ghobadian, A. Gallear,D.N. (1996) Total Quality Management in SMEs. International Journal of Management Science. Vol. 24, Issue 1.

Hall, G. (1991) Non-Financial Factors Associated with Insolvency amongst Small Firms in Construction. Working Paper No. 214. Manchester Business School.

Hall, G. (1995) Surviving and Prospering in the Small Firm Sector. Routledge.

Harvey, R. C Ashworth, A. (1998) The Construction Industry of Great Britain (2nd Edition). Laxtons. Oxford.

Hornaday R. W. (1990) Dropping The E-Words From Small Business Research: An Alternative Typology. *Journal of Small Business Management* (Oct) (from database of Small Business Research).

Hughes, W. Gray, C. Murdoch, J. (1998) Specialist Trade Contracting-a Review. Construction Industry Research and Information Association. Special Publication 138.

Jennings, P. Beaver, G. (1995) The Managerial Dimension of Small Business Failure. *Journal of Strategic Change*. Vol. 4, No. 4.

Jennings, P. Beaver, G (1997). The performance and competitive advantage of small firms: A management perspective. *International Small Business Journal*. Vol. 15, Issue 2.

Kotey, B. Meredith, D. D. (1997) Relationships among owner/manager personal values, business strategies and enterprise performance. *Journal of Small Business Management*. Vol. 35, Issue 2.

Latham, M. (1994) Constructing the team. HMSO, London.

Lefebvre, L. Mason, R. Lefebvre, E. (1997) The Influence Prism in SMEs: The Power of CEOs' Perception on Technology Policy and Its Organizational Impacts. Management Science.

Malmberg, A. Solvell, O Zander I. (1996) Spatial Clustering, Local Accumulation of Knowledge and Firm Competitiveness, Geografisca Annaler, Vol. 78B, Issue 2.

Nomis, (1997) Statistical Data 1995 for Industrial South Wales

O' Farrell, P. N. Hitchins, D.M.W.N. (1988). Alternative Theories of Small Firm Growth: A Critical Review. Environment and Planning. Vol.20, pp1365-1383

Pathway to Prosperity (1998) A New Economic Agenda for Wales

Penrose, E (1995) The Theory of the Growth of the Firm. Oxford University Press.Perrow, C. Organisational Analysis: A Sociological View. Tavistock Publications.

Porter, M. (1990) Competitive Strategy: Techniques for Analysing Competitors, Free Press, New York.

Saad, M. Jones, M. (1998) Unlocking Specialist Potential. Reading Construction Forum.

Silverman, D. (1985) Qualitative Methodology and Sociology, Gower, Aldershot.

South Glamorgan TEC. (1996) Labour Market Assessment for Industrial South Wales Report, (eres).

Small Business Research and Consultancy. (1996) The European Observatory for SMEs Report, European Network for SME Research.

Stanworth, M.J.K. Curran, J. (1976) Growth and the Small Firm-an Alternative View,. *Journal of Management Studies*. Vol.13, no.2

Stockport, G. Kakabadse A (1992) Using Ethnography in Small Firms Research in Small Enterprise Development-Policy and Practice in Action. Eds Caley, K. Chell, E. Chittenden, F. Mason, C. Lancashire Enterprises.

Stokes, D. (1995) Small Business Management (2nd Edition). BPC Hazell.

Storey, D. J. (1994) Understanding the Small Business Sector. Routledge.University of Reading (1988) Building Britain 2001. Centre for strategic studies in construction, London.

Westhead, P (1990) A Typology of New Manufacturing Firm Founders in Wales: Performance Measures and Public Policy Implications. *Journal of Business Venturing*. Vol. 5, pp 103-122.

Woo, C.Y. Cooper, A.C. Dunkleberg, W.C. Daellenbach, U. Dennis, W.J. (1989) Determinants of Growth for Small and Large Enterprises Start Up. Paper presented at Babson Entrepreneurship Conference.

Wynarczyk, P. Watson, R, Storey, D. Short, H. Keasey, K. (1993) Managerial Labour Markets in Small and Medium Sized Enterprises. Routledge.

Yin, R. K. (1994) Case Study Research. Design and Methods (2nd Edition) Sage Publications.

A CASE FOR PROMOTER PARTICIPATION IN THE DEVELOPMENT OF FAIR CONDITIONS OF CONTRACT IN UGANDA.

A. TUTESIGENSI and K. MOODLEY
School of Civil Engineering, University of Leeds, LS2 9JT, Leeds, England.
cenatu@leeds.ac.uk

Abstract

Conditions of contract affect harmony and levels of profit in the construction industry. The positive contribution of conditions of contract to harmony and profit in the construction industry can degenerate into a dysfunctional one when one or both of the following are exhibited in the system: lack of considerable understanding of the conditions of contract by at least one of the participants; and lack of trust and belief in the conditions of contract by one or all of the participants.

This paper reviews the current situation in Uganda as far as conditions of contract are concerned. Uganda has an active construction industry operating with a variety of conditions of contract. The paper highlights the fact that promoters in the Ugandan construction industry do not get a fair deal from the conditions of contract in use.

The paper indicates that equity is desperately needed to aid the development and competitiveness of the construction industry in Uganda. The paper proposes the organisation, training, participation and harmonisation (OTPH) framework as a useful tool in turning around the status quo and bringing about the desired development and improved competitive edge of the Ugandan construction industry. The framework may have application in other developing economies.

Keywords: Conditions of contract, harmony, participation, promoters, Uganda

Introduction

Conditions of construction contract govern 'the game' of construction contracting. They lay out obligations and liabilities between the promoters and contractors. Conditions of construction contracts have a tremendous effect on harmony and levels of profit to all the participants in the construction industry. They affect harmony in the sense of laying out clearly the distribution of risks among the contracting parties. It is important that this distribution of risks is fair and understood by all the parties. Effective distribution of risks is facilitated by addressing issues such as: who can best control the events that may lead to the risk occurring; who can best control the situation if the event occurs; who should carry the risk if it cannot be controlled; what premium is to be charged by the party accepting the risk; and could the person accepting the risk actually sustain the consequences of the risk.

Conditions of construction contract affect levels of profit or benefits to either party because they set out delivery and payment procedures in an environment of uncertainty. Effective conditions of contract in this aspect would be those that take into consideration all the uncontrollable factors for each of the parties and strike a deal that is unbiased. This may be easier said than done, but considerable success can be achieved with involvement of all the parties in the development, evaluation and modification of conditions of contract.

Ideally, conditions of contract are meant to be functional in bringing about harmony and benefits between the contracting parties. However, the extent to which they go in accomplishing this function varies from project to project, organisation to organisation and indeed country to country. The reasons behind the differences in accomplishment levels can be traced from cultural, economic, training and other aspects. However, the functional contribution of conditions of contract to harmony and profit can degenerate into a dysfunctional one when one or both of the following are exhibited in the system: lack of trust and belief in the conditions of contract by one or all of the participants; and lack of considerable understanding of the conditions of contract by at least one of the participants.

Lack of trust and belief in the conditions of contract arises from two main sources. The first source is lack of universal involvement in the formulation and/or introduction of the forms of contract. In the lead author's experience in the construction industry in Uganda, the major forms of construction contract in use (with or without modification) are the following:

- The FIDIC Conditions of Contract, produced by the International Federation of Consulting Engineers in association with the European International Federation of Construction;
- The ICE Conditions of Contract, produced in the UK by the Institution of Civil Engineers and sponsored by the Institution of Civil Engineers, Association of Consulting Engineers and Federation of Civil Engineering Contractors;
- The JCT Conditions of Contract, produced by the Joint Contracts Tribunal in UK whose composition includes bodies of contractors, engineers, architects, surveyors and promoters;
- The East African Conditions of Contract, produced by the East African Institute of Architects with sanction of structural engineers and building contractors; and
- Other conditions of contract in use include purpose written conditions of contract, drafted on the basis of the terms of engagement of the various professionals as published by the respective professional bodies.

The main observation to be made about all these conditions of contract is that most, if not all have received little if any input from the Ugandan promoter in their development. Generally speaking, the conditions of contract are considered alien to the majority of the indigenous public and private promoters. The promoters therefore find it hard to trust and believe in the conditions of contract.

The second source of lack of trust and belief in the conditions of contract are the inherent weaknesses of the conditions of contract in use.

The most prevalent weakness is the unilateral nature exhibited by most of the conditions of contract in use except the JCT conditions of contract. This arises from the fact that apart from the JCT conditions of contract, the rest are exclusively produced by construction professionals' bodies without the input of promoters' bodies. Many of them give the construction professionals almost judicial immunity and extra-ordinary powers to the construction professionals much to the detriment of the promoter. Because of this, a considerable section of indigenous promoters in Uganda conceives the majority of the conditions of contract as construction professionals' tools of exploitation.

Lack of understanding of the conditions of contract arises mainly from lack of familiarity with the conditions of contract. This is particularly so in Uganda because, as can be seen from the list of conditions of contract in use above, many of them are 'imported'. They originate from a variety of sources, having been designed for different social, cultural, political, legal and economic backgrounds. In using 'imported' conditions of contract therefore, several parts of the jig saw are clearly missing and the end result is a less than optimal achievement.

Another important contributor to the problem of lack of understanding of the conditions of contract is the 'legal' language in which many of the forms are written. Legal language has been a source of controversy in conditions of contract in the United Kingdom and other Commonwealth countries. There has been argument against the use of 'legal' language and calls for the use of plain English in drafting conditions of contract, [Cutts and Maher (1986)]. The basis of the argument is that the use of 'legal' language results in increased costs, distortion and oversight in the process of the so often necessary translation and re-translation. Clearly, if these problems are exhibited in the United Kingdom where the first language is English, they can only be worse in Uganda where English is the only official and yet not the first language.

The nature of the problems with current standard conditions of contract are summarised by the fact that the draftsmanship of the available standard conditions of contract in all countries is of the poorest kind, [Duncan Wallace (1986)]. In the UK and Commonwealth the conditions of contract are largely derived from very old precedents, often in archaic language and drafted by lawyers with little or no experience of the background and needs of a construction project. The draftsmen (lawyers) take their instructions from clients who are often unable to analyse or explain the needs of the construction project to the draftsmen, [Duncan Wallace (1986)]. These observations tend to suggest that the way forward can only be for all direct project participants to equally take a more proactive role and get personal involvement in designing clearer and economical conditions of contract.

Problems related to conditions of contract in Uganda

Uganda is a former British colony in East Africa that gained its independence in the latter quarter of 1962. It is a developing country and member of the Commonwealth. Uganda exhibits the typical characteristics of most developing countries. In particular for purposes of this paper, the Ugandan legal system is based on that of Great Britain and English is the national and official language (even though it is not the first language of almost all Ugandans).

In Uganda, construction projects can be grouped into two groups namely, the internationally and locally funded projects. International funding tends to come with its own terms and conditions. International funding bodies like The World Bank, International Monetary Fund, African Development Bank, European Union, etc. fund most of the infrastructure projects in Uganda. The funding is most often provided with the funding body's own choice of procurement procedures and conditions of contract. Some of these include FIDIC Conditions of Contract, ICE Conditions of Contract and World Bank Conditions of Contract. A number of projects have had problems with the procedures imposed by the funding bodies. The problems range from cost and time overruns to termination of contract. In the last 18 months for example, Uganda has seen two major contracts namely the Hydro-Electric Power Project at the Owen Falls Dam in the industrial centre of Jinja and Workers House Project in the capital city, terminated. These problems arise from one or a combination of the following: lack of understanding of the conditions of contract by the participants; modifications to the standard conditions of contract which often have to be made; and inherent weaknesses of the standard conditions of contract. The procedures and conditions of contract in the above cases tend to be biased towards serving the purposes of the funding body rather than the promoter's. However in some cases, arguments have been advanced to the effect that these conditions of contract and contract procedures are imported because there is nothing to use on the ground.

Most locally funded projects are in the private sector. Private sector promoters often look to the construction professionals for services because of the lack of in-house capacity to handle construction projects. In most cases, the promoters lack knowledge about the management of the construction project. Unfortunately, the construction professionals have exploited this ignorance in a way that has hitherto not encouraged harmony between themselves and the promoters. Locally funded projects tend to use purpose written conditions of contract, normally modelled around the standard conditions from other areas around the world depending on the backgrounds of the professionals involved. There are many projects in which standard conditions of contract from Europe have been adjusted to suit the Ugandan situation. The problems with these copies of conditions of contract are:

- Ill-considered modifications often distort the closely knit terms in the standard conditions of contract;
- Modifications could lead, and have often led to effects not originally intended by at least one party; and
- In most cases, these modifications are never acknowledged which tends to mislead some of the parties especially those well conversant with the in-built balance of obligation and liability in the standard conditions of contract.

The results of the above tend to include increased levels of claims and escalation of project costs, stagnation and sometimes complete abandonment of the projects.

The consequences of these problems can be clearly seen from the findings of a recent research effort. It was found that the average construction cost of building projects in

Uganda's formal building sector for the period 1991 to 1996 was approximately US $0.5m, [Tutesigensi (1997)]. In the same study, two building projects in Kampala each with a construction cost in excess of US $2m, well above average size, exhibited the mistrust in the conditions of contract.

The developers actually executed their projects to completion using direct labour methods as an alternative to the formal contracting procedures. These particular individuals indicated that the perceived unfairness of the conditions of contract in use was the main reason for the steps taken in executing their projects. In their view, it was best that disputes in the project be handled as managerial rather than legal problems. Whether these individual points of view were right is a moot point. The points of view however, demonstrate the failure of the contracting practice to perform the functional role of bringing harmony in the construction industry. This then brings into question of whether optimal performance is being achieved. The simple answer to this is that optimal performance is not being achieved and something needs to be done to improve the situation. The following section of this paper will help in responding to this urgent cause.

From the preceding revelations, it can be said that the situation in Uganda as far as conditions of contract are concerned, can be attributed to the following facts: the formal construction industry is young, undeveloped and disorganised; the conditions of contract in use are complex, not well understood and not trusted; and fair procurement practices have not been ensured. It is difficult to find any developing country in which the set of facts above is completely inappropriate. The situation in most other developing countries is more or less similar to the Ugandan one. Therefore, any mechanism that improves the situation in Uganda may well find some application in other developing economies.

The way forward

General

In proposing a way forward, it is important that lessons are learnt from those who have been where developing economies are at the moment. Research in the UK suggests that promoters of all types tend to demand simple contract forms. There is also a need to have terms in the conditions of contract that are practical and whose application can be comprehended by all the parties, [Smith and Wearne (1995)]. Harmony can be achieved if promoters and contractors collectively agreed on what the essential general terms are and end the practice of adaptations and modifications for each project to meet what might well be illusions about the special need in the project. There is overwhelming recognition that the promoter should decide on the priorities, allocate the risks, propose the terms of contract, select the contractor and manage the uncertainties, [Smith and Wearne (1995)].

The key point from the UK experience is that a successful construction sector will depend on increased promoter involvement and competence. It is therefore important that in Uganda, promoters get increasingly involved in all areas of the industry; conditions of contract formulation, monitoring and evaluation inclusive. This task however demands that the promoters possess the required competence to handle the tasks. Competence is a function of work-related knowledge, skills and abilities, [Nordhaug, (1993)]. Relevant competence is a necessary (and in some cases both necessary and sufficient) condition for strategic success. Competence may be gained through training and experience in the work place, [Nordhaug and Gronhaug, (1994)]. It is against this premise that the improvement framework below is proposed.

The proposed improvement framework

On the basis of the description of the state of affairs in Uganda and the trend in the current calls for competitiveness to ensure development of the construction industry, the following improvement framework is proposed. The framework, simply referred to as the construction industry procurement organisation, training, participation and harmonisation (OTPH) framework is illustrated in Figure 1 below. The framework is aimed at creating a strong body (bodies) of construction industry promoters, who are well trained and who participate in the formulation, audit and re-formulation of policies that affect them to bring about harmony with all its benefits to the construction industry. The framework is composed of four main activities namely organisation, training, participation and harmonisation proceeding in the numerical order shown in Figure 1. It is however, often necessary with real world problem solving approaches to revisit previous activities before proceeding. This is catered for by the double arrows in opposite directions in the figure. The following sub sections describe the activities in the OTPH framework.

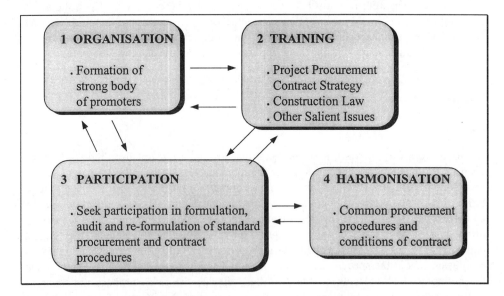

Figure 1. The Construction Industry Procurement OTPH Framework

Organisation: Organisation refers to the act of bringing together individuals or individual groups to build a structured whole that represents common aspirations and goals with more strength and effectiveness. Organisation is the beginning of a struggle for any worthwhile cause. In this particular case of improvement of construction industry procurement, organisation must involve the formation of strong pressure group(s) to foster the needs of the promoters. This will help in making sure that the promoters speak with one voice, thereby creating more chances of being heard. The promoters need to come together under one umbrella and speak with a common voice to government and construction professional bodies in order to get more involved in the issues that affect the promoters. Strength is derived by creating a strong institutional base.

With the implementation of government decentralisation in Uganda in the last five or so years, Local Authorities and District Councils have taken over the role of the majority employer in the construction industry from the Central Government. Two associations namely: The Urban Authorities Association of Uganda and The Uganda Local Authorities Association, are already in place [Tutesigensi and Kiggundu, (1997)]. These could be mandated to fight for involvement in the formulation of harmonious conditions of contract. On the other hand however, the private sector has not taken any visible steps in coming together under one body. They could however, benefit tremendously from the formation of a strong body in fostering their interests in conditions of contract formulation.

Training: Training is the process of providing the individual or group of individuals with an organised series of experiences and materials that comprise opportunities to learn. From previous sections, it is very clear that most indigenous promoters are not well equipped with skills to handle procurement and basic construction contract procedures. Training can be effected quickly and economically when the promoters are organised as a group or groups. The areas that need to be tackled include project procurement, contract strategy, construction law and other salient issues. Training should be aimed at improving the skills and/or attitudes of the promoters and increasing their knowledge and experiences. The outcome of such training is likely to be positive industrial development through one or more of the following: project cost reduction; increased profit in the construction industry; improved quality of construction industry output; and overall harmony in the construction industry.

Training programmes need to be focused, realistic, evaluated, monitored and perpetual for effectiveness [Tutesigensi, (1998)]. They should be designed to begin with clear and measurable strategic objectives. From the strategic objectives, desired levels can be described depending on available resources. An inventory to establish present levels is then taken and the gap between the desired and present levels shortened and eventually eliminated through a well developed, implemented, monitored and evaluated training programme. Evaluation of the training programme leads to identification of necessary modifications and/or new desired levels, thus re-starting the loop.

Participation: Participation refers to active involvement in any activity. When the promoters are organised under a common body/bodies and equipped with knowledge acquired from the training programmes, they are in a better position to get effectively involved in the formulation, auditing and re-formulation of policies about procurement and conditions of contract. The idea of training preceding participation as shown in Figure 1 is important because it ensures that an informed and more competent promoters' body gets hold of its destiny and the destiny of the construction industry since the promoters are the back-bone of the construction industry.

***Harmonisation*:** Harmonisation relates to the creation of rapport between two or more parties. Harmonisation of procurement and contracting procedures requires agreement on standard forms and procedures between the promoters, construction professionals and construction main and sub contractors. This can only be achieved through adequate and effective representation and involvement by all the parties. Agreeing on standard procedure will help to uproot the prevailing mistrust in the conditions of contract and construction project procurement practices.

This will improve the construction industry by ensuring that the practices exhibit wide acceptability, consistency of use, familiarity in use and legal case law precedents.

The OTPH framework above can help render inappropriate some of the arguments for the importation of conditions of contract in Uganda that when there is nothing on the ground to use, one can only borrow from ones' neighbour if he must get on with the job, otherwise nothing is achieved. As can be deduced from the Uganda case, the neighbour's tool is not always perfectly compatible. So, the best approach may well be to develop ones own appropriate tools. The OTPH framework is one strategy that can be employed in developing appropriate and harmonious conditions of construction contract.

Comments about the OTPH framework

The above framework is generic in nature. Thus it can be used at organisational, national, regional and international level. It emphasises involvement of the promoters in the formulation, auditing and re-formulation of policies affecting construction project procurement and conditions of contract in use.

The framework allows iteration between organisation and training, organisation and participation, training and participation and between participation and harmonisation. This ensures that the promoters keep abreast of advancements in theory and practice and are able to modify their approaches or even regroup for more effectiveness. It also allows the promoters to take a proactive role towards knowledge and skills acquisition. The iteration between participation and harmonisation ensures that individual experiences and knowledge acquired are brought into the overall context of procurement and contract procedures. This is bound to create the synergy effect and lead to overall improvement of the construction industry.

While the extent of the problem of unfair conditions of contract cannot be quantified at this point in time, concern must be about its effect on the development of the construction industry in developing economies in terms of ensuring optimal returns, commitment to the industry and human resource development to meet the infrastructure and housing demands of the 21st century. Therefore with careful design of the details in the individual components of the OTPH framework, it could be applicable to most developing economies.

The implementation of the OTPH framework requires commitment of resources for success. It requires resources in terms of finance, technical personnel and administrative personnel. The availability of these resources will be a pre-requisite, and their absence a barrier, to the successful utilisation of the OTPH framework. The main key to success is to get as many promoters as possible involved. The question of finance is not likely to be a problem once the benefits are appreciated. A modest annual subscription should be sufficient to meet the financial requirements. Financial problems might be more pronounced at the beginning but through a few sponsors (government, willing promoters and other organisations), sufficient funds could be raised. The limits to initial funds can only be determined by the organisational capability of the OTPH framework champion. The status of technical personnel in Uganda is a promising one.

Ugandan promoters should be able to benefit from the developing technical skill of the graduates of the construction management studies at the local institutions and other institutions mainly in UK and USA from where considerable numbers of Ugandans have been and continue to be trained. In addition, promoters can benefit from the tremendous wealth of expertise and experience from construction professionals from all over the world who in the last decade have played a substantial role in the Ugandan construction industry, thanks to economic liberalisation. The question of administrative personnel should not be a big problem as long as there is the will to succeed. The framework champion and promoters should be able to recruit from within themselves or even outside sources, effective administrative staff to execute their programmes.

Implementation of the OTPH framework will create a need for change. This may meet the usual tendency for people and organisations to resist change whenever there is a call to shift from the current status to another. Large promoters may take the view that they have managed all this time, therefore they do not need any changes. Small promoters may take the view that they are small after all and any such procedures are not applicable to them. On the other hand, large and small promoters may just be sceptical about the new framework and its benefits as opposed to their own current practices. In such circumstances, the government may need to step in and take the lead and slowly transfer responsibility to the promoters. The current situation in Uganda seems to suggest that while there is growing awareness of the need for change amongst the promoters, there is nobody to champion the challenge amongst the promoters especially because most of them are not institutionally organised. The government could therefore help out here and champion and promote the implementation of the OTPH framework for some time. When the promoters have developed enough courage and motivation, the government can then slowly withdraw and allow the promoters to handle their affairs. The growing awareness of the need for change amongst the promoters and the potential benefits of using the OTPH framework seems to indicate that more and more promoters can become interested and involved, thereby making this particular approach feasible.

Conclusions

This paper has reviewed the current situation in Uganda as far as conditions of contract in use and highlighted a number of issues, which are summarised below.

Most promoters in the construction industry in Uganda do not get a fair deal from the conditions of contract in use. This is because of the complex 'legal' language in which the current conditions of contract are written and the conditions of contract's unfamiliarity and perceived untrustworthiness. Attempts are sometimes made to modify some of the terms but these create other problems including distortion, effects not originally intended and misleading information.

The fair deal is needed to aid the development and competitiveness of the construction industry in Uganda. There are signs of people seeking alternative actions in a bid to avoid conditions of contract completely. This only highlights the disharmony in the industry which needs to be abated by development of promoter-friendly conditions of contract. Promoter participation in the development of such contract forms should be an important consideration.

The proposed organisation, training, participation and harmonisation (OTPH) framework is useful in turning around the status quo and bringing about the so desired development and improved competitive edge of the construction industry in Uganda. It is expected to be effective because it tackles the root cause of the current signs of the disharmony in the Ugandan construction industry.

There are similarities in conditions of contract use between Uganda and other developing countries. Therefore the latter could benefit from the OTPH framework in addressing issues of harmony and increased profit in their respective construction industries.

The implementation of the OTPH framework requires finance, technical and administrative personnel. The absence of these resources could pose formidable barriers to success. The government could take the lead initially to get things going and transfer responsibility to the promoters slowly as and when they become ready to take control.

The paper has highlighted issues of trust and risk management as the main problems in the construction industry in Uganda. The proposed OTPH framework can be useful in building up mutual trust among the contracting parties thereby creating harmony and making risk management more effective with the consequence of increased profits in the construction industry.

References

Cutts, M. and Maher, C. (1986) The plain English story. Whaley Bridge: Plain English Campaign.

Duncan Wallace, I. N. (1986) Construction contracts: principles and policies in tort and contract. London: Sweet and Maxwell.

Tutesigensi, A. (1997) An Investigation into Building Project Identification Practices in Uganda, Unpublished Research Report.

Smith, N. J. and Wearne, S. H. (1995) Contractual Arrangements in the European Union. *In Engineering Project Management* (ed. Smith, N. J.), pp. 267-278. Oxford: Blackwell Science.

Nordhaug, O. (1993) Human Capital in Organisations: Competence. Training and Learning. Oslo: Scandinavian University Press.

Nordhaug, O. and Gronhaug, K. (1994) Competences as Resources in Firms. *The Interna tional Journal of Human Resource Management,* Vol. 5, No.1, pp. 89-106.

Tutesigensi, A. and Kiggundu, B. M. (1997) Conditions of Contract Application in Uganda. *Proceedings of the 1ˢᵗ two day seminar on construction contract law*, Kampala, Uganda, July 9-10, No pp.

Tutesigensi, A. (1998) Manpower for Industrial Development; Which Way Uganda? *Pro ceedings of the 3ʳᵈ Uganda National Technology Conference*, Kampala, Uganda, April 15-17, No pp.

UNDERSTANDING CONTRACT DOCUMENTS AND ITS RELATION TO THE PERFORMANCE IN CONSTRUCTION PROJECTS

ADNAN ENSHASSI

Civil Engineering Department, P.O.Box 223, Islamic University of Gaza, Gaza Strip-Palestine. enshassi@mail.iugaza.edu

Abstract

Management is the mobilization and development of resources, human, material and financial, in order to produce goods and services. In the absence of effective management, resources are wasted or underutilized and economic and social development goals are not met. A plentiful availability of people with management perspective and skills is therefore essential. At the enterprise and organization level in Gaza Strip competent management is more necessary than ever before.

Recently, Palestinian territories have become increasingly aware of the need for a systematic approach to the management of development projects in order to improve the effectiveness and the performance concept in regulations, specifications and contractual procedures. This demand arises from the complexity of such projects and the need for effective coordination and cooperation between many different governmental agencies, international donors and construction firms. The aim of this paper is to study the level of local contractors` understanding of contract documents and its relation to the performance concept in construction projects in Gaza Strip.

Keywords : Contract documents, construction projects, local contractors, performance.

Introduction

With the incredible growth rate of large scale projects in Gaza strip and West Bank, more and more local construction companies are placing emphasis upon understanding contract documents and regulations. Indigenous project managers will have to undergo more training in contract systems than international project managers. International project managers should place emphasis upon the understanding of social and cultural issues.

To understand the necessity for this added training, consider the following response made by executives to the question" what are some of the major problems your company faces with World Bank projects in the Gaza Strip". Some of the pitfalls in trying to manage projects effectively and maximize projects performance are as follows:

- Problems associated with misunderstanding of contract documents, especially
  general and special conditions have a tremendous impact on project performance.
- Cultural variations and difference in value systems present problems to foreign
project people.

This paper will begin by presenting the project management environment in Palestine, prior to discussing the contract documents. The methodology of study will also presented. The purpose of this paper is to study and analyze the problems associated with the lack of the understanding of contract documents` by local contractors and its relation to performance.

Project management environment in Palestine

In an attempt to understand the constraints under which Palestine projects are managed, one must briefly examine the political and economic environment.

Political climate

Following the peace agreement and transfer of authority to the Palestinian people, international support for economic and social development in Palestine was mobilized. The international community felt strongly that there was an urgent need to deliver tangible benefits to the Palestinian population to reinforce the momentum towards peace. The success of the investment projects require political stability in Palestine, steady progress in bilateral and multilateral negotiations, and maturation of internal political process. Instability in the area would hinder implementation. Program success in providing tangible benefits to Palestinian population could help and speed the implementation.

Economic development

The economical situation of the Gaza Strip and the West Bank is mainly service-oriented with agriculture accounting for 30% of Gross Domestic Product (GDP), industry about 8%, construction about 12% and services the remaining 50%. Private sector activity dominates the economy of the West Bank and Gaza Strip which account for about 85% of the Gross Domestic Product. A striking feature of the Palestinian autonomy is its heavy dependence on the Israeli economy.

The economy of the Palestine territories is currently in turmoil. Income levels have stagnated over the past decade. Unemployment and underemployment are rising rapidly. Public infrastructure and social services are grossly overstretched. The fragile natural resource base is threatened with irreversible damage. Above all, the economy remains highly vulnerable to external development, as shown vividly by the economic hardship being experienced in the aftermath of frequent border closure with Israel.

Growth started slowing down with the end of the regional boom in the early 1980s, and decline set in after 1987. Between 1980 and 1987, real GNP per capita increased by 12% and the real GDP per capita increased by 5%. Export growth also stagnated during that period. The situation was exacerbated after 1987 with the intifada, which caused disruptions in economic relations with Israel. Increasingly periodic territories closures adversely affected employment and trading activity.

Imbalances of the Palestinian territories economy are manifested in several areas, as for examples, heavy dependence on outside sources of employment for the Palestinian labor force. There is low degree of industrialization. Then, the trade structure is heavily dominated by trading links with Israel and with a large trade deficit. Finally, there is inadequancies in the provision of public in infrastructure and services.

Project environment

Recently, Palestinian territories have become increasingly aware of the need for a systematic approach to the management of development projects in order to improve the effectiveness and the performance concept in regulations, specifications and contractual procedures. This demand arises from the complexity of such projects and the need for effective coordination and cooperation between many different ministries, agencies and international donors. Figure 1 shows the typical project environment in Gaza. This organization must relate to all appropriate agencies and ministers as well as to the internal relationships common to conventional project management. These external and internal relationships define the organizational project environment (Enshassi, 1996).

In this environment, as a project moves from initiation to implementation, the principal responsibility for management and decision making changes. For example, a consulting firm may do the preliminary project analysis, the Palestinian Economic Council for development and Reconstruction (PECDAR) may make the evaluation and selection with the World Bank, and an implementation agency may manage the design and engineering. Since the quality of the work in each of these steps affects later phases of the project, the interfacing at the points where responsibility shifts become very important (Enshassi,1997).

International firms are becoming more and more involved in the process of managing and implementing infrastructure projects in Gaza Strip. This require special consideration in order to adapt to new project environment. The foreign managers must understand the work environment of the Gaza Strip and be able to cope with it. The specific factors that must be considered are social and cultural, governmental and political, economic, technical, and operational.

Contract documents

Contract documents are legal forms in which the scope and the requirements of a construction project are comprehensively laid down. In these instruments the obligations and responsibilities of the parties to the contract as well as the engineer's powers, duties and functions which flow from such a contract are defined. The engineer is not generally a party to the contract and therefore has no contractual rights or obligations under it (Hasewell, 1982). Contract administration is a natural association of duties for the project manager. He is thoroughly familiar with project and its contract documents and is in the best position to carry through the contractual terms.

The tender documents issued to prospective tenderers are in effect pro-forma contract documents. When the blanks in these documents are filled in by a tenderer, and an offer made thereon is accepted by the employer, these documents together with any amendments, modifications and amplifications that are agreed upon before the final acceptance of the tender become the contract documents governing the performance of the object or objects expressed in the contract (Bent,1989).

A contract is an agreement between two parties. Normally it is a written document but can be binding even as an oral agreement. In general, when we agree to certain items in a contract, one party cannot unilaterally change this. Construction contracts are unique because of the changes clause which is contained in virtually all of them. The changes clause allows the owner to make changes within the scope of the contract. The contractor will be compensated for the changes, but must perform them or be considered in default (Trauner, 1993).

Composition of contract documents

The contract documents adopted for civil engineering construction normally comprise the following parts (Bent,1989 - Kerzner,1984).

- Instruction to tenders.   -   Form of tender.
- Form of agreement.   -   The general conditions.
- Special conditions.   -   Specification.
- Bill of quantities.   -   The plans or drawing.

The instructions to tenders is normally the first part included in the bound volumes of the tender documents. The purpose of this part is to ensure that all tenders are properly prepared and delivered so that they may be evaluated on an equal basis. The form of tender is the section of the document the tenderer is required to fill in to make an offer to execute the works. The form of agreement is the formal document confirming the existence of a contract between the employer and the contractor (Abrahamson,1979).

The conditions of contract incorporated in contract documents are generally one of the standard forms of contract such as the Institution of Civil Engineers (ICE) conditions of contract, conditions of contract for overseas works mainly of Civil Engineering Construction prepared by the Association of Consulting Engineers and the International Conditions of Contract (FIDIC), and conditions of contract prepared by the World Bank (Sawyer, 1977). The conditions of contract that are used in Gaza Strip are FIDIC, the World Bank Condition of Contracts, and mixtures of Israeli, Egyptian and Jordanian conditions of contract.

The special or supplemental conditions of contract are used to supplement, modify , or expand the general conditions. They are used to tailor a standard contract to a specific project or to add items which might not have been included in initial drafts of the contract. The scope of works together with the general and technical requirements of a project is

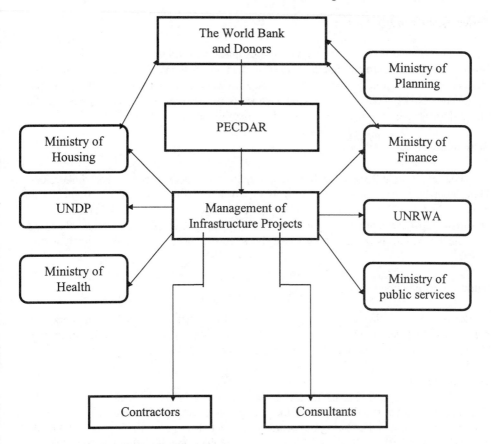

Figure 1 : Project Environment in Gaza Strip

described in the specification. The contract drawing are the drawings on which the works are to be carried out and on which the contract price is based .

Significant contract clauses

The most significant contract clauses that should be carefully considered are: Changes, Differing Site Conditions, Scheduling, Time Extensions, Disputes, Liquidated Damages, Termination, and Exculpatory Language (Trauner, 1988-Trauner, 1993).

Changes clause

The changes clause in a construction contract may be the most important clause. This clause makes the construction contract unique from other forms of contracts. The owner may, at any time, without invalidating the contract, and without notice to the contractor, make changes within the scope of the contract which may either increase or decrease the contract price and/or duration.

Differing site conditions clause

Differing site conditions are also considered changed conditions or concealed conditions. Most construction contracts contain a clause that addresses them. This clause addresses the situation in which a condition is discovered that was not anticipated based on the bid documents. No contract adjustment which results in a benefit to the contractor will be allowed unless the contractor has provided the required written notice.

Scheduling clause

The contract should have a well-written requirement for a schedule for the project. Unfortunately, most construction contracts do not have well-structured requirements for project schedules. The contractor could submit a bar chart or a CPM schedule to satisfy the requirements of the contract. During the planning phase, the owner must decide what kind of project schedule is needed, how it will be effected, and then draft the corresponding clause.

Time extensions clause

While the changes clause may refer to an extension of time for additional work, the mechanics of a time extension are normally addressed in a separate clause. The drafting of this type of clause depends on how time extensions are to be evaluated and granted.

Disputes clause

The disputes clause should clearly define the method that will be employed to resolve disputes should they arise during the course of project. There are many options that could be used, including litigation, arbitration, dispute review boards, administrative boards , neutrals, and mediation. Recently, many contracts are adopting procedures for alternative dispute resolution. The reason for this is to stay out of the courts and try to avoid the associated legal fees.

Liquidated damages clause

Clauses that address delays caused by the contractor or subcontractor are common in construction contracts. The most common form is a liquidated damage clause. If the project is delayed due to the fault of the contractor, the owner has the right to be compensated for the additional costs that it incurs. The purpose of the clause is to reimburse the owner , not penalize the contractor.

Termination clause

The termination clause in a construction contract is a necessity. In essence, it states that should the contractor be found in default of the contract, the owner has the right to terminate the contract for cause. The termination clause should specify what constitutes default of contract. The most common causes cited include bankruptcy, failure to directives of the owner.

Exculpatory language clause

The exculpatory clause protects one party by putting the risk on the other party to the contract. No claim for damages or any claim other that for extensions of time shall be made or asserted against the cause beyond the control and without the fault or negligence of the contractor.

Methods

An exploratory survey was first conducted to identify the contract system used by local contractors in Gaza. Also, efforts were made to identify those significant contract clauses which usually affect the relationship between contractors and owners and has impact on the performance of projects. A total of 103 local contractors in the Gaza Strip were interviewed in order to find out the problems that they face from the contract system and contracts language used. The contractors are implementing several types of projects, i.e. Housing, schools, roads, and water and sewerage projects (see table 1). These projects are located in several places in the Gaza Strip. Understanding contract documents variable was measured in terms of the understanding of the English language which is used in contract documents and the interpretation of contract clauses. Performance of construction projects variable was measured in terms of projects' delay and projects' cost. The data were analyzed using cross-tabular analysis and T -Test to examine the relationship between understanding of contract documents and the performance of construction projects.

Results

The result of T-Test proved that there is a strong correlation between the misunderstanding of contract documents and performance of construction projects in Gaza Strip. The T function is found to be equal 11.889 < 4.303 (the rejection area).

Table 1: The Location of Projects

Type of Project	Number of Contractors in the Sample	Location of Projects	Contract Value for each Project (US$)
Housing	32	Gaza City, Nuseriat, Rafah	1M - 2 M
Schools	38	Gaza City, Jabalia, Khan Yunis	500,000 - 1.5 M
Roads	17	Jabalia, Gaza, Khan Yunis, Absan	500,000 - 2 M
Water and Sewage	16	Jabalia, Gaza, Middle Area, Rafah	800,000 - 2M

These results indicated that the English language is used in most contract documents distributed to local contractors in Gaza Strip (see table 2). A few contract documents are written in Arabic language. It has been found that 85% of local contractors` engineers do not understand contract documents written in the English language very well. The language of contract documents has created a number of problems to contractors` engineers as they do not understand the English language well. They have problems in understanding and interpreting contract clauses. This situation encouraged disputes and arguments between the contractors and employers. This has led to a waste of valuable time and to a decrease in performance of the work.

Table 2: Types of Contracts and Languages of Contract

Types of Projects	No. of Contractors	Type of Contract Used	Types of Conditions of Contract Used	Languages Contract
Housing	32	- Bill of Quantity - Lump - Sum	- FIDIC - ICE - Mixed	- English - Arabic
Schools	38	- Bill of Quantity - Lump - Sum	- UNRWA - FIDIC - ICE	- English
Roads	17	Bill of Quantity	- World Bank - FIDIC	- English
Water and Sewage	16	Bill of Quantity	- World Bank - FIDIC - UNRWA	- English

A large number of projects (90% of the sample) were found to be behind schedule. Delays of some projects are caused by events beyond the control of the contractor, e.g. border closure (see table 3). This posed a severe problem for the local contractors and prevented them from importing the materials required to implement their projects from Israel. This delays has led to additional expenses incurred by the contractors. The prices of material, have been also dramatically increased which has an adverse effect on the performance of the local contractors. If the contractors experience delays to completion of the project which are unenforceable and without the fault or negligence of the contractors, the time allowed for performance of the work may be extended, but no additional compensation provided.

The results showed that the conditions of contract documents which are used in Gaza Strip are the following standard forms of contract:

- The World Bank contract.
- The International Conditions of Contract (FIDIC).
- The Institution of Civil Engineers (ICE).
- Mixture of several standards, e.g. Israeli, Egyptian, and Jordanian.

Although some contractors are familiar with the World Bank contract, they feel that this type of contract is in favour of the owner and is unjust to the contractors. A number of contractors prefer to work with the FIDIC conditions.

Table 3: Delays in Construction Project

Types of Projects	Number of contractors projects	No. of Delayed Projects	% of Delayed Projects
Housing	32	29	91 %
Schools	38	31	82 %
Roads	17	17	100 %
Water and Sewage	16	16	100 %
Total	103	93	90 %

A number of problems that have been observed which caused difficulties for local contractors, and have been attributed to delays of projects, e.g.:

1- Changes clause: as is understood from this clause, the owner may, at any
    time, without invalidating the contract, and without notice to the contractor make changes within the scope of contract. In the Gaza Strip, as the political situation is still not stable, and all the construction materials are imported from Israel, the prices of material increase; 100% in some situations. This result in deficit for the local contractors, and affect their performance.

2- Modern construction equipment are not available to the local constructors in the Gaza Strip because of difficulties in import procedure. This has affected the quality of work and ultimately the performance of projects.

3- Lack of ability in studying carefully the drawing details and the materials specifications for construction projects has adversely affected the performance of the work.

4- Delays in receiving the installments from the owner of the projects created difficulties for the continuity of the work.

5- The supervisory teams employed by local contractors not have sufficient experience in monitoring and controlling projects.

Conclusions

A construction project is a special kind of activity. It has a beginning, a life, and end, in other words it has a life cycle. The performance of the work is the part of the project that most people think of when they think of a project. Basically, it consists of doing the work and reporting the results. Doing the work includes understanding contract documents and language, directing and coordinating other people and controlling their accomplishments so that their collective efforts achieve the project's objectives.

Contract terms are often written in a standard format by the employer or the owner and applied indiscriminately, sometimes standard terms are unfair or unsatisfactory to the construction firm because of the project's special nature and environment like the situation in the Gaza Strip and the West Bank. For Example, the frequent and sudden border closure make it difficult for the local contractors to import the required materials as scheduled. This causes delay for most construction projects which create a negative impact on the performance of projects. In addition, the sudden and extreme increase in materials' cost pose a great problem for the contractors which also has adverse impact on the performance of projects .

It is recommended that some standard terms should be revised or removed through negotiation before the contract is signed. Also creation projects need contract terms that are non-standard and that will appear in the contract only if the project manager thinks that they are necessary. Project managers should understand the contract language and the terms of the contracts for their work in order to ensure:

- that the contract terms will not so hinder them that they cannot perform their
   projects; and

- that contract terms will induce the other parties to behave in ways that will aid
   them in performing their work.

Among the contract elements that are of concern to local contractors and should be clearly understood and clarified are:

1- Language and law clause. The language of the contract and the law governing the
   contract should be distributed in Arabic language.
2- A change clause, i.e., a provision that deals with how changes to the scope of work
   are to be handled.
3- A clause on all allowance costs, i.e., the provision that tells the contractor what
   costs the owner will pay for.
4- A clause on excusable delays.
5- A clause dealing with termination for convenience and for default.
6- A clause dealing with disputes.
7- A clause dealing with liquidated damages.

Project managers need to work under conditions that help rather than hinder their jobs. Therefore, project managers need to negotiate the terms and conditions of their assignments before agreeing to take them. This may result in an increase in the performance of construction projects.

It is also recommended that the contractors should receive their installments without any delay to speed up the process of construction. Regular meeting between the owner and the contractors are encouraged to resolve any problem or misunderstanding between both parties. Standardization of design of building construction components should be encouraged for healthy economy and to secure a high quality. The Arabic language should be used for contracts whenever possible to enable local contractors to administer construction contracts more effectively. This will lead to an improvement in construction performance. Special training courses should be organized for local constructors in order to familiarize them with the international standard forms of contract. This will ensure harmony in procurement and result in profit to all parties

References

Abrahamson, M.W., (1979), Engineering law and the I.C.E. contracts, 4th edn, Applied Science Publishers.

Bent J, and A. Thumann, (1989) , Project Management for Engineering and Construction, Lilburn, Georgia: Fair mount Press, Inc.

Enshassi, A., (1996), A monitoring and controlling system in managing infrastructureprojects, *paper published in the International Journal of Research, Development and Demonstration*, Vol. 24, No. 3 .

Enshassi, A, (1997), Site Organisation and Supervision in housing projects in Gaza Strip, paper published in the International Journal of Project Management, Vol. 15, No. 2.

Hasewell, C.K and de silvan, D.,S., (1982) Civil Engineering contracts, Practice and Procedure, Butterworths, England.

Kerzner, H., (1984), Project Management A systems Approach to Planning, Scheduling, and Controlling, Van No strand Rein hold, U.S.A.

Sawyer, J, and C. Gillot, (1977), The FIDIC Conditions- design of contractual Relationships and Responsibilities, Thomas Tel ford Ltd.

Trauner, T., (1993), Managing the construction Project: A practical Guide for the project Manager, John Wiley and Sons, Inc., UK.

Trauner , T., and M. Payne, (1988), Bidding and Managing Government Construction, R.S. Means company, Inc. New York.

NOVATION CONTRACT AND ITS APPLICATION IN HONG KONG

ALBERT PC CHAN
Department of building and Real Estate, The Hong Kong Polytechnic University, Hung Hom, Hong Kong. bsachan@polyu.edu.hk

TONY YF MA
School of Geoinformatics, Planning and Building, The University of South Australia, North Terrace, Australia. tonyma@unisa.edu.au

Abstract

Novation contract has been adopted in Australia and the UK for several years as an innovative alternative to design-build system. The Contract Journal (1995) reports that there is an increasing trend to the use of novation contract in the construction industry. In 1994, about 36% of all design-build projects utilised the concept of novation in the UK. Despite its popularity in the Western world, novation contract is a brand new idea in Hong Kong with only limited application. The only reported case using novation concept is an abattoir project, which is funded and developed by the Architectural Services Department (ASD). Some cynical comments were obvious amongst the practitioners in Hong Kong. For example, many practitioners claim that quality of this procurement system is inferior, clients use the novation contract as stealth to the contractor, design team lacks genuine commitment, and the contractors are not competent enough to take part in the design process, etc. This paper examines the current practice and appraises the future application of novation contract in Hong Kong.

Keywords: Design-build system, novation contract.

Introduction

Time, cost, quality and statutory requirement are the performance measures in a modern construction project (Chan 1997). In order to maximize the investment value and usage value of the construction, professionals are always attempting to minimize the construction time and cost, but still fulfil the statutory requirements for different departments and accomplish satisfactory quality.

Various procurement systems can contribute to the achievement of the above objectives. Kwayke (1990) states that the choice of an appropriate procurement system depends on:

1. Size and nature of the building program
2. Size and complexity of the works
3. Time available for project completion/occupation
4. Degree to which the client's requirements can be defined before the contract is let
5. Degree to which the client's requirements are likely to change during the construction of the works

6. Availability of in-house or consultant skills
7. Whether a client's financial commitment is known of fixed before the contract is let
8. Logistical problems likely to influence a contractor's risk, e.g. labour supply, remote location, funding, unfamiliar technology, etc.

Many forms of procurement can accommodate these factors. However, there is a growing trend towards the use of alternative contract strategies in place of the traditional lump sum (or fixed price) building construction contract delivery system (Chan, 1995a). These include Construction Management, Project Management, Design/Construct and a host of other closely related variations (Chan and Lam, 1994; Chan and Tam, 1995; Lam and Chan, 1996)

Recently, a new method of project delivery has been added to the ranks, namely, Novation. This construction procurement system has already been used on a number of major projects throughout Australia, and more recently in Adelaide (Chan, 1994; Chan and Tam, 1994; Chan, 1995b; Chan and Lam, 1995). Some of these projects include the Southern Grandstand of the MCG and the new Commonwealth Centre in Melbourne, the Entertainment Centre and Exhibition Centre in Adelaide, and the new Telecom buildings in both cities.

The primary drive towards these alternative delivery systems are the clients' desire to obtain a truly "fixed price" for their purchase. In the past, they have had to face continual escalations in price due to a variety of reasons, many due to the barriers and inefficiencies inherent within the building industry. Novation, therefore, is one attempt to transfer the risks associated with the building process from the client to the contractor, whom, in theory, is in better position to manage the risk.

In 1994, about 36% of all design-build projects utilised the concept of novation in the UK (Contract Journal, 1995). Despite its popularity in the Western world, novation contract is a brand new idea in Hong Kong with only limited application. The only reported case using novation concept is an abattoir project, which is funded and developed by the Architectural Services Department (ASD). Some cynical comments were obvious amongst the practitioners in Hong Kong. For example, many practitioners claim that quality of this procurement system is inferior, clients use the novation contract as stealth to the contractor, design team lacks genuine commitment, and the contractors are not competent enough to take part in the design process, etc. This paper examines the current practice and appraises the future application of novation contract in Hong Kong.

Definition of novation contract

Novation is not a new concept or process but rather it is a contracting process, which dates back to Roman Civil Law. When novation is used in a legal sense it is not found to be complex, however, when referred to the construction industry the nature and extent to which it is used needs to be clearly defined. Since novation is used across a wide range of contracts and in a variety of industry sectors, it is therefore difficult to determine an all-embracing definition. The Oxford Dictionary defines novation as "The substitution of a new obligations for the one existing" (Allen, 1990). In simple terms it is the process of contract substitution, where one contract is discharged and replaced by another.

In legal sense, the concept of novation has been established in the case of *Sruples Imports Pty Ltd v Crabtree & Evelyn Pty Ltd (1983) 1IPR 315 at 320* where novation is considered as a contract between three parties, the obligee, the original obligor and the substituted obligor. The effect of which contract is that in consideration of the obligee releasing the original obligor from his obligation, the substituted obligor promises the obligee that he will assume responsibility for the performance of the obligation (Chitty, 1989).

In the context of the construction industry, RAIA (1991) defines novation contract as a form of Design and Construct agreement, in which the proprietor (client) initially employs the consultant team to carry out design and documentation to the extent that the client's needs and intent are clearly identified and documented. On the basis of these documents, tenders are called and a building contractor selected. The proprietor (client) then novates the consultant agreements to the contractor who takes responsibility for the project to completion.

The novation procurement process

Novation contracting is primarily the process by which the client employs an architect, and design consultants who prepare a design brief of the works to be executed. Once the brief has reached the stage of clarity of what the client requires on construction completion, tenders are called, a contractor is then engaged to carry out both completion of the design and the execution of the works. Novation can in fact have the effect of transforming a traditional contractual pattern into a design and construct pattern. The client, contractor and design consultants must however enter into a contract (a novation contract) under which the original agreements between the client and design consultants are replaced by new contracts between the contractor and the design consultants. Waldron (1993) defines novation process into two distinct stages, namely, pre-novation stage and post-novation stage.

Pre-novation stage

The client initiates the project by commissioning design consultants to develop brief and commence design work in a manner very similar to the traditional system. The terms of the engagement of the consultants are given by means of a contract established between the client and the consultants. The role of the design consultants is to complete the designs to a stage where all the requirements of the client are adequately described to a level of legal clarity. The requirements are normally defined, drawn and specified in the range of 30-80% of the overall design. Once documentation is completed to a stage that legal clarity has been achieved, the client is in a position to call for tenders from contractors to undertake the completion of the design and construction process. The contractual arrangement at the pre-novation stage is quite similar to the traditional system.

Post-novation stage

Once the contract has been novated, the contractor has a direct contractual link with the client and the design consultants. The contractual arrangement is similar to that of the traditional design and construct arrangement. The contractor has become the designer and is responsible for all the design work as well as the construction. The contractor instead of the client pays the consultants, and their first point of loyalty is to the contractor. The sharp distinction drawn between novation contract and design and construct contract is that the contractor must employ the designated designer who has carried out the preliminary design for the client. The contractor has no choice to take their preferred consultant as they can enjoy in the design and construct contract (Swindall, 1993).

The involvement of the contractor in the detailed design phase allows the contractor to implement changes to the design that suit his particular construction practices and equipment. The contractor can even modify the design to suit the availability of materials or skilled workforce. The contractor is required to keep the client informed on design matters while still maintaining prime responsibility of meeting the performance criteria set down in the design brief. By including the appropriate clauses in a novation contract, the client can retain the right to monitor and comment on the design process. However, the client should not do anything that could be perceived as an acceptance of responsibility for the design.

As the contractor is contractually responsible to complete the design, consideration should be given to professional indemnity insurance to cover that responsibility. On the other hand, the client must warrant the design in that it complies with all codes and regulations current at the time of novation and must accept post-novation changes to such codes and regulations as variations (Siddiqui, 1996). Responsibility for authority fees must be clearly noted by the client if he wishes to transfer these costs to the contractor.

Case study ---- Sheung Shui Slaughterhouse

Despite the political uncertainty caused by the return of Hong Kong to China, the property and construction sector continues to prosper. Its contribution to the GDP has averaged over 24% since 1980 (Rowlinson & Walker, 1995). Hong Kong's spectacular economic growth and its growing importance as an international port have given much support to its building and infrastructure development. The demand for more prestigious buildings and sophisticated infrastructure systems results in construction projects of larger scale and complexity (Mo and Ng 1997). Under these circumstances, Hong Kong is going through a phase whereby the so-called traditional method is no longer the only answer for the procurement of a building project. The Hong Kong Exhibition Centre, Kowloon Station Extension and North District Hospital are cases in point, which use Management Contracting and Design & Build respectively.

Novation Contract, a brand new procurement method to the Hong Kong construction industry, was adopted in the Sheung Shui Slaughterhouse project, which is the first government-funded project adopting Novation Design & Build in Hong Kong. Different parties in this project were interviewed in analysing the current practice of Novation Contract in Hong Kong (Butt, 1998). Table 1 shows the details of the persons interviewed.

Table 1:   Details of persons interviewed

Name	Host Company	Role in the project
Mr. Y. S. Lee	Architectural Services Department	Client's Project Manager
Mr. Raymond Ho	China Overseas Building Construction Ltd.	Contractor's Project Director
Mr. Johnnie Lai	Wong & Ouyang (HK) Ltd.	Project Architect
Mr. Francis Leung	Architectural Services Department	Senior Quantity Surveyor

Background of the project

The project concerned involves the construction of a new slaughterhouse in Area 2B, Sheung Shui, New Territories. The design of the new slaughterhouse is based on the operational requirements of a modern slaughterhouse incorporating the latest and technologically – advanced slaughtering equipment. The design objectives are derived from consideration of cost economy, operational efficiency, energy conservation, waste recycling, animal welfare and most important of all, environmental awareness. The new slaughterhouse will operate a daily killing of 5,000 pigs, 400 heads of cattle, and a maximum of 300 goats.

The project site covers approximately 57,807 m^2 and is almost triangular in shape. On the southeast side, the site sits on a straight edge neighboured by the Shek Wu Hui Sewage Treatment Works. On the northeast, it includes an existing formed path next to the Ng Tung River (River Indus). On the west, it is bounded by the K.C.R.C. railway sidings with a curved boundary line. Chuk Wan Street is leading to the northeast corner of the project site.

The site chosen in Sheung Shui area benefited from the close proximity to mainland China where most of the livestock come from and adjacent to the railway siding. The chosen site is also next to Shek Wu Hui Sewage Treatment Works where the processed wastewater can be easily collected and treated. Moreover, the site is far from the residential areas so as to minimize nuisance to the surrounding dwellers.

Project details

Table 2 summarises the general information of the Sheung Shui Slaughterhouse.

Table 2: Project details of Sheung Shui Slaughterhouse

Attributes	Details
Total Gross Floor	59,000 m^2 including Lairage (33,500 m^2), Slaughter hall and associated area (12,000 m^2), Administrative and amenity area (4,500 m^2) and By-product plant/underground wastewater pre-treatment plant (9,000 m^2)
Plot ratio	1.03
Covered area	29,500 m^2
Site coverage	51%
Total building height	20m (4 storeys excluding basement)
Tendering procedure	Selective tendering according to government
Contract Sum	About 1600 million
Time for completion of the project	1070 calendar days (including General Holidays)
Time for completion of construction	804 calendar days (including General Holidays)
Amount of Bond	10% of Contract Sum
Method of payment	Milestone Payments
Form of contract	General Conditions of Contract for Design & Build Contracts, May 1992 Edition (issued by Hong Kong Government)

The novation process of the slaughterhouse project

Pre-novation stage

The project manger of the ASD pointed out that they first employed the Wong & Ouyang (HK) Ltd as their consultant to carry out the feasibility report for the Sheung Shui Slaughterhouse. Upon completion of the feasibility report, they employed Wong & Ouyang again as the design consultant for better efficiency.

Being employed as the design consultant, Wong & Ouyang soon discovered that there were many specialised design requirements in this slaughterhouse project, and even their company did not possess all the necessary expertise in designing the slaughterhouse. Therefore, the ASD invited some overseas experts to assist in the design of the slaughterhouse. These experts came from UK and Netherlands who had experience in slaughterhouse design. They mainly assisted in the design of water treatment work and the detail design of the slaughterhouse.

Wong & Ouyang signed the novation agreement with the ASD to guarantee that they would accept the novation of the contract to the main contractor and would comply with the clauses and terms issued on the novation agreement. The design consultant would complete stages A and B of the RBIA's Plan of Work (brief design of this project), and complete 20% of stage D before calling for tender.

Tendering process

Invitation to tender is restricted only to the tenderers who have been short-listed in a prequalification exercise. An Assessment Panel is formed to recommend the award of the contract. The assessment is to ensure that the contract will be awarded to the best tender, which meets the Employer's Requirements in terms of time and quality; and at the best price. The form of contract used for this project is the General Condition of Contracts for Design & Build Contracts. Ultimately, China Overseas Building Construction Ltd. was appointed as the main contractor for this project.

Post-novation stage

After the appointment of the main contractor, Wong Ouyang was transferred to and employed by the China Overseas in accordance with the novation agreement. Wong Ouyang continued to complete the design with China Overseas. At the same time, China Overseas was obliged to employ a design checker under Section 4 of the Special Condition of Contract. The design checker would directly report to the ASD for the design work and was paid by China Overseas.

Typical uses

The structural interviews with the various project participants revealed that Novation Contract is most suitable for complex and sophisticated projects. Table 3 shows project particulars, which make Novation Contract suitable for Hong Kong.

Table 3:   Project criteria suitable to use Novation Contract

Project criteria suitable to use Novation Contract
1. *High complexity of the project*
2. *High specialization of the design requirement*
3. *Client has high degree of control to the design*
4. *High cost certainty*
5. *Project with tight time limit*
6. *Large project scale*
7. *Single point responsibility*
8. *Minimize the risk of the client side*

Future application of the novation contract in Hong Kong

Although parties of the "Slaughterhouse" expressed satisfaction with the Novation Contract, they were concerned whether it could be adopted widely in the local industry, particularly in the private sector.

Lengthy approval process

The interviewees expressed concerns whether private developers can use the Novation

Contract successfully. They pointed out that the main problem was the control from the various government departments. They stated that one of the advantages of the Novation Contract is the certainty of the construction time. In the case of Sheung Shui Slaughterhouse, the objective of time certainty could be achieved mainly because it was a government project. Obtaining approval is not a problem with government projects because they can have exemptions from the statutory bodies. Should it not be the case, the benefits of exemption from statutory bodies will be gone and the approval process can be time consuming and beyond the control of the main contractor. Hence it will induce more risks to the contractors. Contractors may be reluctant to bid for the contract because the ability to accurately assess the completion time is far too uncertain. Contractors will find it difficult to complete the project within the stipulated time.

Inadequate design experience of the contractors

The project manager of ASD also held negative views about the future of Novation Contract. Since most of contractors in Hong Kong mainly practised as general contractor under the traditional procurement systems, they have little design experience in the construction process. Many contractors have no experience in property development and their design experience and management is not as good as the overseas competitors. With Novation Contract, the contractor is to manage the design consultants at the post-novated stage, contractors who have inadequate design experience may find it difficult to competently head the design team.

Recommendation for future application

The interviewees opined that Novation Contract may be successful if the following criteria are met:

1. The brief is comprehensive
2. Construction is to proceed prior to completion of the documents
3. Builders invited to tender have the necessary expertise to undertake the work
4. The building industry is relatively busy and unlikely to "buy" jobs
5. The project has some unusual or special features where building delivery innovation would provide cost benefits

The management of the risks is an integral part of delivering any project. Should the client wish the builder to carry these risks then it should be acknowledged that a loading or premium will be inclusive in the cost of the works.

Conclusion

It is undeniable that the use of the Novation Contract is at its infancy stage in Hong Kong. The Sheung Shui Slaughterhouse provides a very good opportunity to test whether the Novation Contract can be successfully used in Hong Kong. In fact, the Novation Contract has been used in a number of projects in the UK and Australia. The feedback of using Novation Contract has been favourable.

How about its development in Hong Kong? Up to the time of writing this paper, the Sheung Shui Slaughterhouse is still under construction and it is too early to draw any precise conclusion on the performance of Novation Contract. However, the feedback drawn from the key project participants is positive. Different parties of the "Slaughterhouse" project expressed satisfaction with the Novation Contract. The project up till now shows no delay in construction time and being within the construction budget.

The Novation Contract seems to be more suitable in government project. It is because most of the design approvals can be exempted from government departments and so time and cost can be more easily controlled. Whereas in private project, the situation is not so bullish because of the lengthy approval process required by the various government bodies. Therefore, it is predicted that the Novation Contract in Hong Kong may not be as popular as in other developed countries.

Acknowledgements

The authors gratefully acknowledge the Hong Kong Polytechnic University for providing funding to support this research effort.

References

Allen, R.E. (1990) The concise Oxford dictionary of current English. Oxford, England, Oxford University Press.

Butt, CW (1998) Review of novation contract in Hong Kong, Unpublished BSc Thesis, Hong Kong Polytechnic University.

Chan, APC (1994) Evaluation of novation contract, National construction and Management Conference, Sydney, Australia, 129-142.

Chan, APC (1995a) Towards an expert system on project procurement, *Journal of Construction Procurement,* Vol.1, No.2, 111-123.

Chan, APC (1995b) Novation contract - an innovative way of procuring building projects, *The First Pacific Rim Real Estate Society Educators' Conference*, 333-340.

Chan, APC (1997) Measuring success for a construction project, *The Australian Institute of Quantity Surveyors – Refereed Journal*, Volume 1, Issue 2, 1997, 55-59.

Chan, APC and Lam, PTI (1994) Construction management: a new role to building contractor, First International Conference - Changing Roles of Contractors in Asia Pacific Rim, 145-162.

Chan, APC and Lam, PTI (1995) Novation contract - an innovative variation to design-build, International Congress on Construction, Singapore, 197-203.

Chan, APC and Tam, CM (1994) Design and build through novation, CIB W92, Procurement Systems - Symposium, Hong Kong, 27-36.

Chan, APC and Tam, CM (1995) Project procurement in Hong Kong, National Conference of the Australian Institute of Project Management, 575-582.

Chitty, J. (1989) Chitty on contracts. London: Sweet & Maxwell.

Kwakye, AA (1990) Fast track construction, CIOB Occasional Paper No.46, Ascot.

Lam, PTI and Chan, APC (1996) Construction procurement systems in Asia-Pacific countries, Australian Project Manager, Vol.16, No.2, June, 12-17.

Mo, JKW and Ng, LY (1997) Design and build procurement methods in Hong Kong – an overview, CIB W92 Symposium, Procurement – A Key to Innovation, Montreal, 453-462. pp.27.

RAIA (1991) The use of novation contracts in building delivery, *The Royal Australian Institute of Architects,* August.

Rowlinson, SM and Walker, A (1995) The construction industry in Hong Kong, Longman Asia Limited.

Siddiqui, AW (1996) Novation: and its comparison with common forms of building procurement, *CIOB Occasional Paper* No.60, Ascot.

Swindall, W. (1993) Understanding novation, Building design and supplement, *Building*, 30 July,

Waldron, B.D. (1993) Design and construction through novation, Construction Project Law (International), Seminar Paper presented on 27 April.

4. Procurement Systems and Strategies

EMERGING ISSUES IN PROCUREMENT SYSTEMS

JASON MATTHEWS, STEVE ROWLINSON and **FLORENCE T.T. PHUA**
Department of Real Estate and Construction, Hong Kong University,
Hong Kong. hrecsmr@hkucc.hku.hk

PETER MCDERMOTT
Research Centre for the Built and Human Environment, The University of Salford. U.K.

TOM CHAPMAN
Hong Kong Productivity Council, Hong Kong.

Abstract

This paper commences with a brief review of the work of CIB Working Commission 92 to date which will pick out past themes and major developments and will culminate in a discussion of likely future directions for the working commission, in particular, and procurement systems research in general. The theme of the W92 symposium in Montreal, Innovation, will be addressed and the emerging issues in procurement systems analysed and discussed within this context. The aim of the discussion and analysis will be to try to answer the question "where is procurement systems research going to?"

Keywords: Culture, IT, organisational learning, partnering, procurement systems, sustainability.

Introduction

The purpose of this paper is to explore and explain the strategic issues of construction procurement that have arisen through the published work of CIB W92 - Procurement Systems. The contributions have been made from very different theoretical and cultural perspectives; the contributors have come from all continents, and from different economic, legal, and political systems. The significant strategic issues prevalent in the work of the commission have been identified as follows: - competition, privatisation, development, culture, trust, institutions, procurement strategies, contractual arrangements and forms of contract. Emergent themes - those which have begun to emerge in recent years and are expected to grow significantly have been identified as - organisational learning, culture, developmentally orientated procurement systems, the influence of IT and sustainable procurement.

Much of the work of W92 complements and is complementary to the work of other CIB working groups or task groups. This is particularly true in the areas of organisation and management, culture in construction, environmental sustainability, building economics, conflict in construction and construction in developing countries. This breadth is an indicator of the cross-disciplinary influences within CIB W92. Wide-ranging subject areas are covered; the working group should recognise this as a strength but seek to focus on the manifestation of the phenomena being studied within the context of procurement and contractual relationships in construction. For examples, the study of the significant role that governments can play even in market economies, the comparison and benchmarking of the procurement regulations of the various trading blocks and international agencies, and, researching and debating roles for institutions charged with construction industry development are all legitimate areas of work for the group. This paper focuses on five key themes which are currently developing - sustainable construction, organisational learning, partnering, IT and culture – and briefly reviews their significance.

Sustainable construction

The effect of construction activities on the environment is now well recognized. The construction process, its products and the eventual usage of such products can have adverse environmental impacts and they relate to, and interact with, the environment for many years. Apart from achieving economic development, the construction industry plays an important role in protecting and enhancing the environment now and for the future and developing industry and business.

Understanding sustainability

Sustainability is a multi-faceted concept and infinitely variable in its interpretation. Whether it is sustainability, sustainable development or sustainable construction, it is a process which will generate a profound influence on the construction industry.

One of the main issues discussed in the Kyoto Protocol on Climate Change held at Kyoto in December 1997, was a call for global awareness of the importance of sustainability for the environment brought about by the execution of construction activities and its related processes. However, it has been found that when discussing sustainable construction, there was a big difference in opinions between the people from developing countries compared to those from developed countries (Baba, 1998). The underpinning factor that results in the divergence of opinions lies with the differing principles and concept of development held by developing and developed nations. Many developing countries viewed development as 'industrialization' which is essentially what developed countries had experienced during the development of their own countries.

Supported by experiences of rapid economic modernization from Japan and other Asian countries (i.e. Hong Kong and Singapore), many developing nations believe industrialization is the most effective way to bring about development and economic growth to their countries (Baba, 1998). The drive to achieve further economic development has prompted, among other things, many nations to build more and to build quicker. Very rarely are environmental issues considered carefully. Within the next few years, an enormous amount of construction projects will be carried out, particularly in Asian countries. These developments will substantially affect the global environment and it is high time for the concept of sustainable development to be incorporated into the overall strategy for economic growth in these countries.

Further gaps in opinions also exist locally between different construction businesses with regard to their organizational culture and management strategies. Gilham and Cooper, (1998) commented that the various dimensions of culture affect the decision-making priorities of different organizations and professionals. For many construction businesses, adopting strategies for construction sustainability may well appear to be impracticable and too far from their current business reality (Gilham, 1998).

Having said this, with the increasing emphasis on the importance for protection of the environment internationally, clients within the construction industry are seeking real value-for-money products and several governments are taking the initiative in encouraging the incorporation of sustainability in their building regulations. This results in a stronger need for efforts on using sustainable building material and processes. However, sustainable development means change and the construction industry should transform the demand for change into an opportunity, being innovative, to create and access new markets.

The challenge for the construction industry

The industry is risk-averse, as are the majority of its clients, it is highly regulated, competitive, fragmented and dominated by contracts which regulate the involvement of the client in the construction process. For practitioners in construction projects, the demand for sustainable development brings about increasing complexity. Changes and innovation are complex and emotive issues which can be obstructive in the working environment.

However, largely propelled by the demand for greater efficiencies and to be client focused, there is evidence from the UK construction industry that practices such as Partnering and Value Management are providing a platform for the development of sustainability strategies. Central to the effort for sustainable development is the adoption of appropriate long-term goals and strategic views by the organization and management systems, which involve the client, professionals, government and construction activities. The need for active integration among the fragmented systems of the construction industry is therefore crucial.

Finally, the choice of procurement system can greatly influence the outcome of construction projects and consequently the issue of sustainability in the construction industry can be better addressed. As already apparent in practice in some countries, it is possible to integrate sustainability into construction projects when there is a stimulus for it (Tij huis, 1998). This stimulus will differ from country to country depending on the set of values and culture held by its people. For instance, in an environment like Hong Kong, the incentive would almost always be financially driven whereby there are substantial short-term benefits.

If the issue of sustainability is to flourish, environmental issues must become an integral part of the management systems of the organizations involved. Strategically based objectives and goals and an integrated approach aiming at achieving environmental performance improvement permits the focusing of both organization and industry-wide effort for sustainable development to be realized.

Organisational learning

Organisational learning is a key issue in all organisations but is of particular relevance to construction projects which operate within the concept of temporary multi-organisations. A series of recent Australian studies into construction time performance revealed that neither the building construction companies nor the client undertook any meaningful organisational learning activities to ascertain what could be learned from the experience of managing building and civil engineering construction projects (Walker & Lloyd-Walker, 1998). Organisational learning has the capacity to add value to an organisation through codifying the experience gained by both individuals and teams during project realisation. As such, innovations can be disseminated throughout the organisation, with the outcome that the organisation becomes more effective. Post project evaluation exercises are essential for this organisational learning process. Walker & Lloyd-Walker state that clients must demand and fund the harvesting of knowledge from experience so that this knowledge can be used next time. In this context they identify partnering within the procurement process as a mechanism for encouraging organisational learning.

Organisational learning has a number of objectives:

Increasing effectiveness by making the right decisions;
Increasing efficiency by adopting appropriate methods;
Increasing satisfaction by providing an environment which encourages innovation and knowledge accumulation.

With the acceptance that the construction process is undertaken by temporary multi-organisations the concept of organisational learning takes on heightened importance, unless we wish to rely purely on individual's knowledge and experience. Additionally, the industry's clients are demanding ever higher levels of service, forcing organisations to revise the level of in-house expertise. Thus, organisational learning becomes a business imperative and, within this environment, organisational learning can become a key element in the procurement process; partnering approaches in particular have a role to play in this respect. Partnering allows both clients and construction industry participants to develop learning organisations but also, when used in the construction process, allows learning and technology transfer to sub-contracting organisations that may themselves be too small to undertake this process unaided.

Partnering

The efficient use of resources within the procurement of construction projects has become an even more important issue to the UK construction industry in recent years. Contemporary reviews by Sir Michael Latham (Latham, 1994) and Sir John Egan (Egan, 1998) have identified that the UK construction industry needs to improve its performance and reduce the occurrence of confrontation commonly found between parties, especially between the main contractor and its specialist subcontractors and suppliers. Both Latham and Egan have advocated the use of partnering to meet these improvements.

The contribution of specialist and trade subcontractors to the total construction process can account for as much as 90% of the total value of a project (Gray & Flannagan, 1989). The result of this, and other factors, is that main contractors are concentrating their efforts on managing site operations rather than employing direct labour to undertake construction work. However, the increase of building complexity, (Jamieson et al, 1996) the over supply of subcontractors, conflicting organisation objectives (Matthews et al, 1996a and 1996b) and declining construction output have cultivated adversarial tendencies that are having a negative effect on the main contractor – subcontractor relationship.

The Reading Construction Forum (1998) in its seminal works *"The Seven Pillars of Partnering"* defined partnering as:

> *"a set of strategic actions which embody the mutual objectives of a number of firms achieved by cooperative decision making aimed at using feedback to continuously improve their joint performance."*

Walter (1998) revealed how important partnering has become to British contractors. 'Blue Chip' construction companies such as AMEC, Amey Construction, Bovis, HBG Construction and Alfred McAlpine have all stated that majority of their current projects were partnered. Some of these companies expect their partnering workloads for 1999-2000 to reach more than 50% of their total output. Moreover, the situation seems similar in the USA. Maloney (1997) reported that the use of partnering in the USA was spreading and would continue to spread. Maloney concluded that although there was no up-to-date data available on the use of partnering, anecdotal evidence does support the notion that it's application is increasing in the USA.

Research has shown that partnering, in all it various guises, does achieve substantial benefits for its stakeholders. Weston & Gibson (1993) reported that of 120 partnered projects undertaken by the Arizona Department of Transport the following savings were accrued: average time saving of 12.33%; cost savings (% of bid amount) of 18.49%; value engineering savings of 2.80%; and total cost savings (% of bid amount) of 19.50%. Larson (1995) undertook research to overcome the anecdotal nature of reported partnering benefits. Larson analysed the results of 280 construction projects (including partnering projects). Larson reported that the findings indicated partnered projects achieve superior results in controlling cost and technical performance, and in satisfying customers. Other notable research in partnering, but not limited to reporting its benefits are: Cook and Hancher (1991) [partnering as a contract startegy]; Wilson et al (1994) [partnering and it's interaction with organisations]; Lazar (1997) [overview of research into partnering]; and Crowley and Karim (1995) [partnering defined in the context of organisations].

Within the United Kingdom the Reading Construction Forum (1995) quote that strategic partnering can achieve savings of 30% over time, whilst project partnering can achieve savings between 2 – 10%. Moreover, RCF's most recent publication (RCF, 1998) states that companies adopting 'Second Generation Partnering' (partnering with a greater degree of sophistication than that previously used and with a higher level of strategic decision making) can achieve cost savings in the region of 40% and time savings in the region of 50%. Furthermore, RCF advocate that 'Third Generation Partnering' (partnering approach adopted across the industry that requires clients and customers to work in collaboration to develop comprehensive packages of products and supporting services) can deliver cost savings in excess of 50% and 80% in terms of time savings.

Although many publications exist detailing the benefits of adopting partnering in the United Kingdom and the USA, very few concentrate specifically on the relationship between a main contractor and its subcontractors. Baxendale and Greaves (1997) discuss 'Competitive Partnering' between main contractors and subcontractors. Baxendale and Greaves conclude that there are significant doubts whether the true principles of partnering will ever become widespread between the main contractor – subcontractor when the relationship is characterised by a history of conflict, litigation and adversarial practices. However, Matthews (1996b) developed an approach to partnering that was a manifestation of project partnering and was adopted between a main contractor and its subcontractor on live construction projects throughout the UK. Specific benefits identified by using this approach to partnering included:

- improved contractual situation;
- improved communication and information flow;
- increased understanding;
- improved efficiency of resources;
- improved financial position; and
- improved quality.

It can be seen that partnering can assist both main contractors and subcontractors, however it must not be seen as a panacea to all the problems. Partnering requires for many organisations a 'mind set' change for it to be successfully utilised. The most significant test for those employing partnering will be not to revert back to traditional adversarial approaches when problems arise. Also, the long-term influence of partnering will be measured against the benefits and problems recorded and disseminated by industrialists. It is only with knowing these characteristics that partnering, in whatever form, can contribute in the long-term to the development of improving relationships and in turn the services offered by the UK construction industry. This is not to say that the construction industry's clients have not got a role to play also. Egan (1998) pointed out that in order for his recommendations to be implemented, major clients must commit to fulfil their responsibility in order to improve efficiency and quality of construction.

IT and procurement systems

IT has the potential to act as a facilitator/enabler of change in the construction procurement system over the next five years. For example, IT has made the establishment of virtual projects a reality and from this stems a series of possibilities for the radical re-drawing of the procurement system boundaries. Walker and Rowlinson, 1998, discuss this issue in detail in relation to the use of IT by construction organisations in Hong Kong and Australia, placing particular emphasis on the opportunities afforded by the world wide web (WWW). Web technologies are merely tools that can, with proper training and support, provide the means for effective communications, particularly as it is a medium which creates a presence that can be accessed 24 hours per day. Construction professionals can use the web to gather estimating and forecasting information. They can also undertake other research activities on the web. Most companies are already using information transfer such as CAD drawings or project plans using floppy disks. Information transfer can be more quickly exchanged through the web using file transfer technologies or as attached email files. The most significant change due to the WWW may well come with the use of collaborative design. This would allow distributed organisations to meet in cyberspace and design facilities from their remote locations. With the use of visualisation techniques and object oriented programming the design can quickly be seen by all participants, including the client, and each could ascertain from the virtual model all the necessary details, such as cost plans, construction details and construction sequences. This technology will have the greatest impact on the procurement systems adopted.

Another major driving force for change in the field of procurement systems and IT usage will be the client body. The construction industry's major clients are increasingly focusing their attention on Information Management Systems and automating much of their own production processes. In order for the construction industry's systems to remain compatible with those of the major clients there must be changes in the way that the industry operates. Thus, a demand derived from the clients will be a major change agent, as indicated in the example below. At the CIB W·78 symposium in Stockholm in June, 1998 in his keynote speech Matthew Bacon, of the British Airports Authority, emphasised the need for the construction industry to take on board two key messages from the clients. These messages were:

"Standards are very important but accurate definition of client requirements is even more important. This means that if suppliers are to deliver value to their clients they must understand their clients, business processes - to help them make the right decisions - and provide them with the information that they need.

Modern business is concerned with integration of business information to make informed business positions. The Integrated Data Model (IDM) must facilitate this. This means that a wider definition of the IDM needs to be established. It must enable the client to view data from their perspective too - a perspective probably quite different from the project team."

The pressure is on the construction industry to adapt its processes to the needs of the client, in particular the client business and the whole project life cycle are key issues to be addressed. Hence, the industry must focus on not only construction but facilities management. The opportunity is provided through IT to document and to manage the whole life cycle and these are key issues for the client organisation. Thus, the focus of attention is on the product and the process as well as the link to facilities management. IT is the enabling technology which allows all of these to be brought together in a holistic approach to the procurement system.

Culture

Culture is one of the themes of this symposium and a CIB task group is studying this topic. There are various elements and conceptualisations of culture. It can be distinguished into types approaches and traits approaches which investigate the nature and content of the dimensions of culture. The existence of a set of replicable cultural dimensions is a prerequisite to making meaningful comparisons across organisations. For example, in the UK typical cultural dimensions (according to Hofstede, 1984) are low power distance, low-medium uncertainty avoidance, high individualism and high masculinity. In Hong Kong, cultural dimensions are reflected in high power distance scores, low uncertainty avoidance, low individualism and medium scores on the masculinity index. There is an observed tendency for industrialised societies to have low power distance scores and therefore a more consultative style of management than non-industrialised societies.

Research indicates that national culture has a greater effect on employees than does their organisational culture. Western organisations may need to learn to work with Asian professionals by accepting their low sense of individualism and problem solving, higher levels of respect and communicate in a high power distance context, tending to treat people formally or differently when there are large differences in age or social standing. Seeking intrinsic motivation rather than extrinsic rewards and such intentional collaboration may hold important clues to future developments as each nation's professions learn to adapt their cultural values in the light of the experience of working together.

Within the Far Eastern contracting firm, project co-ordinators and foremen (subordinates) fear confronting their project managers (superiors) and dislike reporting confrontation to them even when these confrontations are problem solving scenarios. Westwood (1992:353) and Adler (1986:136) describe this type of behaviour as a situation accepting culture, which is accepting the current situation and tolerates it without confrontation. This attitude is operationalised in precepts such as the acceptance of falling object accidents as natural events which are tolerable without need to follow up and correct.

The project manager of the future will need to be able to identify and deal with these cultural differences. However, this task cannot be left solely in the hands of the individual. This is a procurement system issue which the system itself must be able to deal with and adapt to. Hence, in formulating the strategic decision concerning choice of procurement system a thorough understanding of culture is vitally important.

Conclusions

The theories and concepts detailed in this paper are not mutually exclusive. They are all multi-faceted and impinge, to a lesser or greater degree, on one another. In discussing sustainability, one of the key factors highlighted was the difference in opinions between those from developing countries compared to those of developed countries. Perhaps this difference can also be attributed to the differences in culture, a point highlighted within this paper.

Collaboration in procurement was also a common theme throughout this paper. With the modern global economy where collaboration between organisations from different countries is becoming more common place it is imperative that parties understand the differing cultural back grounds of one another. Also, collaboration was discussed in the context of partnering. Partnering can be seen as a theme of modern procurement. The discussion held on partnering, and collaboration, viewed these as ways of developing 'sustainability strategies'; facilitating learning between organisations; and improving the relationships between project participants, especially between main contractors and sub-contractors. Clients within the construction industry can also be seen as a common theme throughout this paper. Whilst discussing sustainability it was highlighted that clients were seeking real value for money from products. The needs of clients was also discussed and how IT will facilitate / enable this.

References

Adler, NJ, International Dimensions of Organisational Behaviour, Kent, Boston, 1986

Baba, K. (1998), "Necessity of common understanding of sustainability in construction in Asia", *CIB 98 Proceedings on Construction and the Environment*, Gavle, Sweden, June.

Baxendale, T., Greaves, D. (1997), "Competitive Partnering – a link between contractor and subcontractor. Procurement – A Key to Innovation." *CIB Proceedings Publication 203. CIB W92 Symposium on Procurement,* Montreal, Canada. pp 21-28.

Crowley , L. G., Karim, M. A. (1995), "Conceptual Model of Partnering." *Journal of Management in Engineering,* Vol. 11, No. 5. Sept/Oct. pp 33-39.

Egan, J. (1998), "Rethinking Construction" Department of the Environment,Transport and Regions. *Http://www.construction.detr.gov.uk/cis/rethink/3.htm#chap.*

Gilham, A. and Cooper, I. (1998), "Exploring the cultural dimensions of construction procurement – dealing with difference to achieve sustainable development", *CIB 98 Proceedings on Construction and the Environment,* Gavle, Sweden, June 1998.

Gilham, A. (1998), "Strategies for change – understanding sustainable development from a construction industry perspective", *CIB 98 Proceedings on Construction and the Environment,* Gavle, Sweden, June 1998.

Gray, C., Flanagan, F. (1989), "The Changing Role of Specialist and Trade Contractors." *The Chartered Institute of Building*, Ascot, England.

Hofstede, G, Culture's Consequences, Sage, London, 1984

Jamieson, M. J., Thorpe, A., and Tyler, A. (1996), "Refocusing Collaboration Technologies in the Construction Value System." *Proceedings of the CIB W78 Conference, Construction on the Information Superhighway*, Bled, Slovenia, April, pp 29-289.

Latham, M. (1994), "Constructing the Team." *HMSO*, London.

Larson, E. (1995*), "Project Partnering: Results of study of 280 Construction Projects."Journal of Management in Engineering. Vol. 11, No. 2,* March / April, pp 30-35.

Maloney, W. F. (1997), *"Improvement Example From The 1990'S: Increased Use of Design-Build as a Project Delivery System."* Transfer of construction management best practice between different cultures. *CIB Proceedings Publication 205. W65 Organisation and Management of Construction.* pp. 183- 187

Matthews, J (1996a), "A Project Partnering Approach to the Main Contractor - Sub Contractor relationship." *Unpublished Ph.D. Thesis,* Loughborough University.

Matthews, J., Tyler, A. (1996b), "What main contractors want from their subcontractors." North meets South-Developing Ideas. *CIB Proceedings Publication. CIB W92 Symposium on Procurement Systems, Natal, South Africa.* Vol. 2, pp. 343 – 352.

Reading Construction Forum (RCF) (1995), "Trusting the Team: The Best Practice Guide to Partnering in Construction." *Center for Strategic Studies in Construction*, Reading, England.

Reading Construction Forum (1998), "The Seven Pillars of Partnering." *Partnering Task Force of the Reading Construction Forum,* Reading England.Tij huis, W. (1998),

"Procurement and sustainability in construction industry: Tendering got durable relationships with process and projects", ", *CIB 98 Proceedings on Construction and the Environment,* Gavle, Sweden, June 1998.

Walker, DHT & Rowlinson, S., **"**Use Of World Wide Web Technologies and Procurement Process Implications" in Rowlinson S & McDermott P (eds), Procurement Systems - A guide to Best Practice, *SPONS,* London, at press, 1998

Walker, DHT & Lloyd-Walker, BM, "Organisational Learning as a Vehicle for Improved Building Procurement" in Rowlinson S & McDermott P (eds), Procurement Systems - A guide to Best Practice, *SPONS,* London, at press, 1998

Walter, M. (1998), *"The Essential Accessory." Construction Manager (CIOB),* Feb., Vol. 4, No. 1. pp.16-17.

Weston, D. C., Gibson, G. E. (1993), "Partnering - Project Performance in the U.S. Army Corps of Engineers." *Journal of Management in Engineering, Vol. 9, No. 4,* pp.410-425.

Westwood, RI (ed), (1992) Organisational Behaviour-South East Asian Perspectives, Longman Far East

Wilson, R. A., Songer, A. D., Diekmann, J. (1995), "Partnering: More than a Workshop, a Catalyst for Change." *Journal of Management in Engineering, Vol. 11, No. 5,* pp.40 – 45.

PROCUREMENT AND DELIVERY METHODS RESEARCH PROGRAM

TIINA TANNINEN-AHONEN
Confederation of Finnish Construction Industries (RTK), Unioninkatu 14, 00130 Helsinki, Finland. tiina.tanninen@rtk.fi

Abstract

Confederation of Finnish Construction Industries (RTK) has managed the research program " Procurement and delivery methods" since 1995. The program has had a dozen research and pilot projects with a total budget of about 3$ million. This paper will describe the goals and results.

The goal of the research program is to promote new procurement and delivery methods for building and to promote their introduction. The program has developed the procedures between both clients and contractors, and subcontractors. Introduction of the most interesting procurement methods has been promoted through pilot and demonstration projects. These have been implemented by industry in cooperation with researchers.

The research program consists of the following parts:

1. Criteria to select procurement methods
2. New procurement and delivery methods.
3. Advanced procurement and development of cooperation between main contractor and subcontractor

To form criteria to select procurement methods international and domestic knowledge about new procurement and delivery methods has been studied and described in two separate study.

Procurement and delivery methods where design and building phases are more integrated; like design-build and bridging has been studied and tested in implementing pilot projects. Also the rules and document models for design-build-competition in Finland has been developed. In addition the competitive tendering on technical solution has been developed and tested in co-operation between client, designers, engineers and competitors.

The contractor's procurements constitute 50-90% of the value of the total contract. Procurement as part of production management has been described. A new contract model to manage subcontractors has been developed. Even more important has been the development of cooperation between main contractor and subcontractor (partnering).

Keywords: Advanced procurement, cooperation between contractors, design-build, new procurement methods

Introduction

Confederation of Finnish Construction Industries and Technology Development Centre have developed the building process in their "Procurement and Delivery Methods" research program launched in spring 1995. The goal of the research program is to promote new procurement and delivery methods for building and to promote their introduction. The basis, goals and means of the research program are showed in figure 1.

More than ten development projects had been implemented under the program by the beginning of 1998 of a total value of about 3$ million. Said R&D projects have examined the criteria for the selection of the procurement methods of building projects and developed new methods and rules for applying them. The forms of cooperation between the general contractor and subcontractors have also been developed and tested.

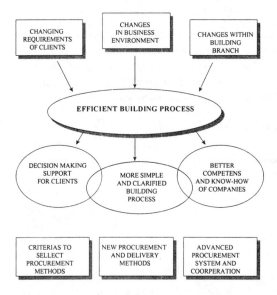

Figure 1 The basis, goals and means of the research program

This presentation combines the key results and conclusions drawn from the R&D projects implemented so far under the program.

Criteria for selecting procurement methods

There are various procurement methods available for building projects. They differ from one another mainly in how the liability for design and procurements is assigned, how tenderers are selected, and what is the contract pricing system. The most crucial factor is who assumes responsibility for project design: the owner or the main implementer (Nykanen, 1997; Pernu,1998). Procurement methods where the main implementer is liable for both design and implementation are gaining ground for several reasons.

The selection of a procurement method has a significant impact on the tasks and scopes of liability of the various parties. Thus, it is important for the owner to be able to distinguish between alternatives and make an informed selection. A procurement method unsuited for the project and owner and unclear scopes of liabilities easily lead to problems that are manifested in quality, upset schedules, extra costs and unnecessary disagreements (Nykanen, 1997).

In the case of traditional procurement methods, the owner assumes responsibility for supervision of design through in-house or hired project management. In such instance, the contractors and suppliers have little chance of influencing design as they are brought in at such a late stage that the key design decisions have already been made. Responsibility for quality is divided vaguely between designers and implementers. Also, the incompleteness and defectiveness of designs in the tender phase has caused practical problems with, for instance, lump-sum contracts.

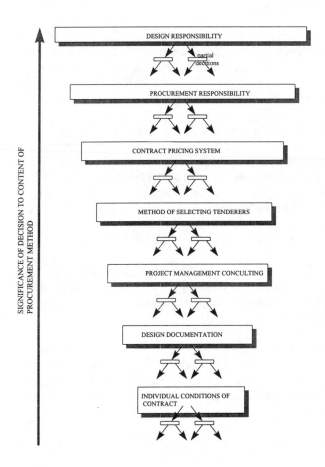

Figure 2 Clients' decisions concerning building project paths (Pernu, 1998).

In the case of separate contracts the owner may employ tens of parties to the project. Yet, the owner bears much of the responsibility for quality and the process.

In construction management and management contracting projects it is in principle possible to have fruitful cooperation between design and production. The owner bears the cost risk. The problem is that projects are often divided into too may deliveries and contracts are entered into at such a late stage.

In design-build and turnkey projects the implementer is liable for design. It is the owner's responsibility to present design criteria and to approve design solutions. As designers and contractors are able to work side by side in the project possible quality, scheduling and cost problems can also be prevented.

Procurement methods involving design

The owner may distribute design liability in various ways. If he decides to assign design responsibility to the contractor, various contract forms that include design are available. On the other hand, design responsibility may also be divided so that the owner is responsible for supervision of general designs while the contractor is liable for the selection of technical solutions and production control.

Rules and documents of alternative design-build methods

A working group representing the various parties to construction has developed rules for Finnish design-build competition which take into account both the concerns of the competition organizers, designers and tenderers. A publication that came out in December 1997 presents three competitive tendering methods incorporating three parallel levels of design responsibility. Methods that emphasize high quality, greatest value for money and lowest price can be selected on the basis of the goals set for the building project.

Competition on quality is suitable for projects that are architecturally sufficiently demanding, value for money comparisons suit regular projects and price competition is fitting for uncomplicated projects. Functional requirements should always be expressed clearly as stated in the study (Pernu et al., 1998).

In design-build competition the solution suggested by the tender group for the owner's spatial needs is central. The tender includes a fixed design solution, price, and quality-level of structural elements. In design-build competition the pre-selection of tender groups is an important phase which ensures that the most competitive participants, as regards personnel resources and technical and financial aspects, are selected.

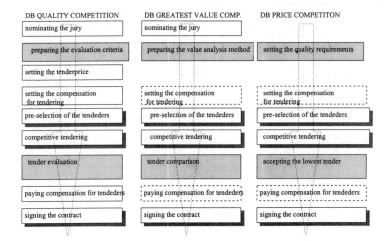

DB QUALITY COMPETITION	DB GREATEST VALUE COMP.	DB PRICE COMPETITON
nominating the jury	nominating the jury	
preparing the evaluation criteria	preparing the value analysis method	setting the quality requirements
setting the tenderprice		
setting the compensation for tendering	setting the compensation for tendering	setting the compensation for tendering
pre-selection of the tendeders	pre-selection of the tendeders	pre-selection of the tendeders
competitive tendering	competitive tendering	competitive tendering
tender evaluation	tender comparison	accepting the lowest tender
paying compensation for tendeders	paying compensation for tendeders	paying compensation for tendeders
signing the contract	signing the contract	signing the contract

Figure 3 Phases of the quality, the greatest value for money and the price design-build competitions.

When responsibility for design is part of the contract, the decision making of the organizers of the competition focuses on the programming phase where the majority of project costs are established. The design brief to be attached to the invitation to tender can be characterized as a detailed project program. In general, the success of a design-build competition requires that the organizers are highly competent professionals.

The aim of the optional and clearly organized standard forms of invitations to tender is flexibility and ease of use. The competition program spells out all the essential rules applicable during the competition which are of no significance after the contract is signed.

The contract program includes the conditions of contract as usual. The contract program of the report only dealt with the conditions of contract characteristic of design-build methods—other supplementary conditions must be compiled from the general conditions of contract for the sector.

The examples of design briefs provide one model for organizing—and a space-based mode of presenting—design requirements, goals and information. The goal has been to present the necessary data of an invitation to tender only once.

Good experiences from competitive tendering on technical solutions

Competitive tendering on technical solutions falls somewhere between traditional competitive tendering and design-build competition. It has been tried out on a housing project of the city of Helsinki. The research/pilot project involved the owner, the architect and four contractors. Each contractor offered a different frame solution. It was possible to submit a tender on equal terms independent of the frame structure (Pernu, 1998).

The owner was responsible for the general design of the project, i.e. functional and visual design, whereas the design of the frame was assigned to the contractor. Quality requirements were set according to the combined building, HVAC and electrical works specification which described only project-specific requirements without committing to any structural element solutions as to the frame.

The firm that submitted the lowest tender won the competition. The deviation in tender prices was only 7 percent.

The competition proved that different technical solutions are mutually competitive and that competitive tendering based on continuous product development by a company that produces alternatives, has wider application possibilities. The competition can easily be extended to, for instance, the mode of heat distribution.

Advanced procurement system and closer cooperation

The development of procurement and cooperation forms is important for construction companies since contractors' procurements account for 60-90 percent of their turnover and subcontracted works constituted over 50 percent of total works in 1996 (Sarkilahti, 1996).

Planning of procurement as part of the production-control system

In the 1980s, construction companies developed actively their quality, accounting and production-control systems; after the mid-1990s determined development of procurement was in turn. One of the results of the latter activity is the presented modern way of planning and supervising the procurement by construction companies. The key changes compared to the old practice are the planning of procurement at the tender calculation phase and the emphasis on the planning of individual tasks (Sarkilahti and Toikkanen, 1997).

The planning of tender phase procurement is based on the company's procurement policy and the basic-production solution devised for a project. The planning of implementation-phase procurement leads to the project's procurement plan comprising a list of purchases for delivery, a procurement schedule, procurement-related liabilities and a logistics plan. The key function of the procurement plan from the viewpoint of project management is to form procurement "packages", i.e., supplier-specific subcontract and material procurement tasks. The planning of individual tasks ensures implementation on schedule (Sarkilahti et al., 1997).

A subcontract is highly significant as a tool of production control. The developed new subcontract form is a standard document to be filled out—it is submitted precompleted in connection with an invitation to tender. The form is complemented by appendices that are generated on the basis of task planning schedule (Sarkilahti et al., 1997).

Towards long-term relationships in subcontracting

The key subcontractors are also to be involved in the development of project management. It is essential that common focuses of development are found and that two-way feedback on the cooperation occurs: a system has been developed for that purpose. Effective project-specific cooperation is a good starting point for sustained collaboration in subcontracting.

The efficacy of subcontracting can be evaluated by various indicators developed and tested in actual projects. The developed joint operation procedure enables stronger development of the activity. Even if tight project-specific competition is not organized, a better end result and lower costs are attained over the long term through improved cooperation (Romppainen et. al., 1998).

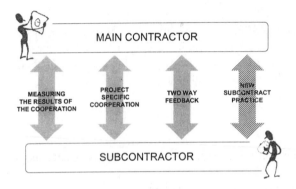

Figure 4 Means towards long-term relationship (Romppainen et. al., 1998).

The cooperation between material suppliers and construction companies has also been developed in the early 1990s through the "Construction Logistics" research program supervised by RTK (Wegelius-Lehtonen et. al., 1996).

The research program continues

The emphases and developmental goals selected at the beginning of "The Procurement and Delivery Methods" research program have been proven right, but the reaching of set goals requires heavier than planned investment and more comprehensive cooperation between the parties to construction. Consequently, Technology Development Centre (TEKES) launched the national "ProBuild—Progressive Building Process" technology program in spring 1997 which focuses on four specific areas:

1. Definition of owners' needs and requirement
2. Procurement of development, design and construction services
3. Implementation of design and construction
4. Development of the quality and flexibility of buildings and construction

The program entity called "Procurement and Delivery Methods" supervised by RTK extends until the year 2001. The future main themes are:

• organizing of projects specifically from the viewpoint of design responsibility to allow natural cooperation within teams
• grounds for selecting implementers with different procurement methods
• strengthening of contractors' capabilities with respect to procurement methods involving design
• long-term collaboration procedures and related rules

Pilot projects are aimed at promoting use of performance criteria in construction projects and the PPP procedure (Public-Private Partnership) also in building construction. In order to increase Finnish construction and construction products exports, the readiness to apply the general procurement methods of other countries will be improved.

References

Nykanen, V (1997) Toteutusmuodot rakennushankkeissa. RTK/Kehitys & Tuottavuus -sarja nro 47.

Pernu, P. (1998) Talonrakennuksen hankemuotojen kuvaus, TKK

Pernu, P., Markku Arimo, Seppo Eurasto, Pekka Helin, Juhani Kiiras, Heikki Miettinen, Erkki Raimovaara ja Kauko Wasenius (1998) Suunnittelu- ja toteutusmenetelmien pelisääntöjä. RTK/Kehitys & Tuottavuus -sarja nro 52.

Pernu, P (1998) Teknisten ratkaisujen urakkakilpailu , RTK:n Kehitys & Tuottavuus -sarja nro 54.

Sarkilahti, T. (1996) Rakennushankkeen alihankinnat 1996. RTK/Kehitys & Tuottavuus -sarja nro 35.

Sarkilahti, T. and Toikkanen, S (1997) Hankintojen suunnittelu ja valvonta. RTK/Kehitys & Tuottavuus -sarja nro 41.

Sarkilahti, T., Toikkanen, S. and Kankainen, J. (1997): Työmaan aliurakkasopimusmenettely, RTK/Kehitys & Tuottavuus -sarja nro 42 A ja B

Romppainen, M., Pahkala, S. and Wegelius-Lehtonen, T. (1998) Alihankintayhteistyö rakennushankkeissa. RTK:n Kehitys & Tuottavuus -sarja nro 53.

Wegelius-Lehtonen, T., Pahkala, S., Nyman, H., Vuolio, H. and Tanskanen, K. (1996) Opas rakentamisen logistiikkaan. RTK/Kehitys & Tuottavuus -sarja nro 38.

PROCUREMENT SYSTEMS' SHARES AND TRENDS IN FINLAND

P.J. LAHDENPERÄ, K.K. SULANKIVI and V.E. NYKÄNEN
VTT Building Technology, Technical Research Center of Finland, P.O. Box 1802, FIN-33101 Tampere, Finland. Pertti.Lahdenpera@vtt.fi

Abstract

The project procurement system is a key means by which the owner creates preconditions for successful realization of a project. Thus, it is of interest to discover the different systems used in different instances.

This paper aims to increase general knowledge in this area by introducing the results of a statistical analysis of data including over eighteen thousand projects that have been implemented in Finland during the last nine years. Thus, a very significant share of the nation's new-building construction volume is covered.

The paper presents the changes in the use of different procurement systems for various building types. The paper also aims to clarify the situation by shedding light on the underlying key factors and, especially, distinct trends.

Keywords: Building types, Finland, procurement systems, shares, statistics, trends

Introduction

The procurement method is a key factor in enabling successful implementation of a building project. The right method may help avoiding problems and be the key to the attainment of project-specific special goals. These goals may include, for instance, quick project completion, low acquisition price, certain distribution of risk between the parties and the owner's desire to affect the details of the design solution and the amount of his own work, etc. Different procurement methods are typically used for different projects.

Several selection methods have been developed internationally to facilitate decision-making by the client with respect to certain goals and boundary conditions. Corresponding numerical or similar systems have not been developed in Finland, but domestic literature does also introduce different procurement methods, favourable conditions for their use, their advantages and disadvantages as well as clients' motives in decision-making (e.g. Nykänen, 1997; Pekkanen, 1998).

The actual decisions of clients and the way in which various procurement methods are used in different situations has, however, received less attention. Even a good theory is worthless as long as it is foreign to practice. For this reason, this review aims to shed light on the share of different procurement methods used in Finland in recent years in different types of construction projects.

The analysis is based on quite broad empirical material that covers a significant share of Finnish new building construction over the nine-year period 1989 - 1997. A total of 18,135 projects were included—a similar review has not been done previously.

Used procurement methods

The implementation of a building project involves the co-operation of many parties: the owner, various designers, contractors and suppliers. There are numerous procurement methods for establishing the division of labour between the parties, contractual relations, and the rules of the game. In this study the breakdown is the following:

- Management-type procurement (MT), where a separate project management organization supervises the project and implementation is realized through numerous partial contracts (construction management, etc.).
- In-house construction (IH), where the client designs or commissions the design, supervises site works and possibly performs part of the technical construction work (e.g. municipalities, speculative building by contractors).
- Design-build (DB), where a contractor under contract to the client is responsible for the project's design and implementation (e.g. turnkey projects, negotiated projects with a contractor's lot, competitive design-build).
- Comprehensive contract (CC), where client assumes responsibility for design and the project is implemented on the basis of a single contract (i.e. traditional design-bid-build method); no subsidiary contracts are use.
- Separate contracts (SC), where the client assumes responsibility for design, and construction divides between a minimum of two, but generally more parallel contracts (i.e. traditional-like design-bid-build with subsidiary contracts).

The procurement methods are only presented on the roughest level. The classification is based on, for instance, the research material, which is dealt with in more detail in the next chapter.

Research material

Initial data of research

The analysis is based on a continuously maintained project database (Rakennusfakta, 1998) and cross-sections of its data representing different time periods. The database is maintained by a private company and project data is generated mainly to serve the direct marketing needs of various construction sector firms. The database contains data on, for instance, the parties and their contact information, building types, key structural solution, schedule, scope and location and, naturally, applied procurement method.

Building-project data is acquired from owners. The starting-point is data on building projects disseminated through various media and continuous contacts with a significant portion of the clientele. The material is not restricted to a certain sub-area of building, but the aim is to cover the entire field. Another goal is to achieve national coverage, which appears to have been realized to a large extent.

Limitation of material

The projects for the procurement-method analysis were selected on the basis of construction launch time. The calculations are thus based on projects started during the review period and no advance information has been used. Due to the timing of the material's annual updates, the review period ends at the end of September of each year; therefore the data is not based on the calendar year, but, for instance, 1997 projects consist of those whose construction was launched between October 1996 and September 1997.

The study is confined to Finnish new building construction. Moreover, all single-family houses (also as a group), vacation homes and small agricultural buildings have been excluded from the study. Projects amounting to less than MFIM 1 (The abbreviation MFIM in the paper refers to one million Finnish marks [FIM]. One million Finnish marks is equal to USD 170,237 or about XEU 193,517 based on the average 1997 exchange rates) were also eliminated so as to better be able to describe the actual professional building practice.

Coverage and reliability of material

An attempt has been made to assess the coverage of the material by comparing the number of projects in the database assigned to 1997 (launched between Oct 96 and Sep 97) to the number of starts of official statistics [Statistics Finland (1997)] while considering the above limitations. The comparison was done prior to eliminating the under-FIM 1 million projects and the results were as follows:

- In housing construction the coverage is up to 64% when comparing apartments and row-houses.
- In the case of industrial buildings (incl. warehouses) and business premises (commercial and office buildings) the figures were 13% and 12%, respectively, whereas the group "other buildings" had 17% coverage.

All in all, the sample covers about 30% of the starts in said building types. The projects included in the material are roughly in line with the construction volumes of various years, so that coverage would seem to be at least as high in other years.

The procurement-method descriptions of the material correspond in general to the above-established division. However, management-type procurement, which was not an existing alternative in the early years, presented another problem.

Use of various procurement methods

Bases of calculation

This chapter reviews the use of various procurement methods in 1989 - 1997. The results will be presented separately for the three most important building types by volume—housing, industrial buildings and business premises. In addition, a summation for all other building types will be presented.

In building type-specific charts (Figures 1 - 4) the number of projects implemented by various procurement methods is the basis for computing shares. The corresponding cost share-based computation is also explained.

When examining time series, we should remember also the changes in economic conditions, which obviously have had an impact on the procurement methods used. The record new-building construction volumes of 1989 - 90 resulting from the economic boom at the end of the 1980s were followed by a sharp decline of over 50% within three years. Only in 1996 - 97 have volumes started to grow again.

Housing

Housing construction, which represents the largest share of the total building construction volume, will be presented first (Figure1):

- In-house construction consist mainly of the development activity of construction companies, where the number of projects diminished significantly as the recession hit at the turn of the '90s.
- The share of traditional comprehensive contracting was emphasized during the recession—it accounted for about half of all housing construction. Since, its share has continued to decline. As production volumes fell, the relative share of state subsidized housing construction grew which is likely also underlie this phenomenon as a result of the need to prove tough competition.
- The share of design-build has grown steadily. Methods that combine design and construction (in-house construction & design-build) account presently for half of all production, the rest deriving from traditional production capacity-based competition.
- Management-type procurement methods are hardly used in housing construction, although a few exceptions can be found in the material (not sufficient to show in figure).
- In calculations based on monetary value the share of design-build is much larger, and that of in-house construction correspondingly smaller, than in the calculations of the figure which are based on project numbers. Thus, the former are also larger, and the latter smaller, than average projects.

Industrial buildings

The following can be said of the procurement methods for industrial buildings (Figure 2):

- The traditional separate contracts method has been by far the most popular in industrial building. Its has held a constant share of at least 50% of the total volume.
- In-house construction has lost share to separate contracts during the review period. The owners of this type of building are industrial firms and the like, and here in-house construction actually means implementation by them—not the development activity of construction companies.
- Management-type procurement also occurs, but its share is very small also in the industrial sector. The projects are larger than average, i.e., a review based on monetary value gives this sector a slightly larger share than an analysis based on numbers.

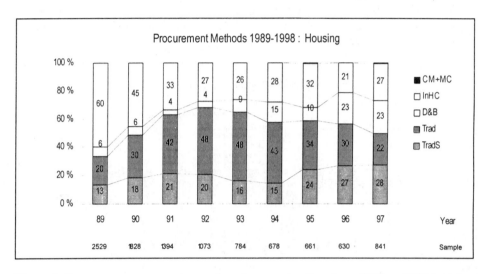

Figure 1. Procurement systems in new housing construction from 1989 to 1997 based on the number of projects.

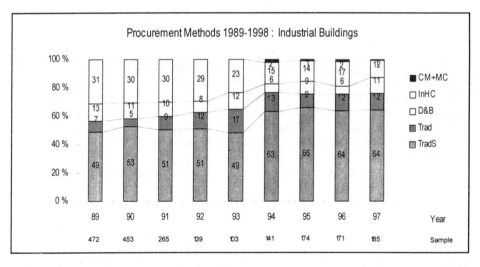

Figure 2. Procurement systems in new industrial construction from 1989 to 1997 based on the number of projects.

- The use of other procurement methods seems reasonably stable and relatively scarce.

Business premises

The trends in business-premise building are as follows (Figure. 3):

- The share of in-house construction has decreased essentially in the review period, but unlike in housing construction, the decline in business premises started only when the recession was well underway, not immediately as the country was caught in it. In this sector, in-house implementation involves both development activity by construction companies and projects of other companies.

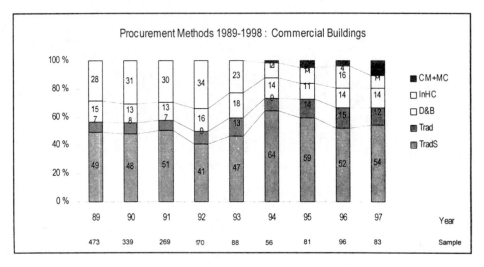

Figure 3. Procurement systems in new business premises construction from 1989 to 1997 based on the number of projects.

- The traditional separate contracts is the single most used procurement method for business premises. Its share has continuously been about half of the total volume.
- The most significant area of application of management-type procurement would appear to be the construction of business premises where it has won a moderate share in recent years. Measured by monetary value, the share of this procurement method has been manifold in some years compared to the numbers of the figure—the projects are significantly larger than average.
- The use of other procurement methods seems relatively stable.

Other buildings

The group "other buildings" covers various public buildings (e.g. schools, churches, old-age homes) and transport and communications buildings (post offices, parking garages, railway stations, etc.). Changes there have been quite small in the review period (Figure 4):

- The traditional separate contracts procurement method has been used the most also in this group. Its share has been growing slightly especially when measured by monetary value.
- Other procurement methods hold a smallish, but stable share, although the share of the traditional comprehensive contract, especially on the basis of the value of the production, has been slightly larger in the middle of the review period than in other years.

Concluding comments

The paper presented the use of various procurement methods in Finland in the last nine years. The analysis is based on quite extensive material and the depictions of trends, especially, can be considered reliable. However, no material on management-type implementation in the early years exists.

As to the results themselves, it can be stated that the comprehensive contract was used most in the middle of the review period while Finnish construction was doing especially poorly. It is obvious that competition solely on price was emphasized then. As the economy has revived, the shares of design-build and management-type procurement have increased possibly due to the faster implementation and lot acquisition, etc. that they make

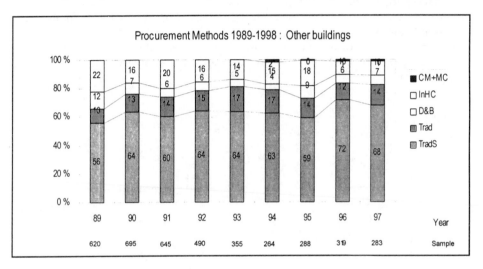

Figure 4. Procurement systems in other new construction than housing, industrial and business premises, from 1989 to 1997 based on the number of projects.

possible. Thus, the comprehensive contracts method would seem to be the choice especially in a recession—although only indirectly as a result of the market structure.

The share of in-house construction has decreased sharply in the review period. The most important reason is that development by construction companies has dropped significantly, especially in housing, but also in business premises due to the recession early in the period. At that time builders accumulated abundant unsold space-which had fateful consequences for many-and understandably left them cautious. They are still wary even though the market situation is now much more favourable than a few years ago. Today's markets appear to be based on "direct demand", especially as concerns business premise construction-there is hardly any speculative activity in the sector.

At the moment, the shares of design-build and separate contracts seem to be increasing. The focus of public debate and development work on design-build methods is at least partly responsible. Management-type implementation is also increasing its share, which is nevertheless minor measured by the number of projects. The clearest loser appears to be in-house construction, although the traditional comprehensive contract is also headed in the same direction.

References

Nykänen, V. (1997) Toteutusmuodot rakennushankkeissa [Procurement systems in construction]. Rakennusteollisuuden keskusliitto RTK. Helsinki. Kehitys & tuottavuus 47. 67 p. (in Finnish)

Pekkanen, J. (1998) Rakennuttamis-, suunnittelu- ja rakentamispalveluiden hankkiminen [Procurement of Consulting, Design and Construction Services]. Helsingin teknillinen korkeakoulu. Rakentamistalouden laboratorion raportit 163. 111 p. + app. 34. (in Finnish)

Rakennusfakta (1998) Rakennusalan projektitiedosto [Construction sector project data bank]. Rakennusalan projektitiedosto. Espoo. (updatable data bank)

Tilastokeskus (1998) Rakentaminen ja asuminen. Vuosikirja 1998. / Construction and housing. Yearbook 1998. Tilastokeskus / Statistics Finland. Helsinki. (in Finnish/in English)

PUBLIC PROCUREMENT OF PROFESSIONAL SERVICES IN CONSTRUCTION IN ITALY. THE WAY FORWARD.

N. COLELLA
Dipartimento di Processi e Metodi della Produzione Edilizia, Via S. Niccolo 89/a, 50135 Firenze, Italy, n.colella@extero.it

Abstract

The domestic market of professional services in construction is changing rapidly. As compulsory competitive selection of consultants replaced negotiated appointment, a concerning weakness, hidden for decades, emerged from the whole sector. The EC directive pushing toward transparent selection procedures and open market opportunities unveiled the inadequacy of the market to cope with the quality challenges of modern construction industry. While on the demand side additional efforts are asked in order to adequately manage the selection/award process, architects, engineers and design firms must align their organisational structures with higher international standards. The threats of a global market and the need to assure value for money pushed both the demand and the supply side to a radical re-thinking of their role and management methods.

The paper critically reviews the meaning of "value" of design services in the light of legislative and market transformations. While an overview of key changing factors is given, a proposal for a best practice guide to the procurement of professional services in construction is outlined.

The work is part of a three year PhD research.

Keywords: Best practice, design services market, EC legislative harmonisation, procurement of design services, value for money,.

Design services market in Italy, the need for change

Until 1995, when Italian law implemented the Services Council Directive 92/50/EEC, appointment of design consultants was based on the juridical principle of *intuitu personae* i.e. a selection based on the confidence the client bestowed on the professional. Municipalities (comuni), regional governments (regioni), central departments and the government itself procured professional consultancy in both design and engineering fields using negotiated procedures.

Traditionally, the public client is not a demanding one. Public institutions do not push design consultants to innovate, to challenge traditional construction methods, to improve the efficiency of the design process or to cut cost while improving quality. The public is also a political client looking for electoral support when commissioning business to consultants and firms. As a consequence, design services have always suffered from high favouritism, poor quality requirement, and scarce attitude to innovation.

Competitive selection became compulsory when a nation-wide corruption scandal was discovered and also right in the middle of a profound legislative renovation, in the field of public procurement. Soon demand and supply sides understood the inconsistency of the whole appointment system. Public clients faced the emergency determined by the lack of the basic technical skills to define needs, to draw up technical documents, to prepare project briefs and to select consultants. Architects, engineers and design firms realise their lack of organisational and management skills, especially if compared to foreign companies. Their financial standing was poor, the average size of professional organisations was very small, the efficiency of their operative and decisional processes was totally inappropriate to assure quality standards to more demanding clients.

In 1995 when the whole sector had to converge toward the new regulatory framework and both demand and supply sides had to compete in a larger market, it became clear that the global market was not imaginary but a real threat.

Demand side today and the role of small tendering authorities

One of the biggest problems in the Italian public construction sector is the fragmentation of the demand market. In Italy there are a striking number of public agencies. Central and local administrative bodies, state-owned companies, ministries and government departments, the army, universities, local health agencies and so on, amount to several thousands tendering authorities. To date, *Comuni*, the smallest local administrative bodies procure more than 70 % of the whole demand of public design services (see figure 1). Other weighty tendering authorities are *Provincie* and *Regioni*, the central government and the national health care local authorities.

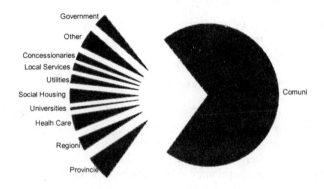

Figure 1. Design and engineering services market share per institutions (tenders in year 1997)

To date *comuni,* especially small and very small ones are at the very forefront of the boom of the design and professional consultancy market.

Decades of slack procurement policy and unrestrained provision of public financial resources caused tendering authorities to suffer from inadequate organisational structures [EC Commission (1997)] and ineffective managing methods. In-house technical resources and skills are often insufficient to execute design activities, to properly identify needs, to outline requirements and to efficiently manage the selection and contract award phases. While the opposite is recurrent in many other large private and public clients, as a result of conscious outsourcing strategies, this sounds illogical in Italy if one considers that the government resolutely requires design services to be delivered by the technical staff of the public tendering authority. Law n.109 of 11/2/1994, art.17 in fact states: "Preliminary, final and detailed design shall be done, with foremost priority, by internal technical staff of the tendering authority".

Evidence suggests that while promising outgrowths take place in the regulation framework, many public client organisations, mostly city and small town administrations (comuni), still suffer from cultural and organisational frictions. This causes an irremediable loss of the advantages in terms of value for money. Among all functions performed by *comuni*, procurement is especially impacted from this evolving scenario.

Supply side: different suppliers same market, same rules

Today's supply side is divided between two main groups: engineering firms and professional practice (engineers and architects).

Engineering and technical-economic consulting organisations, which are usually referred to as "engineering companies", are complex organisations providing technical services related to investments in industrial installations, infrastructures and engineering works.

Since 1950's, these organisations have considerably grown due to major investment projects in developing countries in the iron and steel industry, motorway construction and petrochemical and hydroelectric sectors. Such organisations have specialised in providing turnkey engineering installations, carrying the contractual risk of the projects. Thus, under Italian law, engineering companies are *not* considered as professional firms. With the decline of large foreign projects, engineering companies turned back to the domestic market. Now they share the public contract market with smaller design firms even if their financial standing and average size is much bigger.

In most cases professional design practices include no more than one architect/engineer with a couple of employees, a secretary and a CAD drafter. A recent survey (Censis, 1998) on architect and civil engineering professional practices showed that 43.9% of chartered engineers perform their professional practice as individuals (*libera professione*). This percentage rises to 68.8% in the case of architects. Only one graduate in architecture out of four is employed in a design firm.

The reasons for such a small and fragmented supply scenario are rooted both in a 1939 law, proscribing the creation of professional firms, and in a tightly conservative policy implemented by professional associations. Consequences have seriously hindered the competitiveness of the whole professional consultancy sector. Architects and engineers have never been challenged to efficiently implement and display to potential clients their design management capabilities, simply because a real market never existed.

The design itself suffered from this distorted scenario. Architects and engineers have adopted a generalist approach to the practice spanning from urban to industrial design, because they understood that narrowing their market niche was a "business suicide" with such an unqualified demand. University and professional associations also vigorously preserved this status from change.

To date, neither engineering companies nor professional practice adequately satisfies the need of a growing professional service market. Both groups show too much opposition to change, because their concern about loosing achieved economic benefits far exceed the need to efficiently turn into mature providers of design services.

Today's design market scene

Competitive selection of design consultants is actually booming. Tables below (Osservatorio OICE, 1998) give a bright idea of the marked increase since 1995.

Table 1 Public tenders issued monthly (average)

Contract value		1995	1996	1997	Jan, Feb 1998
Below threshold	EC	60	329	539	460
Above threshold	EC	8	15	28	24
Total monthly		68	344	567	484

Table 2 Yearly public tender per project class

Service category	1995	1996	1997
Housing	156	1133	1776
Healthcare	47	92	151
Education	20	210	375
Civil engineering	176	1360	2227
Technical consultancy	108	524	1112
Architectural contests	48	100	196
Urban planning	50	200	256
Industrial plants design	51	509	705
Total	656	4113	6798

Tenders advertised in 1995 when the new regulation was adopted is tenfold increased in 1997 amounting to a value of nearly 1000 billion Liras. In March 1998 public contracting activity has grown at an average monthly rate of 8,5% while total contracts value increased of 21,5%. Recent survey states that average contract value will grow regularly. However, statistical data are less amazing if compared to other European countries (see Figure 2).

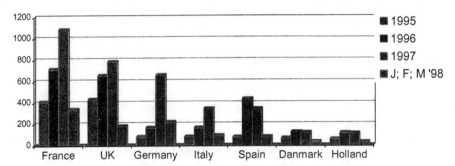

Figure 2 Design and engineering services contractual activity in Europe

Actually, Italy shows a moderate tendering activity. Tenders exceeding the EC threshold are more in France, UK and Germany while Spain slightly exceed the national contractual activity.

Despite statistical evidences, professionals and consulting engineering companies recurrently complaint about:

- limited advertising of contractual opportunities;
- scarcity of big sized contracts;
- narrow technical consultancy contracts, far exceeding complete design services;
- negative variance between the yearly planned expenditure and the actual contractual activity;
- little percentage of awarded contracts (78% of tenders).

Changing factors on demand side

Starting from 1992, the joined effect of internal changing factors (corruption scandal called *bribesville* and the Maastricht parameters convergence) together with the EC legislative harmonisation process, brought substantial changes both to the role of the public and to the ways services are procured and delivered today. The principles underpinning this profound transformation of the public sector are [Technical Secretary of the Ministry of Public Works (1997)]:

- "the State shall directly provide facilities and services *only if* the public intervention is proved to be more fitting than the private one";
- public facilities and services shall be provided to citizens by the closest public entity i.e. *comuni*, while Regions and Central Government shall only act at planning, co-ordinating and control levels.

The first principle accounts for the straightforward privatisation policy, the second inspires a determined decentralisation process to improve local financial accountability.

While the long term objective is the general improvement of the country competitiveness, in the short period the government wants to achieve a substantial change of distorted "modelli di spesa" (founding and procurement policies and procedures). The government resolutely wants chaotic public founds allocation, inconsistent development policies, and frequent use of emergency procedures, to come to an end.

A solid regulative framework has been shaped to foster the change process, notably:

1. A new law regulating public works contracts planning and awarding procedures. The law on public works (Legge quadro sugli appalti pubblici, n°109/94 and following modifications), in its most recent release, regulates both the programming and procurement activities of central government, public bodies and local authorities. It is generally considered the law underpinning the renovation of the whole public procurement market. The Law, called "Merloni" after his early drafter, promotes design as the milestones of a thoroughly restoration of the construction sector. The idea behind this is that if public administrations go out for tender with thoroughly detailed drawings, changes will be avoided and contractors constrained to achieve agreed results. Its complete promulgation (fall 1998) is expected to re-establish transparency and rigour in the field of public works contracting.

2. A complete set of laws leading local authorities toward a management approach inspired by the private model. In 1990 Law n. 142 set the basis for a complete renovation of *comuni* and *regioni*, based on efficiency and economy principles. The first article states: "The administrative activity [...] shall be inspired by economy, efficiency and suitable publicity criteria". Further on, some new operative tools have been shaped: firstly the establishment of the "responsabile del procedimento" (a kind of project sponsor); stakeholders involvement in the project planning; free access to administrative documentation; lastly the "conferenza dei servizi"(a public administrations roundtable). A consistent legislative body, following Law n. 142, has thoroughly defined minimum standards to achieve:

   - simplification of administrative procedures;
   - decentralisation of administrative functions;
   - re-engineering of public agencies core processes.

The thorough renovation of the regulatory scene, laid the foundations for a rapid bridging of the concerning gaps. *Comuni* and *regioni* are committed to lead the country to win the next millennium global challenges, providing quality services to satisfy their taxpayers needs and aspirations while fostering competitiveness of local private enterprises. They own the necessary operational tools and the political authority to become consistent with the philosophy and the needs of market economies.

Changing factors on supply side

Design practice and professional technical consulting in construction undergoes a concerning identity crisis. The whole sector from academic institutions to chartered organisations understands the failure of the entire service supply system, and its plain unsuitability to face the challenges of a market economy.

The rapid evolution imposed by the Services Council Directive 92/50/EEC severely impacted the design professional community. Its unyielding conservative policy proudly preserving the intellectual core nature of the design practice has been also harmed by the joined attack of EC Courts (Brivio, 1997), Antitrust (the Italian Competition Authority), the National Association of Economic Consulting Engineering Companies (OICE) and the government itself. In less than two years few milestone of the national design professional system, such as minimum fee threshold (defined by law), restricted access to professional registers, illegality of advertising campaign, and a widespread elusion of competition rules, have been thoroughly criticised. The Italian Competition Authority has launched the most uncompromising attack. In 1997, Antitrust, issued a fact-finding investigation on professional institutions (Antitrust, 1997), to assess presumed abuse of dominant position. This is, in short, the verdict of the controversial survey:

- design professionals shall be recognised as trade or commerce firms with no concern of their individual status as they achieve economic results;
- the "ordini professionali" (professional institutions) shall be re-defined so that public and private aims should be kept separate (Malatesta, 1996);
- minimum fee thresholds shall be unconditionally removed as no explicit connection has ever been shown with the final quality of the service;
- restricted access to the professional practice creates economic disadvantages to clients, mainly to those interested in "average" quality;
- advertising professional services ostensibly reduce the costs sustained by the client to survey the available market offer. It also gives a valuable opportunity to young professionals entering the market.

The international scenario also supports results of the fact-finding investigation. Large European countries, such as France, United Kingdom, Denmark, Norway and Sweden, show open professional practice. Countries with some kind of restrictions like Germany and Belgium, have neither limited access to the practice nor compelling professional registers. Concerning fees, similar restrictions have been found in Spain (art.4 Law n.7 of 14/4/1997, (1997)), Germany, Holland and France, but in all those countries fee value has only reference purposes.

Cultural evolution, a pre-requisite to take advantage of new market opportunities

Today's market mirrors the paradoxes of a transitory phase. In 1997 a survey from a leading research institution (Quasc, 1997), outlined main features of the newborn design market in Italy. According to the report, titled "design services: booming of an invisible market", the domestic market opening to transparency and equal opportunities go together with new dramatic issues. These prevent advantages of competitive selection from taking place. Compulsory competitive selection, according to authors, does not drive sound competition itself. Likely it suggests ways to elude regulations if not coupled with a comprehensive cultural transformation.

To date increased accountability of public officers has not been coupled with a through improvement of people's skills and attitudes yet. This causes a considerable distortion of the procurement process in terms of:

- tendering authorities focusing on accuracy of the procurement process rather than it's efficiency and effectiveness;
- mismatching criteria and information provision used within same service category procurement or, on the opposite, same criteria used within distant service categories;
- experience in similar projects far exceed other selection criteria;
- contracts total value split in order to keep it under the Community threshold;
- same opportunities given to state owned firms taking advantage of public funds;
- misunderstanding about the differences between ability of providers and quality/cost of services;
- boundaries between work and service contracts still blurred.

Under the facade of transparency and equal opportunities, groups' interests are vigorously defended. The real thing is that, disregarding community needs both demand and supply sides are striving to find the most suitable way to innovate *while* keeping their interest safeguarded. So while the controversy is confined to juridical topics, such as correctness of the procedure, evidence of pre-requisites satisfaction, tender documentation format, submission deadlines and so on, value for money in public procurement is irremediably lost.

Research objectives

In 1997 a PhD research developed within the Department of Processes and Methods of the Building Process, Faculty of Architecture, University of Florence, firstly approached the problem of the newborn design service market in Italy. At the end of the first year fact-finding survey it was clear that best results depend on research's ability to:

1. focus on the demand rather than supply side;
2. bring the debate to its inner disciplinary field i.e. the management of the design process;
3. focus on the early phases of the procurement process.

Preliminary investigations also revealed that Comuni still perform their procurement functions with little or no concern about:

- what happens before and after the contractual "event";
- consequences a defective procurement policy has both on taxpayers and consultants;
- value for money.

Thus, they act like mere *buyers* rather than *clients*. Evidences suggested that demand side mostly suffer from the new legislative background. Local public administrations such as *comuni* in recent years have been consistently empowered. The main part of the overall transformation process, relies on their ability to both satisfy community needs and foster the development of the local contractors. The study understood that this was an arduous mission if arranged with mere procedural "weapons". Most urgent needs were:

- a suitable organisational and management framework to fully exploit new available procurement tools, and
- tailored best practice tools to support the appointment of consultants in construction.

The research aimed at a comprehensive re-design of procurement function within the public administration. Intelligible connections with both the external environment i.e. the social, natural, economic factors, and the client organisation, were to be consciously set so that performing procurement activities become a *means* to achieve long term objective (Connaughton, 1994), not an end itself. Also, suitable tools to efficiently identify needs, draw requirements, select design consultant in construction were needed to significantly improve the judgement ability of the public administrations, preventing these from running risks of opportunistic or unskilled consultants.

Methodological approach

The overall methodological approach is outlined in Figure 3.

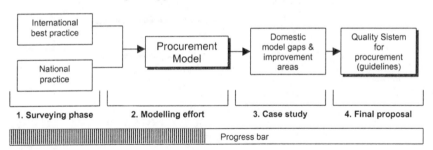

Figure 3 Methodological research flow

To date a preliminary model of the international procurement has been drafted. It outlines main elements of the procurement system with relevant links to the inner and outer environments. The model will act as the base for next phases while allowing a "map" of the regulatory framework to be identified and "overlaid". The map is used to distinguish "constraint free" areas i.e. zones of the domestic model where the regulatory pressure is slack and scopes for improvements are larger.

Case studies, starting presumably in fall 1999, allow the model to be applied on a "real" public organisation. Negotiations are progressing with the municipalities of Florence, Bologna and Naples. Improvement areas for use in case study have been identified in the fields of:

- Strategy definition: to help client organisations to set fitting procurement objectives and policies.

- Management of the procurement process: To provide public administration with procedural and operational tools to set requirements, select consultants and award contracts so that client's objective are successfully achieved.
- Client organisational structure: supporting clients to define roles, responsibility and information management so that requirements are satisfied with value for money.

Quality management: continuously improve the system.

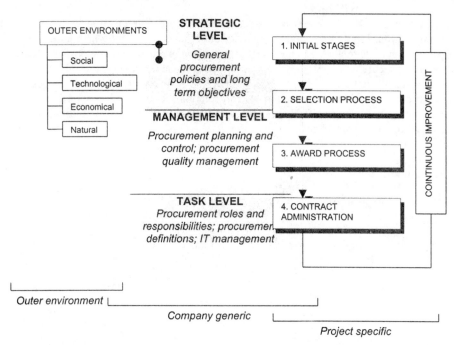

Figure 4. Draft model

Conclusions

Quality of design and competitiveness of the national professional design practice increasingly relies on the maturity of the public demand. Today small local administrations have been empowered to implement the European legislative harmonisation process while improving the cost-effectiveness of public expenditure and the quality of professional service providers. While developing sound procurement practices is compulsory to achieve fair and transparent tenders a major cultural change is to be achieved to enhance the competitiveness of national design supply system and people's quality of life.

Alongside new regulatory framework, local authorities shall be provided with consistent best practice to deliver value through their procurement activities. Public administration's technical staff needs cultural and operative tools to perform procurement tasks with a greater understanding of criteria underpinning the value of the service and community needs in terms of quality of the built environment.

References

EC Commission (1997) The competitiveness of the construction industry. *Communication from the Commission to the European Parliament,* Bruxelles, June 1997, pp.18-20

Censis (1998) 31° Report on work availability in industrial years 1997-98

Osservatorio OICE, Informatel (1988), Design service tenders updated march 1998. Rome, 1998

Technical secretary of the Public Works Ministry, (1997) Per restare in Europa: le infrastrutture fisiche tutte quelle che servono, solo quelle che servono, *Draft document,* Rome, 18/7/1997

Brivio, E (1997) La corte Ue abbatte le tariffe degli albi. *Il sole 24 Ore,* Milano, 19/6/1997, pg. 29

Antitrust (1997) Relazione dell'autorità garante della concorrenza e del mercato al termine dell'indagine su ordini e collegi professionali. Roma, October 1997

Malatesta, M (1996) Professioni e professionisti. *Storia d'Italia,* Torino, 1996

Art. 5 Law 14/4/1997 n. 7, De medidas liberalizadoras en materia de suelo y de Colegios profesionales. Madrid, 1997

Quasco, ANCPL, (1997) Progettazione; il mercato che non c'è. *Quaderno n.27* Bologna 1997, pp. 4-19

Connaughton J and Langdon Consultancy (1994) Value by competition. A guide to the competitive procurement of consultancy services for construction. Ciria, London, 1994, pp.14-18

PROCUREMENT AS A LEARNING PROCESS

ROBERT NEWCOMBE

Department of Construction Management and Engineering, The University of Reading, Whiteknights, P.O. Box 219, Reading, RG6 6AW, United Kingdom.
r.newcombe@reading.ac.uk

Abstract

The way in which people learn from experience and conduct experiments in self improvement has been analysed by many writers including Argyris and Schon (1978), Kolb (1984), Senge (1990) and Pedler (1991). In particular, Kolb's learning cycle and the single and double loop learning postulated by Argyris and Schon has been used to explain how learning takes place in organisations. Creating the 'learning organisation' is a current preoccupation of management theory and practice.

With few exceptions (Lowe and Skitmore, 1994; Ogunlana, 1991), little of this accumulated knowledge about learning in organisations has been applied within the project context and even less within the construction project situation.

This paper will argue that different procurement paths encourage or discourage different types of learning. For example, the traditional design-tender-build method based on competitive tendering does not allow original design concepts to be challenged by the builder and therefore prohibits double loop learning. Equally, learning from experience can only occur where there is opportunity to reflect on those experiences and compare them with mental models which is then followed by experimentation; design and build methods may allow this to occur as may some of the management methods of procurement. Partnering is clearly a vehicle for experiential learning.

It is vital that we understand the link between learning and procurement if we are ever to improve the efficiency and effectiveness of construction projects.

Keywords: Construction, learning, procurement, projects, organisational learning.

Introduction

I have argued in a previous paper (Newcombe, 1992) that the selection of a procurement path for a construction project is rather more than just the establishment of a contractual relationship or the allocation of roles. That paper postulated that different procurement paths not only created a unique set of social relationships (Tavistock 1965, Cherns and Bryant 1984) but also a power structure within a coalition of competing or co-operating interest groups. In that paper the Traditional design-tender-build approach was contrasted with the Construction Management (CM) path in terms of power distribution and use.

In this paper it is proposed that the procurement path adopted creates a 'learning arena' with opportunities to establish a 'learning organisation' or 'learning network'. I shall argue that some procurement paths provide better opportunities for learning than others and further, that some procurement approaches positively discourage learning.

In the construction industry the failure to learn is illustrated when the same mistakes are made on successive projects. The published reports about the UK construction industry in the last 40 years are largely repetitive in their criticism and comments on the structure and functioning of the industry [Emmerson (1962); Banwell (1964); Tavistock (1966); Wood (1976); NEDC (1978); British Property Federation (1983); Department of Industry (1982); NEDO (1983) (1985) (1987); Construction Management Forum (1991); Latham (1994)]. These reports reinforce the impression from well publicised examples of project failure that the participants involved in construction projects are not learning from their mistakes or successes for that matter. Why is this so?

Reasons commonly given are that the prototype nature of construction projects makes learning futile, the temporary nature of construction teams is blamed and the fact that there is no financial incentive for people to learn. Few construction organisations have systems for systematically conducting post-project analyses or have any interest in doing so (Phillips, 1996).

That is not to say that individuals do not learn from projects and retain these lessons in mental maps and tacit knowledge. But construction organisations rarely capture this knowledge and transfer it into the corporate memory; it is lost when the individual leaves the company or practice.

The advent of partnering and longer-term relationships between team members in recent years nullifies some of the reasons for failing to learn from projects given above and makes learning an imperative (Bennett and Jayes, 1995; 1998).

It has been demonstrated that learning takes place at three levels (Inkpen and Crossan, 1995):

a) the individual
b) the group or team
c) the organisation

As stated above, individuals often learn from projects, sometimes teams learn, particularly in long-term relationships, but the satellite organisations surrounding the project and supplying personnel rarely learn (Inkpen and Crossan, 1995; Phillips, 1996).

Pre-requisites for learning from projects

Serious research into projects as vehicles for organisational learning seems to be restricted to joint ventures and alliances (Dodgson, 1993; Hamel, 1991; Simonin and Helleloid, 1993; Inkpen and Crossan, 1995).

Hamel's seminal study proposed an inductively derived model of inter-partner learning which consisted of three main concepts as determinates of organisational learning: intent, transparency and receptivity.

Learning intent is a prerequisite to learning - if a partner believes they have nothing to learn from other partners then it is unlikely that learning will occur. This is what Inkpen and Crossan (1995) memorably describe as 'I'll see it when I believe it' rather than the usual expression, 'I'll believe it when I see it'.

Transparency is the openness and willingness of the partner firm to share its embedded knowledge. Clearly there can be no learning unless there is an opportunity to observe partner skills and motivation and ability on the part of individual participants to absorb new knowledge and skills.

Receptivity is defined by Hamel as the learning firm's ability to absorb skills from its partner. Even if a partner is open and accessible with its skills, individual managers must have the motivation and ability to notice the differences. At the individual level, receptivity is closely linked with individual interpretation and, therefore, transparency and receptivity overlap as learning determinants.

In terms of procurement paths it can be argued that learning intent, transparency and receptivity are lowest in traditional design-tender-build contracts because of the separation of design and production and the hard bargaining nature of the contract agreement. Design and build contracts often rely on repeating variations on a standard design for a series of projects so that learning intent and receptivity could be predicted as low; transparency within the design/build team will be high but the nature of the lump-sum contract will normally preclude transparency to the client. Conversely, these three concepts are often cited as prerequisites for Construction Management approaches (Construction Management Forum, 1991; Gray, 1996) and Partnering (Bennett and Jayes, 1995; 1998).

An inability to absorb new information (receptivity) may be confused with low transparency; manager's may blame their lack of learning on their partner's unwillingness to share information whereas the real reason is their inability to perceive differences because their knowledge is context-bound. Again, this is more likely to be true of Traditional procurement paths because of the separation of design and production but may also occur within a joint venture or partnering arrangement (Inkpen and Crossan, 1995; Phillips, 1996).

Levels of learning

According to Inkpen and Crossan (1995) the major areas of disagreement among organisational learning theorists are:

a) whether organisational learning occurs at the individual, group or organisational level
b) whether learning refers to cognitive and/or behavioural change

c) whether learning refers to content or process

d) whether learning should be tied to performance.

Their study of a major international joint venture project led them to conclude that learning in project organisations of this type:

a) occurs over three levels - individual, group and organisation;

b) involves both behavioural and cognitive change;

c) involves a process of change in cognition and behaviour where the changes may be viewed as the outputs or content of learning;

d) should not be tied directly to performance enhancement.

A full discussion of Inkpen and Crossan's conclusions is outside the scope of this paper but point (a) above merits further consideration within the context of procurement approaches.

They argue that different learning processes are at work at each level.

At the *individual* level the critical process is *interpreting* and the outcome is a change in individual beliefs and behaviour.

At the *group* level the critical process is *integration* and the product of the group process, manifested in co-ordinated group actions, is shared beliefs and concerted actions.

At the *organisational* level the critical process is *institutionalising,* reflected in new strategies, systems and routines.

Clearly, learning at the individual level can take place under any of the procurement paths if the participants have learning intent and receptivity to partners' knowledge and skills. This will enable them to interpret what is happening and change beliefs and behaviour. For example, a project manager who attended a course at Reading on the Construction Management approach was able to apply, with some success, the more egalitarian and trusting philosophy of CM to the Traditional JCT project that he was managing.

At the group or project team level whether integration occurs will depend largely on the transparency of the relationship between the participants, although learning intent and receptivity are also important (Tavistock, 1966; Cherns and Bryant, 1984; Latham,1994; Bennett and Jayes, 1995). Integration within the team is less likely to occur under traditional approaches where fragmentation, confrontation and conflict are the norm. Design and build and CM paths are much more likely to generate the 'shared beliefs and concerted actions' which are the output of group integration (Gray and Newcombe, 1996) and thus encourage learning.

As stated earlier, the parent organisations rarely learn from projects because they do not have the mechanisms or interest which would allow knowledge to be captured (Inkpen and Crossan, Phillips). According to Inkpen and Crossan the essential pre-requisite for organisations to institutionalise learning are systems which routinely and regularly collect intelligence from projects and partners. This is most likely to occur where partners have a long-term relationship and see advantages in setting up systems and adopting strategies which maximise their continuing learning. This could happen under any of the procurement approaches where teams are kept together on successive projects but is more likely to occur where the environment creates mechanisms for learning as in CM and partnering.

Experiential learning

Honey and Mumford (1986) have popularised Kolb's Learning Cycle (1984) as a model of individual learning. The phases of the learning cycle - hands on experience, reflection, abstract conceptualisation and active experimentation - have been tested extensively and been shown to accurately model individual learning. Kolb's unique contribution was to develop the concept of learning styles from the learning cycle model which predict the ability of people to learn from different phases of the cycle depending on their natural learning style. Honey and Mumford's recasting of Kolb's learning styles as Activist, Reflector, Theorist and Pragmatist reflect the emphasis that individuals place on the phases of the learning cycle. *Activists* learn best where there are new experiences/problems/opportunities from which to learn. *Reflectors* learn best where they are allowed to watch/think/cogitate over activities. *Theorists* learn best from activities where what is being offered is part of a system, model, concept or theory. *Pragmatists* learn best where they have an opportunity to try out and practice techniques with coaching/feedback from a credible expert.

The implications of this for procurement are twofold:

a) New experiences alone within a project setting are not sufficient unless (i) time is allowed for reflection, (ii) mentoring encourages conceptualisation and (iii) a non-threatening atmosphere encourages experimentation. Both Traditional and design and build paths are unlikely to provide the environment for this to happen, whereas CM and Partnering have the potential to do so. Schon's (1983) study of the way in which architects learn from practice is particularly insightful in this respect.
b) The recognition of learning styles suggests that learning experiences need to be carefully designed and tailored to individual needs if learning is to occur. Again this is more likely to occur in a situation where learning intent, transparency and receptivity are present (Hamel, 1991).

Single and double loop learning

Finally, Argyris and Schon's (1978, 1996) classic analysis of organisational learning has particular implications for projects and procurement paths.

Their definition of learning is the detection and correction of errors. This can be achieved through a single-loop or double-loop learning process. They point out that organisations are good a single-loop learning but poor at double-loop learning which requires a more extensive rethink of underlying norms, policies and objectives.

Individual versus organisational learning presents an interesting paradox. In one sense organisations know less than the individuals that comprise the organisation but conversely, organisations are collections of individuals and therefore organisational learning is potentially the sum of individuals' learning and is therefore greater. Further, organisations learn through individual experiences and action.

All organisations have a 'theory of action', which comprises an 'espoused theory' - what members *say* are the organisations values and norms, and a 'theory-in-use' - the observable action and experience of members of the organisation. Each member constructs his or her own image of the theory-in-use of the whole organisation. This picture is always incomplete but organisational members strive continually to complete it and to understand themselves in the context of the organisation; they try to describe themselves and their own performance insofar as they interact with others. In addition there are public representations of organisational theory-in-use, called organisational maps, to which individuals can refer. These are the shared descriptions of the organisation which individuals jointly construct and use to guide their actions.

Single-loop learning occurs when *'members of the organisation respond to changes in the internal and external environments of the organisation by detecting errors which they then correct so as to maintain the central features of organisational theory-in-use'* (Argyris and Schon 1978). In their latest edition of the book, they refine the differences between single and double loop learning as follows:

> By *single-loop learning* we mean instrumental learning that changes strategies of action or assumptions underlying strategies in ways that leave the values of a theory-of-action unchanged.

> By *double-loop learning,* we mean learning that results in a change in the values of theory-in-use, as well as in its strategies and assumptions. The double loop refers to the two feedback loops that connect the observed effects of action with strategies and values served by strategies (Argyris and Schon, 1996).

Both single and double loop learning can occur at all three learning levels - individual, group or organisational. For double-loop learning to occur there must be a change in the individual, group or organisational theory-in-use.

The implications of this theory for procurement as a learning process are considerable.

The prototype nature of construction design means that, whichever procurement route is used, there is a large element of trial and error during construction as the design evolves and therefore potential opportunities for learning are significant. The difference is the context created by different procurement paths. The Traditional approach makes design errors the subject of claims so that conflict rather than learning is the likely result. Design and build which uses standard designs is less prone to errors but, paradoxically, presents fewer opportunities for learning. The transparency which should be present in CM and Partnering will encourage openness which allows people to admit mistakes and look for constructive solutions thus emphasising learning. The complexity of these types of projects will offer greater opportunities for learning.

The network of firms that provide inputs to a construction project has been described as one of 'interdependent autonomy' (Tavistock, 1966), stressing the problems created by autonomous organisations participating in an interdependent construction process. Under Traditional contracts individuals can and do learn, often at the client's expense! Teams do not stay together and there is therefore little group learning. As stated previously the parent organisations learn very little from projects and the tacit knowledge of individuals is lost when they leave the organisation. In this sense, most construction organisations know less than their individual members because they do not have the systems which would allow access to this tacit knowledge. Where companies systematically collect information by debriefing sessions then the learning of the organisation will be greater than that of any single individual.

Traditional procurement systems encourage the development of an 'espoused theory' which is captured memorably in the Tavistock (1966) report's description of an initial project meeting at which formal positions are stated that none of the parties, except the client, actually believe. The distinction between espoused theory and theory-in-use is the main criticism of most of the industry reports from Emmerson (1962) to Latham (1994).

Unthreatening (CM) or long-term (Partnering) relationships should enable participants to develop 'shared descriptions of the organisation which individuals jointly construct and use to guide their actions' expressed in the form of organisational maps of the evolving theory-in-use.

The difference between single-loop and double-loop learning is particularly evident in the various procurement paths.

Traditional and design and build approaches both inhibit double-loop learning and promote single-loop learning by prohibiting any changes to the original design concept - the 'values of a theory-of-action'. In the Traditional contract the design is completed before the contractor is involved and is not expected to be challenged - a clear case of single-loop learning. Equally, the design under design and build is expected to conform to standard criteria and changes are minimised; single-loop learning is the norm.

Only CM offers the opportunity for double-loop learning where the open, transparent environment encourages all participants, especially the Specialist Trades Contractors, to challenge the underlying design concepts, norms, and values of the current theory-in-use.

The advantages claimed for CM stem from a change of attitude that may be characterised as 'a change in the individual, group and organisational theory-in-use'.

Deutero-learning

Bateson (1972) defines deutero-learning as second order learning or 'learning how to learn'. Argyris and Schon (1978) amplify this definition as follows:

> When an organisation engages in deutero-learning, its members learn, too, about previous contexts of learning. They reflect on and inquire into previous episodes of organisational learning, or failure to learn. They discover what they did that facilitated or inhibited learning, they invent new strategies for learning, they produce these strategies, and they evaluate and generalise what they have produced. The results become encoded in individual images and maps and are reflected in organisational learning practice.

Deutero-learning represents the ultimate in learning endeavour - a study of the learning process itself in order to accelerate the learning curve. This can occur under any of the procurement paths but is more likely to arise in contexts which encourage learning.

Conclusion

We clearly do not know enough about the way in which the selection of a procurement path establishes a learning arena which may encourage or inhibit learning from projects.

This paper has attempted to pioneer the application to construction projects of some of the theories of learning to improve our understanding of the link between procurement and learning.

The importance of Hamel's pre-requisites of learning intent, transparency and receptivity cannot be overstated, whichever procurement path is used.

Inkpen and Crossan's analysis of the levels of learning has long-term implications for the transfer of tacit knowledge and skills from individuals to the corporate memory.

Understanding how experiential learning occurs and designing learning experiences that match individual learning styles as Honey and Mumford argue is only possible where the procurement path creates a non-threatening context.

Moving from single-loop to double-loop learning as Argyris and Schon suggest is made possible by recent trends towards more participative and open forms of contract, where all the participants are involved at an early stage in the project process.

Whilst it is recognised that learning can occur under any of the procurement approaches, it seems that deeper and more widespread learning and deutero-learning requires conditions for its cultivation and growth that are unlikely to be present under Traditional contracts. The 'Holy Grail' of long-term improvement in the efficiency of the construction industry will depend on learning at all levels which CM and Partnering seem to engender.

A research agenda

Much research is now needed which compares different procurement paths in terms of their learning potential.

Research projects comparing procurement paths for construction projects could look at:

a) The extent to which learning intent, transparency and receptivity are present.
b) The effective transfer of learning from individuals to teams to parent organisations.
c) The design of procurement contexts which match learning experiences to individual learning styles.
d) The extent of single-loop learning and the development of double-loop learning within procurement paths.
e) The process of deutero-learning on projects and the learning curve.

References

Argyris, C. and Schon, D.A. (1978), *Orgnizational Learning: A Theory of Action Perspective,* Addison-Wesley.

Argyris, C. and Schon, D.A. (1996), *Organizational Learning II: Theory, Method, and Practice,* Addison-Wesley.

Banwell Report (1964), *The Placing and Management of Contracts for Building and Civil Engineering Works,* London: HMSO.

Bateson, G. (1972), *Steps to an Ecology of Mind,* San Francisco: Chandler Publishing Co.

Bennett, John and Jayes, Sarah, (1995), *Trusting the team,* Centre for Strategic Studies in Construction, University of Reading.

Bennett, John and Jayes, Sarah, (1998), *The seven pillars of partnering*, Thomas Telford.

British Property Federation (1983), *Manual of the BPF System,* London: British Property Federation.

Cherns, A.R. and Bryant D.T. (1984), 'Studying the Client's Role in Construction Management,' *Construction Management and Economics 2,* pp. 177-184.

Construction Management Forum, (1991), *Report and Guidance,* Centre for Strategic Studies in Construction, University of Reading.

Department of Industry (1982), *The United Kingdom Construction Industry,* London: HMSO.

Dodgson, M. (1993), 'Learning, trust, and technological collaboration', *Human Relations,* 46, 77-95.

Emmerson Report (1962), *Survey of Problems before the Construction Industries,* London: HMSO.

Gray, C. (1996), *Faster, better value construction,* The Reading Production Engineering Group, University of Reading.

Gray, C. and Newcombe, R. (1996), *Design of rapid team integration for multi-organisational teams,* Final Report on EPSRC project GR/J48320.

Hamel, G. (1991), 'Competition for competence and inter-partner learning within international strategic alliances', *Strategic Management Journal,* **12,** 83-104.

Inkpen, A.C., and Crossan, M.M., (1995), 'Believing is seeing: Joint ventures and organisational learning', *Journal of Management Studies,* 32:5 September.

Kolb, D. (1984), *Experiential Learning,* Englewood Cliffs, N.J.: Prentice Hall.

Latham, M, (1994), *Building the Team,* HMSO.

Lowe,D. and Skitmore, M. (1994), 'Experiential learning', *Construction Management and Economics,*12.

Mumford, A. and Honey, P. (1986), *The Manual of Learning Styles.*

National Economic Development Council (1978), *Construction for Industrial Recovery,* London: HMSO.

National Economic Development Office (1983), *Faster Building for Industry,* London: HMSO.

National Economic Development Office (1985), *Thinking about Building,* London: HMSO.

National Economic Development Office (1987), *Faster Building for Commerce,* London: HMSO.

Newcombe, Robert (1994) *Procurement Paths - a Power Paradigm,* paper presented at CIB W92, Procurement Systems Symposium in Hong Kong, December.

Ogunlana, S.O. (1991), 'Learning form experience in design cost estimating', *Construction Management and Economics,* 9-2, 133-150.

Pedler, M., Burgoyne, J. and Boydell, T. (1991), *The Learning Company,* McGraw Hill.

Phillips, R, (1996), *Organisational Learning and Joint Ventures - A Case Study,* Unpublished MSc dissertation, University of Reading.

Schon, D. (1991), *The Reflective Practitioner* ,Basic Books Inc.

Senge, P. (1990), *The Fifth Discipline, The Art and Practice of the Learning Organization,* Doubleday.

Simonin, B. L., and Helleloid, D. (1993), 'Do organisations learn? An empirical test of organisational learning in international strategic alliances'. In Moore, D. (Ed.), *Academy of Management Best Paper Proceedings 1993.*

Tavistock Institute (1966), *Interdependence and Uncertainty,* London: Tavistock Publications.

Wood Report, (1976), *The Public Client and the Construction Industries,* London: HMSO.

CONSTRUCTION MANAGEMENT OF LARGE SCALE BUILDING PROJECTS. THE CASE OF COLOMBO CENTRE PROJECT, LISBON

PROFESSOR FRANCISCO LOFORTE RIBEIRO

Instituto Superior Técnico, Departamento de Engenharia Civil, Secção de Estruturas e Construção, Av. Rovisco Pais 1, 1096 Lisboa Codex, Portugal. loforte@civil.ist.utl.pt

Abstract

This paper aims at contributing to clarify the major problems of project delivery method for large scale building projects and inputting the lessons learned during the development and construction of the Colombo Centre in Lisbon. Traditional project delivery methods are ill suited for delivering a large size building project. Issues such as developing a project delivery strategy and contractual relationship, project structure, selecting project participants and improving constructability are crucial to deliver a large size project on time and under budget. Colombo project used a project deliver approach that combines the strengths of the construction management with design-build methods. This project delivery approach and the management tools described in this paper lead to the overall success of the Combo project, including meeting owner's goals related to cost, schedule, and quality.

Keywords: Constructability, construction management, partnering, project delivery methods

Introduction

Construction industry, due to its important economical role and as a large scale employer, grew to become an area that can not be neglected by any government or economical group. This sector is under pressure to reduce project delivery times and costs despite increased uncertainties and complexities that surround today's construction projects. A host of new project delivery methods and management techniques have been promoted to help achieve this. Delivering a large size building project on time and under budget is still an increasingly complex and risky business. Issues such as developing a project delivery strategy and contractual relationships, project structure, selecting project participants and improving constructability are crucial to deliver a large size project on time and under budget (Kluenker 1996). The project delivery solutions used in the Colombo centre are new and they worked well. The project's owner developed and implemented a project delivery approach that combines construction management and design-build with partnering, concurrent construction to deliver the project on time and within the budget. This paper tries to contribute to the body of knowledge on construction management with lessons learned with the Colombo Centre project in Lisbon.

Methods of project delivery

There is no perfect project delivery method for every building project. Also, there are no absolutes in project delivery methods, just variations along a spectrum (Kluenker 1996). Traditional projects use design-bid-build delivery process and fixed-price contract type, with little opportunity for designer/builder interaction. This approach entails the owner engaging separate organisations for three key services: design, cost advice, and construction. Design-bid-build delivery is ill-suited to managing and delivering large scale building projects with multidisciplinary jobs for the following reasons: (i) each phase of the project is managed separately; (ii) information generated and used in each phase is often held by separate organisations and not shared by all the project participants; (iii) no direct link is maintained between decisions made in each phase and the objective functions of the project; (iv) the emphasis is normally on control rather than on proactive management of the project or continuous value addition; and (v) the management emphasis is on contract administration.

To overcome these problems, the construction industry is moving towards non-traditional project delivery methods such as design-build, construction management, bridging, fast-track, and concurrent construction, and some other similar methods of project delivery have been practised widely in recent years with varying degrees of success (Newcombe 1996, Jaafari 1997). Design-build is a form in which one firm or joint venture is responsible to the owner for both design and construction, setting up conditions for good co-operation between them. Although design-build approach has been used successfully on large and complex projects, this method is a solution that depends upon a design-build contractor having all the necessary design and construction skills within one organisation - a situation that would be rare for large size building projects with multidisciplinary jobs.

Bridging is a variation of the design-build method. Bridging puts side by side designers and construction managers with clients and contractors to serve clients better, improve business and deliver cost-effective solutions. It attempts to secure all benefits of design-build yet retain control for the client whenever the client needs it (Kluenker 1996). Construction management is a method of project delivery that differs significantly from design-bid-build and design-build. In the construction management method the client contracts directly with a series of trade contractors, thus eliminating the role of general contractor. The construction management method is a solution that depends upon a project and construction management organisation having all of the necessary engineering, procurement and construction skills within one composite team (Hughes 1997).

The project management method has been around for many years in the construction industry. This method is justified when there is a need for overall project management and an integration role with other delivery methods, even with the design, management and construction management. In the fast track method the project is executed in a phased manner except that design and construction activities are merely overlapped to save time. This practice is claimed to cause sub-optimal design, cost and time overruns and lack of teamwork between designers and contractor (Fazio et al 1988, Mohamed and Tucker 1996, Jaafari 1997).

Concurrent construction (CC) may be defined as an integrated approach to the planning and execution of all project activities, from the planning stage through to the handover of the facility. Under concurrent construction, all project activities are integrated and all aspects of design, construction, and operation are concurrently planned to maximise the value of objective functions while optimising constructability, operability and safety (Jaafarri 1997).

Partnering has received significant attention in the project and construction management literature (Abudayyeh 1994; Albanese 1994; McGeorge and Palmer 1997). Partnering attempts to change the nature of the adversarial relationship between owners, contractors, architects, and other project team members by getting them to commit to common objectives, formalising improved communication, and preventing disputes (Gardiner and Simmons 1998). However, partnering is not prevalent in the Portuguese construction industry. This is manly due to the lack of commitment, by some parties, to the non-adversarial win-win attitude that is an essential element for the partnering process. Combination projects are those using a combination of two or more approaches, including partnering, design-build, and/or formal constructability programs (Pocock et al. 1997).

Research methods

The objectives of the research were met through an intensive literature review, interviews, and a case study. A real construction project utilising a combined project delivery approach was chosen to study the major problems of delivering a large scale building project. A total of 32 interviews were carried out with project owner's staff, construction management contractor's staff and trade contractors senior staff. The interviewees, who had worked on the case study project throughout the construction process, were asked to identify: (1) issues regarding the selection of project delivery strategy and contract relationships for delivering a large scale building project; (2) advantages and disadvantages in the construction management method combined with design-build method at trade contractor level; (3) key factors that should be considered in applying partnering and concurrent construction with the construction management method; (4) critical schedule and cost implications associated with the project delivery solutions. The writer also solicited the lessons that the project managers learned using the construction management delivery method together with design-build methods in the actual project and how those lessons can be applied in subsequent projects. All 32 interviews were recorded on audio cassette, and were subsequently transcribed and evaluated. The data collected were enhanced if necessary through further contact with the original respondent and by carrying out interviews with additional project personnel .

Qualitative analysis

The study was first carried out by analysing the interviews and then studying the project. The analysis of the interviews are summarised into two topics: (1) project delivery strategy; (2) critical schedule and cost implications associated with the project delivery strategy.

Project delivery strategy

The following question was asked:

1. Is the traditional project delivery method -design-bid-build- suited to meet the performance requirements of a technically complex and large size building project?
2. Why choose a combination of construction management with design-build?
3. Does partnering contribute to reduce the conflicts and disputes related to the design-build method together with the fixed-price contract type?
4. Does partnering help to improve integration and interaction among project parties?

Critical schedule and cost implications associated with the project delivery strategy

5. Does construction management with design-build method at trade contractor level help to reduce construction time and costs?
6. Does concurrent construction contribute to maximise the value of schedule while optimising constructability?

To perform the analysis for these questions, a total of 32 interviews were evaluated. The results are summarised and tabulated in tables 1 and 2.

Table 1: Summary of Yes/No questions

No. of – Inter- views	Is the traditional project delivery method – design-bid- build- suited?	Does partnering contribute to reduce the conflicts and disputes?	Does partnering help to improve integration and interaction among project parties?	Does construction management with the design-build method at trade contractor level contribute to reduce construction time and costs?	Does concurrent construction contribute to maximise the value of schedule?
YES	6	20	18	28	25
NO	26	12	14	4	7

As can be deduced from the table 1 for the first question 81% of the respondents answered that the traditional method- design-bid-build is not suited for technically complex and large scale building project while 19% answered yes. 63% of all respondents agreed that partnering contribute to reduce the conflicts and disputes within the construction management method.

Table 2: Summary of second question

	Why choose a combination of construction management with design-build?				
	Reduce management layers	Improve constructabil ity	Provide the owner with an early price guarantee	Provide opportunities to reduce time and cost	Owner could keep a closer monitoring of the design and construction process
No. of – Interviews	27	17	28	23	29

Case study: construction management at colombo centre in lisbon

This section provides a description of the project delivery framework used for the Colombo Centre project.

Colombo project characteristics

Colombo Centre is a commercial complex consisting of retail, office and leisure areas. Its total construction area is on first phase approximately 400,000 m2 over six floors, and two twin 16 story office towers, the final construction area will exceed 450 000 m2. The centre includes more than 6800 car parking places, 400 shops, 10 cinemas, a carting track, a roller coaster, a jogging circuit, a golf driving range, swimming pools, 12.000 m2 for entertainment areas and 7.000 m2 of open air restaurants and coffee shops on the top floor. The building structure consists of an orthogonal 8x8 mesh of concrete columns and 7 slabs, from -3 to +3 levels. The Colombo project also includes the construction of three flyovers, the longest being approximately 400 meters long, more than 4 km of roads and one tunnel road as counterparts to the city's administration. The capital investment for this project was 50.000.000.000 Escudos (about 167 million pounds), construction works of the first phase started in November 94 and finished in September 1997.

Project delivery approach

Delivering the Colombo project on time and within the budget was a major goal set up by the owner in the early planning stage. The owner had to consider the trade-off between issues such as: assigning and sharing of risk, conflict interests, designers/contractors as agents, designers/contractors as vendors, partnering, and selection of the project deliver strategy, contract types, project structure and project participants. The Colombo project required a combination of different of project-deliver solutions which cloud provide means to delivery the project in time and within the budget at minimum risk. Thus, it used a project delivery framework (figure 1) which combines construction management with design-build, partnering, concurrent construction and constructability at trade contractor level to achieve the goals set up by the owner.

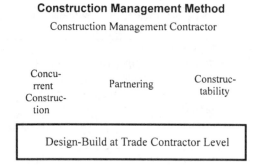

Figure 1: Project delivery approach used for Colombo project

The experience on large scale building projects with a high degree of multidisciplinary jobs has led to a move to reduce the number of management levels and increase the degree of decentralisation (Harrison 1992). The construction management delivery method was justified because it provides a way to reduce the management levels by eliminating the role of main contractor and to approximate the owner to the trade contractors. The absence of the main contractor placed the trade contractors in a direct relationship with the client's management contractor, rather than through an intermediary. The construction management contract was awarded to a consortium of three firms each one with experience in architectural, engineering, procurement and construction fields.

Design-build was chosen as the delivery method of the different parts of the project (work packages). In such arrangement the trade contractors were not given a complete design, but to complete the design and construct the work package put under bid. Why choose design-build? The value of design-build within the construction management approach shown in figure 1 comes in: (i) the integration of the design and construction of each work package within the same contract arrangement, which can then bring construction technology and experience; (ii) lump-sum contract and design-build providing the owner with an early job price guarantee; (iii) design-build offering opportunities to reduce construction time and costs.

Implementation of the project

The design-build method followed at trade contractor level combines the strengths of the design-build process and the fixed-price contract type. Following this approach the client had opportunities for reducing time and cost required to deliver a fixed price work package. The trade contractors were selected by competitive bid based on a concept design and well defined requirements. Also, the contract price was fixed and enforceable. The client selected the trade-contractors on the basis of their qualifications, delivery and financial record, reputation, and negotiated in every instance a firm price at concept. This process allowed the trade contractor (design-builder) to collaborate in completing the design with the best technology and cost-effective solutions.

The Colombo project was divided into 60 design-build packages of work according to a work breakdown structure developed for the project. Each one of these package corresponded to a distinct area of work which was awarded to one trade contractor and in some cases to a consortium of trade contractors by competitive bidding on the basis of the design-build delivery method. These packages were contractually managed in an integrated manner. The construction management contractor assisted the client in defining, co-ordinating, reviewing and managing the project designs, establishing a master schedule and an achievable budget. The first step in its work was to follow the execution and co-ordination of the several design projects. Once the works had enough definition, a master schedule and a general budget were developed. This budget served as upper limit and guideline for the prices of the different packages during the negotiations with the contractors. It is extremely important to have a master schedule and a budget as accurate and detailed as possible from day one.

Figure 2 presents a diagram showing the procurement process at trade contractor level. Partnering as had a great impact in taking the best advantage of the opportunities provided by the combined approach while reducing the impacts of the problems related to the design-build method and the fixed-price contract type. Thus, partnering was as used as a managing conflict tool throughout the project.

Contract relationship between owner and trade contractors

The contract relationship between the owner and trade contractors chosen was the fixed-price contract type, more precisely the lump-sum contract between the client and the trade contractors. According to this contractual arrangement the trade contractor or the joint venture of trade contractors was required to establish a stipulated sum for the completion of defined scope of work. In such a contract arrangement the trade contractor of the joint venture agreed on completing a defined design-build package of work for a fixed sum of money. Risks were allocated primarily towards the trade contractor. Hence, trade contractors bear all the financial burdens of cost overruns if the scope or site conditions do not change.

The project management information system

The project delivery approach shown in figure 1 needed a unified, information management system to hold all key information relevant to the management of the design, procurement, and construction processes of the project. Thus, an integrated information structure was set up as the core the project delivery system and all project management functions linked to it. Figure 3 shows how this framework works. The execution of a given project management function commanded the system to access all the information of the project relevant to the function under consideration stored in the project database, assemble this information, analyse or evaluate the information, and produce reports. Each time a new item of information was entered or an existing piece of information was changed it would be possible to re-evaluate the status of the project by executing all project management functions on an integrated basis. All information relevant to the development and implementation of the project was stored in a relational database management system. On the top of this relational database were the project management system KMES property software) and the project planner PRIMVERA from Primavera Systems, Inc. KMES is project management system which can be easily integrated with other software such as PRIMAVERA

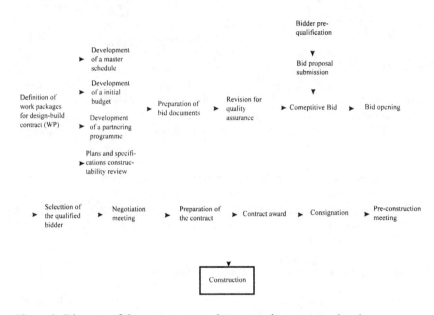

Figure 2: Diagram of the procurement phase at trade contractor level

Conclusions and lessons learned

Delivering and managing a construction project is a complex process and any finished project has something to add to the extensive body of knowledge in construction management. Colombo project has successfully demonstrated that a project delivery approach that combines construction management and design-build methods with other management tools can produce better performance results in terms of schedule, costs and reduction of disputes on large scale building projects with multidisciplinary jobs. Partnering, constructability and concurrent construction are some of the means by which the Colombo project have obtained greater value from the resources committed by the owner to the project. Some of the major contributions of the Colombo project are the lessons learned during construction stage. They fall into the following categories: (i) the project delivery approach; (ii) compensation scheme; (iii) effects of partnering on the construction process; (iv) effects of constructability on the design; and (v) effects of concurrent construction on the construction process. The following sections summarise these lessons.

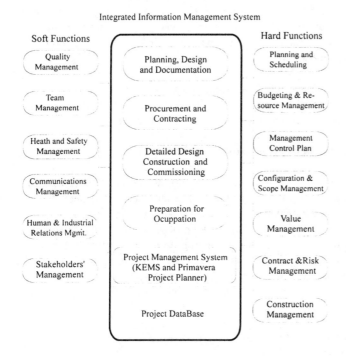

Integrated Information Management System

Soft Functions

- Quality Management
- Team Management
- Heath and Safety Management
- Communications Management
- Human & Industrial Relations Mgmt.
- Stakeholders' Management

- Planning, Design and Documentation
- Procurement and Contracting
- Detailed Design Construction and Commissioning
- Preparation for Ocuppation
- Project Management System (KEMS and Primavera Project Planner)
- Project DataBase

Hard Functions

- Planning and Scheduling
- Budgeting & Re-source Management
- Management Control Plan
- Configuration & Scope Management
- Value Management
- Contract &Risk Management
- Construction Management

Figure 3: Schematic representation of the information management system of the project

Lesson 1: Project delivery approach

Colombo is a technically complex and speedy building project. The combined project deliver approach was ideally suited to meet the schedule and cost goals of the Colombo project and avoid construction claims and cost/time overruns. The construction management method creates contractual relationships and liabilities that are unique in Portugal.

Why choose a project management team instead of a general contractor? A general contractor takes over the responsibility of meeting the schedule and the budget, usually on a lump sum basis (maximum price guaranteed). To cover its risk, the general contractor is entitled to a percentage around 15% of the contract price. In this case, that would mean 7500 million escudos (25 million pounds). Besides the client loses most of its control over the construction process and how the contractor gets the job done. Choosing a construction management team, although an increase of the management costs, the owner could keep a closer monitoring of the design and construction process and was in a better position to drive the project schedule and budget according to his goals. The other consequences were that the owner kept most of the responsibility of finishing the job on time, and the project had to be divided into 60 work packages. Most packages had different contractors, that greatly increased the complexity of the construction management task. The experience has shown that would be more profitable to divide the project into a number less than 60, 35 or forty packages of work. In addition, in the case of large packages of work, it is better to award the whole package of work to a consortium of trade contractors rather than dividing the package into smaller jobs and awarding each part to a single trade contractor, and hence avoiding task interdependence conflicts.

The design-build deliver approach used at trade contractor level gave access to construction knowledge during the design process, offered cost-effective solutions and a early price guarantee prior beginning construction of each work package.

Lesson 2: The compensation scheme: lump sum vs. unitary prices

A lump sum contract type seems an obvious way to control the costs of the project. In a design-build job, the contractor will be responsible for the measurement indicated in his proposal and is not entitled to extra payments for larger quantities. The measurements of the different contractors can be compared during the bid process and adjusted if necessary. This transfers a large amount of quantity surveying work from the construction management team to the trade contractors. However, this procedure also has disadvantages. In the case of multidisciplinary building projects, many entities are involved. This represents a high potential for changes to the original design, often as late as during finishing.

Lesson 3: Effects of partnering on the construction process

Partnering is a co-operative management tool that was used was used on the Colombo project with other tools as a dispute prevention and interaction improvement tool. It represented a change in the way a design-build process is managed. The partnering programme developed and implemented in the Colombo project entailed a series of formal activities and mechanisms that were used to sustain collaborative teamwork and trust between parties from different organisations, and commitment to common goals. These mechanisms typically included: pre-construction team-building sessions, formulation of a project manual of procedures, periodic assessment of adherence to partnering principles, guidelines for resolving disputes in a timely and effective manner, and provisions for improvement and risk sharing. Partnering had three significant benefits. Foremost, it helped parties to approach problems with the attitude of "how can we resolve problems jointly rather than independently". Second, it provided excellent interaction between the construction management contractor and local trade contractors. This was particularly evident in situations involving physical work space, sharing equipment, joining skills to perform complex jobs, finding cost-effective design solutions. Third, it helped to reduce or even avoiding conflict due to: (i) job interdependency resulting from dependency upon others; (ii) differentiation resulting from organisational and technical differences among trade contractors; (iii) different values, interests, and objectives of the various trade contractors and designers.

Lesson 4: Effects of constructability on the design

Constructability related to the efficiency in the construction process, i.e., all efforts to consider how the design of each job should be implemented by a trade contractor without resulting scheduling problems, delays during the construction process, inefficient use of resources, and out-of-sequence work. The construction management was concerned as to the ease with which the inputs of the construction process (design information and data, labour, production equipment and tools, and material and installed equipment) could be brought together by a trade contractor to complete each job in a timely and economic manner.

Lesson 5: Effects of concurrent construction on the construction process

Under concurrent construction project activities were integrated and aspects of design, construction and preparation for occupation were concurrently planned to maximise the value of schedule while optimising constructability and operability. Concurrent construction in Colombo project was based on: (i) integration of design and construction activities at trade contractor level; (ii) establishment of a manual of procedures that integrate all planning, design and construction activities; (iii) formation of composite teams, having a representative from each of the pertinent parties, including designers, engineers, suppliers, trade contractor, tenants, facility manager and in some cases, representatives of the local government and service providers were included; (iv) division of work into separable work packages; (v) integration of total project information over its construction process; (vi) integration of the work of the joint trade contractors into a single design-build contract; (vii) establishment of direct and real-time control system to facilitate the whole process.

References

Abudayyeh O. (1994) Partnering: A team building approach to quality construction management, *Journal of Management in Engineering and Management*, ASCE, 10 (6), 26-29.

Albanese R. (1994) Team building process: Key to better project results, *Journal of Management in Engineering and Management*, ASCE, 10 (6), 36-44.

Fazio P., Moselhi O., Theberge P., and Revay S. (1988) Design Impact of Construction Fast-Track, *Construction Management and Economics*, 6, 195-208.

Gardiner P. and Simmons J.E. (1998) Conflict In Small - And Medium - Sized Projects: Case of Partnering to the Rescue, *Journal of Management in Engineering*, ASCE, 14 (1), 35-40.

Harrison F.L. (1992) *Advanced Project Management A Structured Approach*, Gower Publishing Company Limited, England.

Hughes W. (1997) Construction Management Contracts: Law and practice, *Engineering, Construction and Architectural Management*, 4 (1), 59-79.

Jaafari A. (1997) Concurrent Construction and Life Cycle Project Management, *Journal of Construction Engineering and Management*, ASCE, 123 (4), 427 - 435.

Kluenker C.H. (1996) The Construction Manager as Project Integrator, *Journal of*

Management in Engineering, ASCE, March/April, 17 – 20.

McGeorge D. and Palmer A. (1997) *Construction Management New Directions*, Blackwell Science Ltd, Oxford.

Mohamed S. and Tucker S. (1996) Options for Applying BPR in the Australian Construction Industry, *International Journal of Project Management*, 14 (6) 379-385.

Newcombe R. (1996) Empowering the construction project team, *International Journal of Project Management*, 14 (2), 75-80.

Pocock J.B., Liu L.Y., and Kim M.L. (1997) Impact of Management on Project Interaction and Performance, *Journal of Construction Engineering and Management*, ASCE, 123 (4), 411 - 418.

Puddicombe M. S. (1997) Designers and Contractors: Impediments to Integration, *Journal of Construction Engineering and Management*, ASCE, 123 (3), 245 - 252.

THE DEMAND FOR INDEPENDENT PROCUREMENT ADVICE IN THE NORTHERN IRELAND CONSTRUCTION INDUSTRY

J. G. GUNNING and M. WOODS
School of the Built Environment, University of Ulster, Northern Ireland BT37 OQB.
jg.gunning@ulst.ac.uk

Abstract

This paper presents the results of a 1997 survey into the procurement needs of building clients in Northern Ireland. The requirement for an independent advisory service was investigated by evaluating the levels of project performance and client satisfaction within the local construction industry. By means of interviews and questionnaires, the research gathered information from thirty members of the architectural, quantity surveying and client organisations, which indicated the general views of the local industry regarding procurement advice.

Fifteen projects were studied in detail and found that only two of the projects achieved high performance in terms of speed – both procured traditionally. Half of the projects finished on time and within budget. Unsuccessful projects were found to use both traditional and non-traditional systems. The paper suggests that problems mainly arose because the parties to the projects had not fully considered their specific objectives and the implications of the chosen procurement method.

The research found that clients do require guidance in commissioning a construction project, but that the professionals tended to lack a comprehensive knowledge of procurement issues. Whilst projects generally run quite satisfactorily, those time and cost overruns which do occur are considered to be avoidable if there was increased guidance on procurement from the outset. Clients who were better advised benefit from increased performance, and more educated clients now demand a better service from all parties on their projects. The research concludes that independent procurement advice leads to significant improvement in project performance, and sets out recommendations as to how such an advisory service should operate to provide harmony and profit for all.

Key Words: Building clients, independent procurement advice, Northern Ireland.

Introduction

It is now well recognised that the strategic choices made by clients as to how to organise and manage their projects can have a crucial impact upon subsequent project performance. As building designs become more complex, costs escalate and projects increase in size, client organisations continue to choose their procurement systems from a narrow range of the numerous methods available.

Research reveals that the way in which many clients, and their professional advisors, select the method of procurement can be haphazard, ill-timed and lacking in logic (Masterman, 1992). It has also been established that one of the principal reasons for the construction industry's poor performance is the inappropriateness of the choice of procurement system (Skitmore and Marsden, 1988). It is therefore essential that clients are made aware of all the available procurement alternatives and make their choices by the use of more sophisticated methods of selection than those that are currently used for this purpose.

Procurement options

Although there are many different forms of procurement available, these can be regarded as falling into a number of distinguishable types. These types vary with respect to how responsibilities for design, cost control, planning and construction are defined and allocated (Chartered Institute of Building, 1982). The three main groups of such systems are widely recognised as being Separated/Co-operative, Integrated and Management oriented. Partnering can be employed in conjunction with any of these procurement systems.

Previous research (McMullan, 1996; Gunning and McDermott, 1996) has identified the following methods of procurement as those which are most commonly used in Northern Ireland (in descending order of frequency);

1. Traditional
2. Design and Build
3. Management Contracting
4. Construction Management
5. Design and Manage

Whilst we now have an array of procurement options available, construction clients, especially those who are inexperienced, still rely extensively upon advice given on the most suitable method of procurement.

The British Property Federation's 1997 survey of major UK clients reveals that:

• More than a third of major clients are dissatisfied with contractors' performance in keeping to the quoted price and to time, resolving defects, and delivering a final product of the required quality;

• More than a third of major clients are dissatisfied with consultants' performance in co-ordinating teams, in design and innovation, in providing a speedy and reliable service and in providing value for money.

A recent survey by the Design Build Foundation, quoted by Egan (1998), shows that:

• Clients want greater value from their buildings by achieving a clearer focus on meeting functional business needs;

- Clients' immediate priorities are to reduce capital costs and improve the quality of new buildings;

- Clients believe that a longer-term, more important issue is reducing running-costs and improving the standard of existing buildings;

- Clients believe that significant value improvement and cost reduction can be gained by the integration of design and construction.

Without guidance a client is likely to be attracted to that approach which has impressed its image most strongly upon him, usually the traditional method. The result of this may be that client satisfaction is not optimised, due to a mismatch of need to capacity (Baldry, 1997). Despite this requirement there is a lack of expertise amongst architects, quantity surveyors and engineers in the area of procurement advice (Bowen et al, 1997). The increasing use of project management services overcomes this problem to a certain extent, but for those who do not employ the services of a project manager there is a need for an independent procurement advisor who is suitably trained and experienced to advise upon and implement a set of arrangements that achieves all the objectives of the client within a balance of priorities.

Procurement Advice

Procurement advice can be given by both consultants and contractors. Standard building contracts with contractors do not contain provisions for obtaining procurement advice, yet the contractor is permitted to offer such advice. Under a design and build contract contractors are faced with extracting a proper brief from the client. If the client and project objectives are not suited to the design-build arrangement the contractor is not in any way obliged to advise the client on an alternative procurement route.

The usual contracts for architects (RIBA 'Architect's Appointment'), consulting engineers (ACE) and quantity surveyors (RICS) do not refer directly to procurement but advice can be offered in the course of reviewing design and construction arrangements with the client. A direct responsibility for procurement advice is only found under "other services" clauses or within project management contracts.

At present the situation within the construction industry is extremely complex. The selection of appropriate contractual arrangements for any but the simplest type of project is difficult owing to the diverse range of options and professional advice available. In the absence of a recognised independent advisory role much of the advice is contradictory and lacks a sound research basis for evaluation. For example, a design-build contractor is unlikely to recommend the use of an independent designer. Such organisations believe that the full integration of design with construction is likely to achieve the best long-term solution both for the project and the client. The professions which provide only a design service generally take the opposite viewpoint.

A project management approach is normally required as a separate function only on large and complex projects. In Northern Ireland the majority of construction projects are of a more modest nature where the architect performs the task of the client's project manager. Through such an approach the client will depend on the architect and quantity surveyor to provide guidance throughout the project. These professions are, however, employed to provide design and cost consultancy services, with neither party directly responsible for issuing procurement advice.

Quantity surveyors already consider procurement advice as part of their job but only offer it as part of their full range of services and not as an independent service. The problem in Northern Ireland centres around the fact that many clients approach the architect in the first instance, and so may not receive impartial procurement advice. The onus is therefore on the quantity surveyor to market their skills and ensure that clients choose the most appropriate procurement strategy and maximise their satisfaction with their project.

In Northern Ireland the architectural profession has not pursued the project management role as vigorously as quantity surveyors. Despite this the architect is still the first contact for many construction clients. This raises a question regarding the architect's advice towards the procurement of a building – will the architect remain impartial and consider what is best for the client and his project, even if this involves redirecting the client to another organisation, or will he recommend the traditional procurement system, through which he receives a greater fee?

Research sample

"The Irish Database of Environmental and Architectural Literature" holds details of a wide range of construction projects in Ireland on which articles have been written in various journals. Information can be called up by selecting a project, a location, architect/designer etc. Hence a list of suitable recent projects was compiled.

Through use of the referenced article on each project the researchers were able to select fifteen projects which were suitable for investigation. The chosen projects represented a range of project and client types from various locations in Northern Ireland. From this sample the researchers decided to target ten members of each party (client, architect, quantity surveyor). Each article provided details on the organisations involved with the project therefore enabling the researcher to identify the initial sources of contact.

Initial contact was made with each specific organisation to make a general enquiry as to whom it would be best to contact concerning the project. Each individual was then forwarded a letter and a copy of the research proposal setting out the aims and objectives of the research, the research method to be employed and the time involved. This was followed-up with a telephone call to confirm a willingness to participate and to arrange a time for interview or for forwarding of a questionnaire.

Due to the small number of questionnaires involved, the researchers had the opportunity to contact all concerned and therefore received personal guarantees of co-operation in completing and returning the questionnaires. A number of reminder telephone calls were required; however all questionnaires were returned and used for comparative analysis by simple manual calculations.

The response verifies that most construction clients from both public and private sectors seem to order a variety of work across a range of different building types. Thirty three per cent of projects were of the commercial/retail type followed by offices, making up 27% of the sample. These projects were undertaken mainly by developers (24%), owner-occupiers (27%) and investors (20%). The majority of developers (57%) used the design-build method of procurement with three out of four stating that 'timely completion' was the principal reason for using this system. One client expressed that the use of the design-build option offered great rewards by way of additional revenue through earlier leasing dates.

Client objectives

Of primary importance in the selection of procurement systems is the identification of the client's primary and secondary objectives (Hashim, 1997). All clients placed a great emphasis on time and price certainty, with each client placing these criteria within their top three priorities. The cheapest cost of building was given a relatively low priority by all clients, with 65% stating that they preferred to be given a reliable estimate rather than a low price resulting in an unexpected cost overrun. Despite this, 25% of traditional clients chose this method to ensure that a competitive price was obtained through the traditional tendering procedures. The main priority of the traditional client lay with the quality of the building, while minimising project time took a lower ranking.

A surprising factor in the ranking of client criteria was the low priority awarded to "the introduction of variations'. This low rating is in contrast to the nine projects which suffered time and cost overruns due to design variations and additional work introduced during construction. Evidently clients and their advisors do not consider the impact of such variations on the time, cost and quality schedules of traditional and design-build projects. In saying this, it is also apparent that the users of the management systems do not select this method for its flexibility in allowing variations as they rank it a low sixth. This indicates that clients and advisors do not anticipate any late changes to the design and therefore do not believe this factor to be of significance when considering their priorities for the project.

Project performance

Data was collected on estimated time and cost schedules and actual costs and durations of each project under consideration. These measures were used as the main performance variables to distinguish between projects that finished under/on/over programme and price. Analysing these figures proved that most projects performed extremely well. Only 27% of projects finished over the planned duration and 43% over price. Such success rates provide an excellent argument as to why the traditional method of procurement is still so widely used in the Northern Ireland construction industry.

The excellent success rate of projects in Northern Ireland was evident with 50% completed within both time and cost schedules. Twenty-three per cent of projects overran cost targets but were both completed on time whereas only 7% met cost schedules when running over time. This low rate is common as project costs increase as the project duration lengthens. In addition to this, 20% of respondents stated that they would pay extra to ensure that the project was completed on time.

To understand why 50% of projects failed to achieve complete success, the respondents were asked to give the reasons for delays to the programme and increased costs. The most common problem appears to be the additional work requested by the client, resulting in time and cost overruns in 40% of the unsuccessful projects. From the nine projects which experienced time and cost overruns due to variations and additional work, five regarded an inadequate brief as the underlying problem. Other reasons included insufficient input from consultants (11%) and genuine late changes during construction (33%).

It was evident that the majority of projects with high performance requirements for speed or cost were managed by the non-traditional forms of contract Only two such projects were procured traditionally, with one using the accelerated version to improve time performance. It is clear therefore that clients who require faster and cheaper projects are looking to the alternative methods of procurement; it remains to be seen if these forms of procurement actually achieve better results.

In order to compare the performances to the clients' expectations it is necessary to create definite categories which will define the cost and speed achievements as high; medium or low. These categories were decided upon by considering the cost and speed statistics of the entire sample of thirty. The researchers then decided upon boundaries for each category as illustrated in Table 1.

The speed of construction is base on the ratio – Gross area (square foot) per week, during construction time. Cost is similarly calculated using the ratio – Cost of construction per square foot.

Table 1. Speed and cost performance categories

Speed performance	
High	> 1200 sq. ft per week during construction
Medium	500-1200 sq ft per week during construction
Low	< 500 square foot per week during construction
Cost performances	
Low	< £50 per square foot
Medium	£50-85 per square foot
High	> £85 per square foot

Table 2 illustrates project performance by cross tabulating the speed and cost levels of performance. The table clearly shows that the number of projects which achieve high levels of performance in respect of speed of construction is extremely low - only two projects achieved this. Both of these projects were procured traditionally. Project five was procured using an accelerated version of the traditional method.

The performance of project ten was surprising due to the fact that neither speed nor cost were identified as high rating priorities. Price and time certainty were stressed as number one priorities; however the client claimed that the high success rate was due to an excellent standard of workmanship and a strong design team. This project was actually one of those which experienced problems during construction resulting in time and cost overruns; however, the project team delivered a successful project regardless.

For a number of projects the outcome was 'as planned', for example, project three achieved a low cost and moderate speed performance whilst project eight was completed at a low cost but at a slow rate.

Table 2. Project performance – cost and speed

Cost	Speed		
	High	Medium	Low
Low	NIL	(3)	(8) (11) (15)
Medium	(5) (10)	(4) (7) (14)	(1)
High	(12)	(2) (6) (9) (13)	NIL

Design and build procurement options seemed to record good rates of construction but often at a higher price. This was also the case for the management contracts, therefore backing up the statements that speed is often obtained at a cost premium. Project one was a major refurbishment project which resulted in a high price due to the complex nature of the project. Design and build is not usually considered for such projects, however, this company was highly experienced in the construction industry and in the use of this procurement method.

Client satisfaction

In an overall response, clients appear to be very satisfied with regards to the achievement of the project objectives; time, cost and quality. However, it is important to ascertain if clients consider the original project objectives when stating their levels of satisfaction, or do they merely claim satisfaction with a completed project despite problems which may have occurred during the design and construction.

Eighty-eight per cent of traditional clients claimed to be very satisfied/satisfied with the time outcome despite 37% of projects going over the planned duration. More surprisingly, 94% of the same clients were satisfied with the final costs of the project, seemingly disregarding the fact that 44% finished over the contract price. However, the cost and time overruns were, in many cases, due to client variations and additions, and so clients could not fairly express dissatisfaction with the outcome. A number of other factors may also explain this. Often clients allow a contingency in their budgets for the construction phase. More strikingly, thirty per cent of respondents agreed that if the client has built up a rapport with the building team and is informed regularly and accurately of time and cost implications throughout the project, then he will not be dissatisfied with overruns.

The time performance of the design-build projects satisfied all the clients; only 12% were dissatisfied with cost performances despite 38% of projects failing to meet the contract price. Results were similar for management contracts with 83% of clients satisfied with the time performance. In this case the dissatisfaction with time (17%) directly correlated with the number of projects failing to meet the planned duration (17%). With regard to the cost schedule, all the clients expressed satisfaction with the outcome despite 50% of projects running over contract price. Once again this illustrates the risk which management clients accept when undertaking this type of procurement procedure. A firm price cannot be guaranteed prior to commencement and will not be known with certainty until the project is completed.

Thirty per cent of the client body stated that they only seek procurement advice occasionally, when 'circumstances are exceptional'. In seeking this advice clients seem to approach the quantity surveyor more than the architect with 64% of the quantity surveyors' clients 'always or occasionally' seeking advice compared to only 45% of the architects' clients requesting this service. Fifty per cent of clients state that they do not request any advice. However 70% of these clients commented that all construction clients should be offered procurement advice by the professions even if they do not request it.

It is clear that both professions claim a comprehensive knowledge of the traditional procurement system (100% of architects and quantity surveyors) and its negotiated form (90% of architects and 100% of quantity surveyors). However, in considering the alternative methods of procurement, the proportion of respondents who regard their knowledge as 'very good' or 'good' falls dramatically. Despite the fact that design-build is growing in popularity in Northern Ireland, (Gunning and McDermott, 1997) only 50% of architects and 60% of quantity surveyors have a good knowledge of the system. The situation worsens in respect of the management forms of contract where, on average, only 35% of architects and 45% of quantity surveyors have a good knowledge. This highlights the main problem in the construction industry regarding procurement advice – regardless of their willingness to provide advice, how are the industry's professions, as "experts", to educate the client when they lack a comprehensive knowledge of procurement systems?

Conclusions

The majority of projects in Northern Ireland are procured in the traditional manner by a client body which is predominantly 'inexperienced' in the procedures of the industry. The principal reasons for the utilisation of this method of procurement lie with the primary objectives of the clients; these were the necessity for time and cost certainty before project commencement and the requirement for high quality levels. In general, the construction client in Northern Ireland is less concerned about minimising project time, with many unwilling to risk uncertainties in time and cost by using non-conventional methods of procurement.

As predicted, the use of this method often resulted in time and cost overruns, therefore reflecting poorly on the ability of the system to guarantee time and cost schedules. However, this rate is much less than originally expected as it was discovered that half the projects surveyed finished on time and within budget. Further analysis revealed that these overruns were, in many cases, the result of additional work ordered by the client during the construction process. Such client inputs significantly impede the progress of the project and were found to occur frequently as a result of an inadequate client brief.

These findings point, not to a failing of the actual procurement system, but to a lack of understanding of the importance of the procurement decision. Clients and their professional advisors are not awarding sufficient time to the initial stages of the project when full implications of all decisions should be discussed. However, the problem continues into the construction stage of the project when work is not carried out in accordance with the procedures dictated by the selected procurement system. The research suggests that this problem occurs because the parties do not fully consider the implications of the procurement system they have chosen and have not considered their objectives sufficiently.

Recommendations

From the findings of this study the following recommendations are made regarding the provision of independent advice in the Northern Ireland construction industry.

- Professional members of the industry should attend educational courses to improve and up-date their knowledge on the subject of procurement. If time permits, employers should encourage their employees to undertake a course in Project Management.
- Architects and quantity surveyors should develop a systematic approach in selecting the procurement system. This may involve the use of 'information systems' or a more simplified alternative, but would ensure that all clients received the same higher standard of service.
- Once the skills have been developed, the professionals should promote this service and offer it in an independent manner. It is essential that clients are made aware of the importance of such a service.
- The client's advisor must consider the characteristics of the project objectives and procurement methods, decide on best fit solutions, and assess the levels of risk and the availability of appropriate skills.

In commissioning a building a client should consider the following recommendations:

- A clear brief should be prepared including both an indication of the service required and the completed building. The client should ensure that the professional advisor understands these expectations and should not be afraid to question any suggestions.
- Client organisations should become more involved in the management of their projects, particularly when using methods within the management-orientated procurement system category. The procurement decision process should involve various members of the client organisation including the stakeholders, users of the building, etc. This would result in a comprehensive brief and reduce the possibility of client variations once construction has commenced.

This investigation has indicated that the problem lies not only in the selection process but throughout the implementation of the procurement strategy; that is, throughout the course of the project. The following recommendations would assist in improving the performance of the procurement system:

- Roles and responsibilities should be clearly defined with a clear chain of communication and decision-making procedures within the project team to ensure progress is made and problems are overcome.
- A comprehensive project strategy and timetable should be determined at the outset and taken absolutely seriously. Any variations to the time schedule or budget should be noted and adoptions made to the master plan if necessary.
- Design management is essential. Establish a clear design leader, co-ordinate all design activities and monitor progress, organise and manage the flow of design information.
- The use of Co-ordinated Project Information should be applied in all projects to ensure that all participants are receiving the necessary information and are working towards a mutual goal.
- Regardless of the procurement system in use, all designs should incorporate a high degree of buildability. Consultants should also inform clients of the benefits to be gained by applying the concept of value engineering to their project design.

In summary, the performance of building projects in Northern Ireland could be greatly improved through the availability of independent advice. It can be concluded that this service may not increase the number of projects which are procured in non-traditional manner, but it should improve the way in which all procurement systems are implemented and encourage clients to continually seek profit with harmony.

References

Baldry, D (1997). Building the Image – A Study of the Performance Factor. *CIOB Construction Paper No 72,* CIOB London.
Bowen et al (1997). The Effectiveness of Building Procurement Systems in the Attainment of Client Objectives. *Proceedings of the CIB W92 Symposium on Procurement, CIB Publication No 203,* Montreal, pp 39-50.
Chartered Institute of Building (1996). Code of Practice for Project Management for Construction and Development, 2nd Edition, Longman, Essex.
Gunning, J G and McDermott, M A (1997). Developments in Design and Build Contract Practice in Northern Ireland. *Proceedings of the CIB W92 Symposium, CIB Publication No 203,* Montreal, pp 213-222.
Egan, Sir John (1998) Rethinking Construction – DOE TR, London
Hashim, M (1997). Clients' Criteria on the Choice of Procurement Systems – a Malaysian Experience. *Proceedings of the CIB W92 Symposium on Procurement, CIB Publication No 203,* Montreal, pp 273-284.
Masterman, J W E (1992). An Introduction to Building Procurement Systems. E & F N Spon, London
McMullan, J (1996). The Procurement of the Building Services Industry in Northern Ireland. BEng (Hons) Dissertation, University of Ulster.

Skitmore, R M and Marsden, D E (1988). Which Procurement System? Towards a Universal Procurement Selection Technique. *Construction Management and Economics, Vol. 6*

ANALYSIS OF GENERAL CONTRACTOR'S FINANCIAL STRATEGY UNDER DIFFERENT PROCUREMENT METHODS

TING-YA HSIEH
Associate Professor, Department of Civil Engineering, National Central University, Chung-Li, Taiwan 32054, R. O. China. tingya@cc.ncu.edu.tw

HSIAU-LUNG WANG
Lecturer, Department of Civil Engineering, Van-Nang Institute of Technology, Chung-Li, Taiwan 32054, R. O. China

Abstract

This paper intends to highlight the varying impacts of major procurement methods on the general contractor and to infer the relevant sustainable financial strategy. It first reviews characteristics of major procurement methods and examines the financial structure of general contractors by comparing the general contractors as a sector with other three representative sectors. Then, the financial implications of different procurement methods on a general contractor's financial strategy are discussed. Finally, some summary remarks are given.

Keywords: Contracting, financial management, procurement method, ratio analysis.

Introduction

General contractors are the major player in the construction industry. They basically act as the middleperson between a client and the resource market. The client delivers the needs for building space or functional infrastructures; and the resource market, which consists of mainly subcontractors and material vendors, supplies the products and services essential to the client's needs. With this basic framework, the function played by the general contractors, therefore, is to provide the know-how and technical service of putting the right products and services together within the required time frame and in accordance with quality standards.

Consequently, a general contractor does not have the direct need to maintain material inventory or to employ permanent workforce, as everything he needs, including product and service, can be procured or outsourced in the resource market. The only question is when and how the general contractor will be paid by the client for his periodic project expenses. Therefore, the financial burden of a general contractor is mostly derived from the need for revolving cash for project execution. The degree of the burden depends on how and when the contractor is rewarded as agreed upon in the contract. Further, the financial need of a general contractor is likely to be for the short term and directly correlated to the frequency of project progress measurement and valuation.

This paper intends to show how a general contractor's financial strategy is affected by the procurement method adopted by the client. First, some basic understanding of major procurement methods will be provided. To contrast the unique financial environment faced by general contractors, several financial indicators are employed for sector comparison. Based on the financial background of major public construction companies in Taiwan, some remarks regarding sustainable financial strategies for general contractors are given.

Procurement methods and financial environment in construction

The most traditional procurement method used in Taiwan is lump sum based on comeptitive tender. In cases where the exact bills of quantity cannot be fully determined before the tendering stage, a modified measurement-based lump sum method may be used instead. This particular method has gained its popularity among most pubic agencies in recent years. In yet more complicated situation, the client may opt for the unit price method, in that no bills of quantity is provided. In such cases, a set of working specification is prepared for and quality standard of materials to be used by the general contractor.

It is the design and build method of procurement which has received most critical attention in recent years (Turner, 1995). However, this method still awaits more serious practical implementation. With this method, the general contractor may be responsible for partial planning, design, and construction to fulfill the client's need. Since the project scope is effectively extended beyond construction, the financial concern would therefore become longer-term. On the other hand, the client is likely to provide some prepayment to the contractor to accelerate the project start-up.

An even more varied procurement approach is the Build-Operate-Transfer method. The basic idea of BOT method is that a private owner participates in building and operating infrastructure without incurring any financial burden on the part of the client. This has to mean that whoever builds and operates the facility will also fully absorb the financial burden and risks which may exist. Since the contractor has to encompass both construction and operation into his scope, the borrowings are likely to be much longer-term and enormous. To a great extent, the role taken by the contractor in a BOT project is little different from being a typical owner. In such a case, the borrowing is for investment rather than cash revolving.

The construction output in Taiwan is in the range of 6% to 10% of Gross Domestic Product (Chang, 1994), which is approximately an annual spending of 24 billion US dollars for the whole region or 1120 US dollars per capita. For this construction output, there are more than seven thousand construction firms together sharing the market. Out of the seven thousand firms, less than ten percent are considered financially well established. More than 60% of them are small-size firms which, according to regulation, can only take part in small-scale projects such as renovations, repairs, maintenance, etc.

The focus of this paper is on the former group of general contractors. Among them, some are with property development background and are deemed an integral part of a large corporation or consortium. In this case, several different contracting firms may in fact share the same financial base. A good percentage of the rest exert most of their effort in contracting government projects. For contrasting the special financial features of general contractors in Taiwan, the paper incorporates another three industry sectors for comparison, including the property development, textile, and electronics. Property development sector is selected for its similarity and strong relationship with the general contractor sector; textile sector for representing a sunset industry; and electronics for representing a booming industry.

1. Current assets

According to the securities data obtained from Taiwan Stock Exchange, the general contractors usually have a higher volume of current assets as well as current debts relatively. This confirms the argument that the financial attention of the general contractors is in fact on the short term. A high ratio of current assets to total assets indicates that the firm favors high flexibility of transferring its assets into cash or the equivalent. In Figure 1, it is apparent that general contractors, property development and electronics all have very high ratio of current assets to total assets in the range of 70% to close to 90%, while the textile industry is consistently below 50% for the years examined.

2. Current debts

A high ratio of current debts to total assets would mean a firm has imminent pressure in paying back the loans in the short term, mostly within a year. It also reveals that the firm's financial structure is highly leveraged and the risk of bankruptcy is relatively high. As shown in Figure 2, the general contractors have a much higher ratio of current debts to total assets among the four sectors. The closest being the property development sector is still at least 10% lower. The reason is that general contractors often acquire the needed cash by costly short term loans only, while other sectors are able to obtain loans of longer terms, in which case the firm's land, plant, and equipment can be used as the collateral.

3. Current ratio

Another interesting aspect of a general contractor's financial structure is the ratio of current assets to current debts or current ratio. A firm's ability to pay back all loans to be realized in the short term depends mostly on the current ratio. The current ratio is heavily influenced by the level of inventory and the amount of receivable. Generally, a healthy current ratio would range slightly above 200%. In Figure 3, it is observable that the general contractors have the lowest current ratio, indicating their relative deficiency in this aspect. In 1990 and 1991, the general contractors' overall current ratios are less than 100%, verifying the fact that the construction industry faced deep recession in those two years.

4. Total asset turnover

Total asset turnover is defined as the ratio of annual net sales to total assets. In a sense, this is to evaluate a firm's ability to generate income through the total assets it owns. In Figure 4, the general contractors' total asset turnover is better than the property development sector and the textile sector but is much below the electronics industry. This fact indicates two possible situations: (1) the general contractors are energetic enough to carry out sales equivalent to its base, namely, the total assets; and/or (2) the general contractors limit their asset base, i.e., economic scale, to improve efficiency. In the latter case, the firm may want to maintain its scale approaching to the perceived efficiency frontier.

5. Return on assets

Return on assets is defined as the ratio of earnings before tax to total assets. Since total asset turnover only reveals how productive a construction firm is, a more direct way to reveal the firm's profitability will be the return on assets. Among the four sectors, the general contractors are as marginal as the textile industry, while the electronics industry is more than three times that of the general contractors.

Based on the above five indicators, the financial environment of the general contractors is outlined and discussed. In addition to the information shown in the preceding text as well as the figures, some further observations are listed below:

1. The general contractors are concerned with short term financing much more than the borrowing of longer terms. The financial need is derived mostly from the execution of a project, or more specifically, for affording the interest of revolving cash for disbursements of services and products procured for the project before being paid by the client in the form of progress payments.

2. Since general contractors do not borrow money for the purpose of investment, they will only favor the least expensive source of finance available to them. In other sectors such as the property development sector, the same concern may not be as critical and far reaching, as long as the cost of capital does not exceed the minimum attractive rate of return.

3. Since most of the financial characteristics of the general contractors are attributable to considerations of project execution, the procurement method used by the client would have strong influence on a general contractor's financial strategy. In a sense, a general contractor's financial strategy serves the purpose of achieving the least expensive cost of capital. In retrospect, the performance of a project may also affect in the long run whether a general contractor is capable of assessing the most favorable source of financing.

In short, the financial concerns of the general contractors are short-term, project-execution based, non-investment oriented and highly leveraged. One of the major driving forces for all the concerns cited here is the procurement method adopted by the client. At one extreme, a general contractor would be least bothered financially, if each payment made by the general contractor to others is reimbursed in time by the client. At the other extreme, the same contractor would have to exert substantial financial strength, if he is to be reimbursed only when the built facility is transferred to the client. The next section attempts to elicit the risk implications of mainstream procurement methods on general contractors' financial strategy.

Risk implications of different procurement methods

The basic elements which distinguish a procurement method from others are (1) the content of the contracting service, (2) the awarding mechanism, and (3) the contractual enforcement. For the first element, the client may ask the general contractor to procure all materials needed for the project and provide all technical service necessary to the project completion. In other less common cases, the client may directly supply certain material items and/or call in other collaborating parties such as direct-hire labor, subcontractors, specialists, etc., to the project. In either instance, the general contractor would bear a unique set of financial risks. This in turn affects the general contractor's financial strategy.

For the second element, the awarding mechanism, the general contractor is concerned with the frequency of progress payment from the client, the percentage of earned value actually paid by the client for products procured and service subcontracted, and the quality standards as well as the related awarding procedures. This element in parallel with the project schedule decides the financial burden incurred to the general contractor in various points in time of the project.

The third element, the contractual enforcement, is for protecting the client from all potential contractual losses throughout the project life cycle. There are several major methods of enforcement which prevail in the construction industry, including the bid bond, the performance bond, the payment bond, the progress payment retainage, and warranty. Discussion about the purpose of each is beyond the scope of this paper. However, it is clear that the contractual enforcement also imposes substantial financial burden to the general contractor. The level of contractual risks for different procurement methods are compared in Table 1.

Table 1. Contractual Risk Implications of Different Procurement Methods

		Traditional Lump Sum Lump Sum with Measurement	Unit Price	Design Build	& B. O. T.
Clarification of Content	Contract High	Medium	Low	Medium Low	to Low
Preference of Mechanism	Award High	Low	Low	Medium Low	to Low
Degree of Enforcement	Contract High	High	N/A	High	N/A

Opinions compiled from interviewees (Hsieh, 1998)

After construction commences, the general contractor may face many sources of construction risk, including (1) fundamental risks such as war damage, (2) pure risks such as fires or storm, (3) particular risks such as collapse, subsoil subsidence, and (4) speculative risks such as change of underground conditions and unexpected poor weather conditions. With respect to the degree of involvement to the project, the general contractor will face different levels of the four sources of risk. At one extreme, in the case of traditional lump sum, the general contractor is free from all sources of risks except for the speculative risks such as poor weather. At the other extreme, such as BOT projects, the entire project is regarded as a package deal to the general contractor. And he is supposed to absorb all sources of risks under any circumstances. The risk spectrum of different procurement methods is summarized in Table 2.

Table 2. Construction Risk Implications of Different Procurement Methods

	Traditional Lump Lump Sum with Measurement	Sum Unit Price	Design & Build	B. O. T.
Fundamental Risks	Low	Low	Low	Medium	High
Pure & Particular Risks	Low	Medium	High	High	High
Speculative Risks	Medium	High	High	High	High

Opinions compiled from interviewees (Hsieh, 1998)

Analysis of the general contractor's sustainable financial strategy

From the above discussion, it is clear that the procurement method adopted by the client will fundamentally affect a general contractor's financial strategy. Since most general contractors are concerned with the source of short-term revolving cash, there is an obvious need for them to maintain the most efficient and least expensive source of short-term capital. In terms of short-term loans, the most essential means is accumulating credits through the execution of various projects. This consideration basically decides the method of banking which may attract general contractors. Alternatively, some major general contractors are also seeking public equity as company capital. Since equity capital is generally less expensive than borrowed capital, "going public" or being publicly owned is regarded by many the ultimate means for maintaining a construction firm's financial competitiveness. If going public is a general contractor's ultimate goal, profit maximization may become a less important objective than maintaining the long-term, stable growth. The main source of capital for a general contractor is compared in Table 3. In recent years, the volume of public investment on infrastructure in Taiwan is quick to reduce. Financially inefficient contractors have been most heavily impacted and strongly sensed the necessity of being able to acquire the needed capital from the public. For those contractors already gone public, the portion of capital from the public is on an average of 46.4%. Since, for these contractors, the average operating capacity, i.e., the ratio of total sales to total asset, is in the range of 100% to 200%, it is perceived that the equity capital acquired from the market may only be used as a contingency or for non-core business investments.

In a seperate study (Hsieh, 1998), the writers have also identified some key criteria for evaluating the general contractor's financial performance. Out of ten financial criteria, including Total Debt Ratio, Long-term Capital/Fixed Assets Ratio, Current Ratio, Quick Ratio, Receivables Turnover, Inventory Turnover, Return on Assets, Return on Equity,

Table 3. Preference of Capital Source for General Contractors

Type of Source	Source of Capital	Preference
Equity	Public: Major owners and market share holders	1
	Private: Single ownership or partnership	2
Debt	Long-term: Collateral-based loans	3
	Short-term: Credit, special loans or non-banking borrowing	4

Profit Margin and EPS, four of them show strong positive correlation to the general contractor's financial performance; in the order of degree, they are:

1. Inventory Turnover
2. Long-term Capital/Fixed Assets Ratio
3. Quick Ratio
4. Earnings Per Share (EPS)

Based on this finding, the following financial strategies are further suggested:

1. The general contractor should restrict the volume of construction-in-progress from being too much relative to his total assets. This is to reduce the level of inventory and improve the inventory turnover.
2. The general contractor should pay special attention and exert every effort in taking the company to public. This is the best way to secure long-term source of funding with least cost and highest level of flexibility.
3. The general contractor should try to maintain his quick ratio in the level of above 100%. This improves his ability in short term solvency. He should also be very critical upon cash equivalent, such as accounts receivable.
4. To upkeep the company's EPS and to avoid its dilution, the general contractor should exercise special care upon the issuance of new stocks.

Summary

Conventionally, the lump-sum method has been prevailing among major clients in Taiwan. Partially due to this practice, most general contractors are overwhelmingly concerned with short-term capital as the major source of revolving cash for project execution. Since different procurement method may impact a general contractor's financial strategy in a different way, it is gradually important for major contractors to pay attention to modern procurement methods, such as design-build and BOT methods. These innovative methods will certainly require a more prepared and longer-term oriented financial strategy. It appears that setting the goal for "going public" is a worthwhile exercise for most general contractors for obtaining the least expensive and most stable source of capital, for both the short term and the longer terms. Further, the procedure required for transforming a privately owned construction company into a public one is a key element in general contractor's financial strategy. If the trend of going public continues to evolve, the following two tracks of evolution may be predicted:

1. In the process of expanding the capital base through public funding, general contractors will begin to develop multiple core businesses beyond pure construction. Turnkey or BOT projects are all qualified as a stepping stone.
2. After establishing a larger capital base, general contractors will then come back to the issue of long-term borrowing. Since collateral is the key means to long-term borrowings, wise general contractors will have to pinpoint certain specific banking system which provides the most favorable collateral valuation.

Acknowledgment

The writers would like to thank Dr. Jung-Hua Hung in the Department of Financial Management, National Central University, Taiwan, R.O.C. for his instrumental remarks on the paper.

References

China Credit Information Service , LTD. (1996), *The financial overall analysis of industry and commerce in TAIWAN region*, September. (in Chinese)

Chang, L.-M. et. al. (1994), "Survey of Current Automation Needs in Taiwan's Construction Industry", Technical Report RT83030, Ministry of Interior, Taiwan, December. (in Chinese)

Hsieh, T.-Y. and Wang, H.-L., (1998) "Preliminary Analysis of General Contractor's Financial Strategy Under Different Procurement Methods", Third International Conference on Construction Technology, Singapore, May, 1998.

Hu, W.-L., (1998). The experience of construction management, Black Stone Publishing Inc. , January. (in Chinese)

Ross, S. A., Westerfield, R. W. and Jordan, B. D. (1995), *Fundamentals of Corporate Finance*, Richard D. Irwin Inc., New York, N.Y.

Turner, D. F. (1995), Design and Build Contract Practice, Longman Scientific & Technical, England.

Wu, Z.-Z. (1997), "The new tendency of diversified operation," Money Guide (Special issue of construction and steel stocks), August. (in Chinese)

Wu, Z.-Z. and Huang, Z.-Y. (1997), "The prosperity of construction industry is revivable and expectable," Money Guide (Special issue of construction and steel stocks), August. (in Chinese)

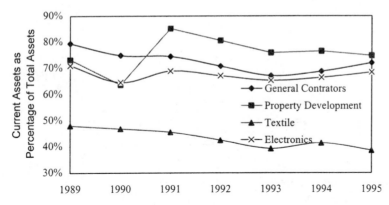

Figure 1 Comparison of current assets to total assets among four sectors

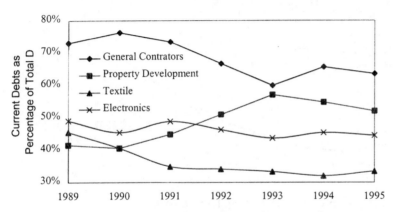

Figure 2 Comparison of current debts to total assests among four sectors

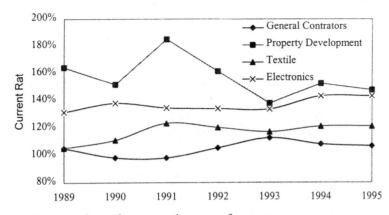

Figure 3 Comparison of current ratio among four sectors

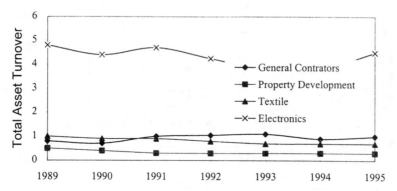

Figure 4 Comparison of total asset turnover among four sectors

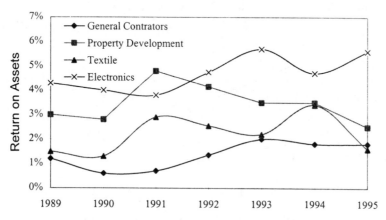

Figure 5 Comparison of return on assets among four sectors

EMPLOYMENT OF VALUE MANAGEMENT FOR DIFFERENT PROCUREMENT SYSTEMS IN THE CONSTRUCTION INDUSTRY OF HONG KONG

C. M. TAM
Department of Building and Construction, City University of Hong Kong, Hong Kong.
BCTAM@cityu.edu.hk

ALBERT P. C. CHAN
Department of Building and Real Estate, Hong Kong Polytechnic University, Hong Kong.
bsachan@polyu.edu.hk

Abstract

Although the traditional competitive tendering is still prevailing in the construction industry of Hong Kong, other types of procurement systems aiming at incorporating the technical know-how and management skills of constructors have been gradually adopted by the industry in the recent years. Hence, traces of Design and Build, Turnkey, BOT, Project Management, Construction Management, Management Contracting can be found in Hong Kong.

The authors have recently conducted a survey to investigate the problems in applying Value Management (VM) to different procurement systems. Results show that VM can be applied irrespective of the types of procurement system used. However, much resistance is experienced from the design teams in the traditional system in which architects tend to resist altering their original design according to the contractor's VM findings. Project management is found to be the ideal system in applying VM. The study presents a detailed comparison in applying VM to different procurement systems.

Keywords: Contracting, construction management, design and build, management project management, procurement systems, value management.

Introduction

Value management (VM) is a philosophy and set of techniques that has its origins in the US manufacturing industry (Male and Kelly 1989). It has been instituted since Larry Miles originated the concept in the late 1940s and early 1950s. However, the concept was not known to the construction industry until mid-1960s.

The delivery of a building involves many complex entities and contains an array of people including the architect, engineers, government officials, main contractor, subcontractors, suppliers and many other parties. The separation of design and construction requires a tremendous effort in co-ordination and communication. Hence, "value for money" in construction always poses a question mark for the clients. This has forced many clients considering to apply "Value Management". Value management is a structured, systematic and analytical process which seeks to achieve value for money by providing all the necessary functions at the lowest total cost consistent with the required levels of quality and performance.

The concept of value management is relatively new and has been introduced only for three years in Hong Kong. Its benefits and costs still wait to be proven. This paper details a study of applying VM in the different procurement systems for the construction industry of Hong Kong.

Objectives of the study

1. To describe how value management can be applied in different contract procurement systems.

2. To point out the advantages, disadvantages and difficulties when implemented VM in different contract procurement systems.

Timing of value management

Value management can be undertaken at any time but best results are obtained early in a program or before a process of change. It is because of the earlier application of VM that the greatest saving potential exists. If VM is applied at a later stage, the cost-saving has to be offset against the cost of re-design and possible delay in a building project.

Therefore, owners and designers should realise this and initiate studies early enough to realise maximum savings with minimum redesign efforts. In fact, the optimum time is at the stage where there is enough substance in the planned action or design and sufficient cost information or procedure timetable to pursue realistic alternatives. Therefore, the timing should be compatible with the extent of information available to the VM team (Patrick,1995). Common target points in the building process for a VM exercise were found to be:

-   pre-brief (user requirements)
-   concept (client brief, 10-20% design)
-   schematic (sketch design, 35% design)
-   production drawings (working drawings, 90% design)
-   post-contract or construction phase
    (Male and Kelly, 1989)

Procurement systems in Hong Kong

Non-traditional contract procurement systems like management contracting, project management, construction management and design and build, aim at incorporating the experience and management skills of the constructor into the design decision-making phase, reducing costs of construction and the time of delivery. Value management is a useful tool helping achieving these goals.

Different types of procurement system ask for different roles to be played by the constructor and have different effect on time and cost. Therefore, the application of value management is also different. In fact, these new procurement systems encourage the adoption of value management because the expertise of the contractor can be incorporated into a project's design decisions through VM studies. The contractor possesses the field experience, the ability to visualise construction methods, and the pricing information which can add credibility and expertise to the VM methodology, thereby providing the employer with a more valuable project.

In order to unveil the popularity of different kinds of procurement system in Hong Kong, a questionnaire survey to 20 large construction firms was conducted. The chart below summarises the findings:

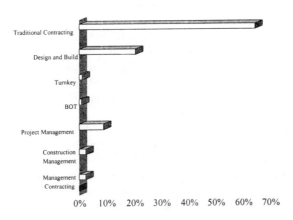

Figure 1: Average percentages of the respective procurement methods that respondents' companies have adopted in the past five years.

The above diagram shows that the traditional contracting approach is still most popular in Hong Kong. Other procurement methods are still new and their full acceptance is waited to demonstrate.

Another finding of the survey is that the respondents, in general, agreed that VM can be applied to different procurement systems, no matter whether they are designer-led or contractor-led types of approach.

Traditional contracting

The traditional contracting follows the procedure that the client produces a brief and sets a budget for the project which establishes the general scale and standard of a building. Then, the client will employ an architect to produce outline proposals and a sketch design. The quantity surveyor will develop a cost plan after received a sketch design. After the design work is completed, the client will select the contractor and to supervise the work till completion.

In this method, an architect is employed solely for the design and the contractor solely for his input in construction. The design is completed by a design firm before contractors are invited to bid.

Reasons for using VM in the traditional contracting

In the traditional contracting the outline proposals and sketch design are produced by architects. The constructor cannot make his contributions on the knowledge of buildability to the design. However, architects or designers may not be fully aware of the construction methods and generate a lot of unnecessary non-value-added costs in the design. If the cost plan exceeds the targeted budget, the design needs to be altered to bring the costs into line.

If the cost exceeds the budget marginally, the architect will tend to resist altering the basic design concept, and therefore, the redesign will usually involve down-grading the specification. However, if the cost exceeds the budget substantially, the design concept may have to be rethought completely. However, down-graded specifications frequently lead to compromised functional performance while complete redesigns involve a considerable amount of abortive work.

Even if the cost plans indicate that the project is within budget, this does not in any way guarantee that the client is getting value for money. It has been observed that designers or architects tend to rely heavily on experience when deciding on a design concept. Assuring value for money requires the consideration of a wide range of design options. Currently, this tends not to happen unless a rethink is triggered by a significant over-budget cost plan.

From the traditional design process, the feedback loops are costly and often devalue the building and delay the design process. Initiate VM can minimise these disadvantages. It is because VM team in the design phase can review the design and come up with suggested savings, a redesign is almost always involved. This can be done by in-house personnel of the owner or an external VM specialist or quantity surveyor as proposed by some practitioners.

Procedure of VM

a) Design stage VM

Value engineering can be carried out in a two-phase process for this procurement system. The first phase is to initiate a VM study in the design phase. The traditional design methodology incorporating VM encourages the designers/architects to consider a wider range of technical options, preventing a fixation on a single solution. Therefore, the VM team usually plays the role to reviews the design and come up with suggested savings.

The multi-disciplinary VM team also ensures that the various specialists involved in the design of a modern building are fully involved at the start in a parallel, rather than serial design process. Hence the influence of design decisions taken in a particular area on the overall design and project cost can be appreciated earlier.

However, there is a paradox in that it is difficult to have accurate formation in the early stage of a project. Some practitioners said that the end of the sketch design stage is the most suitable time while some advocated it is best in the briefing and conceptual design phases. In fact, the maximum benefit of VM is gained during early design stage. VM has the potential of being the first step to success or disaster, depending on how it is executed. Although at the initial procedure, there are lack of complete drawings and specifications. We should bear in mind that VM should be a continuous process throughout the duration of the project.

b) Construction stage VM

A continuation of the VM concepts, principles and techniques into the construction stage can be facilitated by the contractor, who is conversant with construction methods, techniques and costs. The roles of the contractor are to explore all value-added options within given constraints, such as specifications, budget and schedule, and analyse the constraints without sacrificing the intent of the project. The benefit of construction stage VM is that contractors can make use of their construction knowledge to help shaping the design of a project. As a result, the construction cost and future running cost and construction time can be reduced.

In order to achieve the goal of construction stage VM, clients and contractors must reach a mutual understanding on client's needs, preferences and requirements so that the contractor can have a clear target to follow.

Design and build

In this kind of procurement, the contractor provides the design and construction under one contract. The procedure is initiated by the client preparing his requirements on the finished building which are sent to a selection of suitable contractors. When a suitable contractor is selected, the client will leave the rest to the contractor to design and build. The client only needs to appoint an agent to supervise the works and ensure the contractor's proposals are complied with.

The contractor can initiate a VM study at any stage he wants. In fact, the client welcomes or actively encourages contractors to carry out VM studies because the contractor's tender price can be lowered and the period of construction shortened. Meanwhile, the design-builder can increase their competitiveness over other contractors. Moreover, a VM study can ensure the client getting value for money.

However, the design stage VM must be carried out at an early stage because any variations from the original design are discouraged by the contractor and, if allowed, are expensive. In the construction stage, the contractor can apply VM to input his construction knowledge. A research done by Akintoye in 1994 found that VM study is one of the factors that design-builders claimed to reduce overall project costs compared to the traditional method.

Consequently, VM studies by the design-builder can be done through:

- alternative solutions to engineering problems.
- early cost appraisal on efficiency of plan layouts.
- selection of alternative materials and methods of construction.

Management contracting

This kind of procurement method provides an alternative way of bringing construction knowledge and experience into the design process. The management contractor does none of the construction work himself but it is divided into work packages which are sub-let to sub-contractors. The management contractor provides a construction management service to organising, co-ordinating, supervising and managing the construction works in co-operation with the client's other professional consultants. Moreover, the management contractor also provides and maintains site facilities.

The management contractor can initiate and monitor VM activities. Value management program can be implemented at a minimal cost and within minimum disruption to normal operation through close communication with all disciplines. Co-operation of all parties is essential to good value engineering in this process.

Construction management method

Construction management is the provision of a professional management service to the owner of a construction project with the objective of achieving high quality at minimum costs.

The construction manager is responsible for the organisation and planning of the construction work on site and arranging it to be carried out in the most efficient manner.

The construction work itself is normally carried out by a number of contractors, each of whom is responsible for a defined work package.

In construction management projects, the construction manager's responsibility is to work with the designer during the design period. He can provide value management input. Therefore, in the organisation of construction management, it is very common that the construction manager owns a VM team.

Project management

The project manager is referred as the client's representative, with due authority to organise, plan, schedule and control the field work and is responsible for getting the project completed within the time and cost limitations. His duties involve working with the foremen, co-ordinating subcontractors, directing construction operations and keeping the work progressing smoothly and on schedule.

The project management approach provides an ideal vehicle for applying VM from the beginning to the completion of the project. The project manager can initiate and monitor VM activities at any time because he has the overall responsibility and broad authority over all elements of the project.

Moreover, in the large projects the project manager normally is a member of the firm's top management or who reports to the senior executive of the company. VM studies normally apply to large projects; therefore, if project manager believes that VM is beneficial to the project and willing to introduce VM to all development projects, it can be initiated without much difficulties.

In addition, VM studies can be successful because:

■ the project manager has full control of the job and provides the single voice that speaks for the project.
■ the project manager and the field superintendents work very closely together; therefore, a project team can assist in the VM study. The creative experience of an intensive value engineering workshop can do much to breakdown the professional barriers and foster a team spirit within the project.
■ the project manager's prime task is to co-ordinate client's requirements such that clear instructions from a single source can be provided to the other parties involved. Therefore, the project manager can concentrate on the functional requirements to materialise client's value for money.

Conclusion

All procurement methods encourage the adoption of value management; no matter whether they are designer-led or contractor-led. To use VM or not, the issue greatly depends on the client's preferences and the attitudes of the project team rather than upon the form of procurement. In fact, VM can shorten the construction time, reduce costs and provide value for money to the customer. It can foster a cooperation atmosphere rather than a confrontational attitude between the design team and the construction team.

Comparatively, applying VM in the traditional contracting system is more difficult than management-based contracting systems. It is because the value management services applied to the traditional contracting system create an adversary relationship between the architect and the constructor. Due to the pride of authorship of the design, designers often dislike any criticism on it. Moreover, if the design fee is a percentage of the project cost, lower project costs mean a lower fee to the design team.

While in the management-based procurement methods, there is an advantage of implementing VM. One of the reasons is the absence of an adversary relationship between the architect/engineer and the general contractor. Therefore, when the designer criticises the construction manager, he can equally be criticised by the manager. The equal positions provide a healthy climate for valuable ideas to be generated. Moreover, the management-based procurement systems facilitate buildability in the design.

From the experiences obtained by other countries, management-led contracting systems are more suitable to apply VM. In Hong Kong, however, the traditional contracting method is still prevailing. Alternative procurement systems are relatively new. Consequently, it may be the reason why VM is still not very popular in Hong Kong.

References

Male S. and Kelly J. (1989), "Organisational Responses of Public Sector clients in Canada to the Implementation of Value Management: Lessons for the UK Construction Industry." *Construction Management and Economics*, Vol.7, pp.203-216.

Lam, P (1995) "Application of Value Engineering in the Construction Industry", *Asia Pacific Building and Construction Management Journal*, Vol. 1, 1995, pp.18-24.

ZIMBABWEAN PUBLIC SECTOR PROCUREMENT STRATEGIES: TRIAL AND ERROR COMBINATIONS

P.P. GOMBERA and M.I. OKOROH
School of Engineering, Division of Construction, University of Derby, Derby, United Kingdom. p.p.gombera@derby.ac.uk

Abstract

In its effort to provide efficient public facilities for the next millennium, the Zimbabwean construction industry has had turbulent productivity and performance setbacks. These problems have been aggravated by the fact that modern day projects have evolved in structure, autonomy and arrangement, and do not necessarily conform to a traditional and "conservative" prescribed construction matrix of management. Emphasis on project accountability in the public sector has shifted towards private sector participation, profitability, constructability, environmental and project management issues. The continued use of the traditional procurement system has resulted in project delays, cost-overruns, advserialism, poor quality output and lack of teamwork. Harmony, trust and openness remain elusive in the industry. This paper expounds on the current procurement approaches used in the Zimbabwean public sector. It also evaluates the shortfalls and problems faced by the construction industry with regards to choosing and benchmarking the best procurement system.

Keywords: Affirmative action, economic structural adjustment programme procurement strategies, trust, Zimbabwean construction industry.

Introduction

Scant detail is known internationally about the Zimbabwean construction industry, yet it has the second largest vibrant stock investment market in Africa after South Africa. The Zimbabwean government has implemented dynamic and rigorous policies within the construction industry stretching back from the post-independence era (1980), when a new majority government got into power. This government's policy was to introduce measures that fostered maximum economic growth. At an international level, Zimbabwe is neither an economic nor a military giant and hence is only known to its trading partners and those countries who are members of various international fora such as the Non-aligned Movement, the Commonwealth Heads of Government Conference, Group of Fifteen and United Nations. However, within the African context, Zimbabwe is a major actor in the military, trade, political, industrial, manufacturing, agricultural and finance sectors. The country's total trade with countries in East, Central and Southern Africa was well over 2 billion Zimbabwean Dollars (Z$; 25 Z$ = 1 US$) in 1995. The Zimbabwean construction industry holds a pivotal role in the nation's economy. Over the past two decades the Zimbabwean construction industry has contributed immensely to the country's gross domestic product (GDP), assisted in improving employment levels, helped in economic growth and restructuring the unfavourable balance of payment which has been in existence. The Zimbabwean construction industry represents the biggest institution of fixed capital. It employs in excess of 150 000 people of which 1.3 million families are dependant on it for livelihood Mandaza et al (1996). The Zimbabwean construction industry represents about 4% of the GDP and its total workload amounts to Z$2 billion (25 Z$ = 1 US$). It is closely linked to building and allied industries such as cement, glass, steel, timber, asbestos, bricks and many others. On average as observed by Sozen & Gritli (1987), these industries contribute about 60% towards the industry's turnover. Furthermore, construction services and materials have found a market within the Southern African Development Community (SADC) region and the overseas market. Construction business in Zimbabwe has been on a steady decline during the 70's until the 80's, when the construction industry picked up slightly due to the introduction of government supportive policies.

Current procurement routes

The most popular procurement strategies utilised in Zimbabwe are the traditional (conventional), integrated System (design and build), construction management, private finance initiative (PFI) and labour-based methods. Both non-expert and expert clients have used the traditional system.

Design and build is seldom used in the public sector but is used more often in the private sector. It has been used for housing developments by central government agencies or corporations such as the National Security Authority (NASA), a workers' society for national benefits. The authority has in-house expertise and building teams that are responsible for the designing and building of urban low cost housing schemes. This systems has been in existence in a remote sense in Zimbabwe since the Iron Age (1200-1450 AD), when the famous *Great Zimbabwe Ruins* were built in Masvingo (town). The *Great Zimbabwe Ruins* is one of the oldest Southern-Sahara structures built by men. This structure has a perimeter wall that is 820 metres in circumference and 36 feet high. The ancient Zimbabwean craftsmen made this structure as an administrative capital for social and political courtship within their community. It was executed on what can be termed now as "design and build" arrangement. The craftsmen who built the structure used the abundant natural granite stone that was tailor-made for this project. The most amazing feature of this development is that, it was built without any mortar and is still structurally fit as of today.

The construction management or management contracting arrangement method is not widely used in Zimbabwe. This type of arrangement has been used by a handful of experienced public (expert) clients due to the influence of hired S. African consultants who sometimes are contracted to execute massive projects in Zimbabwe especially in the tourism and mining sectors. Zimbabwe Sun (Zimsun) Hotels, a subsidiary of Delta Corporation, a South. African (SA) conglomerate which also owns the Holiday Inn Hotels, is one of the most famous clients which has used this procurement method. Zimsum has its own in-house expertise that was trained in SA, where this method of procurement is used frequently.   Zimsun does undertake complex leisure and hotel renovations and developments.

Privatisation projects are procured using the build- operate -transfer (B.O.T) or build-operate-own and transfer (B.O.O.T). This works on the basis of allowing the private sector to take stakeholding and equity in public sector projects. The government has now moved dramatically towards the Private Finance Initiative (PFI) which has been helping to reduce the burden on taxpayers to procure public infrastructure. This system has gained the support of the public for the following reasons:

- Zimbabwe has limited financial resources to complete its infrastructure programmes and its major strategic projects such as bridges, motorways, airports and power stations. (foreign finance is provided by the sponsors of the bid)
- The provision of "foreign funds" Naoum et al (1994), into a country (such as Zimbabwe) to implement such projects will result in long-term employment creation and the improvement of Zimbabwe's reserves.

Popular specimen of a BOT project is the Limpopo River Bridge in Beitbridge (the border-post of S.A and Zimbabwe).

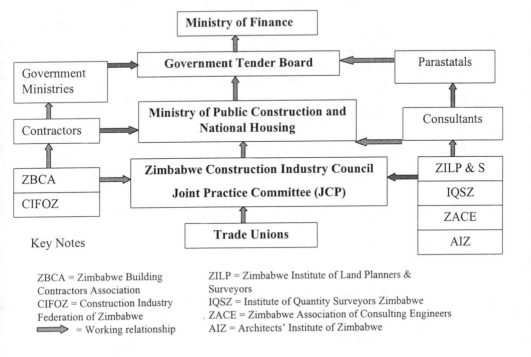

Figure 1: The Zimbabwean Construction Industry Structure

Labour-based contracting has been used in projects sponsored by state departments such as the District Development Fund (DDF), Ministry of Public Construction and National Housing (MPCNH), Agricultural & Rural Development Agency (ARDA). The sponsor provides the supervisory expertise while the local communities provide general workforce. This method has been successful in Zimbabwe to an extent that it is also being extended to urban areas in order to reduce unemployment. In Zimbabwe there are three forms of tendering methods generally utilised and accepted by construction service providers and clients. These are the open, selective and negotiated.

In most procurement systems described above, pre-contract periods are usually short resulting in clients being impatient to see actual site construction started. The modern sector of Zimbabwe's construction industry is similar to that of the former UK construction industry, and the major players include Clients (public) Planners, Architects, Engineers, (Structural, Mechanical, Electrical, and Civil) Land and Quantity Surveyors, Contractors, Sub-contractors, Trade Unions and Suppliers (see fig 1 for industry participants). Their duties and rights on construction projects are outlined in the Standard Form of Contracts supplied by their respective professional bodies and the MPCNH. The Architect or the engineer usually leads the design and construction team.

The Local government and ministries are the largest single employer of most construction entrepreneurs to either procure facilities or to act as consultants to the government. It contributes about 60% of the construction workload carried out in Zimbabwe. It has also another important function of co-ordinating fair practice among construction industrialists and procuring facilities for other central agencies/ministries. The MPCNH integrates and commission construction services for most government ministries. It maintains a register of all qualified building, civil engineering and specialist contractors who are normally invited to tender for government projects. It also keeps a list of various consultants. It has also its own in-house consultants. The MPCNH is also responsible for the commissioning of professional services and the awarding of tenders for major public project such as the proposed Harare International Airport. It is also responsible for the maintenance and management of state owned property both abroad and locally. Firms are registered for the type of projects they would want to execute and are classified by size (estimated project cost) of construction they can handle (see table 1)

Table: 1: General Contractors and Sub-contractors Classes of the MPCNH (1996)

Contractor	Limit	Sub-contractor	Limit
A	Unlimited	A	Unlimited
B	Z$15 Million	B	Z$6 Million
C	Z$9 Million	C	Z$2 Million
D	Z$4.5 Million	D	Z$1Million
E	Z$2.5 Million	E	Z$0.5 Million
F	Z$1.5 Million	F	<Z$0.5 Million
G	Z$0.9 Million		

25Z$ = 1 US$

Expatriate firms who are involved in large public works have dominated the industry. In the urban councils of Harare, Bulawayo, Mutare, Gweru and Chitungwiza the City Architect is responsible for the selection and approval of council contractors who are allowed to tender for council work. His department is also responsible for the design and supervision of all public works within the city council. The City Architect is a qualified Architect registered with the Architects Institute of Zimbabwe (AIZ).

Contract documents

Contract documents used in Zimbabwe are similar to those used in the construction industry in Britain in outlook and sometimes in content. These include architectural and engineering drawings, standard forms of contract, bills of quantities or schedules of rates, specifications (on civil engineering projects) and basic price lists. The contract documents have produced none other than an adversarial atmosphere which is a result of a mixture between dogma, increased project complexity and cost overruns. Although, specifications spell out the final desired outcome whether in building or civil engineering projects, in Zimbabwe they "bear a scant relationship" with the country's climatic and economic conditions.

Contract administration

Contract administration has been the same in Zimbabwe since the colonial era (since 1905). The Architect/Engineer leads the project team and supervises construction works. Poor project delivery mark the order of the day in most projects culminating in the following problems;

- delays in the design and production phases;
- inspection of vital project stages by consultants;
- the issuing of important instructions on projects;
- alterations in the design as consultants differ in opinion due to lack proper working drawings; and
- dispute among participants.

The contractor is also faced with delay problems in honouring interim certificates and a likelihood of project shelving if not sometimes cancellation by the government.

Difficulties with the procurement methods

Not only has the government introduced such policies as the indigenisation decree, which has had its fair share of problems for the past half a decade, but it has made them action plans for most public projects, from where most construction industrialists get most of their construction work. These policies have had a tremendous effect on the construction industry in as far as the practice and procedure of the industry is concerned. The three most important policies includes:

(a) Affirmative Action Policy (1992)

(b) Prescribe Asset Ratio (1990)
(c) Economic structural adjustment (ESAP) (1990)

Affirmative action policy (1992)

The 'Affirmative Action Policy' was introduced to redress the imbalance that existed in Zimbabwe before independence (1980), when the Zimbabwean construction industry was dominated by foreign multi-national and European firms. The main motive of some of these firms was to siphon the huge profits made from construction business to their parent countries of origin whilst leaving the local market in a state of crisis, and the industry not being able to develop itself fully. The majority of the local indigenous entrepreneurs did not have a say in the day-to-day running of the construction industry resulting in lack of indigenous participants. The affirmative action policy aims to encourage indigenous contractors and consultants to bid for public work by offering concessions in the following ways:

* Any government project less than Z$10 million will not be awarded to any construction company which is non-indigenous.

* Any project in excess of Z$10 million will be evaluated on the basis that the bid price given by an indigenous contractor will have 10% (discount) knocked off, to give them an edge over their counterparts who are non indigenous, for example if two bid were Z$30 million, one from an indigenous and the other from a non-indigenous, for comparison sake the indigenous bid would be Z$27 million (30 less 10%) resulting in the award of the project to the indigenous firm which would carry out the work at Z$30 million.

* For any government project awarded to a non-indigenous contractor, at least 30% of the work must be sub-contracted to small indigenous trade contractors.

This situation has caused multi-national corporations in Zimbabwe to acquire share holding in small indigenous firms or using indigenous entrepreneurs who own these companies as "fronts" for the multi-nationals who have a hidden agenda. Although this policy has faced criticism within the multi-national community, it has helped in the harmonising of the whole construction industry.

Prescribe asset ratio (1990)

The 'Prescribe Asset Ratio' allows for the equal distribution of public institution's investment portfolio. The general accepted ratio of prescribed assets is 65% on the money market, and 35% on other investment portfolios. This makes property investment the second choice after the money market. In such a case, most public investors are forced by legislation to invest in construction resulting in a boom in the construction industry. This law allows for fair distribution of wealth and gives advantages to those investing in the property market, which attracts less corporate tax (capital asset).

The economic structural adjustment (ESAP) (1990)

This was a policy introduced to promote and stimulate national and foreign investment in various urban and rural service centres of Zimbabwe after 1980. It was a grant loan afforded to the Zimbabwean government under the International Monetary Fund by the World Bank. The first phase of the grant worth Z$7 billion was invested in the development and improvement of national infrastructure, housing and commercial developments. The grant was used to transform the Zimbabwean economy into a more broad-based economy. The other major objective of this "finance package" was to create home-based, small and medium size industries especially in the rural areas where service centres were to be developed. As a result of this development, the construction sector benefited a lot in the sense that more building projects such as residential development, service centres and public infrastructure were erected to speed up accessibility and greater communication network within Zimbabwe. Between 1990-2 many small and medium scale construction firms were formed to participate in the construction of new projects. This was a result of various easily accessible loan facilities given to contractors below national bank lending rates.

Discussions and analysis

The results from this survey show that architects have complete domination of the project development process and its management. The findings also revealed that the majority of the architects are not willing to relinquish project managerial powers or to accept recent procurement routes namely Design and Build, Project Management, Construction Management, Management Contracting and Partnering.

The traditional procurement method introduced in Zimbabwe over a 100 years ago is still the most dominant. As old as it is, the traditional system possesses a socio-economic project atmosphere that does not exist in the construction industry of Zimbabwe. While some researchers and construction analysts such as Ofori (1985 & 1988), Aniekwu and Okpala (1988) have attributed the slow development pace of the construction industry in emerging economies like Zimbabwe as being due to the lack of proper contract strategies, profitability, innovation and co-operation amongst construction entrepreneurs. Construction business in Zimbabwe is no exception. Rwelamila and Hall (1994) on the hand concluded that successful project management whether in the developing or developed economy, is heavily correlated with construction entrepreneurs " choosing the right method" of working arrangement (procurement) within the environment of that particular economy. The whole construction fraternity is still fighting to come to terms with the issue of opening up to the new era of stakeholder (teamwork) management and the holistic approach to the choosing of proper and dynamic contract strategies for managing projects. It is this issue of teamwork which has been proposed by construction reformats such as Naoum (1989) and Naoum et al (1994), Hibberd et al (1994) and Rwelamila (1992) which would bring about the required "dynamic change" in any construction industry.

Benchmarking for harmonious procurement strategies

The economic liberalisation of the Zimbabwean construction market and the need by clients to achieve value for money on their projects has seen procurement systems become of paramount importance in satisfying clients' needs. The desire to reduce animosity and promote co-operation amongst the project team when procuring facilities, coupled with the need to use economical procurement systems, has made the use of workable and harmonious procurement arrangements the highest priority. This is in relation to time, cost and quality of projects built in Zimbabwe. Various clients (public/private), consultants, contractors and suppliers are still battling to find ways of tackling this serious dilemma faced in industry in order to focus on good contract negotiation, arrangement, management, operation and effective dispute resolution mechanisms on a project. All these has been necessitated by the need to plan and control the project budget taking into account all eventualities (risk apportionment) and finally to observe the stipulated work schedule. Management biased procurement systems have been seen to be flourishing in today's construction industry especially in most developed and developing economies. The idea of using appropriate procurement systems has also been reviewed recently by Sir Michael Latham's (1994) Report (Constructing the team). The Latham Report does not only propose for the use of various contract strategies but calls for a collective "radical decision approach" in selecting any contract procurement system to be used on a construction project.

Conclusions

The traditional procurement (B.O.Q) system still remains the most favoured method of procuring projects by most clients and consultants. This can be explained by the fact that clients are still sceptical about these new emerging methods and would rather stick to the method they are most familiar with mainly due to fear of carrying extra risk involved in other alternative methods such as construction or project management and partnering. Most construction participants in Zimbabwe agree that the "traditional" is a "hard" method of management and has grossly contributed to the adversarial attitude exhibited by the Zimbabwean construction industry. According to the survey, contractors are aware of the benefits of the new procurement strategies like Design and Build, Construction Management, Project Management and the Partnering arrangement. They are also knowledgeable about the benefits the project team could gain from using these arrangements. The rationale behind this argument is that the contractor has input contributions right from the project inception stage. Participation of the contractor in the early stages of construction with a view improves buildability, value engineering, innovation and improved management issues while offering **good value for money** to the client as proposed by the Latham Report (1994). The general attitude of construction participants indicates that there is no unity among the professions and individualism is rife. The construction industry does not speak with one voice due mainly to the fact that each contractor, client and consultant association is heavily interested in furthering the interest of its members. However, the general attitude expressed by the construction industry is that there is a need to use other procurement arrangements which promote *"trust, profit and harmony"* of parties involved in a construction development

References

Abrams, (1974), The Construction Industry in Developing Countries, *International Labour Organisation Vol.* 118, No 3 -5 June 1979.

Andrews J (1969), Construction industries in Developing Countries *1st ed. 169 Longmans (UCL) Publication London.*

Aniekwu, A.N., Okpala D.C., (`1987), The effect of systematic factors on contact services in Nigeria; Construction management and economics, 1988,6, pp.171-182

Edmonds, G.A. & Miles, D.W.T. (1983), Foundations for Change: Aspects of Construction Industries in Developing Countries.; J.T. Publications, London.

Fellows. R. (1989), Development of British Contracts, pp 14-19. *Contractual Proceedings held at the University of Liverpool* 6-7 April.

Gombera. P.P. and Okoroh, M.I. (1998), Zimbabwean Public Sector Procurement strategies: Trial and Error Combinations, Paper to be submitted in *Journal of Construction Procurement.*

Hall. K. & Rwelamila P. (1994), An Inadequate Traditional Procurement System? Where do we go from here? East meets the West. *Proceedings at the University of Hong Kong, CIB Publication* NI. 145.

Heralld. R. B (1995), Project Partnering (Principles and Practice) 1st ed., Thomas Telford, London.

Hibberd. P., Basden. A., Brandon. P., Brown. A, Kirkham. J. Tetlow. S. (1994), Intelligent Authoring of Contracts. East meets the West. *Proceedings at the University of Hong Kong, CIB Publication* NI. 145.

Holt. G., Olomolaiye. P., & Harris, F. (1995b), A Review of Contractor Selection Practices in the UK Construction Industry. Building and Environment, Vol. 30. NO. 4. pp 553-561. Oxford: Pergamon.

Kamaraswamy, M. & Dissanayaka, A. (1996), Procurement by Objectives, *Journal of Construction Procurement.* pp 38-51. Vol. 2. No. 2.

Liu, A. & Fellows, R. (1996), Towards an Appreciation of Cultural Factors in the Procurement of Construction Projects. pp 301. *East meets the West. Proceedings at the University of Hong Kong, CIB Publication* NI. 145.

Mandaza, I, Katerere, Y. & Moyo, J. (1996), Zimbabwe After Independence (An Open Mind on Developmental Issues), International Politics Forum held in Zimbabwe 3-7 Dec 1996.

Newcombe, R. (1994), Procurement Paths - A Power Paradigm, pp 243. *East meets the West. Proceedings at the University of Hong Kong,* CIB Publication NI. 145.

Naoum, S. G. (1989), An Investigation into Performance of Management Contracting and the Traditional Method of Procurement" Ph.D. Thesis, Brunnel University, Middlesex England.

Naoum, S. G., Mustapha, F.H. and Aygun T. (1994), Public Sector Procurement Methods used in the Turkey Construction Industry, *East meets the West. Proceedings at the University of Hong Kong*, CIB Publication NI. 145.

Ofori. G. (1985), The Construction Industry in Ghana, *Construction Management and Economics*, 1984, Vol 2. 1988, 3, pp 22-35.

Ofori. G. (1988), Managing Construction Industry Development, *Construction Management and Economics*, 1984, Vol 2, pp 127 - 132.

Potts, K. & Tommey, D. (1994), East compared with the West: A Critical Review of Two Alternatives Interim Payment Systems as used in Hong Kong and the UK. on Major Construction Works. *East meets the West. Proceedings at the University of Hong Kong, CIB Publication* NI. 145

Potts, K. (1995), Major Construction Works, (Contractual and Financial Management) 1st ed., Longman Scientific & Technical, Essex. UK.

Rwelamila, P.D. (1992):, The Factors Affecting Productivity in the Botswana Construction industry- the case of Gaborone. Botswana *Journal of Technology,* No. 1 pp.23-28.

Sozen, Z. and Grritli, H. (1987), Factor Affecting Construction Productivity, *Construction Management and Economics* Vol.2, No.1 1987.

5. Designing and Building

SELECTING APPROPRIATE PROJECTS FOR DESIGN-BUILD PROCUREMENT

KEITH R. MOLENAAR

School of Civil and Environmental Engineering, Georgia Institute of Technology, Atlanta, Georgia, 30332-0355, USA. keith.molenaar@ce.gatech.edu

Abstract

The use of design-build project procurement is rapidly increasing in the public sector of the United States construction industry. Documented design-build success and recent changes in federal procurement law are indicators of continued large-scale growth. The rapid growth, combined with a lack of experience among many public agencies, spawns a variety of issues that must be addressed. One issue is proper project selection. As new agencies experiment with design-build, appropriate project selection is a primary consideration affecting successful delivery. To date, there is no systematic or formalized method for selecting projects that are appropriate for design-build. This research presents a web-based automated predictive tool that can be used in the earliest stages of project selection to enhance the chances of design-build project success. The hypothesis proven true by this research is that there are public sector projects that are appropriate for design-build and success is predictable. Key project characteristics that are appropriate for design-build projects are identified and a statistically based model is presented. Using multi-attribute analysis, retrospective case study data collection methods, and multiple regression modeling, this research produces a model that predicts the success of the design-build process on future projects.

Keywords: Construction, contracting, design-build, decision support, project delivery

Introduction

Advances in project delivery systems and changing procurement laws are enabling U.S. public sector owners to experiment with innovative contracting techniques for the acquisition of new infrastructure and the built environment. Public policy has historically restricted federal, state and local governments to a traditional design-bid-build method of project delivery, but rapidly changing procurement laws are enabling these agencies to acquire construction via the design-build method of project delivery. Although design-build procurement is not new to the construction industry, it has only recently been adopted by the public sector. Increased use and inexperience in the public sector generates a need for fundamental research of the design-build project procurement process [Songer (1994)]. The need to identify a selection system is illustrated by a call for guidance involving the types of projects that are best suited for design-build ["AIA/AGC" (1994), "Experiences" (1993), and Spaulding (1988)].

Currently, there is no systematic or formalized method for selecting public sector projects that are appropriate for design-build. Projects are selected on a subjectively driven basis. One reason for this subjective selection is the general consensus that no single set of project characteristics dictates the use of design-build ["Design-Build in" (1992)]. While no specific characteristics dictate the use of design-build, this research demonstrates that certain project characteristics affect success. These characteristics can be used to build a formalized selection model.

This paper presents findings of a research project that provides a formalized selection model for public sector design-build projects. At the core of the model is a regression based predictive equation built on 122 retrospective case studies. Data collection was accomplished through a project case study questionnaire derived by a detailed multi attribute analysis. The multi attribute analysis of the problem is used as a framework for the predictive model. Predictions are made on projects' potential for success using design-build. These predictions are the basis for the formalized selection model. The final form of the selection model is a Web-based advisory system. Specifically, this paper discusses the research methodology, data collection and practical application.

Methodology and data collection

The research methodology is simply stated in three main steps; 1) problem definition, 2) proposed solution and 3) validation. Performance data is collected from completed design-build projects, this data is forms the model, and the model predicts the success of design-build on future projects. The analytical tools used in these stages are multi attribute analysis, retrospective case study data collection, and multiple regression modeling. A graphical representation of the methodology, along with the analytical tools employed at each stage, are shown in Figure 1.

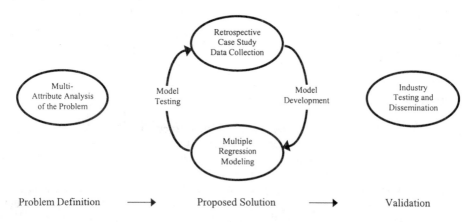

Figure 1. Research methodology

As previously noted, there is a need for a formalized procedure which public sector agencies can employ when selecting projects for design-build. This formalized procedure is referred to as a "model." For purposes of public sector design-build selection, the model must be available for use in the earliest stage of project definition. This stage is what the American Institute of Architects terms "discovery."

There are countless unknowns at the discovery stage of a project. Even with the best planning, there will be many unforeseen events in a construction project (hidden soil conditions, adverse weather, material supplier strikes, etc.). Owners need a model to help select the path with the best chance for success. A prediction model is the most suitable vehicle in determining this path. In this problem, the model will predict what projects have the greatest potential for success using design-build.

To create this formalized model, actual project data must be collected and analyzed. The obvious choice for analyzing this data is a statistical approach. The purpose of a statistical approach is to analyze variables and determine how they affect outcomes. A purely statistical approach works well in cases with a small number of variables and simple definable outcomes, but when selecting projects for design-build, the number of project variables and the complexity of project outcomes becomes too overwhelming for statistical methods alone. Therefore, additional analytical technique of multi-attribute is implemented.

Multi-attribute analysis

To deal with the complexities that arise in the statistical methods, a second analytical tool is added to the model: multi-attribute analysis. Multi-attribute analysis is an approach that compares complex decision alternatives. It is appropriate for analyzing subjective, imprecise information and the interdependency of variables. Multi-attribute analysis is an analytical tool employed to break down large complex decisions into smaller, measurable components. These smaller measures can then be synthesized into one single score representing the complex decisions. In this manner, complex decision alternatives can be compared. In 1966, J.R. Miller pioneered multi-attribute analysis while developing methods to evaluate transportation route alternatives [Miller (1970)]. This research compares various candidate projects for design-build.

A multi-attribute analysis of this problem results in the development of hierarchy of characteristics that effect project success. Figure 2 displays the highest level of the multi-attribute hierarchy and its four main categories.

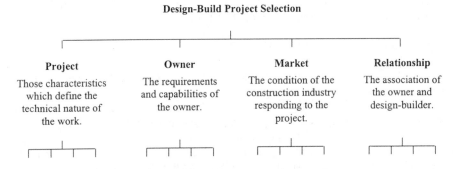

Design-Build Project Selection

Project	Owner	Market	Relationship
Those characteristics which define the technical nature of the work.	The requirements and capabilities of the owner.	The condition of the construction industry responding to the project.	The association of the owner and design-builder.

Figure 2: Highest level of multi-attribute hierarchy

These main categories are broken down until all characteristics are measurable. The final multi-attribute hierarchy of the problem includes 44 project characteristics in the four main sections above. The next section discusses the data collection based on this multi-attribute hierarchy.

Case study questionnaire

The method selected to collect data for the statistical analysis is a retrospective case study questionnaire. The questionnaire format is similar to the multi-attribute hierarchy. It follows the branches of the hierarchy and asks the respondents to measure the criteria. These measurements form the statistical analysis, which produce the weights of the prediction model.

The questionnaire's intent is to find associations and correlation between project variables and project outcome. Questionnaires cannot usually show causal connection between events, but can indicate correlation [Oppenheim (1966)]. Therefore, respondents are not asked their opinions about why projects were successful or unsuccessful, they are asked about objective project variables that may correlate to success. For example, respondents are asked to quantify the project contingency. They are not asked if they think that a larger contingency would have made the project more successful. The next section discusses the actual data collection.

Data collection

The case study questionnaire was prepared and initially sent to 512 public sector owners. Of the 512 owners contacted, 104 completed the survey correctly, 20 completed the survey but the projects were not yet finished, 5 completed the survey incorrectly, and 78 replied that they did not have proper experience to answer the survey or could not complete the survey at this time. This represents a 40% response to the survey. An additional 18 surveys were solicited form other owners to test the accuracy of the model, but these additional surveys were not included in the model construction or in the population characteristics described below.

The predictive model would ideally work on projects of all types and sizes. To accomplish this, a wide range of project types were required for data modeling. Projects of various size, construction type, and from various agencies were collected. The total dollar volume of the projects is $3.1 billion (all amounts are in US dollars). The smallest of these projects was valued at $29 thousand and the largest valued at $780 million. The projects represented various construction types.

The goal of obtaining an evenly distributed sample was met in some respects better than others. A wide variety of agencies at the federal, state and local levels speak well to the diversity of the sample. Also, the wide range of project cost is advantageous to developing a versatile model. The undistributed portions of the population stem from the majority of the cases being federal building type projects. 75 % of the projects were federal and 82% of the projects were of the "building" type. Although theses percentages are representative of the design-build population at large, the data skews the resultant models towards federal building project types. The population characteristics described above relate to the regression modeling presented in the remainder of this paper.

Multiple regression modeling

Regression modeling is extensively employed in fields outside of construction such as psychology, sociology, marketing, and econometrics. The technique is considered underutilized in the construction area and appears to offer promising possibilities [Russell (1992)]. There has been recent success in the area of predicting construction disputes utilizing a similar methodology [Diekmann (1995)]. It is not the purpose of this research to extend the knowledge base of regression modeling, but to further the understanding of its use in the construction industry.

The regression modeling approach is employed when there is evidence for a hypothesis that one variable (*X the independent variable*) causes another variable (*Y the dependent variable*) to change. The definition of covariation is X and Y varying together in a systematic (non-chance) way. Linear regression is a fundamental multiple regression technique used to analyze covariation. In this research, the independent variables are the project characteristics as determined through the multi-attribute analysis. The dependent variables are the measures of design-build project success determined by the authors' previous research [Songer and Molenaar (1997)]. Five multiple regression models are created by the five measures of success: 1) budget variance, 2) schedule variance, 3) conformance to expectations, 4) administrative burden, and 5) overall user satisfaction. The particular form of multiple regression modeling used in this study is the classical linear model. Variation in a dependent variable (Y) is expressed in the form of linear mathematical model of independent variables ($X_1, X_2, \ldots X_{K-1}$). The model is expressed by the following equation:

$$Y_i = \beta_0 + \beta_1 X_{i1} + \beta_2 X_{i2} + \beta_3 X_{i3} + \ldots \beta_{K-1} X_{i,K-1} + \varepsilon_I \tag{1}$$

where:
is the value of the dependent variable;
$\beta_0 - \beta_{K-1}$: are constants or regression coefficients;
$X_0 - X_{K-1}$: is the value of the independent variables;
ε: is an error term;
i: is the index of the individual cases; and
K: is the number of parameters.

Practical application

The remainder of this paper describes the construction and operation of the Design-Build Selector (DBS). The DBS is a Web-based advisory system built to aid in the selection of appropriate projects for design-build contracting. The architecture of the program, data collection, system calculations, and output interface are explained. An example of the advisory system is presented for illustration. The program itself is available via the World Wide Web through the University of Colorado Construction Engineering and Management's web server. The URL for the DBS is:

http://www.Colorado.EDU/engineering/civil/db/

DBS Architecture

The DBS is the product of research conducted into the performance of the design-build method of project delivery. As previously stated, regression models have been constructed to predict design-build project performance in terms of budget variance, schedule performance, administrative burden, conformance to expectations, and overall satisfaction. These mathematical models are the engine which drives the sensitivity analysis for the DBS and in depth interviews with experienced owners provide the advisory information for the system. The mathematical models are quite complex and the advisory information is voluminous. The DBS masks the complexity of the data and produces a simple graphical interface can users receive the only the most critical information for their project.

Input level (data collection)

The DBS is an interactive decision support system. Owners input information concerning potential design-build projects. The DBS responds with an overall score and advice for the owner's project. The first step in the interface is to gather pertinent information from the owners. The DBS accomplishes this through two questionnaires: a *project characteristic questionnaire* and a *success criteria questionnaire*. This information is then utilized by the DBS to generate a predictive score.

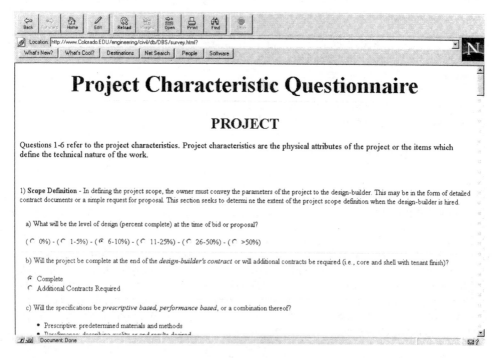

Figure 3: project characteristic questionnaire

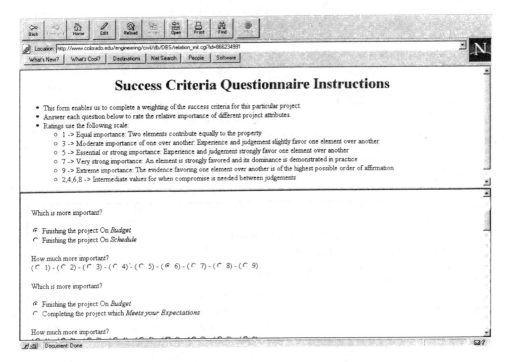

Figure 4: success criteria questionnaire

Retrospective case studies were used in the original data collection used to build the DBS. The project characteristic questionnaire is similar to the case study questionnaire used to collect the original data for the DBS modeling. Whereas the retrospective case study questionnaire collected data on completed projects, the DBS collects data on projects that are in the earliest stages of design for future projects. Figure 3 illustrates the project characteristic questionnaire.

As noted, there are five models used to predict project success: 1) overall user satisfaction, 2) administrative burden, 3) conformance to expectations, 4) schedule variance, and 5) budget variance. The significance of these five models varies on a project by project basis; owner's criteria for project success change. On one project, schedule could be the overriding criteria for project success, and budget or quality can be sacrificed to achieve minimal schedule variance. On other projects, budget might be of paramount importance and the other criteria may be sacrificed. The success criteria questionnaire illustrated in Figure 4 collects weighting data from users for each specific project. The pairwise comparison data interface collects in the same manner as the project characteristic data interface. The input screen is similar, but now has two frames. The top frame is the definitions for the weights and the bottom frame is the actual data collection interface.

The data collected from the project characteristic questionnaire and the success criteria questionnaire is stored in a temporary file for use in the predictive model. Owners may save data from numerous projects for purposes of comparison. The questionnaire may also be viewed from the output screen as discussed later in this paper. The collection of data through the project characteristic questionnaire allows for the generation of a predictive score and the customization of output advice.

Processing level (system calculation)

The processing level is the mathematical engine for the DBS. At this level, the data collected from the questionnaires is transformed into the useable output for design-build project selection. The processing occurs in three categories of processing predictive linear equations, processing success criteria and overall score, and processing the variable analysis. The processing of these functions then generates the output for the user. The user is masked from the tedious calculations required to score and weight the five regression models.

As previously stated, individual projects have different priorities for success. The success criteria portion of the DBS processing synthesizes these different priorities into one overall score. In this manner, different projects with their individual priorities can be compared with one overall score. The method of combing these individual model scores is through a linear weighted model. The weightings for the linear model are derived from the pairwise comparison. The manner in which this weighting of success criteria is accomplished is through the use of a weighted pairwise comparison. The weighted pairwise comparison uses a matrix-base equations to generate a linear equation for combining the method into one overall score. This method has been extensively tested through the Analytical Hierarchy Process (AHP). For a detailed discussion of AHP readers are directed to Saaty (1990).

The true benefit of the DBS is not in the overall score, but in the variable sensitivity analysis. It is through the individual variable analysis that the lessons learned from this research are transferred to the user. The analysis allows the user to quickly access the information that is most critical to their project without being overwhelmed by an excess of information. Often, it is not as important to know which variables cause the greatest change, but which variables are causing a poor score. As previously stated, the DBS allows for the comparison of multiple projects through overall scores, or it allows for the improvement of a single project through the advisory system. The advisory system is linked to the equations through the variables.

Output level

The DBS output interface presents the data processing results. It is divided into three tiers of detail. Tier 1 is the overall project score. This tier is the quickest to use but reveals the least information. Tier 2 is the variable analysis, which contains more information, and has links to the third tier. The third and final tier contains the advice as determined through this research. The actual DBS output interface is shown in Figure 5.

The overall project score is displayed for the user in two manners. First, through individual model scores (as shown in the upper left frame of Figure 5). Second, through one combined overall score (not shown in Figure 5 for clarity). Both scoring displays offer a comparison to an average score as determined through the original case studies. There are benefits to both display methods. The combined overall score allows for quick comparison of multiple projects. The projects with the highest DBS overall scores are the most appropriate for design-build contracting. The individual model scores offer the owner more insight into the project than the overall score. The five model scores are shown graphically through a bar chart. The graphical interface gives a visual reference as to model scores, and to which models have been weighted most importantly. Again, an average project score (with identical success criteria weightings) is offered for comparison.

The variable analysis portion of the DBS is located in the upper right frame of the output interface (see Figure 5). The purpose of the variable analysis frame is to reveal how the individual variables are affecting the overall score. The individual variables are listed in a tabular format. The variables are rated as "good," "average," or "poor." These ratings correlate to the impact on project success. The variable analysis screens are accessible via a pull-down menu located under the overall score graph (not shown in Figure 5 for clarity). There is a separate variable analysis for each of the five models.

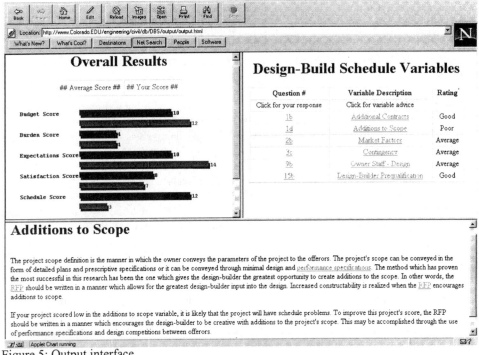

Figure 5: Output interface

The advisory information portion of the DBS contains the lessons learned through this research as previously discussed. The variable analysis format allows quick access to pertinent project advice. The most pertinent information relates to the variables with "poor" or "average" ratings. The user accesses information about the variables through the hyperlink in the variable analysis frame. In the example in Figure 5, the additions to scope variable within the schedule model has a "poor" rating. The user accesses information about the variable and learns why this variable was rated "poor." This is accomplished by selecting the variable itself from the variable analysis frame. The variables are hyperlinks that display the advice in the advisory window. Once selected, the advisory window displays the information regarding the variable score as seen in Figure 5.

The advisory system is a series of textural instructions and comments discovered through this research. These instructions are based on the statistically significant correlations found through the regression equations and in-depth interviews to interpret causal connections. The complexities of the mathematical model and correlations are masked by the user interface. The user only needs to see the overall score and variable ratings to receive the full benefit of the research.

The DBS and an on-line tutorial/demonstration of the DBS system are available at the URL listed in the beginning of this paper. Readers are encouraged to use the tutorial program and explore the html source code that drives the DBS.

Conclusions

Current growth trends and documented success suggest that design-build use will continue to increase. This study extends the understanding of design-build in the construction industry. Specific contributions include predictive project selection and an enhanced understanding of design-build contracting in the public sector. This research provides both theoretical models and interactive tools for project selection.

The results of this research will help owners to start right with the design-build process by choosing the most appropriate projects. It is then up to the owners to stay on the right path by diligently constructing accurate RFPs and making themselves available to the design-builders throughout the lifecycle of the project.

The writers would like to thank all of the public sector owners who contributed their expertise and time to the data collection for this research. Without their willingness to participate, this study would not have been possible. This material is based upon work supported by the National Science Foundation under Grant No. CMS-9410683.

References

"AIA/AGC Recommended Guidelines for Procurement of Design-Build Projects in the Public Sector." (1994) *Rep.*, American Institute of Architects, Washington, D.C.

"Design-Build in the federal sector, a report of the task committee on design-build." (1992) *Rep.*, ASCE, New York, N.Y.

Diekmann, James E., Girard, Matthew J. (1995). "Are Construction Contracts Predictable?" J. Constr. Engrg. and Mgmt., 121(4), 355-363.

"Experiences of federal agencies with the design-build approach to construction." (1993). *Tech. Rep. 122*, Fed. Constr. Council, Consulting Com. On Cost Accounting, National Academy Press, Washington, D.C.

Miller, James (1970). Professional Decision Making, Praeger Publishers.

Oppenheim, A. N. (1966). Questionnaire Design and Attitude Measurement, Basic Books, Inc.

Russell, J., and Jaselskis, E. (1992). "Predicting Construction Contract Failure Prior to Contract Award," *J. Constr. Engrg. and Mgmt.*, 118(4), 791-811.

Saaty, Thomas L. (1990). Multicriteria Decision Making: The analytical Hierarchy Process. AHP Series Vol. 1, RWS Publications, Pittsburgh PA, 1990.

Songer, A. D., Ibbs, C. W., and Napier, T. R. (1994). "Process Model for Public Sector Design-Build Planning," *J. Constr. Engrg. and Mgmt.*, 120(4).

Songer, A.D., and Molenaar, K.R. (1997). "Project characteristics for successful public sector design-build." *J. Constr. Engrg. and Mgmt.*, 123(1), 34-40.

Spaulding, V. M. (1988). "Newport design-build - a study on integrating the Newport design strategy into the NAVFACENGCOM facilities design and acquisitions process." *Rep.*, Naval Facilities Engrg. Command, Alexandria, Va.

THE PROCESS PROTOCOL: IMPROVING THE FRONT END OF THE DESIGN AND CONSTRUCTION PROCESS FOR THE UK INDUSTRY.

MICHAIL KAGIOGLOU, RACHEL COOPER and **GHASSAN AOUAD**
TIME Research Institute, Centenary Building, University of Salford, Peru Street, Salford, M3 6EQ, United Kingdom. M.Kagioglou@dct.salford.ac.uk

JOHN HINKS
Department of Building Engineering and Surveying, Heriot-Watt University, Edinburgh, UK

Abstract

Sir Michael Latham [1994] published a report, which identified fragmentation and confrontational relationships as the greatest barriers to improving quality and productivity in the UK construction industry. It was recognised that manufacturing industry is not as plagued with such problems having introduced a number of improvement initiatives over the past twenty years. One such improvement that has been adopted is the stage gate approach to new product development (NPD). This approach enabled progressive management and monitoring of the 'whole project' lifecycle for all activities involved in the product development with particular attention to the 'front end'.

This paper summarises the main findings of a funded project, which involved a number of industrial partners from the whole spectrum of the UK construction industry. Particular attention is given to the front-end of the design and construction process and a Generic Design and Construction Process Protocol (GDCPP) map is briefly described. IT is also presented as an enabler in undertaking a consistent process and an IT map is presented.

The paper describes how a co-ordinated effort, from a client perspective, at the front-end of the Design and Construction Process could not only improve efficiency but reduce development times, improve quality of the final product and increase predictability.

Keywords: Front end, design and construction, generic, manufacturing, process

Introduction

The technology and knowledge transfer of manufacturing experiences and principles into a construction environment have been the subject of research in the UK for a number of years. Indeed this is further emphasised by the formation of the IMI (Innovative Manufacturing Initiative) and in particular the 'Construction as a Manufacturing Process' sector, which funded the project described in this paper.

There are a number of manufacturing operations which could prove to have a number of similarities with their construction counterparts. The manufacturing industry's continuous mission, towards effective and manageable production and materials handling techniques have been the subject of consideration in the past. The attention has been in utilising material resources and machinery to implement a building, refurbishment or other solutions offered to client. This attention has resulted in moderate elimination of non value-add activities in a construction site and the consideration of resources much closer than was the case in the past. However, the majority of the construction activities are the result of non (strictly speaking) construction activities. These include the identification of the client needs, the formulation of the requirements specifications for the design solution and finally the design and production information. Typically these activities contribute between 10-30 percent of the overall construction budget. Therefore it is fair to say that between 70-90 percent of a construction project's cost is determined during the initial 10-30 percent spend. This inevitably draws the attention at the 'front end' of the construction project.

The manufacturing industry has undergone the same way of thinking as described above, but at a significantly earlier time than the construction industry. Back in the 60's and 70's new management, production, and materials handling techniques have been conceptualised and in many leading companies implemented - particularly in Japan. MRP, MRPII, Kanban systems, JIT, CADCAM and others have initiated this revolution. The focus was to utilise computer technology for undertaking activities faster (CAD), more accurately (robotics), using less stock and plant space (MRP ⯈ JIT), whilst eliminating significant changes during the transition and transformation of a design into a final product.

New product development (NPD) processes have been developed in manufacturing so that the whole development of products from the first steps of identifying a need or capturing a clients' need to the final delivery and replacement of a product can be considered as parts of 'one' consistent process. NPD in its literal sense does not exist widely as a process undertaken by the construction industry but its principles could potentially be applied to the construction industry [Kagioglou et al. 1998]. For the majority of projects undertaken by construction firms the NPD process is enacted once only, each time with a different combination of distribution, production and delivery techniques. That said, it could be argued that an element of commonality exists between each 'one-off' NPD process in construction - as it will be demonstrated. Furthermore, IT can play an important role in utilising and NPD process for construction.

This paper examines the development of an NPD process for design and construction and the use of IT as an enabler within a consistent process.

The project

The Design and Construction Process Protocol project was funded by EPSRC under the IMI 'Construction as a Manufacturing Process' sector. It's two-year duration involved the participation of a number of companies from the whole spectrum of the UK construction industry e.g. clients, contractors, subcontractors, consultants, suppliers and IT specialists. Its main aim was to develop a design and construction process protocol based on Manufacturing NPD and to examine how IT can facilitate such a process. The methodology employed throughout the project involved the distribution of questionnaires within the industrial partners companies, semi-structured interviews with the industrial partners and industry experts, interactive workshops to inform and validate the development and the expert help of a large manufacturing co-operation.

Finding a common ground

Any attempt made to determine the common issues between manufacturing NPD and the design and construction process will have to be based not only on the individual elements of the process but on the undertaking of the process itself in terms of organisation and post project issues. Based on manufacturing industry process development throughout the past three decades the research team concentrated on the following issues in an attempt to find the 'common ground' when developing the design and construction process protocol:

- Overall process structure/framework and decision making mechanisms
- Identification and classification of the steps involved in the design and construction process with particular attention to the front end
- Organisation of project participants to utilise the process
- Learning organisations
- The role of IT as a facilitator/enabler to the process
- Cultural issues

Figure 1 presents those issues in the Generic Design and Construction Process Protocol.

Process structure

Rosenau [1996] notes, that process models are "an effective way to show how a process works" and as a definition:

"A process map consists of an X and a Y axis, which show process sequence (or time) and process participants, respectively. The horizontal X axis illustrates time in process and the individual process activities (and) or gates. The Y axis shows the departments or functions participating in the process..."

Beyond this convention, there appears to be little formality in the method used to represent a process. The research team and industrial partners of the research project have agreed to employ such a method by representing the project/process phases on the X axis, and the project participants on the Y axis. An established process of this kind is proposed by Cooper [1993] which in addition introduces the concept of 'gates' between the stages of the project for decision making purposes. These gates or phase review points review the project and a report is produced (phase review report) which aims to inform the client of the status of the project. In an attempt to streamline the process and eliminate the sequential undertaking of projects, as was traditionally the case, Cooper [1994] introduced the concept of 'fuzzy' gates. These gates are referred to as 'soft' gates in the context of the process protocol. They represent decision points but with the exemption that the project does not necessarily come to a halt but certain activities are allowed to continue. In contrast 'hard' gates require the temporary overhaul of the project until a decision (usually made by the client body/representative) grants the continuation of the project and resources are allocated.

Process stages/steps

According to Hughes [1991], the identification of the steps or stages through which a construction project passes is essential if improvement is to occur. Hughes asserts that:
"...every project goes through similar steps in its evolution in terms of stages of work. The stages vary in their intensity or importance depending upon the project."

The industrial partners to the project and the research team concentrated on the formulation of an agreed description of those steps that were consistent to the process structure whilst at the same time providing for decision making and imply generic properties that could be applied by all project participants. In addition the effort concentrated at the steps for the effective undertaking of the front end. The process of deciding on the process steps involved a number of debates through workshops which were informed by established process steps such as those of the RIBA plan of work [RIBA, 1991]. Furthermore, the steps in the process were classified into broad stages as shown below:

- Pre - project stage:
  - Demonstrating the need
  - Conception of need
  - Outline feasibility
  - Substantive feasibility study and outline financial authority
- Pre - construction stage:
  - Outline conceptual design
  - Full conceptual design
  - Coordinated design, procurement and full financial authority
- Construction stage:
  - Production information
  - Construction
- Post completion stage:
- Operation and maintenance

Project participant organisation

The earlier involvement of the project's participants, throughout the process is a significant development of the conventional approach to building. Traditionally, a construction project's participants are referred to by their professional or expert status. Ball [1988] demonstrates how this may be attributed to the inherent class relations associated with each of the professions and expert groups. As with all class distinctions, the effect that this basis for organisational structure in design and construction has is division. A consequence of this traditional approach, by which even the more recent forms of contract procurement (design and build, management contracting, etc.) are included, is the poor communication and co-ordination commonly associated with construction projects.

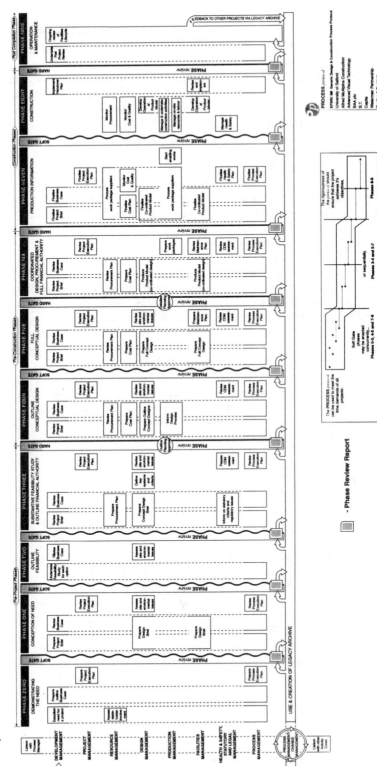

Fig. 1 The Generic Design and Construction Process Protocol

The participants in the Process Protocol are referred to in terms of their primary responsibilities, and are represented on the Y-axis of the Process Model. It is recognised that traditionally, project to project, organisational roles and responsibilities change, resulting in ambiguity and confusion [Luck & Newcombe, 1996]. By basing the enactment of the process upon the primary responsibility required, the scope for confusion is potentially reduced, and the potential for effective communication and co-ordination increased. The Process Protocol groups the participants in any project into 'Activity Zones'. These zones are not functional but rather they are multi-functional and they represent structured sets of tasks and processes which guide and support work towards a common objective (for example to create an appropriate design solution). The Activity Zones for the Process Protocol are the following: Development management, Project management, Resources management, Design management, Production management, Facilities management, Health and safety, statutory and legal management, Process management, and Change management.

The majority of the activity zones are self-explanatory but for the process management. These activity zones are essentially the interface between the Development Management (predominately the client body) and the other project participants. Process Management has a role independent of all other activity zones. A distinction must be made between this conventional view of a project manager and the Process Management role. Process Management, as the title suggests, is concerned with the enactment of the *process*, rather than the *project*. Key to the success of each Phase in the process is the production of project deliverables (reports and documentation associated with each phase). In this respect the Process Management is responsible for facilitating and co-ordinating the participants required to produce the necessary deliverables. Acting as the Development Management's 'agent', it will ensure the enactment of each process phase as planned, culminating with the presentation of the deliverables at each end of Phase Review.

Learning organisation

A learning organisation learns from its experiences, which can either be successful or unsuccessful and applies those lessons to future projects. Furthermore, it can apply those lessons to train future and current employees. To do so an organisation will need to store and maintain the relevant data in an efficient and structured manner. The Process Protocol aims to provide this by the creation, use and maintenance of a 'Legacy Archive'. The structure and undertaking of the process offers itself to such a mechanism due to its consistency through the phase review reports for each phase of the project development. These phase review reports and other relevant documentation can be stored either on a paper driven system or ideally based on IT where the data can be interrogated and presented in different views. Furthermore such a system could provide a communication spine for the whole project team through the creation of Intranet sites and other IT based solutions

IT: an enabler to the process

The legacy archive presented above is only one of the tools that can be created by utilising IT. There are a number of IT solutions that could be applied throughout the lifecycle of a project to undertake diverse activities from modelling of a projects' solution to the production of 'manufacturing' information and operation of the finished facility in a virtual world. The aim is to provide an integrated IT system where IT can be utilised. Attention is drawn in implementing IT from a process focus rather than was the case traditionally where IT solution were and in some cases still are, developed as stand alone tools without a clear place in the design and construction process. Figure 2 illustrates how IT could be integrated in the Process protocol and enable communications between project participants, as well as speeding up the development of a project solution.

Aouad et al [1998] suggests that when IT usage is similar to that shown on the IT map it should be possible for the client to walk through and interrogate various aspects of the designed building, such as cost and specification using VR, and also the information stored in the integrated database. Partnership arrangement could offer long term benefits in that respect.

The future

It is most probable that the process protocol will itself be a catalyst for change and it will require to change flexibly as the industry adopts it partially as part of its change process. The evidence from the prospective cases is that change management and process management are seen as some of the key process changes which need to be put into place first; and that co-ordination of information technology or information systems is another essential requirement. These activities will allow the process to be controlled in a consistent and predictable manner, which will allow the maximum enhancement of capability to accommodate product and process innovation. It will be essential to benchmark the processes as a whole, across the industry as well as the individual organisational processes. This may be a national bench marking initiative which can prove to be a useful mechanism for promoting the process protocol and the changes in individual organisational processes required in order to move to something more harmonised than the existing fragmented system.

It was considered to be an essential element of the process protocol that the efforts of the process team were loaded towards the planning effort at the pre-project and pre-construction phases of the process. This assumption might allow the minimisation of ambiguity and late stage design changes, thereby allowing a greater pre-production planning efficiency and a smoother and more controllable production phase to be achieved. This would also allow the greater predictability associated for production and would also allow the more rapid and cost

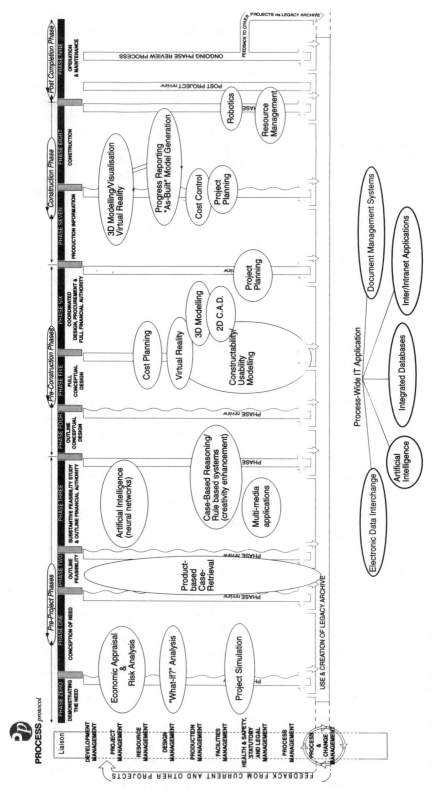

Fig. 2 IT and the Process Protocol

effective introduction of standardisation prefabrication, the current problems of which tend to occur at the interfaces between trades and elements for which pre-planning effort is required to make improvement. It is absolutely essential that risk and risk allocation is considered as an explicit functional role within the early stages of stakeholder involvement and assembly of the virtual design and construction team.

Towards the end of the project it was considered that rather than it being a generic process protocol it may be more representative to consider it as being a prototypical process model; that is a process model which in itself provides a prototype of how the

local customised solutions may be designed in a more typical set of rules than exists currently, and also that this should be done at a strategic rather than operational level.

The effective implementation of the Process Protocol will greatly depend on its ability to effectively translate the strategic to the operational level. To this end further work is needed (and is currently under way at the University of Salford) in examining the sub-processes (Activity Zone) and produce generic maps for those sub-processes.

In such a way the underlying principles and philosophies of the Process Protocol will form the framework for company/project based wide adoption and effective implementation. This is confirmed by the adoption of the Process Protocol by the CRISP Process Group with the comment that sub-level process definition needs to be defined. In addition such developments will address all the issues identified in the Construction Round Table (CRT) 'Agenda for Change'

Conclusion

The principles of the process protocol can therefore be summarised as a model which is capable of representing the diverse interests of all the parties involved in the design and construction process, which is sufficiently repeatable and definable to allow IT to be devised to support its management and information management. A mechanism by which the systematic and consistent interfacing of the existing practices, professional practice and IT practice support tools can be facilitated. Also, simplicity within the protocol, which allows its interpretation and flexible application at a variety of strategic levels across a variety of scales of project using combinations of virtual teams and IT systems should be accommodated. They should all be based in clarity of terms of what is required from whom, when and with whose co-operation, for whom the requirements are to be delivered for what purposes and how they will be evaluated (during phase reviews). Other principles underlying the process protocol where the standardisation of deliverables and roles associated with achieving managing and reviewing the process and the product. The introduction of organisational wide and industry wide co-ordinating process improvement programmes which incorporated the facets of process capability and information technology/information system capability; all of which is grounded in a philosophy of early entry (at the front end) of the maximum proportion of stake-holders and functionaries, with a predominant emphasis of effort on the design and planning to minimise the error and reworking during the construction phase.

The process protocol is divided into a series of sub-stages defined as pre-project, pre-construction, construction and post-completion, and within each of those major stages there are sub-phases which can be operated concurrently or concatenate to make the process more efficient in smaller scale projects.

The process protocol introduces a number of novel principles in a number of areas, in particular the extension of the boundaries of design and construction process into the requirements capture phase of pre-briefing client decision making. Also the extension of the boundary of the process beyond practicable completion to allow the management of use and the learning from performance in use to improve the product and process for future projects. In addition the creation of an explicit process management and change management role to co-ordinate the functionaries and phase deliverables associated with the process, the information that supports the functional roles and is delivered via the creation of products, and a stable platform to allow innovations in process and products and operations to be facilitated in a co-ordinated and repeatable manner.

References

Aouad, G, Cooper, R, Kagioglou, M, Hinks, J and Sexton, M (1998) A synchronised process/it model to support the co-maturation of processes and IT in the construction sector. *Proceedings of CIB Working Commission W78 - Information Technology in Construction Conference,* Royal Institute of Technology, Stockholm, Sweden, June 3rd - 5th , pp.85-95.

Ball, M (1988) Rebuilding Construction. *Routledge*, London, UK.

Cooper, R G (1994) Third-Generation New Product Processes. *Journal of Product Innovation Management*, Vol.11, pp. 3-14.

Cooper, R G (1993) Winning at New Products: Accelerating the Process from Idea to Launch. *Addison-Wesley Publishing Company*. Canada.

Hinks, J, Aouad, G, Cooper, R, Sheath, D, Kagioglou, M and Sexton, M (1997) IT and The Design and Construction Process: A Conceptual Model of Co-Maturation. *International Journal of Construction Information Technology,* Vol.5, No.1, pp. 1-25.

Hughes, W (1991) Modelling the construction process using plans of work. *Proceedings of an International Conference - Construction Project Modelling and Productivity*, CIB W65, Dubrovnik

Kagioglou, M, Cooper, R, Aouad, G, Sexton, M, Hinks, J and Sheath, D (1998) Cross-industry learning: the development of a generic design and construction process based on stage/gate new product development processes found in the manufacturing industry. *Proceedings of the Engineering Design Conference '98 - Design Reuse*, Brunel University, Uxbridge, UK, June 23-25, pp. 595-602.

Koskela, L (1992) Application of the New Production Philosophy to Construction. *Technical Report* 72. Technical Research Centre of Finland, Finland.

Latham, M (1994) Constructing the Team. H.M.S.O.

Luck, R and Newcombe, R (1996) The case for the integration of the project participants' activities within a construction project environment. *The Organization and Management of Construction: Shaping theory and practice (Volume Two), Langford, D A and Retik, A (eds.),* E. & F. N. Spon.

R.I.B.A (1991) Architect's Handbook of Practice Management, *RIBA*, 4th ed., UK.

Rosenau, M (1996) The PDMA handbook of new product development. John Wiley & Sons, Inc., U.S.A.

Acknowledgement

The research team wishes to thank IMI for providing the resources for the research project and the participating companies for their contributions to the project.

DESIGN AND MANAGING FOR SUSTAINABLE BUILDINGS IN THE UK

D. A. LANGFORD, X. Q. ZHANG, I. MACLEOD and B. DIMITRIJEVIC
Department of Civil Engineering, University of Strathclyde, 107 Rottenrow,
Glasgow G4 0NG, United Kingdom. d.langford@strath.ac.uk

T. MAVER
Department of Architecture, University of Strathclyde, 107 Rottenrow,
Glasgow G4 0NG, United Kingdom.

Abstract

This paper discusses the concept of sustainability and particularly its development within the built environment. It looks in details at the principles and indicators for sustainability of buildings and its assessment methods.

Keywords: Buildings, environmental management, sustainability assessment, sustainability indicators

Introduction

Sustainable issues are the subject of a great deal of debate. Growing concern about the threat of the environmental catastrophe has coincided with a downturn in the property market and building industry in the UK at the end of 1980s and in early 1990s. The low rate of global economic growth further moves property owners and managers to choose buildings, which are environmentally friendly and cheap to run (Watts, 1993).

Although a huge amount of literature on sustainability has been produced, the concept of sustainability has been open to wide interpretations in different contexts. The most commonly used definition was provided by the World Commission on Environment and Development in 1987. 'Sustainable development is development that meets the needs of the present without compromising the ability of future generations to meet their own needs' (World Commission on Environment and Development, 1987). On the other hand, the wide interpretations of the term itself also reflect the interest of all areas of professions and sectors in the sustainability issue. Following the 'Earth Summit' in Rio in 1992 and adoption of the international document 'Agenda 21', such interests quickly turn into the government's commitment to sustainable development in various areas of policy and practice (Jacobs, 1993). The building sector bears substantial responsibility in achieving the overall sustainability of the earth. The global concern on sustainability and the government's commitment require the building sector to act differently from the conventional approaches. For all actors in the life cycle process of buildings, techniques or strategies to achieve sustainability within the sector and the whole environment in large are very much high on the agenda.

The concept of sustainability in the wide context and within the built environment

More than two thousand years ago, Chinese philosophers like Lao developed a complex idea about sustainability. Lao's philosophy particularly emphasises that human being and nature is an integral system. Human being must live in harmony with nature. It emphasises the dependence o f human beings on the nature. Human activities should enhance the natural phenomena. *Feng shui* theory was particularly developed to provide a guide that how human construction activities can enhance the natural environment. This could be regarded as the first systematic approach towards how to reduce the human impact on the nature. Thousand years later, this idea of keeping nature and human activities in a balanced harmony was revived on a world wide scale because people realised that our future does depend on the nature, while noticing that the carrying capacity of the earth is limited.

The 1980s rethinking on sustainability and the relationship between human activities and the environment produced fundamental changes in our ways of development in terms of policy and practice. The world has never been so united as in the way that we are concerned with the issue of sustainability. The adoption of 'Agenda 21' was a good start for global action on this issue. Since sustainability concerns both human activities and nature, therefore the issue of sustainability needs to address both the human system and the rest of the environmental system.

Human systems and environmental problems

Human civilisation in its current form has developed since the beginning of the domestication of plants and animals and the creation of settlements. Such activities have met the needs of ease of physical life and facilitated the cultural and social development. This process increases the sophistication of human organisation, for example, increase in settlement numbers and advance in technical skills (The Institute for Policy Analysis and Development, 1992). The development of human skills in using natural resources to serve human purposes gradually challenges the natural environment at the local levels and then at the global level. However, such challenges have not been to a catastrophic level until this century.

According to the Institute for Policy Analysis and Development, the most profound environmental problems we face today include the following (the Institute for Policy Analysis and Development, 1992:27-28):

• *'Soil erosion.* About one third of the world's arable land is being degraded, largely due to poor management. About 7 % of the world's cropland was lost between 1980 and 1990. In spite of this world agricultural production is still increasing. This is because soil losses are being masked by increased levels of agricultural inputs. For example, between 1950 and 1980 fertiliser inputs in developed nations increased from 22.3 kg nutrieds per hectare (n/h) to 16 kg n/h. This means that current increase in agricultural output depend on the continuing availability of inputs. However, most of the inputs are until now depended on fossil fuel. At some point the supply of fossil fuels will start to contract, while the level of human population and hence of demand for food is continuing to grow. This will make the loss of soil fertility a more serious problem' (*ibid*. p. 27).

- *Threat to biodiversity.* There are three levels of biodiversity: generic diversity, species diversity and ecosystem diversity (Blum, 1997). The most often studied is the species extinction. The current extinction rate is probably 100 to 150 species per day. This will change the evolutionary conditions for life on Earth, with unpredictable consequences.

- *Ozone depletion.* Chlorine catalyses the destruction of ozone. About four fifths of the current chlorine loading of the atmosphere has been added by humans acvitities. The destruciton of ozone layers allow more ultraviolet radiation through to surface of the planet, which probably damages crops and marine organisms, thereby reducing yields, as well as causing blindness and skin cancer.

- *Global warming.* As a result of various human activities, additional emissions of certain chemicals are into the Earth's atmosphere. This is likely to cause a number of effects. One effect could be a rise in sea level. A temperature rise to 3.0 °C could lead to a sea level rise of about 1 metre, which would flood about 3% of the coastline land surface of the Earth. Unfortunately, that 3% would contain land for housing the world's many big cities and the world's best farmland. Another effect could be changes n the world's pattern of rainfall, which could result in yet unknown consequences for agriculture and forestry.

- *Population Growth.* For most of human history there have only been a few million people at any one time. However, the population have been growing rapidly for the last two centuries. It is now over 5 billion, which could double again in 35 years, and will certainly continue growing for sometime yet. Unlike early population increase in developed countries, the current increase is mostly in developing countries. Thus rates of population increase, current levels of resource consumption, and existing infrastructure are not in a coherent relationship. This will be particularly problematic in meeting the needs of the additional population.

- *Resource demand.* The economic and population growth put great pressure on resource demand. Given population growth rates, a five- to ten- fold increases in world manufacturing output will be needed if the developing nations have the same living standard of the developed nations by the end of next century. 'This means that similarly large increases in efficiency would be needed just to keep levels of resource demand and pollution out at today's levels. There are also some important cases where resource shortages are likely to cause problems in the relatively near future. There is a considerable potential for conflict. Such issues may become more common. In other words, while most resources will not simply run out, it will be necessary to deal with the pressures created by increased scarcity and growing demand' (the Institute for Policy Analysis and Development , 1992:27-28).

- *Ecological transitions.* 'Life on Earth is an immense ecological system. All ecological systems adapt to change. The process of evolution is one of continuous adaptation and change. It is not realistic, therefore, to think of Earth as having some 'natural' condition to which it would return if humans ceased to exist. What is important, form the human point of view, is that all ecological systems also have a number of points beyond which they cannot recover to the same essential state that they were in before. It is possible that one or some of the changes in the Earth's ecology being brought about such a transition. For example, one possibility is that global warming will start to defrost the Arctic tundra. This would release large quantities of methane, which is a greenhouse gas. This would cause further global warming, and so on. Such positive feedback effects could lead to a situation in which there was no longer any possibility of remedial action. It is not known just how likely such effects might be, but there are clearly some grounds for concern' (*ibid*. p. 28).

The building sector contributes much to the above environmental problems. Buildings are particularly related to the issues of Ozone depletion, global warming and acid rain. The next section will discuss these issues in detail.

Buildings and environmental issues

1. *Buildings and natural environment*. Buildings will inevitably use land. A lot of natural agricultural land has been lost because of buildings. Most of the world's cities are concentrated on the coast areas that are normally the best land for crops and other agricultural production. The consumption of the best agricultural land areas by the construction of buildings inevitably leads people to search alternative and less productive lands for food productions. As a result, grass land and forest was dramatically decreased. The habitat of animals has been affected especially when the site of buildings and cities are located in an environmentally unfriendly way. They can also contaminate the land. Buildings affect not only the surface of the earth but also the atmosphere of the earth.

2. *Buildings and ozone depletion*. The use of compound chloroflucarbon (CFC) is the principal chemical agent causing depletion of the ozone layer. A number of building components use CFCs including some forms of insulation, air conditioning, refrigeration plant and fire-fighting systems. The solution is to avoid or substitute the use of CFCs in buildings. The most significant substitutes could be Hydrochlorofluorocarbon (HCFCs) and Hydrofluorocarbons (HFCs) which both have similar functions but much less negative impacts on the ozone. Meanwhile, the international community seeks to reduce the use of CFC products in the first place. A European Commission regulation bans almost all the production of CFCs by 1997 (Johnson, 1993).

3. *Buildings and global warming*. The earth's low atmosphere is covered with a protective blanket of greenhouse gases, which includes carbon dioxide, methane, nitrous oxide and CFCs. The greenhouse gases absorb some infra-red radiation and keep the surface of the earth warmer and avoid the rapid change of temperature at the surface of the earth, therefore provide a suitable environment for the survival of human and animal life. However, the dramatic increase of greenhouse gases by human activities will also cause problems and make the surface of the earth too warm, which could cause dramatic negative impacts on human beings and animals. Buildings are mostly related to the release of the carbon dioxide and CFCs, which respectively accounts for 50% and 25% of the greenhouse gases. It was estimated that half of carbon dioxide produced in UK is related to the use of buildings (*ibid*.).

4. *Buildings and acid rain*. Acid rain was mainly caused by the emission of sulphur dioxide to the atmosphere. Most of which is from the burning of fossil fuels for the supply of power. Acid rain will cause damages to building materials such as stone and

certain metals. Carbon dioxide can result in problems with concrete. Plastics and paints can be affected by ozone and photo-oxidants (*ibid.*).

Building and sustainability

The study of sustainable buildings immediately concerns the issue of what is to be sustained in the building sector. Sustainability itself is a global issue which involves a wide range of issues and concerns. As far as buildings are concerned, what level of sustainability should be discussed? Or at what scale should the sustainability in the building sector be pursued? At large, The sustainability of buildings should contribute to the ecological sustainability of the earth. But at local levels, many also concern the sustainability of buildings as an economic sector. The sustainability of buildings can involve different objectives or components:

1. *The ecological sustainability.* The sustainability of buildings should contribute to the ecological sustainability of the earth. Sustainable buildings are environmentally friendly buildings. New building activities should enhance the natural environment.

2. *Economic sustainability.* The sustainability of buildings should not affect the economic performance of the building sector and the national economy as a whole. It should increase the employment opportunity within the sector.

3. *Social sustainability.* Sustainable buildings should maintain and improve the existing social structure.

4. *Cultural sustainability.* Sustainable buildings should maintain the cultural richness of human society.

5. *Levels of sustainability.* The sustainability of buildings can be discussed at different levels. It can be discussed at the project level, aggregate building type level, the building sector level, national level and the global level.

Sustainability indicators or principles for buildings

Ecological sustainability

1. *Site planning.* The location of a building has a profound impact on the environment as well as on the building. Buildings should avoid contaminated land and causing potential damages to the surrounding environment.

2. *Orientation.* The choice of orientation should make a building easy for natural air ventilation and ensure an adequate exposure to sunshine for the health of building users.

3. *Building materials.* Reducing the use of non-renewable materials for building use. In order to avoid the use of non-renewable products, the sources of building materials should be specified to ensure that the materials are renewable and the use of non-

renewable is reduced to the minimum. Since the man-made chlorofluorocarbon (CFCs) contribute to both the ozone depletion and global warming. Its use should be avoided. Special attention should also be paid to other environmentally unhealthy materials such as asbestos, solvent-based paints, lead-based paints, timber treatment pesticides and formaldehyde when choosing building materials (*ibid.*). Recycled building materials should be pursued. Attention should also be paid to the embodied energy used in the production of materials.

4. *Energy efficiency in buildings*. Energy is primarily a non-renewable resource. The product of energy mainly comes from the burning of fossil fuels, which again leads to the emission of greenhouse gases and other harmful chemicals to the atmosphere. Energy use in buildings accounts for about half of the UK's emissions of carbon dioxide, of which half are from housing. Through energy efficiency measures, 20-30% of energy and emissions could be saved. Therefore energy efficiency is an important issue for sustainable buildings. New actions have already begun in this area. A comprehensive review of Building Regulations is under way, aiming to achieve energy efficiency and reduce the emissions of carbon dioxide in buildings. Regulations require new energy efficiency standards apply to new buildings and certain adaptation work. The Home Energy Conservation Act requires local housing authorities to identify energy conservation measures for housing and to provide efficient energy services for residents in their area (DETR, 1998).

5. *Waste*. The production and use of buildings should achieve the minimisation of waste production. In 1995, there are some 70 million tonnes of construction waste produced annually (DETR, 1998). A large amount of waste is generated through the process of building use. With regard to waste, the fundamental goal is to minimise the waste production in the first place and to deal with waste in a most environmentally friendly way. A classification of waste is needed for waste management. The unavoidable waste should be considered the possibility for reuse and recycling.

Economic sustainability

1. *Environmental capital*. Environmental capital is a monetary expression of the depletion of natural resources and the deterioration of the natural environment in the process of material production or other human acitivities. The traditional economic theory does not consider the environmental capital. Traditional economic thinking treats natural resources as free gifts. The profits from the treatment of natural resources are regarded as revenue. The sustainable development requires treating the depletion and deterioration of natural resources as capital consumption, which includes any loss of value of natural resources by human activities including the deterioration of environmental quality (Mikesell, 1992). Building activities and maintenance require taking environmental capital into account.

2. *Investment*. Investment in green products should be encouraged. It should also encourage investment in pollution abatement facilities and environmental research and training to achieve a highly qualified environmentally aware labour force. Investment in the use of renewable resources such as energy generated from wind, tide and sun, and the use of CHP (combined heat and power).

3. *Employment*. The building sector should promote employment. Manual labour should be recruited from local labour force in order to reduce the transportation and accommodation supply, which in turn reduce energy and natural resource consumption.

Social and cultural sustainability

1. *Reducing the social cost for future generations*. The activities of past generations have transferred the associated costs to the current generations. Sustainability requires that our current efforts should not yield social costs or should minimise social costs for our future generations. We should leave the natural and human environment to our future generations without or with as little damage as we received from the past generations. Let future generation enjoy equal social and environmental development opportunities as we do.

2. *Quality of life*. Sustainability is a global issue. Our building activities should not transfer our social cost to the poor country. We should not exploit the natural resources from the less developed nations to achieve the sustainability on a narrow national level. It should value natural resources highly and so less developed countries could also improve their life by increasing the price of natural resource products and reducing the rate of depletion of natural resources in turn.

3. *Preserving social and cultural diversity*. The developed countries bear particular responsibilities for this. Any aids to developing countries should not be attached to social and cultural conditions for changes. Social and cultural hegemony is not in harmony with sustainability. At local levels, the preservation of historical and natural heritages and local culture and custom should receive particular attention in building design and construction.

4. *Equal opportunity*. It requires preserving equal opportunities for socially disadvantaged and disabled people. The equal opportunities in distribution of social costs and benefits should be encouraged among different ethnic and social groups.

Assessment of sustainability

Many building related decision-making bodies now require taking into account of the sustainability issues in their business. Therefore, a scientific assessment of sustainability of building projects or use management options becomes increasingly important. Such an assessment could involve the choice of building materials, energy use options and energy consumption, and life cycle analysis of building performance related to environmental issues.

Choice of building materials

1. *Eco-labels*. The eco-label method marks the building materials in terms of levels of their environmental friendliness. It looks at both the initial source and the production methods of the materials. Countries like Germany, Canada and Japan have already

produced operational eco-labelling programmes, which include part of the following elements (European Commission, 1995):

- Materials made from waste paper
  - Materials made from recycled glass
  - Materials made from recycled gypsum
  - Low-pollutant varnishes
  - Asbestoes-free floor coverings
  - Paints low in lead and chromates
  - Insulation from recycled wood-based cellulose
  - Heat recovery ventilators
  - Products from recycled plastic
  - Water-based paints
  - Thermal insulation

The aim of the eco-label schemes is to provide decision makers with information on the impacts of building materials and on how to promote less environmental impact solution during the life cycle of buildings.

2. *Environmental Hierarchy Method.* This method considers the impact of materials on the environment over a building's life cycle. Early work has been done by Woon and Energie in the Netherlands, which concentrates on materials and building components (*ibid.*). However, it does not consider the durability of building materials, which could affect its usefulness. For example, one type of materials during its life cycle has less negative environmental impact, but the life cycle of the material is too short (not durable), it might have higher negative environmental impact than another durable material over the whole life cycle of a building instead of materials themselves. Durability also can reduce the frequency of repairs and replacement and save energy consumed in maintaining materials. Therefore, the assessment of environmental friendliness of materials should include the following issues (*ibid.*):

- durability of materials
- shortage of raw materials
- renewability of raw material resources
- availability of raw materials at local level
- ecological harm by extraction of raw materials
- energy consumed in all stages
- water consumption
- noise and odour pollution
- harmful emissions such as those leading to ozone, global warming and acid rain
- health aspects
- risk of disasters
- reparability
- re-usability
- waste

According to the importance of each above aspect, a points system can be established. By comparing the total score of each material, we can establish a hierarchy of materials in terms of environmental friendliness and provide a choice of environmental preference.

Energy use

Since energy is primarily generated from non-renewable fossil fuels, the efficient use of energy becomes an important element in achieving sustainable buildings. The assessment of energy use in buildings could include embodied energy and energy in use.

1. *Embodied energy.* Embodied energy can be calculated in two different ways. One is energy consumption for each cubic metre of the material in question and another for each tonne of the material.

2. *Energy in use.* Energy in use can be looked at different levels. First, it can look at the total annual energy consumption per space unit (per sq. metres or cubic metres of usable space) for different types of buildings. Second, the annual energy consumption for different building services such as heating, cooling and lighting within the same building type. Third, looking at different components and design and management options and as well as external conditions such as weather in affecting each energy consumption in buildings.

Levels of sustainability assessment

Sustainability assessment can be done at different levels. Sustainability issues can be dealt at different levels – the top level (global level), local level, building industry level and building project level.

1. *The highest level.* The top level is concerned with global issues which includes environmental quality such as the global warming, ozone depletion and pollution; natural resources such as depletion of natural resources and recycling of materials; energy and waste.

2. *The local level.* Local governments concern a lot on local economical sustainability while sustaining the environmental quality. It involves the employment and economic growth, site planning, impacts of buildings on the local environment such as noise and odours from buildings.

3. *The building sector level.* It is concerned with the sustainability of the building industry as a whole. Issues such as adaptive use of buildings, durability of buildings, reuse and recycling of materials, efficient use of energy will have a significant impact on the sustainability of the building sector.

4. *The building project level.* It primarily concerns the indoor environmental health such as ventilation, humidity, lighting, thermal comfort, indoor noise, hazard materials; maintenance and estate management patterns which affect the durability and adaptation of buildings and building materials.

Conclusions

Sustainability is very complicated, which primarily concerns environmental issues. Research into sustainability is mainly to assess the environmental impact of human activities and to search for options, which could have least negative impact on natural environment. The purpose of buildings is to create a human environment. Therefore, the study of building sustainability involves both the sustainability of natural environment and the sustainability of human environment created by building activities. The sustainability of natural environment includes less depletion of natural resources, less pollution and less consumption of energy. The sustainability of human environment aims to achieve a stable and comfortable indoor environment by providing adequate functional space and servicing. With the changes of technology and human activities in mind, the issues of durability and adaptability will be inevitably important, particularly for the sustainability of human environment.

There is no single strategy for building sustainability. What strategies should be used depend on the objectives of sustainability and levels of sustainability. This paper addresses the most important issues, which should be dealt with regard to building sustainability. The principles used in this paper should help to contribute to the sustainability in the process of the creation and maintenance of building environment. The Harmony and Profit symposium in Thailand marks an important step in this area.

References

Blum, E. (1997), Making Biodiversity Conservation Profitable: A Case Study of the Merck/Inbio Agreement, in Owen, L A and T Unwin (eds.) *Environmental Management: Readings and Case Studies*, Oxford: Blackwell, pp. 40-54

DETR (1998), *Opportunities for Changes: consultation paper on a revised UK strategy for sustainable development*, Ref. 97EP072

European Commission (1995), *Environmental Assessment of Buildings*, BRE

Jacobs, M. (1993), *Sense and Sustainability*, London: CPRE

Johnson, S. (1993), *Green Buildings*, London: Macmillan Press

Mikesell, R. F. (1992) *Economic Development and the Environment*, London: Mansell

Watts, T. (1993), foreword, in Johnson, S. *Greener Buildings: Environmental Impact of Property*, London: Macmillan

World Commission on Environment and Development (1987), *Our Common Future*, Oxford University Press.

DESIGN AND BUILD PROCUREMENT OF CONSTRUCTION PROJECTS: HYBRIDS IN SINGAPORE

J. FOO, C. LOW, B.H. GOH, and G. OFORI
School of Building and Real Estate, National University of Singapore,
10 Kent Ridge Crescent, Singapore 119260. bemofori@leonis.edu.sg

Abstract

The design and build (D&B) approach to the procurement of construction projects is gaining popularity in Singapore. Several reasons account for this, including the strong support provided by the government through the use of the contractual arrangement by public-sector clients. The level of utilisation of D&B in the private sector is much lower, although it is growing. It may be said that the approach is only used when the features of the project clearly demand it. Given its relative novelty in Singapore, much experimentation with aspects of the arrangement is taking place.

This paper considers the use of the D&B procurement approach for construction projects in Singapore. It discusses the reasons for its introduction and growth. It then describes some of the main common forms of the arrangement, and uses some case studies to illustrate the dynamism of the situation. It is concluded that whereas clients in Singapore like the convenience and the cost and time benefits which D&B offers, they wish to have some control over the design. It is suggested that contractors should seek to demonstrate clearly to clients their ability to provide value-for-money designs.

Keywords: Construction, design-and-build, hybrids, Singapore

Background

Since the early 1960s, various different ways of promoting and carrying out construction projects have been created and utilised in many countries with the intention of improving on traditional approaches which were no longer found to be satisfactory. Franks (1990) and Masterman (1990) offer good descriptions of the major procurement arrangements for construction projects. They discuss the structure of each arrangement, its evolution, the advantages and disadvantages, and the circumstances in which it is the most appropriate approach to be adopted.

The reasons for the advent of the new arrangements vary from one country to another. The initiator and main promoter also differ. However, in general, the traditional approaches (which also take different forms from one country to the next) were found to be failing to satisfy clients' needs as projects became larger and more sophisticated. They also made the integration of the necessarily different but closely interrelated and interdependent contributions of the participants problematic. Finally, the traditional systems did not easily allow the particular skills of some participants, especially contractors, to be utilised on the project in order to improve upon the overall performance of the design and construction teams, as well as the constructed product.

In introducing new procurement arrangements, attention has focused on reducing the time spent on producing a design and preparing tender documentation, hence enabling construction work to begin sooner. Another important consideration has been to bring the contractor in at an early stage in the development of the design. In traditional arrangements, contractors are rarely involved until the tendering stage, that is, after the design has been fully developed. The increasing complexity of construction projects led to the realization that it was in the interest of clients (and architects) to use the vast amount of knowledge and practical experience of contractors early in the design process. This is to take full advantage of their valuable contribution, especially in terms of design buildability. The D&B procurement approach has mainly been developed with this intention in mind. Essentially, in D&B, the owner enters into a single contract with one entity which is responsible for the architectural/engineering and construction works of the project.

Design and build in Singapore: strategic considerations

Growth of D&B

In Singapore, where the construction industry uses practices and procedures modelled on those of the UK, the design-bid-build approach is the traditional procurement approach for construction projects. The D&B method was used for infrastructure projects such as the construction of roads, highways and bridges prior to the 1990's. It was the dominant procurement approach on the Mass Rapid Transit project, the S$5 billion first phase of which was started in the early 1980s. However, usage of D&B in building construction was uncommon until 1991 when the Housing Development Board (HDB) embarked on a programme to use the D&B method for its public housing projects with the intention of providing variety of design to public-housing residents. After this, the utilisation of the approach grew strongly.

Since 1992, the volume of D&B building projects has risen significantly (CIDB, 1997). Table 1 shows the value of D&B building contracts awarded by the public sector in 1992-1996. In 1995 and 1996, public-sector clients awarded more than S$1 billion worth of D&B building contracts annually. Whereas there are no similar reliable figures for the private sector, there is evidence that the use of D&B is spreading. With this upward trend, it is apt to examine the nature of this mode of procurement in Singapore.

Several studies have been undertaken on D&B in Singapore. Among recent ones, Lim (1994) assesses its usage in a survey and discusses some legal issues; Lim (1997) presented some case studies of its utilisation; and Ling (1997) considered the selection of design consultants by D&B contractors. This paper considers the various structural forms which the D&B procurement approach takes.

Strategic Reasons

The main reasons for the introduction of the D&B procurement approach in Singapore are strategic, and relate to the overall development of the construction industry. The prime promoter has been the Construction Industry Development Board (CIDB), which is responsible for the continuous development of the construction industry of Singapore (CIDB, 1994). Two reasons for the active promotion of D&B in Singapore may be highlighted. First, widespread use of D&B is considered to be a key ingredient of the effort to enhance construction productivity in order to relieve the severe labour problems facing the industry (CIDB, 1992). D&B, where the contractor has responsibility for design, and where there is the potential for the contractor to be involved in the entire design process, would lead to greater attention to buildability during the design stage, which is considered to be a pre-requisite for productivity enhancement (CIDB, 1996; Ng, 1996). Indeed, buildability is considered to be so important that legislation to set a mandatory minimum requirement for all construction projects is being considered (*Construction Focus*, 1997).

Table 1. Value of D&B building contracts awarded by the public sector

Year	Value of D&B Building Contracts Awarded by the Public Sector (S$ million)
1992	73.7
1993	232.9
1994	456.4
1995	1,138.8
1996	1,095.3

Source: CIDB (1997), Design & Build in Singapore, Singapore.

Second, D&B is considered to be desirable as it would widen the capability of construction contracting organisations. The narrow build-only focus of local construction firms has been seen as a limiting factor to their competitiveness both at home and abroad (Economic Committee, 1986). These firms have been advised to endeavour to provide a more comprehensive service in order to reduce their reliance on only a small segment of the construction market. In this way, they would be able to develop the corporate capacity and capability to enhance their competitiveness.

For these reasons, one of the strategic thrusts of the CIDB in its current ten-year plan is (*Construction Focus*, 1994b: 5):

> Strengthen Design-and-Build Capability – Design-and-build has the inherent advantage of single-point accountability in terms of client-builder relationship while encouraging integration of design and construction to find cost-efficient solutions.

The CIDB actively promotes the D&B approach through seminars and workshops, and gives it much publicity in its newsletter. For example, in an article, it notes the advantages of the procurement arrangement:

> D&B brings the builder upfront to be involved in the design process. The builder can contribute ideas on buildability and cost-effective ways of construction. D&B can shorten the project delivery time and lower costs. It also offers developers a one-stop package deal. Developers need to work with only one party who is responsible for coordinating the whole project (*Construction Focus*, 1996: 1).

In another article showcasing a project, the CIDB describes how the D&B arrangement enabled a 26,600 square metre warehouse to be completed in 10 months (including substructure works) through the identification and tackling, at the design stage, of potential construction problems (*Construction Focus*, 1994a). The CIDB has prepared a handbook "to raise the level of awareness and convey recent experiences on Design & Build practice in Singapore" (CIDB, 1997: i).

Forms of D&B in Singapore

The CIDB (1997) traces the evolutionary development of the D&B procurement method in Singapore. It identifies three forms which are mainly used by public-sector clients. These are Traditional D&B, Develop and Construct, and Consultant Novation. As in the literature, the CIDB (1997) notes that the selection of a suitable form depends on the nature of the project and the client's preferred degree of involvement in the design. A brief outline of each of the three forms of D&B is now presented.

Traditional D&B

In the Traditional D&B form, the client approaches a contractor or a small group of pre-qualified contractors at an early stage of the project. The contractor, together with his team of architectural and engineering consultants, proposes an outline design and cost based on the client's brief. Essentially, the contractor is responsible for the design, planning, organisation and control of the construction to satisfy the client's requirements. The contractor is fully responsible for both the design and construction works of the project and the client needs to deal only with him.

Develop and construct

The Develop and Construct method is a variant of the traditional D&B form that is also sometimes referred to as Design Development and Construction. In this case, the client uses in-house professionals or directly employs a design consultant to prepare the design brief which includes the concept and schematic design, specifications and evaluation criteria. Contractors are then invited to submit proposals for the development and completion of the remaining design together with an estimated price for the work. Interested contractors would employ their own consultants, to form a D&B team, to prepare proposals for the detailed design and price. The client normally evaluates the proposals based on price.

Consultant novation

Consultant Novation is another variant of the traditional approach whereby the client employs his own design consultants to develop the concept and schematic design. Similar to the Develop and Construct approach, the next stage involves inviting the contractors to submit their proposals and prices. However, upon the selection of a qualified contractor, the client would "novate" his design consultants to the contractor who then enters into a contractual relationship with them. The role of the consultants is to aid the contractor in developing and completing the detailed design, and to provide assistance to him during the construction stage.

Special features of the different D&B forms

Advantages and disadvantages

The key advantage of all the three different forms of D&B is that single-point responsibility for both design and construction is available to the client. The contractor is solely responsible for failure in both the design and construction. The benefit of having the contractor's early involvement in the design stage, to enhance buildability of the design of the project, is also available in all three forms. Where a list of selected contractors is used, the client is provided with a variety of optimum solutions to choose from.

One of the drawbacks of Traditional D&B is that the client has little opportunity to interact with the D&B team while tenders are being called. This may lead to the team proposing solutions that are not exactly in line with the client's requirements. Moreover, the need to prepare extensive proposals by a number of contractors, only one of which is eventually appointed and paid to do the work, can be time-consuming and costly, especially for the more complex and sophisticated buildings. This may create a disincentive for contractors to submit good quality proposals.

Merits of variant forms

In view of these problems, the two variant forms outlined above are an improvement of the traditional approach. With the employment of consultants to develop the brief up to the schematic design stage, clients would be involved in the development of the design and specification. However, for the Develop and Construct method, if too much detail would have been determined before the contractor is engaged. Thus, it may not offer the full benefits of the contractor's involvement in the design. In this case, it would help if the contractor is able to employ the same team of consultants which has been developing the design to ensure continuity in concepts and in efforts to satisfy requirements. This is a good feature of the Consultant Novation method whereby both the contractor and client can benefit from the design being developed more closely to the brief. However, some contractors may have their preferred consultants, or there may be incompatibilities between the contractor and the consultants appointed by the client. Moreover, the consultants may also experience a conflict of interest owing to the change of employer.

In Singapore, different client organisations which utilise D&B prefer particular forms, which help them to achieve their corporate objectives. The CIDB (1997: 29-40) describes the D&B models used by main public client organisations: the HDB, Jurong Town Corporation, Lands and Estates Organisation of the Ministry of Defence, and Public Works Department. It notes that for each owner, the method described

> ... may not necessarily be the only one that the owner may adopt. The methods are continuously being fine-tuned and improved with time through experience and feedback (p. 29).

The situation in the private sector is even more fluid. To provide a clearer picture of the various permutations of the D&B procurement approach used in Singapore, three case studies were conducted. The focus was on private-sector projects where flexibility and experimentation with procurement approaches are more common, and dynamism in the D&B forms utilised is more evident.

D&B case studies in Singapore

Case study 1

The first case study is on a 19-storey commercial project located in the Central Business District. The total construction cost, inclusive of variations amounted to S$29 million. The client was a publicly-listed property development company with a track record of 35 years. Its operations and existing portfolio of assets span the globe; it is active in several countries in Asia, Australasia, Europe and North America. The company's Projects Department is staffed by qualified and experienced project managers from different professional backgrounds. This department monitors all stages of the client's projects. The project studied was the client's first venture into D&B: its other projects were constructed under the design-bid-build arrangement.

At the inception stage of the project, the consultant quantity surveyor proposed the use of D&B as an alternative to the traditional method in order to achieve the client's objective of an earlier completion. Another consideration was the client's desire to ensure better design management in areas of buildability, choice of construction method and cost management.

To achieve better control over the architectural design, the client adopted a variant of the Develop and Construct approach. The architect was pre-selected from a pool of reputable architects with a good track record in the design of commercial buildings. The requirements stipulated by the client included those in the following areas:

* minimum Gross Floor Area
* minimum Net Floor Area
* performance specifications and technical specifications
* mechanical and electrical services requirements
* maintenance requirements.

These were established jointly by the client's project manager, the architect and the quantity surveyor, and incorporated into the tender documents.

A pre-qualification exercise was conducted to shortlist contractors with experience and expertise in high-rise commercial buildings as well as a track record of good performance on past projects with the client. Two contractors were selected and invited to tender for the project. Each of the contractors was provided with the architect's concept design, client's requirements and some essential statutory provisions. Each was then required to submit a proposal of the structural, mechanical and electrical design, and an estimated price The tenderers were given four weeks to complete their proposals. Upon submission, the proposals were evaluated by the architect and quantity surveyor according to the following criteria:

* deviations from the concept design
* compliance with the client's requirements
* structural, mechanical and electrical design efficiency
* maintainability
* cost efficiency.

A total of four weeks were taken to evaluate the tender returns. Interviews with both tenderers were also conducted to clarify any doubts on their respective proposals.

When the contractor was apppointed, the architect was 'novated' to the contractor. A contractual relationship between the architect and the successful contractor was created, and a single point of responsibility for design established. The quantity surveyor remained on the project team as the client's contract adminstrator. Overall monitoring and co-ordinating of the project was by the client's project manager.

The Employer's objectives of buildability were achieved through the adoption of this hybrid form of D&B. This was because, at the design stage, the contractor was able to provide valuable inputs on the co-ordination of the architectural and structural elements with the mechanical and electrical services. However, owing to changes in the design of the façade to improve the aesthetics of the building (after the contract had been awarded), the objectives of time and cost savings could not be realised.

Case study 2

The second case study was a 671-unit condominium development including a semi-basement car park and recreational facilities. The client was a Singapore-based property developer which owns commercial and residential buildings in Singapore, China, Myanmar, Vietnam and the UK. The company's projects are overseen by a team of experienced project managers. The contract sum was S$144.5 million, a portion of which was allocated to provisional sums for kitchen cabinets, kitchen appliances, wardrobes, ironmongery, fire protection systems and clubhouse equipment. The provisional sums had been set aside for these items in order to ensure that the client would be able to retain some measure of control over the quality of the finished product as well as to reduce the risk exposure of the contractor for any additional statutory requirements for fire protection.

This project began under the traditional design-bid-build arrangement. Mid-way through design development, a decision was taken to adopt the Develop and Construct procurement system to reap benefits of cost and time savings. As the client had previously used the Traditional D&B procurement system without modifications, the transition was accomplished smoothly. The architect was retained as the Client's Representative while the quantity surveyor was entrusted with the task of contract administration. This allowed the client to gain the benefit of the contractor's feedback into structural, mechanical and electrical systems while maintaining control over the aesthetics of the development.

As the architect had substantially developed the design by the time of tender, the scope of architectural design undertaken by the D&B contractor had been reduced. Hence, the contractor was only required to prepare the drawings for Building Plan approval, and co-ordinate the design of the architectural and structural elements, and the mechanical and electrical services. The contractor and his consultants were also responsible for ensuring that the design complied with all relevant statutory requirements. If changes to the architectural design became necessary, the approval of the client's representative would be required.

Three contractors were selected to participate in the tender exercise. They were all registered in the CIDB's highest financial category for General Building, and had track records of good performance on other D&B projects. Again, the contractors were required to submit their proposals for the design. Schedules of materials, finishes and rates were also included in the tender documentation to minimise the risk of any dispute with regards to the quality or cost. Evaluation of tenders was carried out over a period of three months, during which tender interviews were conducted to give both the client and the tenderer the opportunity to clarify aspects of the design. The project manager, architect, and quantity surveyor selected the successful contractor based on the following criteria:

* compliance with the client's requirements
* cost efficiency and effectiveness of structural, mechanical and electrical services design
* track record of the tenderer's proposed consultants
* design management and site management capabilities of the tenderer.

After the contract had been awarded, the task of design detailing and co-ordination of architectural, structural, mechanical and electrical services design was passed on to the successful contractor. In addition, the contractor was required to apply for, and obtain, approvals from the relevant authorities.

The use of this hybrid of D&B ensured that co-ordination between the three design disciplines could be carried out with the contractor's input, thus resulting in buildable design solutions. In comparison with the traditional procurement method, this appeared to be a more efficient use of resources as the co-ordinated construction drawings eliminated abortive work. The project is still under construction and is expected to be completed in the year 2000. Hence, it is not possible to determine the actual time and cost savings at this stage.

Case study 3

The third project was a S$17 million, 7-storey office-cum-warehouse near the Singapore Changi International Airport. The client was an export-import organisation, which was building for owner-occupation, and for the first time. The original intention was to invite tenders for the project under the design-bid-build method. However, at the time of calling for tenders, the architect had only been able to produce a preliminary design. The client wanted the building to be completed within a short period. Due to the urgency of the matter, the client's consultant quantity surveyor recommended a change of procurement method.

The D&B method was proposed. However, the approach adopted did not follow any of the established models of the D&B procurement arrangement described in the literature. Moreover, amendments were made to the arrangement which was selected as it was considered necessary. Some key aspects of the project may be highlighted.

The quantity surveyor prepared a "design brief" based on the client's requirements and the preliminary design. This design brief and the preliminary drawings were given to selected contractors who were asked to quote an 'all-in amount' based on the drawings and the design brief. The contractor was made responsible for the development of the detailed design, which was expected to adhere closely to the preliminary design. The original design team appointed by the client (except the quantity surveyor whose role was to assist the client during the post-contract stage) would be novated to the successful contractor. This was done to reduce the time which the contractor would need to appoint a new team to develop the preliminary design. Moreover, the original design team had a clear idea of the design concept, and the preliminary design which could easily be developed further.

The lowest tender submitted on the basis of the preliminary design exceeded the client's budget. Negotiations were carried out, allowing the contractor to provide input to enhance the buildability of the design and to reduce cost. The major changes made to the preliminary design include the exclusion of the basement car park, which constituted a large proportion of the overall contract sum. The contractor's proposal also saved time as basement construction was eliminated

The project was awarded within the client's budget, and was constructed within a satisfactory time frame. With the design team headed by the contractor, it led to a time saving (decisions on design were communicated faster). However, variations did result as many changes were made to the design brief owing to the client changing his mind at various stages of the construction. A main reason was that at the time of drafting the design brief (when, indeed, the preliminary design had been prepared), the client was still unclear about what he wanted.

Conclusion

The D&B procurement arrangement for construction projects is still in its infancy in Singapore. Various models of the arrangement are used by different clients. Moreover, the same client may use a variety of forms of D&B from one project to another. On some occasions, changes are made to the procurement system in the course of the project. Various prerequisites are necessary if D&B is to be successful, and to provide the advantages and benefits with which it has been credited. One of these is that the design brief should be clear and complete in order to eliminate, or at least minimise, changes to the design at later stages of the project.

Although both public- and private-sector clients in Singapore increasingly prefer D&B, especially because of the single point responsibility which it provides, they still want some form of control over the design. Contractors should prove that they are able to produce good designs which provide clients good value for money, and result in properties which generate high levels of returns in benefits or profits. The CIDB can assist in this process by starting the registration, and systematic and progressive gradation of contractors with a proven record of competence in D&B projects. This would give assurance to clients, and also provide aspiring contractors with targets to aim for in their entry into the D&B market and their development over time.

References

Construction Focus (1994a) Design & build is faster. Vol. 6, No. 4, April-May, p. 3.
Construction Focus (1994b) CIDB's Vision 20004 unveiled. Vol. 6, No. 6, June-July, p. 5.
Construction Focus (1996) D&B for construction today. Vol. 8, No. 4, July-August, p. 1.
Construction Focus (1997) Buildable designs may be made compulsory. Vol. 9, No. 3, May-June, p. 1.
Construction Industry Development Board (1992) *Report of the CIDB Task Force on Construction Productivity.* Singapore.
Construction Industry Development Board (1994) *CIDB 10th Anniversary Commemorative Publication.* Singapore
Construction Industry Development Board (1996) *The CIDB Buildable Design Appraisal System.* Singapore.
Construction Industry Development Board (1997) *Design & Build in Singapore.*
Economic Committee (1986) *The Singapore Economy: New Directions.* Ministry of Trade and Industry, Singapore.
Franks, J. (1990) *Building Procurement Systems,* 2nd Edition. Chartered Institute of Building, Ascot.
Lim, M.M.S. (1994) Design and Build. Unpublished B.Sc. Dissertation, National University of Singapore.
Lim, K.S.G. (1997) Design and Build: A single source of responsibility. Unpublished B.Sc. Dissertation, National University of Singapore.
Ling, Y.Y. (1997) Procuring design services for design-build projects. *Proceedings, CIB 92 Symposium on Procurement: The key to innovation.* I.F. Research Corporation, Montreal, pp. 405-412.
Masterman, J.W.E. (1990) *An Introduction to Building Procurement Systems.* Spon, London.
Ng, P. (1996) Improving construction productivity through buildable designs in Singapore. In *Productivity in Construction: International experiences, Proceedings of the 2nd International Congress on Construction*, Singapore, 5-6 November, pp. 113-121.

A PROTOCOL FOR MANAGING THE DESIGN PROCESS IN THE BUILDING INDUSTRY IN BRAZIL

C. T. FORMOSO, P. TZOTZOPOULOS, M. S. S. JOBIM AND R. LIEDTKE

Federal University of Rio Grande do Sul (NORIE/UFRGS), Av. Osvaldo Aranha 99, 3° andar CEP 90035-190, Porto Alegre – RS, Brazil. formoso@vortex.ufrgs.br

Abstract

The performance of the design process in the building industry has a great influence on the success of subsequent processes in construction projects and also on the quality of the final product. Despite its importance, relatively little attention has been given to the management of the design process, when compared to production.

The main objective of this article is to present the preliminary results of a research project which aims to devise a protocol for managing the design process in house building firms. The protocol will consist of a general plan for developing design activities, including the content of the main activities, their precedence relationships, the role and responsibilities of different actors, and a model of the information flow. The protocol has been developed through case studies, carried out in four different construction companies from the South of Brazil. Each of them is developing a quality system for the design process, in partnership with design offices. Two of the companies are also working towards getting ISO9001 certification.

The development of this research is based on the conceptual framework of the New Production Philosophy (Lean Construction). Some of its concepts and principles have been applied on the management of the design process.

Keywords: building design, design management, lean construction, process management

Introduction

Several studies have pointed out that a large percentage of defects in building arise through decisions or actions in design stages [Cornick (1991)]. It is also widely known that poor design has a very strong impact on the level of efficiency during the production stage [Fergunson (1986)]. In recent years, the increasing complexity of modern buildings in a very competitive market-place has significantly increased the pressure for improving the performance of the design process in terms of time and quality. For instance, it has been fairly common to overlap the design and the production stages in order to reduce project duration and increase the flexibility of product design. Despite its importance, relatively little research has been made on the management of the design process, in relation to the research time and effort which has been devoted to production management and project management in general [Austin et al. (1994); Koskela et al. (1997)]. The relatively small cost of the design process compared to the production costs probably disguises its true importance for the performance of construction projects [Austin et al. (1994)].

To some extent, the fact that design management has been neglected is understandable. Building design is a very difficult process to manage. It involves thousands of decisions, sometimes over a period of years, with numerous interdependencies, under a highly uncertain environment. A large number of personnel are implicated, such as architects, project managers, structural engineers, service engineers and marketing consultants. Each professional category has a different background, culture and learning style [Powell & Newland (1994)]. Moreover, feedback from the production and building operation stage takes a long time to be obtained, and tends to be ineffective.

The design process needs to be planned and controlled more effectively, in order to minimise the effects of complexity and uncertainty. The lack of design planning results in insufficient information being available to complete design tasks and inconsistencies within construction documents. Poor communication, lack of adequate documentation, deficient or missing input information, unbalanced resource allocation, lack of co-ordination between disciplines, and erratic decision making have been pointed out as the main problems in design management [Cornick (1991); Austin et al. (1994); Koskela et al. (1997)].

The nature of the design process

Markus & Arch [1973] pointed out that most descriptions of the design process, both theoretical and empirical, recognise two patterns. One consists of an individual decision making process, usually performed by a single designer, concerned with the creation of alternative solutions. The second one is a managerial process, divided into phases, which develops from the general and abstract to the detailed and concrete. A complete picture of the design method requires the consideration of both patterns. These are discussed below in sections 1 and 2, respectively.

1. Design as a creative process

Designers are traditionally known for the solutions which they produce, rather than the kind of problem they deal with [Lawson (1980)]. In design, the problem is usually poorly defined, i.e. clients sometimes are not able to make their needs explicit. Often, there is no way to develop an adequate solution from existing information, since the client requirements are vague. The solution does not necessarily comes directly from the problem. The attention of the designer oscillates between the comprehension of the problem and the search for a solution [Cross (1994)].

Each designer approaches the design problem in a particular way. One of the traditional ways used by designers for dealing with the problem is to develop quickly a potential solution or a group of potential solutions, which are used as a way to define and understand the problem clearly [Cross (1994)]. A number of models of design as a creative process are presented in the literature. For example, the model proposed by Markus & Arch (1973) is presented in Figure 1. In this model there are four main activities in design: analysis, synthesis, evaluation and decision/communication. One of the main contributions of such models is the fact that they make clear that individual design processes tend to be very unstructured and chaotic. This should be taken into consideration when developing a model for managing the design process. It means that the steps for producing a design solution cannot be pre-established at a very fine level of detail [Lawson (1980)].

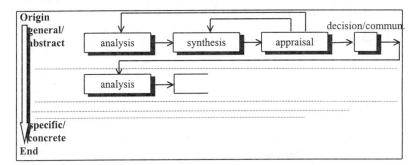

Figure 1: Design as a creative process [Markus & Arch (1973)]

Design as a managerial process

In the building industry, design is traditionally regarded as one of the stages of a project. This is mainly due to the fact that the participation of the design team often starts relatively late in the project and finishes as soon as the production stage starts. In the present study, design is regarded as a process which is present in all stages of the building process, from inception to building operation. In fact, design is one of the most important processes in building projects, since it defines the product to be built and has many interfaces with several other processes, such as production planning, material supply, sales, and building operation.

The way the design process is divided into stages varies considerably in different studies both in terms of content and the names given to each stage [Cross (1994)]. In this research work, the design process is divided into seven stages: (a) Inception and Feasibility, (b) Outline Design, (c) Scheme Design, (d) Design for Legal Requirements, (e) Detail Design, (f) Production Monitoring, and (g) Feedback from Operation.

Each design stage can be further divided into sub-processes, which successively transform information from client requirements and project constraints into product design. In this research project, the design process is analysed from the point of view of the New Production Philosophy [Koskela (1992)], which means that there are four kinds of activities involved: conversion, waiting, moving, and inspection (Figure 2). Only conversion activities are value adding. Waiting, moving and inspection activities are non value adding and should be eliminated, rather than made more efficient. Part of the conversion activities are not value adding, since they cause rework, due to errors, omissions and uncertainty [Huovila et al. (1997)]. This kind of rework should not be mixed up with the iterative process that occurs when a design alternative is generated, as shown in Figure 1, which is an inherent part of design as a creative process.

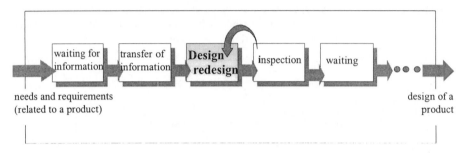

Figure 2: The design process according to the New Production Philosophy [Huovila et al. (1997)]

Based on the view of design as a flow, Huovila et al. (1997) proposes a number of principles and methods for eliminating waste in the design process:

1. Reducing uncertainty, which is one of the main causes of rework, specially in the early stages of design. This can be done by increasing the effort in terms of clearly defining the project restrictions and the requirements of internal and external clients;
2. Reducing waiting time by decomposing adequately the design tasks so that they can be properly planned, and also allow the transfer of information to be made in smaller batches;
3. Reducing the effort needed for information transfer through team work, and by rearranging the design tasks.

Development of the protocol

The protocol will consist of a general plan of the design process which can be used as a basis for devising a model to manage the design process for individual companies. The main elements of the protocol are: (a) the content of the main activities, (b) their precedence relationships, (c) the main inputs and outputs for each activity, (d) tools that can be used for supporting the execution of such activities, (e) the role and responsibilities of the different actors, and (f) a model of the information flow.

The development of a model for managing the design process can potentially bring a number of benefits for design and build organisations, by applying some of the lean construction principles [Koskela (1992)]:

1. The fact that a stable, consensual and explicit model of the design process exists makes the identification of the necessary improvements easy. This makes possible the application of some lean construction principles, such as reducing the share of non value-adding activities, reducing variability in the process, and minimising the number

of steps;

2. All actors involved in the process are able to understand the process as a whole, their roles and responsibilities [Cornick (1991)]. This increases the process transparency, and tends to improve communication between them;

3. It is possible to increase the effectiveness of the information flow, since the necessary information for performing each activity is formally established, as well as the information which must be produced by each activity. This tends to improve both the quality of design and creates the possibility of reducing the duration of the design stage;

4. It becomes easier to devise and implement tools for measuring and controlling product and process performance. Some of these tools can be used for focussing control on the complete process;

5. Continuous improvement can be built into the process, since the design tasks are monitored and registered in a systematic form, including those design related tasks which are performed during the production and the building operation stages. The data collected during those two stages can be used for feeding back future projects and the company strategic planning process.

The protocol has been devised through case studies, carried out in four small sized house building companies from the State of Rio Grande do Sul, Brazil. Their main activity is to develop and construct commercial and residential buildings. Each one of these companies is devising its own model for managing the design process. Two of them are also working towards getting ISO9001 certification. This work has been developed as a partnership between the building companies and NORIE (NORIE is a research group on Building Engineering at the Federal University of Rio Grande do Sul), and is partially sponsored by FINEP, a Federal Government research funding agency. Most of the work in each company is performed by a team of four to five people, named "operational group", which is formed by representatives of different sectors involved in design, including the design manager, and one representative of the research team, who acts as a facilitator. The "operational group" usually meets once a week for two hours. Occasionally the "operational group" is extended by including some of the external designers (architects, structural engineers, service engineers, etc.) who work for the company, forming the "extended group". This is necessary every time there is a need to consider the point of view of the other participants in the development of the model.

Although there have been differences between the companies, the development of the model can be divided into three main stages:

(a) Preliminary investigation: a number of interviews with design managers and external designers were made, aiming to identify their perceptions about the sequence and the content of the main design activities, and about the foremost managerial problems in design.

(b) Design Model: a model of the design process was devised for each company by the "operational group". This model was represented by flowcharts and input-output tables.

6. Design Manual: written procedures and working instructions have been made for some

of the activities of the model. All written information has been put together in a Design Manual, which reflects the culture and working methods of each company. In both companies that have been working towards ISO9001 certification, the design manual is part of the Quality System. The procedures and working instructions have been gradually implemented in the four companies.

Two main tools have been used in this study for modelling the design process. The first one is the flowchart, which represents graphically the process, including the division of the process into sub-processes, making explicit precedence relationships. In order to keep the flowchart of the whole design process as simple and as readable as possible, it was necessary to group information in a hierarchical way. There is a general flowchart presenting the seven design stages, for each stage a flowchart of activities and, for the most complex activities, a flowchart of operations. This form of representation gives a broad view of the design process, since the general flowchart is relatively short, and, at the same time, makes it possible to plan the design process at a relatively fine level of detail. Figure 3 presents the main criteria used for organising the design tasks into stages,

CRITERIA	STAGE	ACTIVITY / ACTIVITY	Operation / Operation / Operation
PRODUCT	• The product of each stage relates to the whole building, in different levels of detail (e.g.: scheme design)	• The product of each activity usually relates to parts of the building (e.g.: selection of structural system)	• The product of each operation usually relates to smaller parts of the building, in relation to activities (e.g.: bathroom wall tiles design)
SEQUENCE	• Linear – dependent tasks • One stage starts when the previous one is finished and approved Dependent tasks (series) A → B	• Linear, independent or interdependent tasks • There might be overlapping between activities. Also, the activities can be looped Independent tasks (parallel) A / B	• Linear, independent or interdependent tasks • There might be overlapping between operations. Also, the operations can be looped Interdependent tasks (coupled) A / B
WHO DEVELOPS	• Involves several participants	• Involves a small number of participants	• Involves usually only one designer
TIME SCALE	• monthly	• weekly	• daily
APPROVAL & VALID- ATION	• Tends to be formal and external • Check lists can be used for control	• Intermediate situation between stage and operation	• Tends to be informal and internal (self control) • Check lists supports decision making
NUMBER	5 to 10	8 to 16	variable
SUB- DIVISION	It is always subdivided	It is often subdivided	It is not subdivided

Figure 3: Criteria for dividing the design process into stages, activities and operations

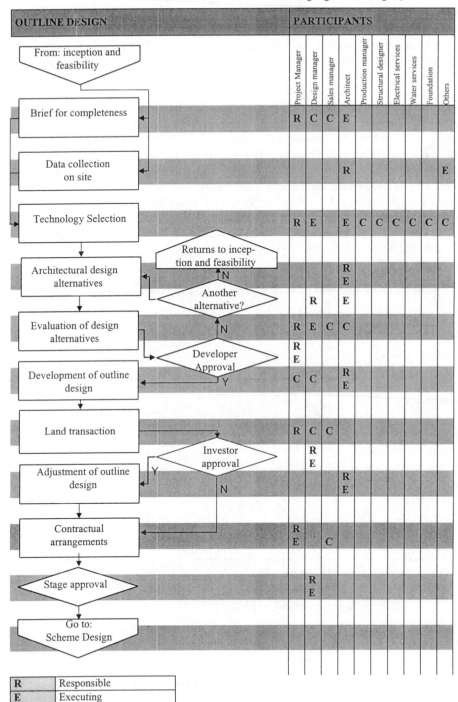

Figure 4: Example of flowchart (outline design stage)

activities and operations and describes three possible kinds of precedence relationships between activities - sequential, parallel, or dynamic interaction. The third kind of precedence relationship occurs when two activities or operations are iterative and it is not worthwhile to model the sub-process at a very fine level of detail, because it is unstructured or too uncertain.

There are limitations in terms of representing the flow of information in the flowcharts. Very often the output of one activity is not necessarily the input for the following one, but corresponds to the input of several different activities. For that reason, not all flows are made explicit in the flowcharts, as this would make the flowchart difficult to understand. Figure 4 presents an example of flowchart for the outline design stage, in which the degree of involvement of the main participants is defined for each activity. The second tool used for modelling the process is the input-output chart. It describes in more detail all activities represented in the flowcharts, by making explicit the inputs and outputs for each one of them. Figure 5 shows an example of input-process-output chart for two outline design activities.

INPUT	PROCESS	OUTPUT
Briefing Data collection from the site; Feedback data (from buildings delivered previously and from production). Definition of the design team; Strategic selection of technologies Initial statement of performance requirements; Regulatory and statutory information.	**ARCHITECTURAL DESIGN ALTERNATIVES**	Outline design alternatives
Outline design alternatives Cost planning and sales price estimate	**EVALUATION OF DESIGN ALTERNATIVES**	Choice of alternative(s) to be developed

Figure 5: Example of input-process-output table

Guidelines for developing models of the design process

Based on the development of the four case studies, a number of guidelines have been proposed for developing models of the design process, which are presented in this section.

The model includes not only design activities but also other activities which are part of project management, such as production of legal and statutory documents, negotiation with the land owner, etc. This is due to the fact that the design process has many interfaces with other processes, which must be made explicit in the flow. The first step for developing a model of the design process is to establish or to make explicit the company competitive strategy. This is an important requirement at the beginning of the design process, since it defines the product strategy (i.e. the market in which the company operates, its clients, their needs, etc.), and the process strategy, in which the main decisions related to the technologies to be used by the company in different projects are set.

There are some design tasks which need to be carried out in phases, developing from the general and abstract to the detailed and concrete, such as the identification of client needs, negotiation of the land, selection of construction technologies, and market and financial analysis. It means that they have to be split into separate activities, which take part in different design stages. Also, some individual activities may occur in different stages of the design process, depending on the characteristics of the company or the project. This is, for instance, the case of land transaction, and design commissioning.

Due to market pressures, the production stage usually starts before the completion of detail design. Thus, it was necessary to identify which documents or information should be available at the beginning of the production process and at some important production milestones. In such a context, planning the design process becomes very important. A detail plan of the design process is made only at the detail design stage. In the initial design stages, the level of uncertainty is very high, and only a rough estimate of the duration of each stage is usually made. Also, some important dates in terms of project strategy are defined at the beginning of the process, such as the date for commercial release and the beginning of the production stage.

There are three main points at the design process in which there is a concentration of efforts for integrating different disciplines. The first one is soon after the first sketch of the structural design and water and electric services design, at the outline design stage. The second one is at the end of scheme design, and the third in the detail design stage. This integration effort is usually represented as a cycle of activities, performed by the design team, and followed by individual adjustments carried out separately by different designers, and an evaluation activity.

The model developed in each company was not based only on its current practices. Some improvements were introduced in the design process, based on the literature review and also on some best practices identified in other companies. The main improvements introduced are presented bellow:

1. Early participation of structural and services designers in the design process, since the outline design stage;
2. Oral presentation of the design to both production and sales people, in order to communicate the design intentions and philosophy clearly;
3. Some activities were rearranged in order to reduce flow activities. For instance, a

procedure was made for collecting a large number of information at the building site in the same visit;

4. There are a number of sub-processes which are cyclic and repetitive, which were made explicit in the flowchart. They are usually started by an event, such as design changes demanded by clients or by production management, repairs demanded by clients during building operation, etc.

5. As in the work developed by Sheath et al. (1996), the model incorporates an approval routine between stages, in order to establish a formal and structured evaluation at the end of each stage.

Final comments

This paper describes the preliminary results of an ongoing research project, in which a protocol for managing the design stage has been developed. The implementation phase of the research has started gradually, and the effectiveness of the model will be evaluated by using performance indicators, related both to the design process and the product. Once the model is fully implemented in each company, an analysis of the information flow related to the design process will be made. Based on that analysis, it will be possible to develop an in depth investigation on the actual content and impact of flow activities in building design.

Acknowledgement

Thanks to FINEP - Programa Habitare and CNPq for the financial support and to the four companies which have been partners in the research project.

References

Austin, S.; Baldwin, A. & Newton, A. (1994) Manipulating the Flow of Design Information to Improve the Programming of Building Design. London, Spon, *Construction Management and Economics*, 12 (5): pp. 445-455

Cornick, T. (1991) Quality Management for Building Design. Rushden, Butterworth. 218 p.

Cross, N. (1994) Engineering Design Methods. Strategies for product design. London, Wiley, 179p.

Ferguson, I. (1989) Buildability in practice. London, Mitchell. 175 p.

Huovila, P.; Koskela, L. & Lautanala, M. (1997) Fast or Concurrent: The Art of Getting Construction Improved. In: Alarcon, L.F. (Ed.) Lean Construction, Rotterdam, Balkema: pp. 143-160

Koskela, L. (1992) Application of the New Production philosophy to Construction. Stanford, CIFE, Stanford University. Technical Report No. 72.

Lawson, B. (1980) How Designers Think: the design process demystified. London, Architectural Press. 216 p.

Markus, T.; Arch, M. (1973) Optimisation by Evaluation in the Appraisal of Buildings. In

Hutton, G.H. & Devonal, A.D.G. (Ed.) Value in Building. London, Applied Science. pp. 82-111

Sheath, D.M.; Woolley, H.; Cooper, R.; Hinks, J. & Aouad, G. (1996). A process for change – the development of a generic design and construction protocol for the UK construction industry. In: INCIT 96. Sidney, Australia. Proceedings.

ENHANCED DESIGN BUILD - AN INNOVATIVE SYSTEM TO PROCURE A HOSPITAL PROJECT

ALBERT P. C. CHAN and ANN T. W. YU
Department of Building and Real Estate, The Hong Kong Polytechnic University, Hung Hom, Hong Kong. bsachan@polyu.edu.hk

C. M. TAM
Department of Building and Construction, The City University of Hong Kong, Tat Chee Road, Hong Kong. BCTAM@cityu.edu.hk

Abstract

Building works in Hong Kong have been mainly delivered in a traditional manner where clients appoint consultants to act on their behalf to produce design and supervise the construction phase. The increasing complexity of buildings, the need to reduce design and construction periods, and the need to improve the project performance have brought pressure to find other ways to deliver the project. In the last decade, "design and build" procurement method has been used extensively both worldwide and in Hong Kong to help dealing with the problems associated with the traditional system. The objective of this paper is to report the findings from a case study of a hospital project which adopted an "enhanced design and build" form of building procurement. The hospital project requires a rapid project delivery and a high degree of public accountability. Twenty project participants representing different organisational interests were interviewed and requested to complete a questionnaire to aid the assessment of the procurement system. The results were put under critical analysis. The perceived advantages and shortcomings of the procurement system were assessed from the perspectives of the client, client's consultants, contractor, contractor's consultants, and contractor's subcontractors.

Keywords: Enhanced design build, hospital projects, procurement system.

Introduction

This paper is based on a research carried out in the Department of Building and Real Estate at the Hong Kong Polytechnic University and involves an investigation into method of procurement for a major hospital in Hong Kong. Despite considerable research into the choice of appropriate procurement systems for construction projects, no generally applicable solutions have been found (Chan, 1995; Bennett and Grice, 1990; Skitmore and Marsden, 1989). It appears that the appropriate solution is in each case largely a function of various contingency factors. Procurement problems are particularly acute in the provision of hospital buildings due to their complexity, long design and construction periods, ongoing developments in health care planning and technology, and the need for a highly accountable approach to procurement by health authorities (Wilkins and Smith, 1994). The objective of this paper is to report the findings from a case study of a hospital project which adopted an "enhanced design and build" form of building procurement.

Methodology

The overall research methodology comprises an in-depth case study of a hospital project which adopted an "enhanced design and build" form of building procurement. The hospital project requires a rapid project delivery and a high degree of public accountability. Twenty project participants representing different organisational interests were interviewed and requested to complete a questionnaire to aid the assessment of the procurement system. The results were put under critical analysis.

The paper presents a detailed discussion of the case study of a hospital project, which adopted "enhanced design build" form of procurement. The paper assesses the selected procurement system from the perspectives of the client, client's consultants, contractor, contractor's consultants, and contractor's subcontractors.

Procurement approaches for hospital project

Recent studies show that variants of the design-build model emerge as a popular procurement system for hospital project (Church and Potts, 1992; Bound and Morrison, 1993; Ndekugri and Turner, 1994). Many researchers advocate that this procurement system can result in a reduction in overall project duration (Pain and Bennett, 1988; Griffith, 1989; Ndekugri and Turner, 1994). In the UK and US, the concept of contractor participation in design is commonplace (Wilkin and Smith, 1994). In Hong Kong, the use of design-build to procure hospital projects is also gaining popularity. The Hospital Authority's first major new capital project, the North District Hospital, adopts a develop/construct approach for its project delivery.

Background of the project

The North District Hospital locates in Fanling; some 30 minutes drive from the Central Business District (CBD). The hospital is a 618-bed acute general hospital with ambulatory care facilities including a 24-hour accident and emergency service, specialist out-patient clinics, and 120 day places including day surgery, psychiatric day hospital and geriatric day hospital. The hospital serves the northern part of the New Territories and part of the Yuen Long area. This is the first project fully devised and managed by the Hospital Authority form inception to completion. Previous hospital projects were controlled by Architectural Services Department.

The project was announced in the March 1993 budget and to be completed by mid-1997 within a total budget of approximately HK$1.3 billion at March 1993 prices. This comprises design, construction, supervision, commissioning, project management, furniture and equipment. Given the tight program and limited budget, the Hospital Authority conducted an investigation of various procurement methods to fulfill these objectives. From the past experience, it was concluded that the traditional procurement method which normally requires three to four years for design development and another three years for construction works, cannot fulfill the time requirement of this new hospital project. In fact, some similar hospital projects took as long as ten years for completion (from inception to physical hand-over) and generated a number of variations due to the rapid changes in medical technology. Therefore, Hospital Authority placed top priority to use fast track method to shorten the period of design and construction of the North District Hospital.

Procurement method – enhanced design build

The Hospital Authority adopted a "enhanced design and build" form of building procurement for this project where rapid project delivery and a high degree of public accountability are required. With this approach, construction work can commence prior to completion of design development thus saving at least one year of time. The standard form of design and build was modified to allow for a greater client control on the design and to minimise the uncertainty of the final product.

The Hospital Authority prepared a comprehensive functional brief with schedule of accommodation, generic room data sheets and furniture and equipment schedules. A consultant team, consisting of architect, structural engineer, quantity surveyor and building services engineer, was appointed to prepare the schematic design in 1:200 scale drawings, performance standard and technical specification. The remaining design development and production of documents required for construction were carried out by the design-build contractor who employed his own consultants for completing the remaining design.

Unlike other design team members, who ceased their design services upon the appointment of the design-build contractor, the structural consultant was novated to the contractor, who was obliged to appoint the same consultant to complete the structural design. The client was happy with this arrangement because continuity and quality of the original design could be maintained. The change of other design consultants resulted in double handling some activities. The new design consultants approached the client for similar queries raised by the original design team members. It was commented that the situation could be improved if novation contract was adopted to avoid duplicating efforts. In addition, the client can assure the quality and performance of the consultants throughout the project.

Form of contract

A modified HK Government form of Design and Build contract was used. The modifications generally strengthen the contract in aspects such as substantial completion and design responsibilities. Bills of Quantities were not used but the contract was on a lump sum basis with a milestone payment system because the Government required a fixed price for the project. A provisional Bill of Quantities for furniture and fittings was provided as a basis for pricing the fitout of the hospital.

Selecting the contractor

The selection of contractor is crucial to the successful completion of a construction project (Chan, 1996b). The schematic design in 1:200 scale drawings, performance standard and technical specification formed the basis of the tender documentation. Prequalification criteria included compliance to Government listing, hospital construction experience, design and build experience and quality assurance track record. The objective was to induce adequate competition and yet at the same time sufficient attraction to the potential tenderers because of the high initial cost of producing design-build bids. Tenderers were, however, invited to make submissions of alternative tenders based either on a revised design, or modifications to the specifications. During the tender evaluation process the Tender Assessment Committee held three meetings. The Committee comprised representatives of the Hospital Authority, end-users and Government. The basic requirements, technical issues, alternative designs and tender price were evaluated. Alternative designs were assessed using a 5-point scale with scores from –2 to +2, with 0 being compliance to the original design, +1 or +2 being better than the original design, and –1 or –2 being worse than the original design.

Project management

A consultant project manager was appointed to take up the project management role on behalf of Hospital Authority. The project manager was to act as a coordinator and to maintain tight control over time, cost and quality for this project. The project manager coordinated meetings to manage the design, construction and commissioning stages of the project. Each of these stages required interaction between the people from the Hospital Authority, design team and construction teams. The Hospital Authority also set up a Steering Committee to coordinate and control the large number of end-user groups. All changes requested by the end-users were managed in accordance with strict procedures and were referred to the HA Steering Committee for final decisions. The organisation chart for the North District Hospital is shown in Figure 1.

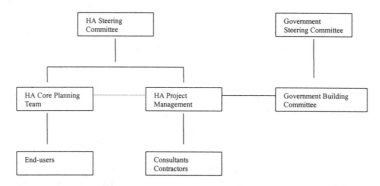

Figure 1: North District Hospital – Organisational Chart
Source: Hospital Authority Procedures Manual, 1994.

A project management system was tailor-made and clearly written down in the HA Procedures Manual which included organisation structure, responsibilities, delegated authorities, reporting system, cost control, time control, payment procedures as well as change process.

The "enhanced design build" form of building procurement for this project required a unique set of procedures for project definition. Because the project would be progressively defined both before and after the award of a building contract, strict controls were placed on the ability of end-users to introduce changes to requirements to ensure that the project would be completed within budget and on time.

Project definition

The project requirements were generally defined in the Functional Brief and Space Program which were developed into 1:200 scale Departmental Plans. The process for the further definition of the project comprises the following steps as illustrated in Figure 2.

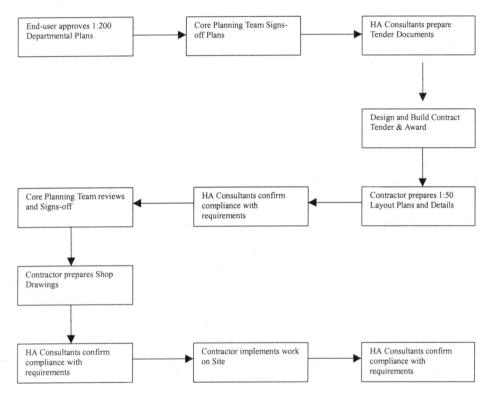

Figure 2: North District Hospital – Project Definition
Source: Hospital Authority Procedures Manual, 1994.

Assessing the enhanced design build system

Twenty project participants representing different organisational interests were interviewed and requested to complete a questionnaire to aid the assessment of the procurement system. The background of the respondents can be seen in Tables 1 and 2.

Table 1: Respondents' professional affiliation and highest academic qualification

Professional affiliation	Percentage	Highest academic qualification	Percentage
Engineer	35%	Diploma/Certificate	10%
Architect	30%	Bachelor Degree	75%
Builder	15%	Master Degree	15%
Quantity Surveyor	10%		
Others	10%		

Table 2: Respondents' year of experience and role in the project team

Year of experience	Percentage	Role in the project team	Percentage
Less than 5 years	5%	Client	15%
5 to 9 years	15%	Project Management Consultant	10%
10 to 14 years	35%	Client's Consultants	15%
15 to 19 years	30%	Contractor's Consultants	20%
20 years or more	15%	Contractor	20%
		Sub-contractor	5%
		Client and Contractor's Consultant	10%

Table 1 shows details of respondents' professional affiliation and their highest academic qualification. Most respondents were engineers (35%) by training. Other major professional affiliations include architects (30%), builders (15%), and quantity surveyors (10%). Most of the respondents (90%) had bachelor or higher academic qualification. Table 2 shows details of respondents' experience and their role in the project team. Most of the respondents (95%) had 5 or more years of experience. Hence it can be considered that the respondents are highly qualified and experienced both academically and professionally. The respondents also portray a balanced representation of the project team, covering the client body, client's project manager and consultants, contractor, and its consultants and sub-contractors. From the interview, the respondents' assessment of "enhanced design build" procurement system is consolidated as follows:

Advantages of "enhanced design build" system

1. A fast track method contributed to completion of project on time.
2. The client can ensure conformance of the basic design to his requirements by retaining control of the initial design.
3. The amount of documentation and detailing compiled were less demanding in comparison with traditional procurement method.
4. Cost savings and speedier construction can be achieved by incorporating the issues of access, working method, buildability into subsequent detailed design.
5. Due to price certainty on accepting proposal and lump sum price from the tenders, the problem of cost overrun was minimised.
6. Direct communication between the client and the contractor.
7. Better working relationship of the project team.
8. The single-point responsibility of the contractor reduced the client's risk on undefined origin of defects.

Disadvantages of "enhanced design build" system

1. There is a limited number of companies with a proven record of both designing and constructing.
2. The contractual risk accepted by the contractor was too high.
3. The tender price was inevitably inflated to cover the additional risks of accepting the consultants' design.
4. The quality of project was lowered due to inadequate supervision of the project.
5. The opportunity for client to make changes or variations was restricted.
6. The opportunity for disputes was increased due to incomplete documentation.
7. Design quality was inferior to those produced under traditional procurement method.
8. Discontinuity of design team created unnecessary duplicating efforts.
9. The project team members were unclear about their roles and duties played in this procurement form.

The perceived advantages and shortcomings of the procurement system were then assessed from the perspectives of the client, client's consultants, contractor, contractor's consultants, and contractor's subcontractors. The respondents were asked to express their degree of agreement with the listed advantages and disadvantages on a 5-point scale, 1 being strongly disagree, and 5 being strongly agree. The ranking by each respondent was transformed into a matrix as the input data for the calculation of the Kendall Coefficient of Concordance using the SPSS package. The results of the Kendall Coefficient of Concordance computation and the ranking were tabulated in Tables 3, 4 and 5. The Kendall Coefficient of Concordance, W, for the ranking of advantages among all

Table 3   Kendall Coefficient of Concordance, W for the assessment of "enhanced design build" system

	Advantages			Disadvantages		
	All	Client's group	Contractor's group	All	Client's group	Contractor's group
No of cases	20	10	10	20	10	10
Kendall Coefficient, W	0.21	0.26	0.36	0.30	0.26	0.52
Chi-square	29.60	18.20	24.94	48.15	20.76	41.32
Degrees of Freedom	7	7	7	8	8	8
Significance	0.00	0.01	0.00	0.00	0.00	0.00

Table 4   Ranking of advantages of "enhanced design build" system

Advantages	All respondents	Client's group	Contractor's group
A fast track method	1	2	1
Conformance of the basic design	2	1	4
Better buildability	3	4	3
Single-point responsibility	4	7	2
Direct communication	5	6	5
Less demanding documentation	6	3	7
Minimise problem of cost overrun	7	5	7
Better working relationship	8	8	6

Table 5   Ranking of disadvantages of "enhanced design build" system

Disadvantages	All respondents	Client's group	Contractor's group
Limited competition	1	1	4
Discontinuity of design work	2	2	5
Increased dispute due to incomplete documentation	3	3	3
Excessive contractual risk to contractor	4	7	1
Inflated tender price	5	6	2
Unclear roles and responsibilities	6	4	6
Inferior design quality	7	8	7
Less flexible	8	5	8
Poor workmanship	9	9	8

respondents is 0.21; among the client's group (the client, the client's project manager, and the client's consultants) is 0.26; and among the contractor's group (the contractor, contractor's consultants, and contractor's subcontractors) is 0.36. Similarly, W for the ranking of disadvantages among all respondents is 0.30; among the client's group is 0.26; and among the contractor's group is 0.52. The computed Ws are all significant at 0.01. The null hypothesis that the respondents' rating are unrelated to each other has to be rejected. Thus, it can be concluded that there is a significant amount of agreement among the respondents on the ranking of the advantages and disadvantages of the "enhanced design build" procurement system. This conclusion is made at 0.01% level of significance.

Discussion of results

It is stressed that the ranking exercise is based on perception; not an objective assessment. A subjective assessment of the ranking results is made to analyse the perceived relative importance of factors. The fact that this subjective assessment does not provide any absolute value on the ranking position is recognised. Emphasis is only given to factors that are placed as the most important and the least important.

Advantages of "enhanced design build" system

In general the respondents perceived "fast track delivery process" as the most significant advantage and "better working relationship" as the least important advantage. "Conformance of the basic design" and "better buildability" were also considered as important advantages. However, factors like less "demanding documentation", and "minimise problem of cost overrun" were not considered as significantly important. When dividing the respondents into client's group and contractor's group and assessing their corresponding perception on the advantages of the procurement system, the values of Kendall Coefficient in both cases increased. It indicates that a better agreement is reached when the assessment is done within the client's group and the contractor's group separately. The client's group perceived "conformance of the basic design" as the most significant advantage, and "better working relationship" as the least important advantage. Whilst the contractor's group perceived "fast track delivery process" as the most important advantage and "less demanding documentation" and "minimise problem of cost overrun" as the least important advantages. The major disparity between the client's group and the contractor's group lies on factors like "single-point responsibility" (ranked 2nd by the contractor's group and 7th by the client's group); and "less demanding documentation" (ranked 3rd by the client's group and 7th by the contractor's group). To the client's group, it is a great relief to dump the tedious work of preparing documentation to the contractor's design consultants. Hence it is understandable why the client's group put the factor relatively high on the list.

Disadvantages of "enhanced design build" system

Generally the respondents considered "limited competition" as the most important disadvantage and "poor workmanship" as the least significant disadvantage. There is a marginal decrease in Kendall Coefficient of Concordance when assessment is made to the client's group alone (W changes from 0.30 to 0.26), but there is significant improvement when the ranking exercise is done only within the contractor's group (W changes from 0.30 to 0.52). Whilst the client's group maintained that "limited competition" as the most significant disadvantage and "poor workmanship" as the least significant disadvantage, the contractor's group considered otherwise. They considered "excessive contractual risk" as the main disadvantage (ranked 7th by the client's group) and "less flexible", "poor workmanship" as the least significant disadvantages. Since the additional contractual risk only affects the contractor, it is logical that the client does not have the same concern. Interestingly, "inflated tender price" was ranked 2nd by the contractor's group, but only 6th by the client's group. It may be interpreted that the client is prepared to pay additional premium for transferring risks to other parties and therefore they do not consider it as a major draw back.

Conclusion

This paper has examined the enhanced design and build method of construction procurement – identifying its contractual form, highlighting its project management and project definition, outlining its advantages and disadvantages, and assessing its performance from the perspectives of the client, client's consultants, contractor, contractor's consultants, and contractor's subcontractors.

A detailed case study of North District Hospital is described to illustrate the process of this procurement system. All the respondents generally agreed that the hospital project was successful in meeting the time, cost, quality, functional and safety requirements set by the client. As the first project adopting the "enhanced design build" procurement system in Hong Kong, the benefits of applying this innovative procurement system were demonstrated.

The innovative procurement system can integrate the construction practices more effectively, as well as to achieve the client's objectives in terms of time, cost and quality.

Acknowledgement

The authors gratefully acknowledge the Hong Kong Polytechnic University for providing funding to support this research effort.

References

Bennett, J and Grice, A (1990) Procurement systems for building, in Quantity Surveying Techniques, New Directions, (ed. PS Brandon), BSP Professional Books, Oxford.

Bound, C and Morrison, N (1993) Contracts in use. Chartered Quantity Surveyor, December/January 1993, 16-17, Royal Institution of Chartered Surveyors, London.

Chan (1996a) Architectural Services Department, 10th Anniversary, Professional Building, June, 1996, 39-48.

Chan (1996b) Determinants of project success in the construction industry of Hong Kong. Unpublished PhD Thesis. University of South Australia.

Chan, APC (1995) Towards an expert system on project procurement, Journal of Construction Procurement, Volume 1, Number 2, November 1995, 111-123.

Church, R and Potts, K (1992) Trends in design and build contracting. Chartered Quantity Surveyor, February 1992, p.ii-iii, Royal Institution of Chartered Surveyors, London.

Ganesan, S, Hall, G. and Chiang, YH (1996) Construction in Hong Kong, Avebury, 1996.

Griffith, A (1989) Design-build procurement and buildability. Technical Information Service Paper No.112, Chartered Institute of Building, UK.

Ho, TOS (1995) Managing for the future management contracting – KCRC Kowloon station renovation and extension project. Asia Pacific Building and Construction Management Journal, Vol.1, 1995.

Lam, PTI and Chan, APC (1995) The recent trend of construction procurement systems in Asia-Pacific world, National Conference of the Australian Institute of Project Management, Adelaide, Australia, 520-527.

Ma, TYF and Chan, APC (1997) Is design-build leading the trend? A study of procurement systems in Australia. Leadership and Total Quality Management in Construction and Building, Singapore, 123-129.

Naoum, SG and Chan, H (1994) The potential use of management contracting in the Hong Kong construction industry – attitudinal survey. CIB W92 East meets West, Procurement Systems Symposium, Hong Kong, 235-242.

Ndekugri, I and Turner, A (1994) Building procurement by design and build approach. *Journal of Construction Engineering and Management*, Vol.120, No.2, 243-256, ASCE.

Pain, J and Bennett, J (1988) JCT with contractor's design form of contract: a case study. *Journal of Construction Management and Economics*, 6, 307-337, UK.

Rowlinson, SM and Walker, A (1995) The construction industry in Hong Kong, Longman, 1995.

Salmon, MJ (1990) Project procurement in Hong Kong and China,. *Proceedings of the symposium on construction project management in China and Hong Kong*, E2.1-E2.23.

Skitmore, RM and Marsden, DE (1988) Which procurement system? Towards a universal procurement selection technique, *Construction Management Economics*, 6, 71-89.

Tam, CM (1992) Discriminant analysis model for predicting contractor performance in Hong Kong. Unpublished PhD Thesis, Loughborough University of Technology.

Wilkins, B and Smith, A (1994) Procurement of major publicly funded health care projects. CIB W92 East meets West, Procurement Systems Symposium, Hong Kong, 307-314.

6. Private Finance Projects

'INTENTIONAL UNCERTAINTY' FOR PROCUREMENT OF INNOVATIVE TECHNOLOGY IN CONSTRUCTION PROJECTS

TOMONARI YASHIRO

Department of Civil Engineering, University of Tokyo, 7-3-1 Hongo Bunkyo Tokyo 113, Japan, yashiro@ken-mgt.t.u-tokyo.ac.jp

Abstract

Construction has the nature of project base economic activity where research and technology development are made after a project is started. The paper indicates examples how technology developments are made during project processes in Japanese construction projects; one example is procurement of claddings where capable and competitive specialist contractors take major roles. Instead of prescriptive design, performance-based contracting is used in which architects create 'uncertainty' intentionally at the stage of conceptual design in order to procure innovative technology from specialist contractors. The other example is some of Japanese conventional version of design-build projects. The alliance of clients and design builder creates 'uncertainty' at the earlier stage of projects in to get better adaptability to unforeseen conditions relating to project circumstances.

Based on the lessons from these examples, the paper proposes the idea of 'intentional uncertainty' strategy as a potential method to procure innovative construction technology by encouraging invention during project process. In order to enjoy the benefit, controllability of 'intentional uncertainty' is essential. The paper discusses about the methodology to control 'intentional uncertainty'. These are continual interdependency management, scope management and team work development.

Keywords: adaptability, design-build, innovation, performance contracting, project, uncertainty

Introduction

As Steven Groak pointed out, construction has nature of 'a project-based (or at least project led) economic activity, with its arrangements and disposal of resources induced by those projects and borrowing across technological bases, almost unpredictably ' [Groak (1994)]. The idea of 'construction as project' is, in a sense, opposite with the idea of 'construction as industry'. An industry has definable boundaries and it uses specific technical skills and specific resources. Contrarily a project induces its own demand chain, its needs and resources, its own process and consequential processes and its own specific organization. It thus creates new patterns of connections constantly between sources of expertise and technical know-how [Groak (1994)].

For the discussion on the method to procure innovative construction technology, Groak's remarks are quite suggestive. In case we stand on the concept of 'construction as industry', technical skills and resources applied in the project need be predictable at the stage of the projects' initiation. However, these ideal situations are far beyond the reality of many construction projects, especially that of long term and/or large size construction projects where innovative technologies are applied. A sort of turbulent environment of these projects requires unpredictable but inventable configurations of supply industries and technical skills. Innovative technologies are invented during the projects driven by a variety of factors and agents, some are identifiable, some are not.

This nature of construction, termed as 'technological paradigms' by Groak, generates unpredictability in terms of innovative inventions during whole construction processes. This unpredictability is a sort of uncertainty for project managers and for design team leaders. If project managers prefer to minimize these kinds of uncertainties and to set prescriptive scope definition about the applied technology, inventions during project processes would be discouraged while probability to miss the opportunity to apply the most appropriate technologies would increase. Contrarily, if the project managers would set only outline scope about the technology, the risk of failure in quality of projects' product would increase while better chance for innovation is available.

Thus, we need well-balanced setting of 'uncertainties' to enjoy the benefit of innovation during the construction process. The paper defines these kinds of uncertainties as 'intentional uncertainties' and discusses about the methodology to procure innovative construction technology by managing 'intentional uncertainties'.

Lessons from Japanese practices

Procurement of cladding from well matured specialist contractors

Claddings of buildings are composed of various prefabricated elements such as sashes, cladding panels, curtain wall components and so on. Curtain walls, mainly applied to high rise building, are supplied by major cladding fabricators.

In Japan, these cladding fabricators perform as specialist contractors. Their roles are far beyond those of trade contractors and building material suppliers. They perform integrated services from design and fabrication of components to installation of components in sites. It is reported that cladding fabricators in Britain are performing similar roles [Connaugton et. al. (1994)].

Cladding fabricators are competing with others. They are making effort to upgrade technical potential and are making intensive investment on technology development in order to overcome competitors. Eventually, each cladding fabricator has its unique technical know-how which is different with these of competitors. In another word, it is usual that every cladding fabricator offers distinct detail design with others even if the appearance and performance are similar.

Since the first skyscraper in Japan was constructed in 1968, Japanese architectural design firms have got empirical knowledge; each cladding fabricator has distinct technical know-how with others. Architectural design firms are not the top runners of technological knowledge about claddings but some cladding fabricators are.

Consequently, capable architectural design firms tried to involve cladding fabricators to the process of conceptual design of large size tall building, because technical feasibility of cladding is an influential factor to whole building design. However, it is general that any official status is not given to the cladding fabricators for their involvement in concept design of building. There are no guarantees that the fabricators who provide informal service are appointed as subcontractor by the main contractor of the project, in case of 'traditionally' procured projects where architectural design firms and main contractors are appointed separately.

Japanese architectural firms took various attitudes to deal with the informal service by cladding fabricators. There are two typical attitudes. One attitude is that they input distinct prescriptive details offered by specific cladding fabricators into contract drawing in order that the main contractor is obliged to nominate the cladding fabricator who provided informal service, because only the cladding fabricator has exclusive technical know-how to realize the details at reasonable cost. The other attitude is that architectural design firms only specify the appearance and performance of claddings in contract drawings in order to make the quality of cladding independent from selection of cladding fabricators by the main contractor, and more significantly, in order to encourage technological development by cladding the fabricator after nomination.

The former attitude lost the popularity. One of the reasons was that the distinct prescriptive details are only one of the criteria for main contractors in the selection of subcontractors; main contractors have the other criteria including the nomination of specific cladding fabricators by clients. Thus, it is usual that the other cladding fabricators are nominated even though the contract drawings involve distinct prescriptive details offered by the fabricator who provided informal service. In those cases, the main contractor requests architectural design firm to accept change of details' design, or alternatively main contractor requests the appointed cladding fabricator to realize the prescriptive detail by developing brand new technical know-how for the fabricator. In most of the cases, the appointed fabricators responded to the expectation by the main contractor. The other and more significant reason why the former attitude lost popularity is that architectural design firms themselves learned through the experience that the prescriptive details offered by specific fabricators are not always the best technical solutions; it is not beneficial to exclude other solutions in the early stages of design development.

Therefore, the latter attitude, a sort of 'performance-based design' by architectural design firms, is now 'ordinary' way to procure claddings for tall buildings in Japan. 'Performance-based design' is the design that architects only define performance at the stage of contract drawing.

Introduction and dissemination of performance-based design have made unexpected impact. It decreases technical potential of architectural design firms in terms of cladding design. Today it is not surprising that in the case of performance-based design, most architectural design firms need help by cladding fabricator for setting appropriate performance level prior to completion of contract drawing.

It is understandable that cladding fabricators would complain to doing services of uncertain scope without certain prospect of payment for their services. However, until now the complaint have not been demonstrated explicitly, simply because these informal services are believed to bring some benefit for cladding fabricators from the aspect of marketing based on long terms 'good' relationship, under the circumstance of expanding construction market.

Now, the construction market is shrinking in Japan so that the informal services by cladding fabricators are losing its motivation. Thus, it needs formalization of these services by cladding fabricators in Japanese procurement system.

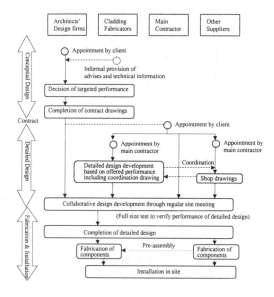

Figure 1. Typical roles of related agents in the process of performance-based cladding design

Architectural Institute of Japan (AIJ) edits and maintains Japan Architectural Standard Specification (JASS), which has been the most influential standard specification in Japan. Section 14 (JASS14) is for definition of claddings' design. Different with other sections, JASS 14 defines performance standard since the first version published in 1985. JASS 14 indicates performance criteria and the methodology to indicate the performance. Through practices based on JASS14, Japanese practitioners got the following lessons about the performance-based contracting.

1. Fixation of prescriptive details in earlier stages of construction is not the best way to

procure most innovative technical solutions.

2. In another word, it is better to provide a sort of 'uncertainty' intentionally at earlier stages of design development in order to utilize technical potential of specialist contractors by encouraging technology development during whole project process.
3. Performances described at contract drawing provide the scope of the technology development, which makes 'uncertainty' controllable by design team leader or project manager.
4. Performance contracting in terms of cladding is applicable both for
   'traditional' procurement projects where architectural design and construction services are separately appointed.
   conventional Japanese design-build procurement where potential general contractors provide whole packages of architectural/engineering(A/E) and construction service.

It is quite usual in Japan that meeting for design development of tall building cladding is held at site offices. The meeting is usually composed of architectural design firms, main contractor, cladding fabricator and related material suppliers such as glass fabricators, sealants' suppliers, stone fabricators and soon. The meeting is held at regular basis (weekly or biweekly). Based on performance drawing, detailed designs are developed and determined through the discussion the detailed design drawing proposed by cladding fabricators. At the meeting, the following issues are discussed until the completion of detailed design.

1. Correlation between defined performance and proposed detailed design
2. Interfaces between proposed details offered by different fabricators (eg. cladding fabricators, glass fabricators, sealants' suppliers, stone fabricators and soon)
3. Need of full size performance tests to verify the performance of completed detailed design

Performance-based design method is being disseminated from cladding to the other parts of building; for instance, building services' specialist contractors are providing integrated service from design to installation of building services.

Adaptive procurement by some Japanese clients

Japanese large size clients tend to provide imperfect project brief, especially in the case of conventional design-build procurement projects. It is simply because those clients are accustomed to 'tacit' communication with the specific general contractors with whom they have long term relationship. The clients expect that the 'partnering' general contractor would understand their implied requirements which are needless to say. It can be termed as 'implied requirements' procurement. Until beginning of 1990s, 'implied requirements' procurement has been quite successful procurement method for these clients, because it was effective to mitigate disputes and conflicts by change orders. The clients could finalize project brief to enjoy innovative construction technology by considering deliberately on information input by the 'partnering' general contractor, who has advanced knowledge on project management as well as technical issues. In case of long term projects, it is probable that project circumstances are changed dramatically during project process. Even in these cases, 'implied requirements' procurement worked well for clients because the 'partnering' general contractor was flexible to respond to the change of circumstance to some extent. Even though 'implied requirements' procurement has a sort of stability of cost and time. In most of the cases, 'partnering' general contractors complete the work within promised time and cost, though the cost is supposed to involve certain amount of overhead. For these reasons, Japanese large clients learned that 'implied requirements' procurement policy has, in a sense, high degree of adaptability. Thus, it can be called as 'adaptive procurement' as well.

For example, in heavy engineering projects such as constructing underground industrial facilities, clients recognize that they do not enough technical knowledge what is possible by most contemporary technology. The client only offers an outlined project brief in contract documents and then the client nominates and invites the contractor as design-builder at basically turnkey contract basis. In practice, the client and design-builder collaborate to develop the scope of the industrial facility including its characteristics. In addition, feasibility of alternative approaches and solutions are being discussed between clients and design-builder. Through these communications, clients' requirements are identified for the client itself and design-builder. Then, design-builder starts to arrange ad hoc whatever supply of services, finance, information and products which are necessary for the project. The ad hoc arrangement generates the need of technological development during project progress. Design-builder is also encouraged to organize to implement technological development.

It is quite often that unforeseen event happens during the progress of projects such as unpredictable geotectonic conditions. The changes by unforeseen events enforce the change of design. The coalition of clients and design-builder analyze the extent and the impact of the change to the project. Then they do not hesitate to agree with the change of project scope. A design-builder rearranges process and organization of technology development as well. Moreover, it is not unusual that a client proposes the change of project scope by quickly changing projects' circumstances. If it is reasonably minor, Japanese design-builders do not demand the revision of the initial contract. Thus, in heavy engineering projects, 'implied requirements' procurement has high degree of adaptability corresponding to the project progress in Japanese construction practices

Although Japanese large clients have enjoyed the benefits of 'adaptive procurement', transformation in economy undergoing in Japan would give pressure to change the 'adaptive procurement' procurement policy, because;

1. Many clients are recognizing the irrationality to appoint the specific contractor with long term relationship; present economical situations provide the chance to procure at more cheaper cost from potential competitors with lower risk of dispute and conflict if the project is managed well.
2. There is certain sign that possible misunderstanding of tacit requirements would give serious damage to the client; projects are becoming complex so that analogical understanding of requirements in other project is becoming almost harmful.

Japanese large clients are now trying to create the procurement method for innovative construction technology with reasonable adaptability and with transparent cost under competition of potential general contractors.

The idea of 'intentional uncertainty' strategy

In general, there are number of uncertainties in construction projects because of;

1. Ever-changing project circumstances and requirements during the projects
2. Unpredictability of available resources
3. Unforeseen potential of members of a construction team
4. Insufficient information on constructability at earlier stages of projects
5. Unpredictable speed of technological innovation

These uncertainties are basically uncontrollable for project managers. It is given constrains for them. Some of these uncertainties turned to be explicit risk especially in long term and/or large size construction projects or in case very few similar projects were implemented in recent past in that region.

Lessons from Japanese practices indicate that there are several examples in which clients or architectural design firms as an agent of clients remain uncertainties intentionally to procure innovative technology by encouraging developments during project process. It is implicit, rather than explicit, procurement policy. Here it can be termed as 'intentional uncertainty' strategy.

The expected benefits of 'intentional uncertainty' are;

1. Encouragement of inventions by contractors, specialist contractors and suppliers during project processes under competitive business environment.
2. Improvement in flexibility and adaptability to the change of project scope during project progress.

'Intentional uncertainty' can be distinguished with five kinds of uncertainties listed above from the aspect of controllability. Five kinds of uncertainties are definitely uncontrollable. However, 'intentional uncertainties' could be controllable if it is managed well. Therefore, it is quite significant to establish the methodology how to make 'intentional uncertainties' controllable in order to apply innovative construction technology.

The methodology to control 'intentional uncertainty

Japanese practices indicate that the qualities of following issues are key factors to improve controllability of 'intentional uncertainty'.

1. Interdependency management
2. Scope management

3. Team work development

Interdependency management

Construction projects consist of processes which are dependent each other. One project process usually affects others. It is essential for project managers and/or for design team leaders to identify what kinds of project processes are dependent on 'intentional uncertainty'. It is also requested for them to identify potential interface problems between 'intentional uncertainty' and dependent project processes. If interdependencies and interfaces are not recognized comprehensively, 'intentional uncertainty' could be uncontrollable.

In addition, project managers and/or design team leaders are requested to make sub-package of processes composed of 'intentional uncertainties' and processed which have interdependency to it. They have to set appropriate phasing of project into several sub-packages. In case most of project processes are identified to be dependent to 'intentional uncertainty', it should be improved of interdependent relations between processes. Interdependent processes to 'intentional uncertainty' should be 'enclosed' to reasonable sized sub-package in whole project. For instance, 'intentional uncertainties' in cladding design described in this paper is 'enclosed' in cladding system only. It does not affect on detailed design of other parts.

Project managers and/or design team leaders have to be able to inform project team members what should be fixed now and what can be changed later, by making the phasing of the project synchronized with timing of converting from 'intentional uncertainty' to fixed issues.

If interdependencies are comprehensively identified and processes are appropriately phased, any changes occurred during project progress and turbulence from project environment do not make harmful damage in controllability of 'intentional uncertainty'. If phasing of processes are made successfully, the impact of delay can be manageable, even in the case that the decision of converting 'intentional uncertainty to fixed detailed design should be delayed behind the schedule. Continual interdependency management during whole construction process is essential to improve and maintain the controllability of 'intentional uncertainty'.

Scope management

In the projects where 'intentional uncertainty' strategy is applied, it is not unusual that the project scope, such as description of the project product and its characteristics, is modified during project processes. Clarity of project scope is significant to keep controllability of 'intentional uncertainty'. However, it does not mean that project scope should be fixed at the stage of project initiation. In order to get the potential benefits of 'intentional uncertainty', the project scope should be adapted flexibly to turbulent environment projects. Thus, continual scope management during project progress is essential for enjoying the benefit of 'intentional uncertainty' and for keeping controllability of it. Continual scope management requires continuous communication between clients and project team, because concept development for scope definition has to reflect updated clients' requirement.

In the communication, setting performance criteria and levels would be translation-tool between clients' requirement and project scope.

Team work development

Partnering of potential parties is the key to encourage collaborative inventions during a project which is driven by 'intentional uncertainty' strategy. Adversarial relationship would delete basis of collaboration and lose controllability of 'intentional uncertainty'. 'Good' formation of a design/construction team is indispensable for partnering, where each member has enough potential to accept the situation that the scope of work for each member is not clearly defined at the stage of project initiation. Moreover, project organizations' formation should provide common interests for each member of project organization. In another word, project processes should be designed so that the inventions during project would be accord with common interest of each participating party of the project. It is desired to invent procurement method refined by domestic context, which create collaborative relations among team members, and which provide the prospect that potential benefit would be larger than potential risk. As it is presented, conventional design-build procurement in Japan had the potential to provide a sort of partnership, though the situation is now being changed.

By any means, creating collaboration is essential for encouraging inventions from 'intentional uncertainties' during project.

Conclusion

The understanding of construction as a project based economic activity, which can be termed as 'technological paradigms', provides different idea for procurement policy of construction. For management of the dynamic nature of the processes, static approach is ineffective; rather than that, more dynamic approach is needed.

The idea of 'intentional uncertainty' strategy has the potential to procure innovative construction technology by encouraging inventions and innovations during whole construction projects processes. During project processes whatever available resources and technical know-how are assembled to realize specific project scope. In case there exist competitive potential suppliers, performance-based procurement would have advantage to utilize the potential.

'Intentional uncertainty' strategy also has potential benefit of flexibility and adaptability to ever changing projects' environment. Partnership applied at some conventional design-build projects in Japan would be the example of better practice to induce flexibility and adaptability from 'intentional uncertainty'.

In order to enjoy the benefit, 'intentional uncertainty' needs to be controllable. Different with uncertainties defined by projects' environment, 'intentional uncertainty' is controllable, if project managers and/or design team leaders are capable enough for continual interdependency management, scope management and teamwork development during whole construction process.

The paper does not focus on the relationship between existing procurement method and 'intentional uncertainty' policy. The author is planning to discuss about it in the other research papers. However, the author believes, at this moment, that 'intentional uncertainty' strategy is compatible or applicable to most of existing procurement method of construction services with minor modification in terms of risk allocation and authority for taking initiatives in the process, because 'intentional uncertainty' strategy is mainly related to process management rather than to contract management.

References

Connaugton, J., Jarrett, N., Shove, E. (1995) Innovation in the cladding industry, Department of the Environment, HMSO

Groak, S (1994), Is Construction an industry? Notes towards a greater analytic emphasis on external linkages, *Construction Management and Economics*, vol.12, no.4, pp287- pp293

Yashiro, T (1994), How combine explicit and implicit code in housing practice ?, Proceedings of 4th annual Rinker International Conference Affordable housing; present and, University of Florida

Yashiro, T (1998), Localization of industrialized construction method for blocks of flats, Proceedings of the Sixth East Asia-Pacific Conference on Structural Engineering & Construction, National Taiwan University, pp1029-pp1034

PRIVATIZATION-INDUCED RISKS: STATE-OWNED TRANSPORTATION ENTERPRIZES IN THAILAND

SANTI CHAROENPORNPATTANA and TAKAYUKI MINATO

School of Civil Engineering, Asian Institute of Technology, P.O. Box 4, Klong Luang, Pathumthani 12120, Thailand. minato@ait.ac.th

Abstract

Privatization has been a well-accepted tool for financing public works. It reduces investment expenditure by the government. When it is successful, privatization enhances efficiency and effectiveness of the investment. On the other hand, privatization is very risky. Unless appropriate care is taken, it may result in failure.

In Thailand, privatization is not a new issue. It has been discussed for more than a decade. Recently, the Thai government has begun to accelerate infrastructure development using private finance initiative, and this is recognized to be a potential solution to financial problems in Thailand.

However, in privatization, if risk is allocated to any single party, it may generate too much burden on the organization. Therefore, proper risk allocation is a key to the success of privatization. Risk-sharing is a way of responding to risk. A risk that is allocated to a single party would be shared by several parties according to their abilities. This research explores the identification of privatization-induced risks for transportation profects in Thailand, with the objective of pinpointing which risks could be shared between the public and private sectors.

Keywords: Privatization, risk identification, risk-sharing, Thailand

Introduction

Privatization has been a popular topic for the past two decades. According to the Asian Productivity Organization (APO), "the 1980s was the era of privatization" [APO (1993)]. As many as 50 countries, both developed and developing, opted for privatization of state-owned enterprises (SOEs) from 1980 to 1987. From 1988 to 1996, "there were 2,655 privatization transactions in 95 countries, yielding 271 billion U.S. dollars" [Shafik (1996)]. Many developing as well as some developed countries needed innovative infrastructure financing tools because "their capital pools were decreasing continuously, and they thought that privatization might provide answers" [Walker and Smith (1995)].

On the other hand, there have been several arguments against privatization. For example, "some people consistently believed some particular activities that had been carried out by public sectors, such as transportation, utilities, and education, had to be steadily provided as public services and could not be left to private decisions and market mechanism" [APO (1993)]. Dismissals and layoffs were other arguments against privatization.

In Thailand, privatization has been discussed for a decade. The focus has been on state-owned enterprises in the public utility and transportation sectors. The SOEs, interchangeably called public enterprises, are government owned and controlled entities that are supposed to earn most of their revenues from sales of public services. At present, there are 61 SOEs in Thailand. In the transportation sector alone, there are 14 SOEs playing major roles in infrastructure development. Examples include the State Railway of Thailand (SRT), the Metropolitan Rapid Transit Authority (MRTA), the Expressway and Rapid Transit Authority of Thailand (ETA), the Airport Authority of Thailand (AAT) and the Thai Airways International Co., Ltd. (THAI). In terms of performance, however, many of the SOEs have not lived up to the expectations of the founders. In addition, due to the continuing economic crisis in the country, the government is willing to accelerate "non-traditional" infrastructure development programs to gain benefits from the private sector. As a result, privatization has been a major focus of attention.

Options for privatization consist of *ownership option* and *contractual option*. The ownership option refers to transfer of ownership of a SOE to private hands. It includes total or partial transfer of ownership as well as establishment of new joint venture business. In contrast, the contract option involves provision of services by the private sector through contractual relationships with the government or SOEs. Under this option, government gives concessions to private firms so that they operate and maintain the public services for a certain period of time.

Ownership Option is further divided into *trade sale*, *public offering*, and *buy-out* (management or employee) as described below.

> **Trade sale:** Government chooses to sell SOEs to investors, possibly to a corporation or partnership that has experiences in running the utility and the financial capabilities to improve and expand the services. "The trade sale can be made more quickly than full-blown public offering (floating), which requires months of preparation" [Hyman (1995)].

> **Public offering:** It refers to selling shares to the public through a long and elaborate procedure. Government might have to take a year to fix the utility in order to market the offer. Some public offerings sometimes fail, even after expensive preparations, to the great embarrassment of all those involved.

> **Buy-out:** Government sells assets to management or employees of SOEs instead of to outside third parties with the objective of encouraging employee involvement in the management of the firm.

Contractual option includes *contracting out* and *concession* as described below.

> **Contracting out:** Government participates in services that may be better supplied by private companies. In this form, government maintains ownership of SOEs, and the private company is paid according to the contract terms. This form of privatization could increase efficiency of SOEs by reducing long-term expenditures that are used for, for example, permanent employees.

Concession: Government gives private investors concessions for investment, management, and operation of assets for a certain period. At the end of the concession, the investors normally transfer all assets to the government. Sometimes, the private firms hold rights to continue operation of the assets even after the concession.

In privatization, if risk is allocated to any single party, it may generate too many burdens on the organization. Therefore, proper risk allocation is a key to success. The underlying idea of this research is to employ risk-sharing concept for privatization as a better risk management strategy. For the purpose, this research attempts to identify sharable risks for privatization in Thailand.

General background of Thailand

Thailand is a country covering 198,461 square miles. The population is approximately 60 million, and grows at approximately 1.2 to 1.4 percent per year. It is projected to reach 64 million by the year 2000. In Thailand, the government plays major roles in driving the economy. The economic growth rate averaged 5.7 % per year during 1977-78, while that of other developing countries as a whole was 3.8% only. It was highest in 1988 at 13.2%, and has been declining since 1989.

In 1997, the financial market and the whole economy faced a series of problems such as currency devaluation, downgrades of credit by rating corporations, debts from the property sectors, and budget deficit. All these factors served to hinder the performance of financial institutions in the money markets. These events forced Thailand to request loans from the International Monetary Fund (IMF). A major condition of the loan agreement required the government to accelerate privatization programs in order to solve the deficit problems. Thailand is still a developing country. Poverty is still widespread and the national saving rate is negative. Neither of individuals nor local banks have capabilities to invest in major projects. The economy at this moment is in a bad shape. There is no sufficient budget to invest in the infrastructure development. In order to accelerate economic growth, infrastructure development is essential, and innovative financial solutions must be developed.

Privatization-induced risks in Thailand

To identify risks associated with privatization, historical experiences in the UK, Italy, Malaysia, Japan, Philippines, and Vietnam were initially reviewed from books, journals, newspapers and via internet [Miller (1995), Bishop and Kay (1989), Kohno (1996), Savona et al. (1994), Leeds (1989), Sundaram (1995), Samia (1996), Gupta (1997)]. Based on the review, risks were classified into five categories; 1) political, 2) economical, 3) legal, 4) transaction, and 5) operational as shown in Table-1.

By reviewing newspapers and interviewing experienced people, significant privatization-induced risks in Thailand were identified. The economic and political situation in Thailand has changed significantly in the past few years. For example, a new constitution has been drafted, and the economic crisis that started in 1997 has affected confidence in the economy. Therefore, recent news and interviews were thought to be must reliable for this study. The results showed that attention should be paid on such risks as *internal resistance, labor resistance, political influence, currency devaluation risk, small capital market, uncertainties of government policies and changes in law and regulation, delay in privatization program, small number of interested investors, unfair selection process of investors, unfavorable investment environment, associated infrastructure risks, demand and supply risk, price escalation, and other operation risks.* Description of these Thai specific risks is also given in detail in the following sections.

Internal resistance: Some government authorities are against the idea of privatization. For example, it was reported that the secretary-general of the State Enterprise Relations Confederation said "this idea is to cut good meat for some groups of people supporting politicians and political parties, or for foreigners who want to take over Thailand's profitable public enterprises" [Bangkok Post (July 25, 1997)]. There are also politicians and government officers who are known to be against privatization.

Labor resistance: There is resistance against privatization from various labor unions, too. For example, staff of Thai Airways International (THAI) rejected the management's plan to privatize the airline's cargo business, which they consider neglected their interests [Bangkok Post, (July 26, 1997)]. In general, government employers are opposed to privatization primarily because they do not want to loose benefits such as life-long employment and full health support.

Political influence: Political intervention in public works is a well-known characteristic of Thai culture. According to some interviews, for example, the route of Second Stage Expressway was altered due to political interventions. It was initially planned to reach Bangkok International Airport, but the plan was changed to terminate the route at Chaeng Wattana road.

Devaluation risk: The Thai currency was devalued by almost hundred percent within about six months in 1997. Consequently, many privatized projects in Thailand funded through borrowings of foreign currencies faced difficulties in loan repayment.

Table-1 Privatization-Induced Risks Identified from Experiences of Various Countries

Risk	Country					
	U.K.	Japan	Italy	Malaysia	Philippines	Veitnam
1 Political Risk						
1.1 Internal Resistance	*	*	*	*	*	
1.2 Labor Resistance	*	*	*	*	*	
1.3 Nationalization						*
1.4 Political Influence	*	*	*	*	*	*
1.5 Uncertainty of Government Policy					*	
1.6 Unstability of Government					*	*
2 Economical Risk						
2.1 Devaluation Risk				*	*	*
2.2 Foreign exchange Risk						*
2.3 Inconvertibility of Local Currency						*
2.4 Inflation Risk						
2.5 Interest Rate Risk						
2.6 Small Capital Market	*		*	*	*	*
3 Legal Risk						
3.1 Changes in Law and Regulation					*	
3.2 Inefficient Legal Process					*	*
3.2 Legal Barrier	*	*	*		*	*
4 Transaction Risk						
4.1 Delay of Privatization Program		*	*	*	*	*
4.2 Improper Privatization Program						*
4.3 Incapable Administration Body				*		
4.4 Reluctance to Proceed		*	*		*	*
4.5 Too Small Number of interested Investor						*
4.6 Unfair Process of Selection of Private Investor				*	*	
4.7 Unfair Selection of SOEs to privatize				*	*	
4.8 Unfavorable Investment Environment				*	*	*
4.9 Valuation of Asset	*	*	*		*	*
5 Operation Risk						
5.1 Associated Infrastructure Risk						
5.2 Demand and Supply Risk						
5.3 Incapable Investor						
5.4 Improper Regulation (too loose or too tight)	*	*	*	*	*	*
5.5 Liability Risk				*	*	*
5.6 Management Risk						
5.7 Price Escalation Risk	*	*	*	*	*	*
5.8 Technical Risk						
Sources:	1	2	3	4	5	6

1 Miller (1995), Bishop and Kay (1989)
2 Kohno (1996)
3 Savona et al. (1994)
4 Leeds (1989), Sundaram (1995)
5 Samia (1996)
6 Gupta (1997)

Small capital market: Thailand's capital market is heavily dependent on foreign lenders. After the collapse of the "bubble economy," foreign financial institutions became reluctant to extend loans and liquidity to the local money market. According to Mr. Banthoon Lamsam, the president of the Thai Farmers Bank, "many businesses want loans to lubricate their operations, but commercial banks have no funds available for them" [Bangkok Post (June 23, 1997)].

Uncertainty of government polities and changes in law and regulation: Thailand's had seven different governments for the periods of 1989 to 1998. Each of the governments, composed of different political parties, always changed the policies of their predecessor in order to protect his own interests.

Legal barrier: According to Thai law, state-owned enterprises that are privatized by divestiture of assets in the stock market must have the legal status as a company. However, approximately 70% of state-owned enterprises are not legally recognized as companies. According to Mr. Mahidol Chantrangkurn, the chairman of Thai Airways International (THAI), "the planned privatization would be delayed because of existing legal requirements which stipulated that Thai nationals must own not less than 70% in the national carrier. Therefore, the government needed to amend the law to allow foreign investors to hold a combined stake of up to 49%" [Bangkok Post, (December 20, 1997)].

Delay in privatization program: Many factors influence privatization programs in term of timing. For instance, the process of amending laws is unpredictable and a timetable is difficult to set. Political influence is also a factor that causes delay in privatization.

Ffor it small number of interested investors: One of the difficult problems in privatization in Thailand is the small number of competitive bidders. "This situation can lead to failure to price the assets properly and finally to accusation of corruption and favoritism" [Laohasomboon (1992)].

Unfair process of selection of private investor: Collusion among bidders is common on government projects as political influences play significant role in the process, especially for large projects. One recent example is the bidding of construction of infrastructure for 800,000 telephone lines in Bangkok and the provinces. It is alleged that an influential politician persuaded 13 bidders to form one group instead of two to submit tenders for the project. As a result, a foreign firm had to be screened out by the bid-evaluation committee [Bangkok Post (November 15, 1996)].

Unfavorable investment environment: Current economic crisis raises questions about likely share prices and potential buyers. Local investors have been hit hard by the economic slump while foreigners are waiting for the Thai Baht to stabilize for investment.

Associated infrastructure risk: Transportation infrastructure development always faces land acquisition problems as property owners often refuse to sell their land. As such, costly route charges have to be made. For instance, residents of the Ban Krua community have been against the Urupong-Ratchadamri expressway project for over 10 years" [Bangkok Post (March 9, 1997)].

Demand and supply risk: Factors that directly affect demand and supply of transportation infrastructure are significant. For example, change of import tariff for goods such as cars may affect the number of cars on toll ways. Similarly, higher tariff rate for construction steel may give rise to shortage of the material in the market.

Price escalation risk: Fees of public transportation tend to escalate after privatization. Although privatization may augment efficiency, enterprises that have been subsidized by government in the past are likely to increase fees after privatization.

Other operation risks: The factors that cause operational problems are incapable investors, improper regulations (too loose or too tight), liability risks, management risks, and technical risks.

Identification of sharable risks

The final step in this paper is to assign sharable or non-sharing classifications to the risks identified in the previous sections. The classification scheme consists of two steps: *risk characterization*, and *analysis of characterized risks*.

Risk characterization: In this process, predetermined criteria are assigned to each risk. In this study, six criteria were chosen to classify risks into either 1) Static (S) or Dynamic (D), 2) Fundamental (F) or Particular (P), 3) Government (G) or Private source (OTH), 4) Speculative (SP) or Pure (PU), 5) Financial (FI) or Non-financial (N), and 6) Measurable (M) or Immeasurable (I) [Vaughan (1997), Perry and Hayes (1985), Flanagan and Norman (1993)]. These six criteria were categorized into two groups: main criteria and supplementary criteria. Main criteria are those that give decisive implication in evaluating whether the risk is sharable or not, and consist of 1), 2), and 3). The implication and characterizations of the three main criteria are described in Table-2. On the other hand, supplementary criteria 4), 5), and 6) give no decisive implication.

Analysis of Characterized Risks: Next, attributes such as static, fundamental, and government source were assigned to individual risks. The assignment was rather subjective, depending on evaluation of past research and the authors' view.

Table 2 Implication and characterization of criteria

Criteria	Characterization
1) Static or Dynamic	The key consideration is whether or not the risk will be altered or changed, in terms of direction or magnitude of losses, according to changes in economy or macro public policies. If it would be changed because of changes in economy or public policies, it is dynamic. If it changes independently or its changes are not relevant to the economy or macro public policies, it is *static* [Vaughan (1997)]. **Dynamic risks**: Government is the body that has the ability to manage, control and direct the economy or public policies. **Static risks**: Most static risks are more likely to be managed by the public sector. In some cases, static risks can be handled through insurance.
2) Fundamental or particular	Fundamental risks involve losses that apply to all entities in the system, whereas particular risks involve losses that apply to a particular entity or a group of entities in the system [Vaughan (1997), Al-Bahar (1988)]. **Fundamental risks**: Fundamental risks are more likely to be managed by government body. **Particular risks**: Particular risks are more likely to be managed by the private sector.
3) Government Source or Private Source	Government source risks originate from the actions of governmental entities, whereas private source risks arise from private sectors. Each risk should be allocated to us "creator." **Government Source risks**: should be allocated to the government. **Private Source risks**: Private source risks could be allocated to private investor.

Depending on the reasoning given in Table-2, each attribute of the three main criteria was directed into either of government or private's domain as shown in Figure1. For example, if a risk was evaluated to be dynamic (D), then it indicated that the risk belonged to the role government sector.

The final step in the analysis was to classify risks considering whether they are classified into sharable or not. The following rules were applied. If all the attributes of a particular risk belonged to the same domain of either private risks or government's risks, the risk was identified to be a non-sharable risk. In other words, if all the attributes of a risk did not belong to the same domain, the risk was thought to be sharable. Table 3 illustrates the final results of the analyses.

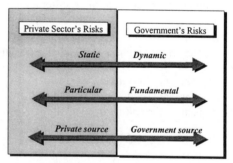

Figure 1. Classifications by main criteria

The results show, for example, "currency devaluation risk" is non-sharable because it is attributed to dynamic (D), fundamental (F), and governmental (G) sources. It is, therefore, categorized to be government' risk. On the other hand, "demand and supply risk," for example, is dynamic (D), particular (P), and originated from other (OTH) sources. It indicates that the risk is not totally assigned to any single organization, or sharable.

Table-3 Final Results from Analysis of Privatization-Induced Risks

Risk	Static-Dynamic	Fundamental-Particular	Government-Private Source	Pure-Speculative	Financial-Nonfinancial	Measurable-Immeasurable	Response	Allocated to	Thai Risk
1 Political Risk									
1.1 Internal Resistance	S	F	G	PU	N	I	Sharable		*
1.2 Labor Resistance	S	F	G	PU	N	I	Sharable		*
1.3 Nationalization	D	F	G	PU	N	I	Non-sharable	Government	
1.4 Political Influence	D	F	G	PU	N	I	Non-sharable	Government	*
1.5 Uncertainty of Government Policy	D	F	G	PU	N	I	Non-sharable	Government	*
1.6 Unstability of Government	D	F	G	PU	N	I	Non-sharable	Government	
2 Economical Risk									
2.1 Devaluation Risk	D	F	G	SP	FI	M	Non-sharable	Government	*
2.2 Foreign exchange Risk	D	F	G	SP	FI	M	Non-sharable	Government	
2.3 Inconvertibility of Local Currency	D	F	G	PU	FI	M	Non-sharable	Government	
2.4 Inflation Risk	D	F	G	SP	FI	M	Non-sharable	Government	
2.5 Interest Rate Risk	D	F	G	SP	FI	M	Non-sharable	Government	
2.6 Small Capital Market	D	F	G	PU	N	I	Non-sharable	Government	*
3 Legal Risk									
3.1 Changes in Law and Regulation	D	F	G	PU	N	I	Non-sharable	Government	*
3.2 Inefficient Legal Process	D	F	G	PU	N	I	Non-sharable	Government	
3.3 Legal Barrier	D	F	G	PU	N	I	Non-sharable	Government	*
4 Transaction Risk									
4.1 Delay of Privatization Program	D	P	G	PU	N	I	Sharable		*
4.2 Improper Privatization Program	S	P	G	PU	N	I	Sharable		
4.3 Incapable Administration Body	S	P	G	PU	N	I	Sharable		
4.4 Reluctance to Proceed	D	P	G	PU	N	I	Sharable		
4.5 Too Small Number of interested Investor	D	F	OTH	PU	N	I	Sharable		*
4.6 Unfair Process of Selection of Private Investor	S	F	G	PU	N	I	Sharable		*
4.7 Unfair Selection of SOEs to privatize	S	F	G	PU	N	I	Sharable		
4.8 Unfavorable Investment Environment	D	F	G	PU	FI	I	Non-sharable	Government	*
4.9 Valuation of Asset	D	P	OTH	SP	FI	M	Sharable		
5 Operation Risk									
5.1 Associated Infrastructure Risk	ST	P	PR	PU	FI	M	Non-sharable	Private	*
5.2 Demand and Supply Risk	D	P	OTH	SP	FI	M	Sharable		*
5.3 Incapable Investor	S	F	OTH	PU	FI	I	Sharable		*
5.4 Improper Regulation (too loose or too tight)	D	P	G	PU	FI	M	Sharable		*
5.5 Liability Risk	ST	P	PR	PU	FI	M	Non-sharable	Private	*
5.6 Management Risk	ST	P	PR	PU	FI	I	Non-sharable	Private	*
5.7 Price Escalation Risk	D	P	OTH	SP	FI	M	Sharable		*
5.8 Technical Risk	ST	P	PR	PU	FI	M	Non-sharable	Private	*

It is interesting to note that most of transaction and operation risks are classified to be either sharable or private source risks while political, economical, and legal risks are evaluated to be non-sharable risks that should be controlled by government. In fact, all transaction risks, except for "unfavorable investment environment," are identified to be sharable. Similarly, sharable operational risks include "demand and supply risk," "incapable investor," "improper regulation," and "price escalation" risks. The other operational risks are evaluated to be long to the private sector.

Suppose that political, economical, and legal risks are labeled as *macro events at country level*, and transaction and operation risks are *micro at individual project level*. It could be concluded that government should take care of most of macro risks, while micro risks are either directed to the role of private parties or shared between public and private sectors (Figure-2).

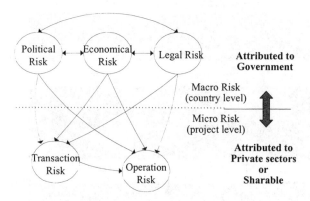

Figure-2. Risk allocation in general

Conclusion

Privatization is a very complicated matter for which successful implementation requires the application of risk management. While most forms of traditional privatization include "transfer of risks" to one party, risk-sharing between the public and the private sectors may improve the efficiency and effectiveness of privatization. In this study attempt was made to identify the major privatization-induced risks and to classify them into sharable or non-sharable risks.

The five sources of political, economical, legal, transactional, and operational risks were classified as privatization-induced risks. The major conclusion obtained through this study was that government bodies should control macro level risks such as political, economic, and legal risks. It was also discussed that transaction risks might be sharable; operation risks might be either sharable or controlled by the private sector. Among sharable risks, it was discussed that there were eight significant risks to be recognized in Thailand: internal resistance, labor resistance, delay of privatization program, small number of interested investors, and unfair process of selecting investors, demand and supply risk, incapacity of investors, improper regulation, and price escalation risks.

In Thailand, privatization of state-owned transportation enterprises is driven by economic needs. Failure to manage business under the changing environments, lack of autonomy, low efficiency and fraudulent intervention by government have been considered the major problems of public agencies, especially state-owned enterprises. In order to accelerate economic growth, infrastructure development is essential to improving efficiencies of SOEs.

Different social environments give rise to different national systems. In the future, extensive study to actually propose successful privatization in Thailand will be necessary. Inputs from experts of related disciplines such as economist, lawyers, engineers in this country as well as from overseas will be needed in order to disseminates insights among the society and find solutions under which risks could be fairly controlled.

References

APO (1993) The Privatization and Industrial Relations. Tokyo.

Al-Bahar, J. F. (1988) Risk Management in Construction Projects: A Systematic Analytical approach for Contractors. *Doctoral dissertation*, College of Engineering, University of California, Berkeley, U.S.A..

Bishop, M.R. and Kay, J.A. (1989) Privatization in the United Kingdom: Lessons from Experience. *World Development*, Vol.17, No.5, pp.643-657.

Flanagan, R., Norman, G. (1993) Risk Management and Construction. *Blackwell Scientific*, Oxford.

Gupta, J.P. (1997) The Privatisation Process: A Survey of State Owned Enterprises in Vietnam. *Sasin Journal of Management*, Vol. 3, pp.1-11.

Hyman, L.S. (1995) The Privatization of Public Utilities. *Public Utilities Reports*, U.S.A.

Kohno, M. (1996) Privatizing State-Owned Enterprises: experiences of Asia-Pacific economies. *K. Yanagi, ed., Asian Productivity Organization*, Tokyo.

Leeds, R.S. (1989) Malaysia: Genesis of a Privatization Transaction. *World Development*, *Vol.*17, No.5, pp.741-756

Miller, A.N. (1995) British Privatization: Evaluation the Results. *The Columbia Journal of World Business.* Winter 1995: pp.82-98.

Perry, J. G. and Hayes, R. W. (1985) Risk and Its Management in Construction Projects. *Proc.Instn Civ. Engrs* 78, June, pp. 499-521.

Samia, A.S.J. (1996) Privatizing State-Owned Enterprises: experiences of Asia-Pacific economies, *K. Yanagi, ed., Asian Productivity Organization*, Tokyo

Savona, P., Lorch, M., Pontecorvo, G., Arzac, E., and Greenwald, B. (1994) Privatization in Italy: Roundtable Discussion with Paolo Savona the Italian Minister of Industry. *The Columbia Journal of World Business* , Spring, pp.72-83.

Shafik, N. (1996) Selling Privatization Politically. *Columbia Journal of World Business*, Vol.31, No.4, pp.20-29.

Sundaram, K.S. (1995) Privatizing Malaysia: Rents, Rhetoric, Realities. *Westview Press*, Malaysia.

Vaughan, E.J. (1997) Risk Management. *John Wiley & Sons*, U.S.A.

Walker, C., and Smith, A.J., (1995) Privatized Infrastructure: The BOT Approach. Thomas Telford, London.

PPI IN THE PRC: TWO CATEGORIES AND SIX MODELS OF PRIVATE PARTICIPATION IN CHINESE INFRASTRUCTURE

R. MACDONALD, T. YASHIRO, T. YOSHIDA, and M. KUNISHIMA
Construction Management and Infrastructure Systems Laboratory, Department of Civil Engineering, University of Tokyo, 7-3-1 Hongo, Bunkyo-ku, Tokyo, Japan.
regm@iname.com

Abstract

Concepts of private participation in infrastructure (PPI) have received increasing international attention in recent years. As governments and the private sector in developing and advanced countries alike experiment with procedures for cooperating in the efficient development of infrastructure, acronyms such as "BOT" (Build-Operate-Transfer) and "PFI" (Private Finance Initiative) have become essential parts of a new vernacular. This paper outlines private participation in infrastructure in the People's Republic of China. It represents one portion of a comparative study on private sector infrastructure initiatives in China and Japan.

The paper divides PPI models used in China today into foreign private and domestic private categories. It describes three models within each category, outlining the merits and demerits of each as well as sources of financing. It goes on to give an overview of Chinese legal and regulatory conditions which affect the development and operation of infrastructure. Finally, the paper finishes with observations on the results of PPI models in China to date, and speculates on future directions of PPI in the PRC.

Keywords: Domestic private, foreign private, build-operate-transfer (BOT), legal and regulatory framework, private participation in infrastructure (PPI), sino-foreign joint venture

Introduction

The paper divides current models of PPI used in China today into two categories: foreign private and domestic private. Within the foreign private category there are three models: Cooperative Joint Venture (CJV), Equity-Only Joint Venture (EJV), and Build-Operate-Transfer (BOT). Within the domestic private category there are a further three models: "red chip", provincial private utility, and "ITICs". The sections below provide an explanation of each model, as well as a description of sources of financing accessed by the two categories.

Each of these PPI models has merits and demerits. Regardless, they are the outcome of attempts by government agencies on the one hand and private business on the other to harness the dual potentials of Chinese infrastructure growth and international private capital flows within an uncertain regulatory framework. In this light, the paper provides an overview of China's regulatory environment regarding PPI, and cites the resulting positive and negative aspects of PPI in the PRC to date.

Foreign private infrastructure participation

Since the first successful foreign private investment in Chinese infrastructure, at the Shajiao B power plant in 1987, there have been a large number of projects in the PRC which have involved the participation of foreign private investors and developers. These ventures have made significant contributions to the provision of physical infrastructure to China, a country with insufficient public funding capacity in the face of immense infrastructure needs.

Models 1 & 2: Sino-Foreign Joint Ventures

Many Chinese infrastructure projects or proposals involving foreign investors / developers in China take the form of Sino-Foreign Joint Ventures. Sino-Foreign Joint Ventures can be further divided into two models: Cooperative JV and Equity-Only JV.

The joint venture models involve a foreign party (Foreign-Invested Enterprise registered in the PRC) in cooperation with an affiliate of a regional, provincial or municipal government agency. Both types of Sino-Foreign JV, as applied to infrastructure projects, feature the foreign partner helping to finance the costs of development and then sharing ownership of project assets and profits for a set period of time. When this period expires, the foreign partner transfers its ownership interests in the project to the Chinese partner. Since the Chinese partner in these deals is invariably a bureau of – or strongly affiliated to – the local government, the Joint Venture models in effect implement a version of Build-Operate-Transfer (BOT) concession structure.

However, an important difference of this model from infrastructure concessions used in other countries or the formal BOT model being formulated in China itself is the restriction on operational control by foreign enterprises. Although foreign-invested enterprises are allowed to engage in Sino-Foreign Joint Ventures to undertake infrastructure development projects, they are prohibited in most cases from holding controlling interests in the project companies that actually operate and manage the facilities [Sorab, B., Allen, S., et al (1997)]. The operator, off-taker (in the case of power or water facilities), main construction contractor and input supplier (again in the case of power or water facilities) are usually all related to the same local government agency. Foreign firms are prohibited from taking a controlling role in any of these contracts, although they may perform advisory services, including construction / project management and O&M services, under sub-contracts to the project company. Therefore, depending on the roles and risks taken on by the foreign partner, its role in Chinese infrastructure via Sino-Foreign Joint-Venture can range from that of active advisor for construction, procurement, and operation as well as equipment or material supplier… to that of passive investor with virtually guaranteed return. In the latter case, the investment can be considered a kind of quasi-debt, with the foreign partner acting as a conduit connecting the local government to international capital at a fixed rate of interest (the minimum rate of return for the foreign investor agreed to in deals of this type).

The fact that the local government takes so many roles in Sino-Foreign Joint Ventures has both advantages and disadvantages: on the one hand, it enables the foreign party to off-load many of the risks associated with facility construction and operation; on the other hand, by forcing the foreign partner to relinquish effective control of the facility, there may be inefficiencies in the operation and management of the project. In cases where the foreign partner is an international utility with extensive know-how in facility operation, losing control of operational management may be seen as a serious drawback to its investment.

In the Cooperative Joint Venture (CJV) model, the project company functions as a distinct legal entity under Chinese law, and enters into all contracts related to the project, including any supply agreements (e.g. fuel, raw water, material, construction), offtake agreements, etc. It is further allowed to directly borrow from foreign and/or domestic lenders, and in doing so can assign its rights as defined in the project documents as security for the loans [Sorab, B., Allen, S., et al (1997)]. One more important feature of the Cooperative Joint Venture is that it allows the partners to determine allocation of profits according to a negotiated schedule: many CJV's are structured to give greater profit to the foreign partners during the early stages of project operation in order to satisfy the strict financing commitments of the foreign investors, with the Chinese partner (local government agency) receiving its share of profits later on. Such flexibility is often mentioned as the most attractive feature of the CJV model to foreign investors in practice.

Under the Equity-Only Joint Venture (EJV) model, there are a few differences from the CJV structure described above. The project company in an Equity-Only JV is not allowed to directly access debt capital or give any kind of security on project assets to either lenders or shareholders. Instead, financing of the project company is entirely in the form of equity contributed by the foreign and Chinese partners [Sorab, B., Allen, S., et al (1997)]. This equity can be raised by whatever means, including corporate or project finance, from whatever sources; however, all debt financing must be limited to the shareholder level. The advantage of this structure, particularly on smaller projects, is that since the project company itself does not have to apply for government approval of external debt, it can skip one time-consuming step of China's project approval process and speed up development of the venture. On the other hand, the inability of the project company to assign security for project assets to lenders or investors inhibits the use of this model for larger scale projects, since few lenders will agree to invest in a project without some solid form of security.

All in all, the forms of Sino-Foreign Joint Venture described above – particularly the CJV model - have been the norm in the way infrastructure deals in the PRC have been structured over the past ten years. More to the point, there has been little alternative!

Model 3: Build-Operate-Transfer (BOT) Concession

The lack of alternatives in structuring foreign private infrastructure ventures in the PRC is being addressed now by the central government with proposal of the Chinese version of Build-Operate-Transfer (BOT). In China, the term "BOT" refers to a specific project model and legal structure being developed by the central government whereby, "*a government authority may, through a franchise agreement and within a specified period, authorize a BOT project to a project company established by a foreign investor particularly for such BOT project, and have the project company responsible for its financing, construction, operation and maintenance.*" [China State Planning Commission]. The key differences from the JV models listed above are that: (1) the concession is granted by international competitive bid; (2) the foreign winner of the bid is allowed to own a controlling interest in the project company; (3) the foreign party is allowed to maintain control of operation of the facility; (4) the government agencies involved are authorized to provide limited support and guarantees of specific undertakings by Chinese government-affiliated corporations involved in the project, including foreign exchange and convertibility; however, they are not allowed to provide any form of guarantee regarding the project's rate of return [China State Planning Commission]. This BOT structure has so far been used on only four projects, one of which (Laibin B power) is close to construction completion while the others (Changsha power, Chengdu #6 water, Puqi power) have been awarded and are presently in financing. The central government has announced that it will prepare a comprehensive set of legal BOT administrative guidelines (the "BOT Law") in the near future based on its experiences in three designated pilot projects (Laibin, Changsha and Chengdu).

Table 1: Samples of CJV, EJV, and BOT infrastructure project deals in the PRC, 1987-1998

Project	Developer	Structure
Toll Roads		
Hui-Ao Roadway	NWI, local govt	CJV
Guangzhou-Zhuhai East-Line Expressway	NWI, local govt	CJV
Guangzhou-Shenzhen-Zhuhai Superhighway	Hopewell, local govt	CJV
Shunde Roads	Hopewell, local govt	CJV
Greater Beijing Region Expressway	GBFE (Hong Kong investors, Bechtel), Hebei Prov. Highway	CJV
Guangzhou City Northern Ring Road	NWI, local govt	CJV
Wuhan Airport Expressway	NWI, local govt	CJV
Shenzhen-Huizhou Expressway	NWI, local govt	EJV
Power Generation		
Laibin B	EDF, GEC Alsthom	BOT
Changsha	National power, ABB	BOT
PuQi	Sithe Energies, Marubeni	BOT
Shajiao B	CEPA, local govt	CJV
Shajiao C	CEPA, local govt	CJV
Xiangci-AES Hydro Power	AES, local govt	CJV
Meizhou Wan	Intergen, Lippo, local govt	CJV
Zhujiang	NWI, local govt	EJV
Water Treatment		
Chengdu #6	Generale Des Eaux	BOT
Dachang Waterworks	Bovis, Thames Water, Shanghai Municipal Water Bureau	CJV
Yueyang Water Plants	CKI, Local govt	CJV

Domestic Private Infrastructure Participation

The examples described above focus on China's efforts to introduce foreign private capital and participation in the development of infrastructure. However, there is also <u>domestic</u> private and semi-private participation in Chinese infrastructure, and this is growing at an increasing rate in recent years. Domestic Chinese corporations, or corporations registered in foreign countries but effectively controlled by domestic Chinese interests, are beginning to directly access the international capital markets as well as China's growing domestic equity and debt markets.

Model 1: "red chips"

The term, "Red chips", has become part of the common investor vocabulary, particularly in Hong Kong where many of these firms have publicly listed shares. These companies are registered outside of mainland China (usually in Hong Kong), and are technically defined as foreign enterprises; however, they are controlled by Chinese interests – state-owned enterprises, government agencies or their recently corporatized bureaus – and their objectives generally include generating profits from international business and channeling investment back into China. The use of red chips began mainly in trading and shipping; however, more recently there are red chips that invest heavily in Chinese infrastructure. Examples include firms such as the China Power International Corporation, the wholly-owned Hong Kong subsidiary of the Chinese state-owned China Power Investment Corporation [Bauer, J., Allen, S. et al (1996)], Beijing Enterprises, and China Merchant, a red chip company that has significant holdings in Chinese toll roads.

Model 2: Provincial private utility

Recently, a growing number of domestic infrastructure-related corporations have been formed as the result of China's movement to separate government from enterprise. Bureaus of local government agencies responsible for development of certain infrastructure sectors are breaking off to form corporations. With few exceptions, the new companies are in a transition stage, maintaining ties of influence with parent government agencies and displaying management styles that are not yet completely "commercial"; however, they are registered as private corporations with the ability to participate in infrastructure development as private parties. They benefit from a ready-made portfolio of existing assets, since formerly state-owned infrastructure facilities (e.g. roads, power plants, etc.) are packaged and transferred or sold to the new corporations when they are formed. This provides the companies with an asset base of revenue-generating projects which can be used to boost their credibility on international and domestic capital markets. The companies are usually focused on investment and development within the same home localities where they were formerly government departments. Examples of this are particularly noticeable in the sub-sectors of transportation – especially roads – and power and include: Jiangsu, Zhejiang, Shenzhen, Anhui, and Sichuan Expressway companies in toll road development, and Beijing Datang Power, Shandong Huaneng, and others in the power sector.

Most recently of all, there have been a few Chinese infrastructure companies which are displaying more commercial management and moving outside of their home localities in search of profitable projects. In doing so, these companies often form joint ventures - indeed, some of these joint ventures are with foreign investor partners. However, there are also examples where purely domestic joint ventures are being formed across regional boundaries to pursue infrastructure development projects throughout China. Similar to Sino-Foreign Joint Ventures, domestic joint ventures can be single project or multi-project holding companies.

Model 3: "TICs" and "ITICs"

Trust and Investment Corporations ("TICs") and their International counterparts ("ITICs") have provided another vehicle for financing Chinese infrastructure. Defined under Chinese law as "nonbank financial institutions", the TICs have been a source of innovative financing and investment opportunity, offering Chinese citizens the opportunity to increase returns on savings compared to bank deposits, and investing funds in a variety of ventures, including private infrastructure. Most of the more advanced Chinese provinces, as well as many municipalities, have set up their own TIC / ITIC to facilitate investment in their locality. TICs and ITICs are often involved as investors or partners in CJV relationships, and are also sometimes involved in backing up Chinese side guarantees on returns or foreign currency availability in joint venture infrastructure projects.

However, recently the high flying investment corporations have met serious troubles, as the Chinese economy slows and ill-advised investments in property come back to hurt the their balance sheets. While TICs and ITICs have provided some bright spots in their aggressive participation in private infrastructure financing, it remains to be seen how viable this model will be for future PRC infrastructure development.

Sources of Funds:

Allowing private investors to enter the infrastructure development arena has enabled China to supplement limited public financial resources with those of private - mostly international - investors.

Foreign private participants in infrastructure undertakings have played a vital role in linking various forms of international private and public capital to PRC infrastructure development. The ability of foreign investors to develop creative methods of accessing and structuring financial resources continues to be a major competitive advantage for foreign private participation models.

At the same time, as Chinese enterprises grow in size and sophistication, their ability to access international capital markets directly on a private basis is growing. Furthermore, as the country continues to make progress in its shift to market economics, domestic capital markets are growing and will provide a potent source of financing for domestic PPI models in the future.

Regardless of the PPI model, there are a number of potential sources of funds for private investment in Chinese infrastructure:

Foreign direct investment. This generally consists of capital contributed by the sponsors of project companies and joint venture. Particularly in the case of foreign private participation models, foreign partners provide "registration capital" - general in cash - when setting up project companies or JVs, and contribute further capital at different stages throughout project development. The foreign companies may in turn source funds by whatever means of corporate or project finance, as they wish.

International equity markets, through publicly listed shares of project holding companies. The Stock Exchange of Hong Kong (SEHK) has been particularly useful in this context, with roughly eight listed Hong Kong-based companies that hold multiple infrastructure projects in China. These listed firms generally use capital raised through initial public offerings (IPOs) of stock to finance further project developments in the PRC. The stock market in Hong Kong has provided not only a useful source of financing, but also an important medium for "exit" by firms wishing to liquidate their ownership interest in Chinese ventures by selling shares through the stock exchange.

From the Chinese side, the pioneers in accessing equity financing through international stock markets have been the "red chips", publicly listed on the SEHK. Within the past two years, there has also been a rapid increase in the number of provincial / regional utilities issuing public shares on the HKSE (known as "H-Shares"), including toll road firms such as the Jiangsu, Zhejiang and Sichuan Expressway companies.

International debt markets, including non- and limited-recourse loans by foreign commercial banks, and corporate bonds issued by the project company or holding company.

Loans can be provided through project financing arranged by syndicates of international commercial banks, and are by definition non-recourse, i.e. banks only have recourse to the assets of a specific project to securitize the loans, with no claims on other assets of the companies or governments involved. However, considering the risks inherent in Chinese law, foreign commercial banks have been unwilling to undertake full project financing to date; instead, they have required some form of guarantee from Chinese government agencies or affiliated enterprises. In this case, syndicated bank loans become a sort of "quasi-" project financing [Sorab, Allan, et al (1997)]: with local government guarantees, the deal is no longer pure project finance; however, considering the relatively high risks associated with local government creditworthiness and actual enforceability of the guarantees, the deal still bears similar risk considerations to a project financing.

Issuing bonds by private placement, usually under US SEC rule 144A, enables project holding companies to access the large body of capital held by sophisticated international institutional investors such as major insurance companies and pension funds [Finnerty, J. (1996)]. There have been several successful private bond placements by infrastructure holding companies in China, including those by Cathay International Infrastructure and Greater Beijing First Expressways [Prospectuses].

Export credit financing and project guarantee support by the export credit agencies (ECAs) of various foreign countries. The ECAs can provide significant amounts of funding, as well as coverage against country political risks, to projects that purchase capital goods or services from their respective countries. Whereas traditionally they have required sovereign backing by the host country, many ECAs are now becoming involved in limited / non-recourse financing in Chinese projects [Sorab, Allan, et al (1997)].

Multilateral agencies, such as the World Bank's International Finance Corporation (IFC) and the Asian Development Bank's Private Sector Group, can lend directly to projects or take a minor equity share, with any contribution usually small in size. However, with the agencies' beneficial impact on project credit-worthiness, "co-financing" offered by commercial lenders can range up to 5 times the amount of the multilateral agency's commitment. Further, the IFC and ADB can help by arranging guarantees for country political risk.

Domestic commercial bank loans may become a key competitive factor in Chinese private infrastructure models in the future. The ability to finance project costs with local currency denominated loans is a key factor in helping infrastructure developers to off-set foreign exchange risks. It has yet to be seen whether domestic private ventures will be given greater access to this potentially large new source of financing than models involving foreign private parties.

Domestic equity markets offer an increasingly attractive source of capital for domestic private models of infrastructure development. While the PRC's bond market is under-developed, its stock markets – in Shanghai and Shenzhen – have been growing in their capitalization and liquidity during the 1990s. Some domestic ventures have begun to pass up listing on foreign exchanges, where market sentiment is currently poor and disclosure rules are strict, in favour of listing on domestic stock markets. One example is Jilin Electric Power, a provincial utility which had initially planned to list on the HKSE, but switched to the Shenzhen exchange instead in 1998 [Morarjee, R., et al (1998)].

Legal and regulatory framework for PPI

With regards to the legal and regulatory framework, China is still in the process of establishing the rule of law in many areas of society... and infrastructure is no exception. Until recently, there has been no specific set of regulations defining the legal role of foreign private entities in the financing, development and operation of infrastructure facilities. In addition, the dynamic tension between central and regional governments in China has a profound influence on the environment surrounding PPI. Decentralization of authority with regards to public finance – including funding and administration of infrastructure – has been an important step in China's transition from centrally planned to market driven economics [Kato, H. (1997)]. There is an inherent conflict between the central government – whose objective is to undertake policies that balance the economic and social development needs of the all China – and regional governments, particularly in the advanced coastal provinces – whose aim is to promote local economic, and sometimes personal, needs via increased investment in local power, water, roads, ports, and telecoms.

The combination of China's fledgling regulatory systems, decentralization of authority, and hazy separation of public and private sector creates the shifting matrix of unclear laws, conflicting regulations, lengthy approvals, rival bureaucracies and opaque decision-making that define the environment for private participation in infrastructure. The models of PPI must adjust to fit into this set of conditions on a case-by-base basis. In Sino-Foreign Joint Ventures, for example, investor / developers comply with a patchwork of laws and regulations, from industry-specific laws such as the new Electric Power Law to cross-industry legislation such as the Sino-Foreign Cooperative Enterprise Laws and the Security Law, to local regulations defined and enforced differently in each locality.

Numerous problems have resulted: in some cases, regulations overlap and conflict on certain issues (e.g. legal enforceability of guarantees given by Chinese parties); in most cases, foreign investors waste significant amounts of time trying to keep track of the different laws and lists of required project approvals; and even Chinese government officials are often unaware of the mix of laws, central government circular notices, and provisional rules that can affect their projects when foreign investment is involved.

Conclusions

The following observations can be made on developments in Chinese PPI since the country's first foreign-invested infrastructure project a decade ago.

Sino-Foreign Joint Venture models have been a breakthrough in enabling local Chinese governments to attract foreign private capital into investments in infrastructure. They have allowed for maximum flexibility at the local government levels, given an imperfect legal and regulatory environment. Depending on the degree of risk taken on by the foreign partner, JV models can be considered as equivalent to variants of "BOT" concession, with local government maintaining ownership interest and operational control.

There are drawbacks to the Joint Venture models. They are conducted in an unclear regulatory environment – often with guarantees of questionable legality, and always with arduous tiers of approvals. They involve local governments in multiple – sometimes conflicting – roles, which could be abused to the detriment of foreign private investors with little legal recourse. Finally, they limit the level of control a foreign investor can have over operation of a project: while this may be beneficial for risk allocation reasons in some cases, it is inefficient to limit foreign partners – some with international technical and managerial expertise in the development and operation of infrastructure – to a role as mere conduits to international capital markets.

The Chinese Build-Operate-Transfer (BOT) model could provide the administrative framework that has been lacking so far in Chinese foreign PPI models. However, the successful introduction of BOT is far from assured. For one thing, the four BOT projects in China to date have been based on a set of draft guidelines published in 1995. Passage of the "BOT Law" itself has been delayed for three years at this point, apparently as a result of resistance at the upper levels of central government. Assuming the law is passed, it may not be welcomed by local governments, since the current Sino-Foreign JV models provide more immediate returns to them while not requiring them to relinquish control over infrastructure facilities. Certain types of foreign investors which have built strong relationships with regional governments to secure CJV/EJV deals in the past may also resist the BOT model, which requires international competitive tender. Despite the potential negative reception, it is hoped that the BOT model is legislated to facilitate creation of a transparent, reliable framework for Chinese infrastructure development in the future.

The rise of domestic models of private participation in infrastructure will increase competition and introduce fundamental changes to the landscape of PPI in China. With direct access to international capital markets, domestic models could completely replace Sino-Foreign JV models where foreign partners are limited to funding conduit-type roles. In the same way, these models could marginalize the BOT model. The efficiency of domestic PPI models in accessing and managing capital and projects will ultimately determine the benefits they can bring to China, as well as the threat they will pose to foreign PPI models. Without this, they will be forced to depend on local connections and preferential bank financing in their home localities, possibly leading to inefficiencies in both investment allocation and infrastructure facility operation.

China's developing legal system has had both positive and negative effects on PPI efforts. Ambiguities of the legal system have allowed flexibility by local governments in structuring deals. On the other hand, a lack of legal enforceability of contracts made under these conditions could potentially hurt foreign investors in the future.

Similarly, the balance of tension between regional and central levels of government also presents good and bad news. Decentralization has enabled local governments the freedom to promote local economic growth and enthusiastically attract foreign investment. This has been a strong factor in accelerating the rate of economic development in many of China's regions. At the same time, localization and short-term economic interests at the provincial and municipal levels of government has motivated many of them to promote the development of small scale projects – under central approval limits – and compete with neighbouring provinces or localities. In some cases, the result has been poor overall coordination and an inefficient national infrastructure network.

As described above, PPI in the PRC presents a wide array of opportunities for foreign and domestic private investor / developers to contribute to meeting China's huge infrastructure needs, while making returns commensurate with the roles and risks undertaken in the process. However, the business is far from easy. With two categories and six models for private participation in infrastructure, as well as shifting legal, regulatory, administrative and economic conditions, prospective investors and developers – not to mention government officials, lenders, insurers, and even academic researchers – are engaged in a learning process that will no doubt continue for years to come.

Acknowledgements

This paper is based on existing literature, interviews with private firms and government agencies in Beijing and Shanghai (Aug. '97, Mar. '98), as well as an internship in a private infrastructure investment fund in Hong Kong (July - Oct. '98). The author wishes to express his deepest appreciation to everyone who cooperated with interviews and tolerated his many questions.

References

Bauer, J., Allen, S. et al (1996) Project & Infrastructure Finance in Asia, Asia Law & Practice Publishing, 2nd Edition, p.113

China State Planning Commission, Articles 2, 20 and 23, Provisional Regulations on Foreign Investment Build-Operate-Transfer Projects, draft

Finnerty, J. (1996) Project Financing: Asset-Based Financial Engineering, John Wiley & Sons, Inc., pp. 157-187

Kato, H. (1997) Regional Disparities and the Shift to Market Economics, *China's Economic Development and Market Transition (chuugoku no keizai hatten to shijouka)*, pp. 137-162

Morarjee, R., et al (1998) , Jilin Power Prefers Domestic Listing, *China Infrastructure Newsletter*, Clear Thinking

Prospectuses, Greater Beijing First Expressways and Cathay International Ltd.

Sorab, B., Allen, S., et al (1997) The Life and Death of an Infrastructure Project, Asia Law & Practice Publishing, 2nd Edition, pp. 162-165, 181-184

Zeng Peiyan, Minister of State Development Planning Commission, Comments on China Infrastructure Investment Program (June, 1998)

ECONOMIC REGULATION OF INFRASTRUCTURE PROJECTS INVOLVING PRIVATE SECTOR PARITICIPATION

H.C. CHO, S. TAKEUCHI, T. YOSHIDA, and M. KUNISHIMA
Construction Management & Infrastructure Systems Laboratory, Department of Civil Engineering, University of Tokyo, Tokyo, Japan. cho@ken-mgt.t.u-tokyo.ac.jp

Abstract

The aim of economic regulation in private infrastructure development is to safeguard consumers from monopolistic power while ensuring that the private sector remains viable and has incentives to develop and operate projects efficiently. Setting an adequate governance mechanism is important because when this regulatory control is too lax, consumers have to bear some risks (e.g., price hike or quality degradation); on the contrary, excessive control may discourage private sector participation in infrastructure development. This paper discusses three alternatives in economic regulation on private infrastructure pricing systems: Price regulation, ROR (Rate-of-Return) regulation, and No-regulation. It analyzes the incentive effects with regards to efficiency and the degree of risk to which private company is exposed under each mechanism. It also presents some experiences in Asian developing countries and finally makes a suggestion for improving economic regulation for these developing nations.

Keywords: Economic regulation, no-regulation, price regulation, ROR regulation

Introduction

Since the 1990s, as the governmental costs of regulating specific sectors of the economy increase, it becomes a global trend that policy-makers view deregulation as a cost-effective strategy for promoting economic growth. This trend is likely to continue in the future as a result of the accelerating globalization of the market. However, not all regulation is on the decline. People in many nations today express a desire for more and stronger regulations in several areas, such as environmental protection, public health and safety standards, which are particularly classified as "social regulation". In the area of privately-promoted infrastructure development, another form of government intervention, called "economic regulation" needs to be contemplated to remove market failures in the provision of infrastructure. Economic regulation refers to controls on prices, quality, quantity, ownership, entrance and exit conditions for specific industries (Jones 1993). In the case of economic regulation, the primary rationale has to do with the companies' incentive for improving efficiency of facility development and operation. Among several subjects of economic regulation, the focus of this paper is on pricing systems under private infrastructure scheme. In general, three main streams of economic regulation on pricing have been developed and promoted by many governments: Price regulation, ROR regulation, and "No-regulation". Figure 1 summarizes the different forms of economic governance in a hierarchical manner.

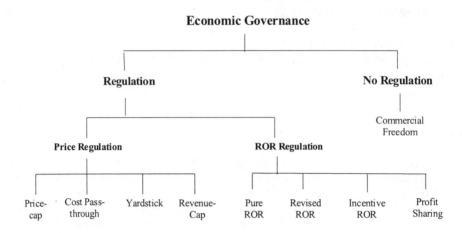

Figure1 Hierarchy of Economic Governance in Private Infrastructure Development

Price regulation

Price-cap regulation

In an original price-cap regulation, an explicit limit is set directly on prices of infrastructure services charged to consumers over a long period of time such that a well-run company can expect to earn a higher rate of return; conversely a company with inefficient management may get a lower return, suffer losses and possibly go bankrupt. Yet, the opportunity to gain theoretically infinite profits gives a company an incentive for improving efficiency. The UK was the first nation to develop price-cap regulation for the privatized British Telecom in 1984. In a generalized form, the price setting rule of price-cap regulation can be written as (Littlechild 1983)

$$\text{New Price} = \text{Previous Price} + RPI - X, \tag{1}$$

where RPI is the changes in Retail Price Index, a standard measure of general annual rate of inflation in the UK, and X stands for the expected gains in efficiency. The regulator decides on the level of annual efficiency gains X at will based on the assumption that only an efficient company can expect to reach this level(Alexander, Mayer and Weeds 1996). Once X is determined, it remains unchanged for the next five years until an adjusted X is decided via periodic price review by the regulator. Generally, the company has an incentive to obtain even higher efficiency gains X than the one assumed by the regulator, because all those additional gains would belong to the firm in the end. However, the company is exposed to the risk of cost-changes which cannot be controlled by the company.

Price-cap with "cost pass-through" regulation

In recognition of the fact that certain cost elements are beyond the control of the company and exposure to such unmanageable variables increases the risk with no benefit in terms of incentive under the pure price-cap control system, a modified form called "price-cap with cost pass-through" was developed for the privatized British Gas in 1986 (Levy and Spiller 1993). This new mechanism allows certain cost changes outside of the company's control to be passed on to the prices charged onto consumers without waiting for the next periodic price review. The level of risk borne by investors becomes much decreased, while the incentive properties of the pure price-cap are not undermined as long as these cost components are truly exogenous from the company's perspective. Under a cost pass-through scheme, by adding Y representing the pass-through of cost, the expression (1) can be rephrased as

$$\text{New Price} = \text{Previous Price} + RPI - X + Y, \tag{2}$$

Yardstick regulation

One demerit of the pure price-cap and cost pass-through regulations is that they create an auxiliary incentive for the company to cut costs by degrading the quality of services: trading profits with quality downgrade. To avoid this problem, a yardstick scheme was introduced, in which the regulator sets a level of expected efficiency gains X as well as benchmarked quality of services. The regulator compares performance in meeting the agreed quality level from several different operators within the same or similar projects. When a company performs above the specified level, it is allowed to raise the prices by adding higher Q, representing the monetary value required for quality improvement. While one who produces poor quality below the benchmarked level is penalized or compelled to lower or remove Q (Shoji and Yamagishi 1997). Under a yardstick scheme, the pricing rule (1) can be rephrased as

$$\text{New Price} = \text{Previous Price} + RPI - X + Q. \tag{3}$$

ROR regulation

Pure ROR regulation

In a pure ROR regulation, which was firstly developed in the US (Alexander and Irwin 1996), the regulator allows the private company to charge prices expected to make profits commensurate with a fair rate of return on the fair value of capital investment. If the company's return falls below the compromised floor level, it can request an approbation for a new set of higher prices; on the other hand, when return rises above the ceiling level, the regulator would demand a reduction in prices to an acceptable level. In this situation, the company bears very little risk, as any unforeseen costs can quickly be passed on to customers. A fair ROR can be estimated based on the normal rate of return for the country, plus a premium based on the systematic risks of the specific industry and project in question. In general, the normal rate of return for the country implies the nation's social interest rate. Due to the lack of risk, the risk premium tends to be extremely low. In theory, if any change in costs can pass through to the consumers automatically, the risk premium should be equivalent to zero. Thanks to the virtually risk-free characteristic, the ROR model can make a stable and safe investment environment for the private sector; however, it delivers limited incentives to the company to be efficient and save costs.

Incentive ROR regulation

To compensate for the problem of limited efficiency-incentive, the Incentive ROR scheme has been developed, in which several criteria to promote particular objectives are set. For example, the regulatory authority of the California Mid-state toll road project determined a target level in several areas including accident rate, occupancy rate, maintenance service, and administration costs: they are designed to provide incentives to the private company to reduce fatality or injury traffic accidents, to encourage car-pooling policy for conservation of resources, to improve quality of maintenance service, and to reduce overhead costs (Enabling Act). When the company satisfies the requirements of these negotiated criteria, it will be rewarded with bonuses from the government, which have no relation to the firm's rate of return. One shortcoming of incentive ROR regulation is that in order to meet the elucidated objectives, the company is likely to spend more money rather than less (e.g., patrolling, advertisement, warning and encouraging sign-boards, maintenance costs), which in the end may raise the total costs.

Profit sharing

According to the profit sharing scheme, instead of fixing the ceiling return as in normal ROR regulations, any excessive profit above the ceiling point will be shared between the government and the company on the basis of negotiated proportions. A good example is the Teodoro Moscoso Lagoon BOT bridge in Puerto Rico (Schnettler 1994), whose scheme is shown in Table 1. The government determined 19% as a reasonable rate of return on equity investment (ROE) for the company, yet instead of fixing the maximum ROE, the regulator permitted the company to be remunerated with a portion of any ROE surplus over the 19%, while the remainder returned to the Puerto Rican government.

Table 1. Profit Sharing on the Teodoro Moscoso Lagoon BOT Bridge

Profit Share	ROE 19% or Below	19% - 22%	22% or Over
Company	100%	40%	15%
Government	0%	60%	85%

No-regulation

No-regulation, also referred to as "Commercial Freedom", is another legitimate alternative to Price regulation and ROR regulation. Two possible conditions, that are quite different from each other, can be considered where the no-regulation selection can be more effective than others. Firstly, selection of no-regulation can be made in a competitive market, based on the perception that only those segments where competition is not feasible would remain subject to economic regulation, and that when competition is freely allowed, market forces would balance demand and supply in deciding the final prices (Crampes and Estache 1997). Secondly, no-regulation can also be chosen when the project involves highly incomplete designs, unproven technologies, or insufficient information that make the capping of prices or ROR too risky, difficult and contentious.

Economic regulation in Asian developing countries

Experiences

China

Officials have attempted to limit the equity return of private foreign investors in power projects to what the regulators regard as a reasonable rate. Company have had to bear the risks owing to any change in government policy, even including sudden adjustments of the limit on ROR by the government. Two early power concession projects in China, the Shajio B and C power plants, ran into serious problems due to the government; unilateral declaration of a 12 percent ROR ceiling on all BOT power plants in the late 1980s, after the projects were undertaken (Huang 1995). Recent events are signaling a transition of focus from ROR regulation to the price control. This trend is shown in the Laibin B BOT power plant. The competitive bidding process for this recent power project required the participants to specify an initial electricity tariff. If a bid involving a pricing tariff is accepted, the bidder will have the opportunity and incentive to elevate efficiency and to generate the profits by reducing the construction and operation costs (Pote and Wong 1996).

India

In India, the government has outlined its plan for private investment in the power sector. Within this plan, the government has asserted a 16 percent return on foreign equity to attract foreign investment (Ullman and Dayal 1996). For example, in the Balagaph project, a 500 megawatt coal fired thermal power station project located north of Calcutta, a 16 percent return was allowed and incorporated in the form of a Power Purchase Agreement (PPA) for a 30 year period. The negotiated tariff for the energy sold is in accordance with the notification issued by the government and is calculated to cover full costs plus the allowed 16 percent rate of return on total equity investment if the project operates at an annual Plant Load Factor (PLF) of 68.5 percent. On top of the 16 percent return, there is an additional incentive of 0.01 Rupee for each kWh generated above the normative PLF of 68.5 percent. Furthermore, the return on equity may be enhanced by more efficient operation and the concessionaire aimed to uplift efficiency by advanced heat rate and innovative design and operation (Interview 1997). This project is a good example of pure ROR regulation that allows the company to be reimbursed for all costs plus a fair return.

Philippines

A typical example of price-cap regulation appears in the concession agreement for BOT of the Manila Grains Terminal (MGT). In this agreement between the Philippines Ports Authority and the private concessionaire. The concept of price-cap with cost pass-through is stated in the Article on Revenues and Rate Adjustment as,

> "The Offer to Bid will establish the rates and charges to be assessed by the concessionaire for services at the MGT. Rates set forth in annex 6 are assumed to establish a ceiling under which the concessionaire may levy its charges. However, as a publicly issued concession, the MGT shall offer its service to the public and shall not discriminate among customers in pricing or level of service. Provided that the

concessionaire is free to levy differing rates for differing services and differing level of service."

The prescribed rates and charges collectible by the concessionaire from the users of the services which the MGT provides shall be subject to adjustments in accordance with adjustment provisions and price policy. For example, when the rate of increase in fuel/power cost or government-mandated wage adjustments is higher by 10% than those reflected in the economic parameters for the period in question, the Philippines Ports Authority will consider ordinary increase in rates.

Pakistan

In a BOO oil terminal project at Port Qasim, the tariff rate was calculated based on the assumption that the cash flow of the concessionaire will provide shareholders with a return of 18 percent in US dollar terms after covering all cash expenses and principal repayments with interests. An 18% return is guaranteed by the government with a pass-through contract, where the tariff structure is allowed to change flexibly in case of any changes in costs such as wages or fuel. Recently, the efforts to implement price regulation can be seen in the power sector. In March 1994, the government issued its "Policy Framework and Package Incentives for Private Sector Power generation Projects in Pakistan", which set forth incentives offered by the government to the private sector to participate in BOO power projects. In this framework, the price regulation scheme is included in a statement that an internationally competitive bulk tariff rate of $0.065 kWh to be paid in Pakistan currency will be controlled (Interview 1997).

Thailand

Thailand's experience of economic governance in the power sector reflects the pure price-cap regulation scheme. In Thailand, the Electricity Generating Authority of Thailand has set a price for the power in the Houay Ho concession power project at US$0.0422 per kWh. The Houay Ho consortium requested an increase to US$0.0435 per kWh, but eventually, the consortium had to back down and agree to the original ceiling price (Anderson 1996). Table 2 shows examples of economic regulation selection in Asia.

Table 2. Economic Governance in Concession Projects

Project	Host Country	Economic Governance
Shajio B & C power	China	Traditional ROR Regulation
Laibin B Power	China	Price-cap Regulation
Balagaph Power	India	Traditional ROR Regulation
Manila Grains Terminal	Philippines	Price-cap cost with pass-through
Houay Ho Power	Thailand	Price-cap Regulation
Fauji Oil Terminal	Pakistan	Traditional ROR regulation

Characteristics

The background for the participation of private sector in infrastructure development in Asian developing countries is quite different from those in developed countries. Most private infrastructure companies in Asia are formed by means of concession arrangement, while in developed countries, many companies have been privatized via sales of government enterprises. In short, seeking "investment gains" has been the more significant rationale for privately-promoted infrastructure projects in Asia. Several critical points that should be contemplated for further institution of economic regulation can be summarized as follows:

- Infrastructure services have been provided at prices well below the direct costs of supply or free of charge, and these services are normally subsidized by the government to achieve social and national development goals under public development plans (Mosley and Linklaen 1995). Therefore, in the absence of subsidies to the private company, the prices are likely to rise significantly in spite of some savings in costs achieved by efficiency

- Demand for infrastructure services far exceeds supply: thus, competition among providers is not yet fully mature or simply non-existing. This makes the option for allowing bankruptcy of companies unrealistic. Once an infrastructure company commences providing services, the sustainability of service production and protection the public are likely to override simple efficiency-based considerations.

- Laws and rules guiding economic regulation have not been clearly written and there is a serious ambiguity in whose authority governs company between the central and local government.

- Information can be asymmetric because of the relatively immature stock and capital markets, and non-uniform accounting rules disable a fair and transparent evaluation of private company performance. Also, nationwide, well-balanced annual inflation rate figures like *RPI* in the UK may not be available.

- Most governments have not yet established a regulatory agency and it may be too early to expect governmental regulatory institutions to be independent and discretionary from political interference because the room for high renting behavior, including nepotism and bribery, is great in these nations.

The foregoing findings conclude that infrastructure market in Asian developing countries can be generally regarded as still primitive, imperfect and distorted compared to those in developed nations.

Suggestions : "Hybrid" regulation

The above study has recognized the existence of dissimilarities in background and current status in private infrastructure evolution between developed nations *vis-à-vis* Asian developing countries. One obvious point is that neither pure price-cap nor pure ROR regulations will work effectively for the Asian developing countries. This leads to the conclusion that some modulations of the two main streams are essential prior to application in this region. For Asian developing countries, a combination of two controls on both individual prices and rate-of-return, is desirable to shelter the powerless consumers and assure the investors simultaneously. Extreme inclination to either one side may be risky and a hybrid device in between the two extremes must be designed.

As one possible regulatory tool, a new concept called "Hybrid Regulation" is proposed in this paper. At first, explicit level of prices, *p1*, that consumers can afford must be determined by the government based on international standard prices, as well as the historical and present domestic prices. These affordable prices, *p1*, are most likely to fail in covering the total costs plus a fair return requested by the investors. Then, real prices, *p2*, which enables the company to earn a fair rate of return should be decided. The disparity between *p1* and *p1* multiplied by q, quantity of sales, would be compensated for by cash-subsidies from the government, represented as *G*. Based on above assumptions, the two formulae can be induced as

$$
\begin{aligned}
P &= R - C \\
&= p2 \times q - (OC + FC) \\
&= p1 \times q + G - (OC_1 + OC_2 + FC)
\end{aligned}
\tag{4}
$$

$$
\begin{aligned}
G &= P + (OC_1 + OC_2 + FC) - p1 \times q \\
&= P + (OC_1 + OC_{2\text{-}1} + OC_{2\text{-}2} + FC) - p1 \times q,
\end{aligned}
\tag{5}
$$

where P is the company's profits; R the revenues; C the total costs; OC the total operating costs; OC_1 the endogenous operating costs to the company; OC_2 the exogenous operating costs to the company; FC the financing costs; $OC_{2\text{-}1}$ the costs uncontrollable by the company but controllable by the government (e.g., changes in taxes, or transaction costs of additional licenses or permits); and $OC_{2\text{-}2}$ the costs uncontrollable by both the company and government as weli (e.g., changes in imported oil prices). From the government's perspective, the lower *G* is desirable and its ultimate goal is to pull the amount of government subsidy to virtually zero. To achieve this, variables in the right hand side must be reduced. Here, P is fixed at the level commensurate with a fair rate of return to the company at the beginning, and FC is almost constant according to the loan agreement agreed with the lenders. Also, q can be regarded stable with less possibility of decrease, based on the characteristic that demand far exceeds supply in Asia. Accordingly, there remain three possible ways to reduce *G*.

First, the company can help to reduce *G* by bringing down OC_1 via efficient management. The most obvious way to realize this is to furnish an incentive to the project company by sharing gains in *G* reductions together between the government and company. In principle, for any change in OC_2, the government will be responsible regardless of types ($OC_{2\text{-}1}$ and $OC_{2\text{-}2}$) and directions (up and down). Yet, under this hybrid model, the government will have a stronger rationale to limit $OC_{2\text{-}1}$, because it will have to pay for any increase in $OC_{2\text{-}1}$ in the end through an increased *G*. Therefore, political risks can be mitigated in itself thanks to the government's willingness to restrict the increase in $OC_{2\text{-}1}$. This system encourages more frequent price review such as biannual or quarterly compared to five years review term in the UK-system, because if time elapses for a longer period, it may become difficult to distinguish exogenous and endogenous cost changes.

Secondly, in the long-run, the government holds an option to raise the affordable prices *p1* gradually to the *p2* level, while contemplating other economic and social environments. The third option for reducing *G* can be achieved by setting the company's fair rate of return, which is represented as company's profit P, at a lower level during the incipient stage of project development. The best and most straightforward way to drop the initial P will be a fair and transparent competitive-bidding process allowing free entry of companies for the tendering in selecting the project company rather at entry stage than a direct and private negotiation with a single company.

There are also several other approaches rather than cash-subsidy that the government can offer to compensate for the disparity in the affordable prices $p1$ and real prices $p2$. Yet, among these instruments, the proposed cash-subsidy system appears to be better for controlling companies in Asia in view of: 1) the transparent and simple calculation of the exact amount of required government support, 2) the incentive-creation to the company to be efficient, 3) the incentive-creation to the government to remove transaction costs, and 4) the achievement of larger "investment gains" by avoiding up-front cash-in as in equity investment, land grants, and subordinated loans. It is also advisable that the government allow a flexible and preferential provision in tax treatment, because the collected taxes from the company will be returned to the company as long as the cash-subsidies are provided, and this take-and-give process would only incur the additional transaction costs. Yet, forming a responsible regulatory agency insulated from political intervention will remain as a critical task to tackle for many years to come. A concept of suggested "Hybrid" economic regulation for Asian concession projects is depicted in Table 3.

Table 3. An Example of Cash-Subsidy Hybrid Regulation

Cash-subsidy G	Incentive to the company when OC1 changes	Incentive to the company when OC2 changes
Required to meet fair return		
If G increases to GH	- (GH - Normal G)	No
Normal G = $(p2-p1) \times q$	No	No
If G decreases to GL	(Normal G - GL)/2	Normal G - GL

A sharing proportion in profits between the government and company is 50:50

Conclusion

Economic theory discovered, based on the analysis of three types of economic regulations, suggests that regulatory differences may cause the degree of risk borne by a company to vary significantly, with an inverse relationship existing between the degree of risk and the level of cost efficiency incentives imposed on a private company involved in infrastructure schemes. Based on this principle, each country should develop its most appropriate regulatory policy for more successful implementation of infrastructure projects involving private sector participation. The proposed "Hybrid Regulation" can function as one guide to improve the prospects for the private infrastructure schemes in Asian developing countries, while other creative ideas must to be contemplated when characteristics of the political, economic, social, and regulatory environment change in the future.

Acknowledgements

Several interview investigations have been conducted with staff at private companies and commercial banks in Japan (Oct.1996-Dec.1998), as well as investment officers at Private Sector Groups in the Asian Development Bank (Aug.1997). The authors would like to express their sincere gratitude to everyone who cooperated with interviews.

References

Jones, L. P. (1993) Appropriate Regulatory Technology. The Interplay of Economic and Institutional Conditions. Proceedings of the World Bank Annual Conference on Development Economics 1993. The World Bank.

Littlechild, S. (1983) Regulation of British Telecommunications' Profitability. London.UK.

Levy, B. and Spiller, P. T. (1993) Regulation, Institution, and Commitment in Telecommunications. A Comparative Analysis of Five County Studies. Proceeding of the World Bank Annual Conference on Development Economics 1993. The World Bank.

Alexander, I., Mayer, C. and Weeds, H. (1996) Regulatory Structure and Risk: An International Comparison. Oxford Economic research Association.

Shoji, H. and Yamagishi, R. (1997) Private Sector Financed Infrastructure Development in Developing Countries. OECF Journal of Development Assistance Vol. 3, No.1

Alexander, I. and Irwin, T. (1996) Price Caps, Rate of Return Regulation, and the Cost of Capital. Public Policy for the Private Sector. The World Bank

Enabling Acts: Midstate Tollway Development Franchise Agreement by and between Caltrans and California Transportation Ventures Inc.

Schnettler, J. S. (1994) Puerto Rico Privatizes The Roads. Civil Engineering, February 1994.

Crampes, C. and Estache, A. (1997) Regulatory Trade-Offs in the Design of the Concession Contract. Working Papers. The World Bank.

Huang, Y. L. (1995) Project and Policy Analysis of Build-Operate-Transfer Infrastructure Development. University of California at Berkeley.

Pote, J. L. and Wong, E. B. (1996) Investing in Electric Power Project in China. Project & Infrastructure Finance in Asia – Second Edition.

Ullman, G. N. and Dayal, A. S. (1996) Financing Power Projects in India. Project & Infrastructure Finance in Asia – Second Edition.

Interview (1997) Interviews with investment officers at the Asian Development Bank.

Anderson, J. A. (1996) The Independent Power Project Market in Thailand. Project & Infrastructure Finance in Asia – Second Edition.

Mosley, M. P. and Linklaen Arriens, W. T. (1995) Toward a Policy for Water Resources Development and Management in Asian and Pacific Region, Issues and Opportunities. A Discussion Paper for the Interdepartmental Water Resource Policy Group of the Asian Development Bank. The Asian Development Bank.

BUILD-OWN-OPERATE - THE PROCUREMENT OF CORRECTIONAL SERVICES

B. CONFOY, P.E.D. LOVE, B.M.WOOD, and D. H. PICKEN

School of Architecture and Building, Deakin University, Geelong, Australia, 3217,
pedlove@deakin.edu.au

Abstract

In an attempt to improve the effectiveness and efficiency of correctional facilities the Australian Government has recognised the need for their privatisation. Consequently, the Victorian Government initiated an 'Infrastructure Investment Policy', which led to the development of a portfolio called the 'New Prisons Project'. The paper presents findings from several prison projects that have been procured using different procurement methods by both the public and private sector. The findings reveal that prisons procured by the private sector using BOO systems are more cost efficient, specifically in relation to construction and operating costs, than those procured by other means. Discussions on the future of privatising correctional services using BOO systems are also presented.

Keywords: Consortium, build-own-operate, risk, cost efficiency.

Introduction

The Australian construction industry has gone through a period of introspection, which has been initiated by various government inquiries such as the New South Wales Royal Commission into the Building Industry (1992), and the Victorian Government Economic Development Committee's report into corruption of the tendering process (1993), and the publication of the 'No Dispute' recommendations (NWPC and NBCC, 1990). Extensive criticism of the construction industry has followed from these initiatives together with the general consensus emerging that reforms and, in particular increases in efficiency are required if major construction projects are to be procured successfully. One of the key elements arising out of the introspection process has been a review of current procurement strategies and how they might be improved, give rise to less disputes and generally made more efficient and effective. Yet before the procurement system can selected a detailed analysis of the risks involved with a major project is required. This will involve the identification, allocation and where necessary, sharing of construction risks. Such an analysis can be use to establish a procurement framework and how the parties' respective risks and rewards are to be measured.

Managing risk is an integral part of the procurement process (Akintoye and Taylor, 1997). The response of the industry to managing risk, such as design risk, which has been traditionally assumed by principals, has been the increasing use of Design Document and Construct or "novated" consultant agreement type arrangements. Thus, the risk of ambiguities or conflicts in detailed design documentation are assumed as a risk by the contractor, who as a consequence of assuming that risk, gains direct access to and control of the design consultants.

There is a shift toward the use of design-and-construct methods in Australia (Love *et al.*, 1998). This is reflected by the increasing demands of clients requiring a single organisation to be responsible for the delivery and sometimes the finance of their project. For example, in the context of infrastructure, the State and Federal Governments have encouraged the private sector to take a financial interest in the procurement of their projects. Consequently, this has led to an increasing number of BOOT (Build-Own-Operate-Transfer) and DBFO (Design-Build-Finance-Operate) methods being used to procure privately funded public infrastructure projects such as the Sydney Harbour tunnel, the Sydney Opera House car park, the Melbourne Trans-Urban Link and the M4 and M5 Motorways.

Since the early 1990s, the State and Federal Governments have encouraged the private sector to take an active interest in the funding of public infrastructure projects. Fundamentally, Governments are trying to obtain value for money by aiming to eliminate the waste and inefficiencies inherent within the public sector. According to Victorian Department of Treasury (1994) and HM Treasury (1995), value for money can be expected to increase as risk is transferred to the private sector to the point where all risks can be effectively managed by the allocated party. Needless to say however that there is a price to pay for the amount of risk a party is willing to accept. In most privately funded infrastructure projects the allocation of risk can be categorised as follows:

- design;
- financial (interest rate changes etc);
- construction (guaranteed maximum price); and
- operational (including maintenance and running costs)

Essentially, the utilisation of the private sector places the formal responsibility for the management of risk on the construction industry (Akintoye and Taylor, 1997).

The current Victorian Government has recognised that the benefits of involving the private sector in procuring public infrastructure, these include (Victorian Department of Treasury, 1994):

- improved value for money through the appropriate allocation of risk (eg., accountability);
- close integration of service need with design and construction;
- a continuing commercial incentive; and
- improved cost efficiency (construction and operating costs).

Bearing in mind these benefits and the need to improve the effectiveness and efficiency of the public sector, the Victorian Government has recognised the need for private investment in infrastructure and service delivery. Consequently, the Victorian Government initiated an 'Infrastructure Investment Policy', which led to the development of a portfolio called the 'New Prisons Project' (NPP). The NPP programme has attempted to strengthen its relationship with the private sector by using BOO methods of procurement. The paper presents findings from several prison projects that were procured using different procurement methods. The findings reveal that prisons procured using BOO systems are more cost efficient, specifically in relation to construction and operating costs, than those procured by other means. Discussions on the future of privatising correctional services using BOO systems are also presented.

The procurement of infrastructure projects

Traditionally, the provision of public infrastructure has been the responsibility the State and Federal governments in Australia, and has typically been financed through taxes. The need and demand for new infrastructure, rehabilitation and maintenance of existing facilities continues to increase. Yet to decrease expenditure, deficit reduction programs to control government debt have stimulated the use of innovative procurement and financing methods. According to Russell and Abdel-Aziz (1997), the emergence of public-private sector initiatives, such as BOOT, DBFO and BOO for procuring infrastructure facilities provides governments with option of satisfying their infrastructure needs and demands by alternative means. Generally, such means involve a user-pays concept, which invariably can be implemented by governments, yet many governments have preferred to execute the concept through the private sector so as to minimise their financial liability (Russell and Abel-Aziz, 1997). It has been suggested by numerous authors such as Tiong *et al.* (1992), Dias and Ioannou (1995), Woodward (1995), Tiong (1996), and Russell and Abel-Aziz (1997) that the private sector also offers the opportunity to improve the quality and efficiency of public projects through the use of their skills in design, engineering, construction, operation, and marketing. In addition, with the use of these methods the government attempts to transfer to the private sector as many project risks as possible, particularly those associated with time and cost overruns.

The procurement of infrastructure projects using those methods identified in Figure 1 (after Russell and Abel-Aziz, 1997) requires both the public and private sectors to change their existing mindsets and adopt new skills, roles, responsibilities and risks so that all the phases of a project's life cycle can be managed effectively. Noteworthy, Russell and Abdel-Aziz (1997:410) state that "the private sector must acquire expertise not only to justify such joint-ventures, but also to be successful in promoting, winning and managing them". This invariably requires the project owner/promoter, (usually a consortium of contractors and operators) to have the necessary technical, financial and legal skills in place if the project is to succeed. For further details pertaining to the attributes needed to procure successful infrastructure projects using concession initiatives such as BOOT refer to Dias and Ioannou (1996) and Russell and Abdel-Aziz (1997).

The Public and Private Sector

Two fundamental attributes for procuring successful infrastructure projects are commitment and trust, which need to come from the both the public and private sectors. In the case of the Government for example, such commitment is explicitly demonstrated by the Victorian Department of Treasury *'Infrastructure Policy for Investment in Victoria'* published in 1994, which sought to bridge the gap between the public and private by encouraging the private sector to invest in State Infrastructure. The policy sets out clear guidelines for promoting greater certainty for those businesses making infrastructure

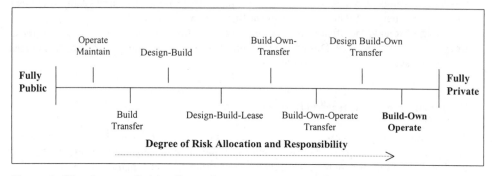

Figure 1. The degree of risk allocated to procurement methods used for infrastructure projects.

investment decisions by highlighting a number of important factors that need to be considered, these include:

• the establishment of guidelines for assessing the merits of proposals;
• examples of appropriate forms of investment; and
• property provisions and confidentiality.

Figure 2 provides an overview of the Victorian Government's guidelines and depicts the attractiveness of various private sector investment proposals that may be submitted.

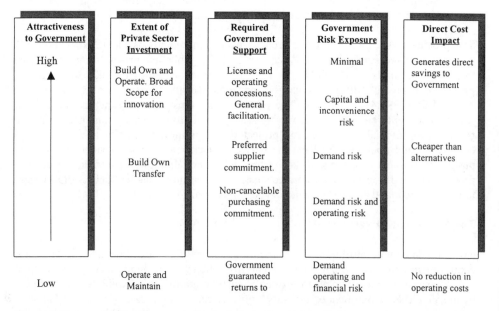

Figure 2. Degree of attractiveness of private investment proposals

The private sector is typically faced with uncertainties when tendering for infrastructure projects. The Victorian Government has developed policies, which demonstrate its commitment to reducing the effect of these matters. By reducing such uncertainties the efficiency of the tendering process, in terms of time and cost process can be improved. Private sector investment is encouraged in both new and existing infrastructure. The policy however requires that a project to be structured in a manner that is value-adding to both the public and private sector, that is, there is on-going benefits for both parties throughout the procurement process. Such an approach to infrastructure procurement can provide a win-win situation and stimulate innovation as well as provide direct cost savings.

Procuring infrastructure projects

The use of BOOT, DBFO, BOT (Build-Own-Transfer), and BOO approaches to procuring infrastructure projects have become ubiquitous in Australia, mainly as a reflection of the current Governments emphasis on the privatisation of publicly owned assets. An important aspect of these approaches is that the government grants a concession to the private sector to own, and operate and therefore implicitly finance an activity, which has traditionally been carried out by a government authority. The suffix "T" at the end of these acronyms refers to "Transfer". Thus, in a number of cases what will happen at the end of the concession period is that ownership of the asset will actually transfer back to the government at the termination of the agreement. Barnett (1997) provides a comprehensive definition of BOOT as the:

> "Government granting to a private sector organisation a concession of franchise to build a specific facility, to own it for a specified period, to operate it and to take the revenue from it, and ultimately to transfer it back to the Government".

Individual BOOT projects tend to differ from one another, but they usually display similar structures as shown in Figure 3 (McCarthy and Tiong, 1991; Woodward, 1995). The vehicle for a BOOT project is the consortium, which may be a trust, joint venture or partnership. In this arrangement the principal enters into a concession agreement with the consortium. This agreement essentially identifies and describes the support to be provided by the principal and the rights and obligations of the consortium. If a consortium develops a project then they often negotiate the sales agreement with the Government and then use it as security to go and raise the equity and debt support with the lenders. The consortium has the responsibility to construct, operate and take revenues from the running of the facility for the duration of the concession period which terminates when the amount borrowed has been repaid or when the concession period expires.

In the context of a BOO project however ownership does not generally transfer back to the government at the termination of the agreement. With this in mind, is a BOO approach synonymous with privatisation? On the one hand it can be considered to be privatisation inasmuch as the concession being granted for procuring a facility equals the economic life of the asset. On the other hand, for privatisation to occur a government authority should issue some wholesale equity so it can be transformed into a listed entity. Although, it is outside the scope of this discussion to address the nature of privatisation, for the purposes of this paper the BOO approach will be considered synonymous with privatisation.

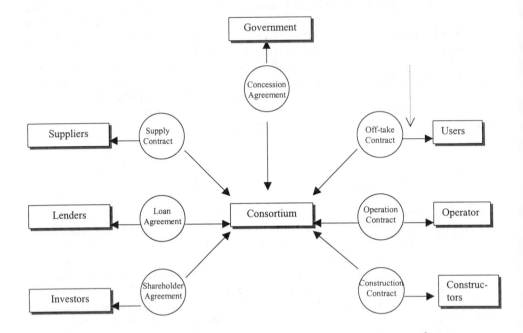

Figure 3. The structure of BOOT project (after McCarthy and Tiong, 1991)

Noteworthy, however, the grantor of the concession does not have to be a government; it could be a private sector organisation. The major difference between financing projects of the BOOT/BOO type and the more conventional approaches is that lenders have only the project's expected cashflows to indicate its economic viability (Woodward, 1995). The consortium of a BOOT/BOO project can take on numerous roles. As well as assuming the role of borrower, through their financial advisers, they have the added responsibility of structuring the financial package in a way that the project will prove attractive to lenders, whilst minimising their risk and potential liability.

The structure of a BOO can be the same as that of a BOOT illustrated in Figure 3. The only major difference relates to the nature of the offtake contract and the allocation of risk, as shown in Table 1, which can be primarily determined by the concession agreement. Build-Own-Operate projects employ a particular form of structured financing. Such projects are complex by way of the number of parties involved and the corresponding number of contracts, which must all interlock. Furthermore, each party is dependent on the performance of all parties to the project.

Under a BOO arrangement the private sector designs and builds the infrastructure, finances its construction and owns, operates and maintains it over a period, often as long as 20 years, which is often referred to as the 'concession' period. Technically this form of arrangement should be Design-Build-Finance-Own-Operate (D-B-F-O-O), but for simplicity we use the term BOO in this paper. Traditionally, such projects provide for the infrastructure to be transferred to the government at the end of the concession period,

Table 1. Risk allocation under the New Prisons Project using a BOO system for a recently completed project

DEVELOPMENT PHASE	Consortium	Government
Planning approvals for Government site	*	*
Other planning approvals	*	
Site contamination etc	*	
Construction:	*	
All risks	*	
Programme milestones	*	
Cost to complete	*	
Changes in legislation and policy not aimed at the projects		
Changes in legislation changed at the project		*
Government initiated variations		*
Conditions:		
Industrial disputes aimed at the Government		*
Industrial disputes	*	
Other Government/Public Authority/ Local Government Approvals.	*	
Increases in Local Government charges	*	
Completion and certification	*	
Commissioning plant and equipment	*	
Finance:		
Interest rate to date of signing	*	
Interest rate change after date of signing	*	
State taxation change	*	
Federal taxation change	*	
OPERATING PHASE		
Cost to operate agreed standard	*	
Industrial relations aimed at Government		*
Other Industrial Relations	*	
Operating performance/outcomes	*	
Deviation from Standards:		
Change in regions prisoner mix and demographic profile	*	
Change from current prisoner numbers	*	
Change in operating standards at Government request	*	*
Maintenance of any building/plant	*	
Periodic replacement	*	
Accreditation:		
Personnel	*	
NPP accreditation	*	
Financial:		
Interest rates	*	
Federal and State Tax legislation	*	
Changes in funding mechanisms applied to all privately funded prisons	*	
Building Defects Liability	*	
Buildings/Plant Durability	*	
Design suitability		
Design performance/operability		
Rates and Taxes	*	
State Legislation and Policy not aimed at the project	*	
State Legislation aimed at the project	*	

though in Australia this is often not required due to the Governments power to renew most contracts. This form of procurement method is currently being used by the Victorian Department of Justice under their New Prisons Project (NPP), which was announced in 1993. Their primary aim is to encourage private sector invest so that cost efficiencies can be obtained in terms of capital and operating costs.

The new prisons project: a commitment toward the use of BOO systems

The Victorian prison system currently consists of 14 publicly owned and operated prisons throughout metropolitan and non-metropolitan areas. The existing system has a number of inadequacies, which require a number of high priority improvements. Amongst these are:

- *antiquated facilities* - Some existing building stock is antiquated and obsolete and urgently needs to be replaced;
- *requirement for further improvement to work practices* - Prisons operating costs are high and despite recent improvements, public sector correctional services work practices have substantial scope for further efficiency gains; and
- *projected growth in prisoner numbers* - The existing capacity of the Victorian prison system is considered inadequate.

The Victorian Government is committed to the implementation of the NPP and has developed the following objectives (Wilson, 1995):

- replace existing and ageing plant with new facilities and increase the capacity of correctional services to meet projected demand
- reduce the costs of infrastructure development;
- reduce the costs of correctional services via the adoption of improved work practices;
- ensure the scope and quality of services to prisoners is maintained and/ or enhanced without affecting security or safety;
- meet Government policy objectives of private sector involvement in prison operations by introducing private equity into Victorian prison infrastructure with consequential transfer of risk to the private sector;
- establish competition amongst private and public sector providers of correctional services; and
- introduce innovative approaches to the design, construction and management of prisons.

Under the NPP programme three prisons were identified:

- a 125 bed women's prison to be located in Melbourne;
- a 600 bed medium security men's prison located in rural Victoria; and
- a 600 bed multi-functional men's prison in metropolitan Melbourne.

These facilities will accommodate over 60% of female and 50% of male prisoners in Victoria. The listed prisons identified have been or are being procured under a BOO arrangement. The reasons for adopting such an arrangement have been based on findings from prisons procured in other states of Australia such as Queensland (QLD) and New

South Wales (NSW). The next section of this paper will present findings from prisons that have recently been procured in Australia and demonstrate how private sector private sector involvement has led to the Victorian Government opting to procure their prisons using a BOO system.

Cost efficiency

Economic rationalisation is the primary driver for privatising prisons in Australia. Yet, the initial decision by the Queensland Government to privatise Bollaron prison was based only on an assertion made in the *Kennedy Report* (1988) that stated "...in some particular areas of prison operations the private sector can do it cheaper and better. Since the privatisation of this prison, New South Wales and Victoria have followed suit. In order to justify the involvement of the private sector and the cost efficiencies that could be attained, financial benchmarks were recommended. The financial benchmark was based on the *net annual cost per inmate per year* (See Table 2).

The most recently commissioned prison being procured through government capital expenditure is a 400-bed maximum-security institution, which cost $90 million to construct in Western Australia (WA). Such funds are simply not going to be available up-front in the foreseeable future. With this in mind, the NSW and Victorian Governments have decided to use the private sector as a mechanism for procuring correctional facilities. Recently, however, the NSW Government procured a 600-bed medium security institution for $57 million using a BOO system, which was approximately half the estimated cost which the State Government itself would have incurred though a direct procurement approach. A net present value analysis using discounted cash flow techniques was applied to the combined construction and management offer put forward by the consortium. Over the full range of both discount rates and analysis periods, the tender by the consortium was found to have a lower net present value than a direct government procurement approach by a significant margin (NSW Parliament Public Accounts Committee, 1993).

Operation costs (recurrent costs)

Operating costs or recurrent costs of the prison system will remain a source of concern to all Australian Governments, particularly at a time when public sector spending is under scrutiny by opposition parties, the public and media. To provide the required level of service and at the same time save money the Government must justify the need for privatising prisons. Yet, there can often be a problem when comparing costs between public and private sector inasmuch as it is often difficult to identify a private and public prison whose operations are comparable. Factors that need to taken into account when comparing prisons include:

- location as this may affect salaries;
- the design of the facility can have a bearing on the number of required staff members,
- the security classification of inmates will determine the extent and cost of their custody requirements; and
- the population size of the prison.

The most important factor that determines the cost per inmate per day is the prison's population size. With this in mind, the following examples, identified in Table 2, compare the cost of operating a public and private prisons in Queensland (QCSC, 1993).

Table 2. A cost efficiency comparison of correctional facilities procured by public and private sector

	Private	Public
Total Direct Costs	$9.3 m	$8.96 m
Average Daily Offenders	237	210
Net Annual Cost Per Inmate	$39 240	$42 870
Central Office Overheads	-	$11 690
Gross Annual Cost Per Inmate	$39 240	$54 560

These prisons were constructed at the same time and have a similar infrastructure and number of inmates. According to the QCSC (1993) the net annual cost per inmate is 9.3% ($42 870) higher in the public prison than in the private ($39 240). The difference in the gross annual cost works out to a total annual cost saving of $15 320 per inmate and a total annual cost saving of $3.63 million for all inmates accommodated. The operation of a prison is a labour intensive activity. A prison must be staffed 24 hours a day and there also

Table 3. Cost comparison between various correctional facilities in Australia between 1990-1993

Correctional Facility	Procurement Method	Operational System	Capital Cost Per Inmate Housed ($)	Operational Cost Per Inmate Per Day ($)
Victorian Prison A	Build Transfer	Public	395 000	160
NSW Prison A	Build Transfer	Public	327 000	180
Victorian Prison B	Design & Construct	Public	271 000	160
NSW Prison B	Construction Management	Public	219 000	140
NSW Prison C	Build Transfer	Public	204 000	140
WA Prison	Design & Build	Public	169 000	160
Victorian Prison C	Design & Build	Public	150 000	150
QLD Prison A	Build-Own-Operate	Private	130 000	150
QLD Prison B	Build-Own-Operate	Private	100 000	110
NSW Prison D	Build-Own-Operate	Private	92 000	120

needs to be a sufficient number of prison officers to direct inmates to ensure internal security and safety. Achieving efficiencies in these areas offers the potential for savings by private prison contractors. This can be achieved by reducing staff levels relative to public prisons without loss of efficiency or security. Table 3 compares the cost of various correctional facilities procured in Australia during the last decade. More importantly, however, this table identifies the cost efficiencies that can be obtained using BOO systems to procure prisons.

Discussion and conclusion

Prison privatisation is a relatively new concept in Australia. To date there has been limited research undertaken in Australia which provides a genuine insight into the performance and cost effectiveness of the private sector. This paper has attempted to contribute to providing this insight demonstrating the cost efficiencies that can be obtained by using BOO systems. There appears to be a future for privatization in Australia as:

- Growth in the appeal of correctional privatisation has proven to be far stronger than even the most ardent privatisation proponents imagined possible a decade ago. For example, the average annual rate of growth in the capacity of private facilities in operation or under construction over the last ten years was 45.04%. Furthermore, with a five-year average annual growth rate of 41.08% and a three-year average growth rate of 38.11%, there is no indication of year-on-year rates of growth is moderating in any significant way. Such robust and persistent growth has encouraged many to predict similarly high rates of growth for the foreseeable future (Thomas, 1997).
- Privatisation is becoming an accepted alternative to a 'business as usual' approach rather than something being viewed as experimental. Evidence of significant operating and construction cost savings is growing rapidly. Evidence that privatisation can yield improved services as well as reduced costs are also growing. The pressure on government departments to provide better public services without a corresponding increase in tax burdens is increasing, which favours privatisation. Thus, this implies continued strong growth for the private correctional services industry.
- Evidence is becoming available which suggests that competition, caused by the private providers of correctional services is yielding improvements in the efficiency and effectiveness of public agencies. The more efficient and effective public correctional agencies become, the lower will be the benefits that privatisation yields – though one should be mindful of the fact that the private firms continue to seek out means of improving their efficiency and effectiveness.

A decade ago the burden was on those who favoured the competition between alternative providers of correctional facilities. No government department risked being criticized and took to the position that correctional privatisation was too new, too unproven, and simply too risky.

Today the tide has turned dramatically as the burden is on those who wish to block the privatization initiative. They are quickly becoming a minority, and are being called upon to defend their negative points of view. According to Thomas (1997) this is how it should be, as the volume of evidence is now growing large enough and positive enough that those who oppose it now have an obligation to consider privatisation as a potentially valuable innovation. The day may be coming when policy makers agree that they have an obligation to provide for the delivery of correctional services that meet or exceed the specified and professional benchmarks and to do so at the lowest possible cost. The day may be coming when policy makers understand that the public or private identity of correctional service providers is irrelevant.

References

Akintoye, A., and Taylor, C. (1997). *Risk prioritisation of private sector finance of public sector projects*. Proceedings CIB W-92 Procurement Symposium, Procurement - The Key to Innovation, University of Montreal, Montreal, 18th -22nd May, Canada, pp. 1-10.

Barnett, M. (1987). Role of the Merchant Banker in Projects. *International Journal of Project Management*, 5(4), pp.197-203.

Dias, A., and Ioannou, P. (1995). Debt capacity and optimal capital structure for privately financed infrastructure projects. *ASCE Journal of Construction Engineering and Management*, 121, (4), pp.404-414.

HM Treasury (1995). *Private Opportunity, Public Benefit*. Progressing the Private Finance Initiative, November, London, UK.

Hodgson, G.J., and Davis, J.A. (1997). Can Value for Money be realised by Using Private Sector Finance for the Procurement of Public Service Infrastructure? Proceedings CIB W-92 Procurement Symposium, Procurement - The Key to Innovation, University of Montreal, Montreal, 18th -22nd May, Canada, pp.285-293.

Kennedy, J.J. (1988). *Commission of Review into Corrective Services in Queensland*. Final Report, Vol. 2, Brisbane, Queensland, Australia.

Love, P.E.D., Skitmore, R.M., and Earl, G. (1998). Selecting a suitable procurement method for a building project. *Construction Management and Economics*, 16, (2), pp.221-233.

McCarthy, S.C., and Tiong, S.L. (1991). Financial and contractual aspects of build-operate and transfer projects. *International Journal of Project Management*, 9, (4), pp.222-227.

New South Wales Government. (1992). *Royal Commission into Productivity in the Building Industry in New South Wales*. Vols. 1-10, Sydney, Australia.

NWPC and NBCC. (1990). *No Dispute – Strategies for Improvement in the Australian Building and Construction Industry*. A Report by the National Public Works Conference and National Building and Construction Council Joint Working Party, May, Canberra, ACT, Australia.

Queensland Corrective Services Commission (QCSC). (1993). *Comparisons between Lotus Glen and Borallon Correctional Centres*. Queensland Treasury Department, Unpublished Report, April, Brisbane, Queensland, Australia.

Russell, A.D., and Abdel-Aziz, A.M. (1997). *Public-private partnership and public infrastructure*. Proceedings of the First International Conference on Construction Industry Development: Building the Future Together, National University of Singapore 9th-11th December, Singapore, pp.410-419.

Tiong, R.L., Yeo, K., McCarthy, S.C. (1992). Critical success factors in winning BOT contracts. ASCE Journal of Construction Engineering and Management, 118, (2), pp.217-229.

Tiong, R.L. (1996). CSFs in competitive tendering and negotiation model for BOT projects. ASCE Journal of Construction Engineering and Management, 122, (3), pp.205-211.

Thomas, C.W. (1997). Private Adult Correctional Facility Consensus. Tenth Edition, March 15th.

Victorian, Department of Treasury (1994). *Infrastructure Investment Policy for Victoria.* State Government of Victoria, Department of Treasury, June, Melbourne, Victoria, Australia.

Wilson, A. (1995). *New Prisons Bulletin.* Department of Justice, Victoria, Australia.

Woodward, D.G. (1995). Use of sensitivity analysis in build-own-operate-transfer project evaluation. *International Journal of Project Management,* **13**, (4), pp.239-246.

RISK IDENTIFICATION FRAMEWORKS FOR INTERNATIONAL BOOT PROJECTS

A. SALZMANN and S. MOHAMED
School of Engineering, Griffith University, Gold Coast Campus, Queensland, 9726
Australia. A.Salzmann@mailbox.gu.edu.au

Abstract

The BOOT procurement scheme requires a firm to (B)uild an infrastructure facility, (O)wn and (O)perate it for a stipulated period of time, and then (T)ransfer it free of charge back to the Government authority at the end of the period. This in combination with the large, complex nature inherent in overseas projects provides a myriad of risks that need to be addressed. Potential investors need then to thoroughly appreciate and investigate associated risks, so they are able to make informed decisions regarding the viability of the project. A number of recent studies in the literature concentrate on risk and critical success factors present in BOOT projects. The listing and categorizing of these factors has varied quite significantly in the literature and little attention has been paid to comprehensively detailing every aspect of a BOOT project where problems may arise. A more general approach that covers all areas of risks inherent in BOOT projects is required to enhance this decision making process. In this paper, two frameworks are presented that comprehensively cover all areas of BOOT project risks. Particularly heightened are those regions neglected by previous research and those with greatest impact on procurement performance and profitability. The developed frameworks are justified by the use of a comparative study with published literature.

Keywords: BOOT projects, international projects, risk identification, success factors.

Introduction

BOOT projects are becoming increasingly common throughout the world as a means for providing infrastructure to a country, without directly impacting on the sovereign finances of that country. The BOOT procurement scheme requires a firm to (B)uild an infrastructure facility, (O)wn and (O)perate it for a stipulated period of time, and then (T)ransfer it free of charge back to the Government authority at the end of the period. Because of the multifarious nature of BOOT schemes, a number of specialist firms, each providing a particular service, forms the hence created consortium. Within this consortium there are many risks that need to be addressed in order to ensure that investment into the project is viable. These risks are heightened in international BOOT projects as they depend on a combination of commercial, political and economic factors. The potential risks and rewards for an international BOOT project are therefore great and are directly related to the consortium capability of balancing the often competing principles of harmony and profit.

Previous research has focussed upon identifying individual risk and critical success factors in BOOT projects that lead to project success or failure. The listing and categorizing of these factors has varied quite significantly in the literature and little attention has been paid to comprehensively detailing every aspect of a BOOT project where problems may arise. It is considered important to undergo this listing and categorising stage so to allow further detailed investigations into the interaction between the identified factors. It is believed that it is a combination of the many factors and their effect on the system through some mechanism that influences a project outcome. The objectives of this paper are to develop risk frameworks that comprehensively map where risk and success factors are present and how they may impact upon the performance of an international BOOT project during its lifetime.

Risk factors – an overview

Risk identification is an important step prior to risk analysis. In order to be able to correctly manage project risks through analysis, comprehensive identification at the preliminary stage is required. Many authors have listed, what appear to be, the more unique and detrimental risk and success factors relevant to BOOT projects. Yet a comprehensive list is required in order for those bidding for a BOOT project, experienced or otherwise, to make sound decisions regarding risk allocation. Each author presents their list based on different definitions, categories and studies. Furthermore, each list is just that, a list under a few categories determined during a particular study in a particular area. In addition, each list has been arrived at in completely different ways. For example, Keong et al. (1997) base their list of critical success factors on three Malaysian BOT projects. Tiong & Alum (1997), on the other hand identify distinctive winning elements primarily through surveys. The background of published research, presented herein provides a broad introduction to BOOT risk and success factors.

Tiong (1990) pioneered the initial research into risk and success factors inherent in BOT projects. He investigated many areas of project risks and securities and introduced the concept that an infrastructure project can be viewed as a relatively high-risk construction project and a relatively low-risk utility project. Three categories of risk present during the entire life cycle of the project that deserve attention are those described as financial, political and technical. The political risk category is considered by Tiong to be the most significant.

Further research undertaken by Tiong et al. (1992) focussed on the critical success factors involved in winning a BOT contract. The authors identify constraints as well as critical success factors in winning BOT contracts through two sources (case studies on BOT projects and interviewing project sponsors and government officials). A comprehensive checklist of critical success factors that must be considered prior to the bidding process was developed. The list included the six main categories of entrepreneurship, selecting the right project, strong team of stakeholders, imaginative technical solution, competitive financial proposal and special bid features. It was concluded that if given proper attention, these six critical success factors will help to successfully overcome the menagerie of risks presented in a BOT project.

Walker and Smith (1995) used the broad categories of elemental and global risks. The elemental risks are described as those risks inherent in the elements of the project, and that are generally retained by the concessionaire. The global risks are those risks outside the project elements but still impact upon the concession, these are generally controlled by the concession itself. Two other categories of risk are also identified as when financing risks occur. They include the pre-completion or construction phase and post-completion or operation phase. These categorisations then lead on to a general, but not comprehensive list of risks that affect the project right throughout its life cycle. Finally, the authors concluded that infrastructure projects are particularly vulnerable to demand risks, that is, the project is at the mercy of those who are willing to patron the facility.

Tam (1995), through case studies of the BOT power industry in South East Asia, identified risks and critical success factors relevant to power station investments in developing countries. The identified factors include political opposition, underdeveloped legal framework, underdeveloped fiscal framework, corruption and political instability. Finally, the author develops the five P's framework of project, partner, pattern, profitability and protection for successful BOT project execution in China. The framework was developed on the basis of observations of, and interviews with, companies involved in a number of Chinese BOT projects.

David and Fernando (1995) investigated the difficulties and risks involved in a BOT project and specifically those identified within the private power sector. The authors divide risks into two distinct groups based on who perceives the risk as most relevant. The two divisions used are the investor perceptions of risk and the host perceptions of risk. The former of the two includes construction phase risks, inflation, foreign exchange and no-take risks. No-take risks can be described as the investor fears the host may fail to purchase the project offtake in future years. The second division of host risk perception includes risks such as take-off risks, complicated negotiations, non-competitive bidding and failure of the investor to maintain the facility. Take-off risks are analogous to no-take risks in that the host is cautious in committing to buying something for a pre-agreed price which it may not need in say thirty years.

Keong et al. (1997) further build on the idea of critical success factors in private sector initiated BOT projects. Through case studies of three Malaysian BOT projects, namely a water supply, a power and a maritime port project, the authors identify country, project and client as most critical to project development success. Some of the most critical factors identified in the country group included, an economically stable country, political will to proceed, well developed local capital market and proper regulatory and legislative frameworks for infrastructure privatisation. The project group should have a steady demand forecast, adequate lender returns and be fairly large in size. Finally, the client group included factors such as creditworthy government signatory, and willingness for project implementation through participation and support.

Tiong and Alum (1997) studied the BOT tender process to determine what distinctive winning elements need to be present in order for a contractor to win the bid during the final selection process. They ascertain that BOT concessions are not awarded on the lowest bid price or tariff alone. Rather it is based on the best overall bid package presented by the potential concessionaire. They add to previous research undertaken by Tiong et al. (1992) by focussing on three critical success factors, namely that of technical solution advantage, financial package differentiation and differentiation in guarantees.

Ma et al. (1998) identify five major risk categories in typical BOT projects. They include political, construction and completion, operating, financial and market and revenue risks. These risks were identified as occurring throughout the duration of the project, but at no specific stage within the project. Ma et al. (1998) acknowledge the fact that physical completion of the facility does not annul the identified risks posed by BOT projects. Therefore, it is suggested by Ma et al. that during the development of a BOT project, risks should be adequately identified, managed and allocated by undertaking feasibility studies. In summary, the authors define success of a project as being characterised by strong, stable government support, stable currency, reliable economic system and a technically and economically feasible project.

Kerf et al. (1998) deal with risk identification in the context of determining which party to the concession should be allocated that risk. Kerf et al. identify eight groups in which risks may be grouped under. These include design risk, construction risk, operating cost risk, revenue risk, financial risk, force majeur risk, performance risk and finally environmental risk. The theory used behind risk allocation is that the party best able to deal with the risk should bear it. However, Kerf et al. explain why this is not as simple as it seems for many reasons, a couple are explained as follows. Firstly, it is not always easy to determine how well a party can protect itself against a particular risk, for example, how can you define the line where a company will pass a risk on to the consumers rather than bearing that risk. Secondly, it is not always feasible to determine which party has the best risk mitigation tactics for particular risks.

Framework justification

From the background, one can see the diverse range of topics covered within BOOT/BOT project risk and success factors and the depth with which they have been explored for the basis of their selection. One of the main drawbacks of previous research is that due to the nature of the studies, only those factors associated with that type of study have been documented. For instance, Tam (1995) primarily utilises power station BOT project case studies and interviews to determine those particularly relevant risk factors. On the other hand, Tiong and Alum (1997) use surveys and questionnaires to identify risk and success factors relevant to the specific area of final bid selection process. Therefore, neither list can be considered to be a comprehensive listing of BOOT project risks. Furthermore, no serious attempts have been made to examine and explain the interactions that exist between the identified factors. The inter-relationships that exist between the factors play perhaps an even more significant role in affecting the projects progress. For instance, political opposition to a proposed BOOT project is consistently identified in the literature as a risk factor. However, the authors believe that political opposition is quite often a response to the presence of the community opposition risk factor. It is believed that before such interactions can be identified, a comprehensive listing of all the risk factors involved in all aspects of BOOT projects must first be developed.

Two comprehensive risk frameworks have been developed in response to the perceived lack of such a listing. The frameworks have been developed so to include every possible aspect where risk may emanate from a BOOT project. Presented below in *Figures 1 and 2* are the developed risk identification frameworks for BOOT projects. It was deemed appropriate to categorise according first to where the factors' influence is first encountered. *Figure 1* details those factors present during the feasibility to (B)uild stages (Development phase). *Figure 2* presents those factors likely to be encountered during the (O)wn, (O)perate and (T)ransfer stages of the project (Operations phase).

Within each framework, the factors have been divided into four 'superfactor' groupings. They are that of:

- Host Country;
- Investors;
- Project; and
- Project Organisation/Management.

A brief description of each superfactor and the associated subfactors and elements is hence given below. Each subfactor is considered not only to present inherent risks but also to highlight those particularly crucial areas where attention should be focussed. Special emphasis is given to subfactors where the concepts of harmony of relationships play a major role in meeting project objectives.

Development phase risk factor framework

The host country superfactor includes some of the most important risk factors posed to international BOOT projects. The political subfactor contains elements such as how the political and legal frameworks are developed to enhance or otherwise, the companies' ease with which it conducts business dealings. The economic and national market subfactors describe all those situations where significant instability and unknowns remain as to the magnitude of the threats posed by them, such as inflation and currency. The social subfactor describes those interactions undertaken in conjunction with the wider community, which if not properly addressed, severely affect project progress. Cultural barriers and their impact on the harmony between construction relationships is highlighted in this subfactor. Chan et al (1996) investigated how the unique management style and financial strength of Japanese contractors working in Hong Kong contributed to their success. Differences in their respective cultures were responsible for the majority of conflicts that might have affected project performance and hence profitability. Baba (1995) also reinforced how differences in cultures provide difficulties or "barriers" to U.S.A. contractors bidding within the Japanese bid system. Baba (1995) suggests these difficulties stem from the fact that Japanese management is founded on harmony and profit, American management on logic and reason, these being directly descended from a country's culture, history and background.

The investors superfactor constitutes a significant area of concern. In large, complex BOOT projects, the ability to finance is absolutely critical to be able to even consider bidding for the project. One needs to consider the characteristics of the investor, such as their financial capability, previous record and market share as a whole. The country of origin may also affect the ease with which the investors may conduct business, such as foreign investment policies. Also critical is prior BOOT experience. This is especially

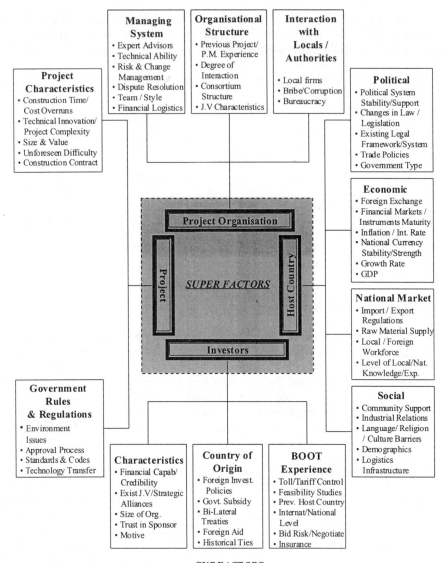

Figure 1 Development phase risk factor framework

important in the feasibility stages as it can help predict the relative financial success or failure of the project. Toll and tariff control requires significant judgement, which also stems from prior experience in the area.

The project superfactor includes all associated avenues of risk connected to the actual project itself. The first category of project characteristics includes the type, size and value of the project itself and any associated features. The second subfactor of government rules and regulations is particularly important to be aware of and adhere to in a foreign country, so to prevent any unnecessary delays in project completion.

The subfactors contained within the project organisation superfactor include that of managing system, organisational structure and interaction with locals and authorities. The managing system describes the risks created by how and who is in control of any 'situation' that may arise during the concession. For instance, inadequate risk or change management techniques or the absence of certain key people, such as expert advisors can create risks. The organisational structure attempts to include those areas of risk that emanate from how the project organisation as a whole is formed and those who comprise it. For example, the consortium may be created as a joint venture or say a partnership. Each structure will present unique and different contractual arrangements and parties that need to be examined for risks. Sridharan (1997) reported that international joint ventures are subject to a very high failure rate (up to 50%in developing countries). This is mainly due to cultural and operational difficulties at both national and organisational levels. The interaction with locals and authorities is important, as a significant amount of liaison is required in foreign contracting. Risks in this subfactor include political bureaucracy and bribe and corruption.

Operations phase risk factor framework

The operations phase is important in BOOT projects as it constitutes a significant component of the total amount of time involved. However, one can see that the number of identifiable subfactors given in *Figure 2* is noticeably less than that given in the development phase. This is primarily due to a downsizing in the amount of activities being undertaken and personnel and companies involved. This does not however, diminish the impact of the remaining risks posing a threat to the project.

The influence of the host country superfactor has diminished slightly at this stage. However, significant threats still remain. Within the political subfactor, changes in laws resulting in expropriation (confiscation), still remain an extreme threat to the project. Economic subfactor elements of foreign exchange, inflation and interest rate fluctuations still remain, but to a lesser degree than the development phase as many safeguards such as insurance can minimise these threats.

The national market elements generally act upon the revenue generating capacity of the project at this stage, such as facility demand and raw material supplies. In a similar vein the social subfactor elements have the potential to affect project revenue generation, such as strikes (industrial relations) or boycotts (community support).

The investors superfactor contains the subfactors of project-dependent and financial in the operations phase. At this stage of the project, risks are created through the investors' control over the subfactor of project-dependent finances. This includes project-dependent revenue generation and revenue-dependent debt servicing. The financial subfactor encompasses all the other areas where the investors may have financial dealings, such as with the government, insurance companies and the investors themselves.

The project superfactor influence in this framework has heightened considerably in the operations phase as operation and maintenance become the key focus of the project. The first of the three subfactors is that of maintenance. Elements include how long the maintenance contract is for, and who is involved in the execution and control of the service. The operation factor elements cover the problems that are likely to arise (such as cost overruns) and the reasons behind these problems (such as contracts, regulations and those involved). The post transfer stage risks and remaining liabilities should also be

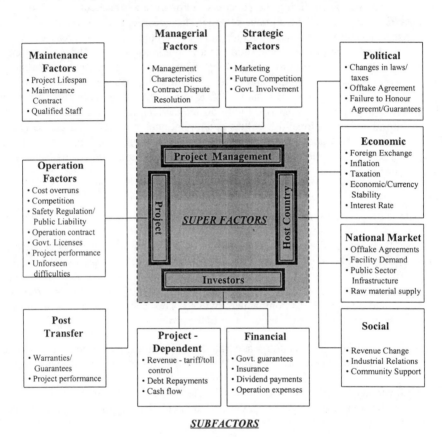

Figure 2 Operations phase risk factor framework

pre-empted so to avoid costly problems. During this stage, consideration should be given to any warranties or guarantees that are to be provided with regards to long-term project performance.

Project management is divided at the operations phase into two divisions, managerial and strategic. Managerial factors are those characteristics that describe how day to day operations are conducted. For instance, managing human resources, quality issues and business operations. Strategic factors describe the threats and opportunities which might affect how the project is operated. Some of the many facets of the strategic management issue included here, are developing business plans, marketing, future competition and the risk of technology transfer.

Framework cross-verification

Table 1 presents in a clearer manner, those factors identified in the overview. This type of presentation allows cross-verification with individual authors, and with those of the presented frameworks. As can be seen, both risk and success factors are tabulated together in *Table 1*, due to the fact that a factor can be described in not only in a negative sense (risk factor) but also in a positive sense (success factor). For example, Tiong (1990) mentions that political risk is one of the most significant of all the risks posed to a

Table 1. Comparison of published risk and success factors

Factor • risk factor ■ success factor ▲ critical element	Tiong (1990)	Tiong et al (1992)	Smith (1995)	Tam (1995)	David and Fernando (1995)	Keong et al (1997)	Tiong and Alum (1997)	Ma et al (1998)	Kerf et al (1998)	Proposed Framework
Development / Construction										
Completion delays/nodelays	•		•				■	•	•	▲
Construction cost overruns/effectiveness	•					•	■	•	•	▲
Change in taxes/laws								•		▲
Complicated negotiations		•				•				▲
(Under)developed legal/fiscal/economic framework	■				•		■			▲
Corruption						•				▲
Feasibility studies							■			▲
Consortium structure					■					▲
Local Partner					■					▲
Project management ability			■		■					▲
Existing infrastructure	•						■			▲
Political (in)stability/opposition (support)	• ■	•	•	•	•	■		•	•	▲
Raw material supply						•				▲
Environmental concerns/least impact							■		•	▲
Bid process		•				•				▲
Technical solution		■					■			▲
Inflation/foreign exchange/interest rate	• ■		•			•		•	•	▲
Financing	•	■				•	■			▲
Force majeur	•					•			•	▲
Operation / Maintenance										
Cost overruns - operation									•	▲
Operation & maintenance failure	•		•			•				▲
Raw material supply	•					•			•	▲
Market competition	•		•					•		▲
Revenue tariff/demand change			•			•	■	•	•	▲
Project performance	•		•			•			•	▲
Foreign exchange/inflation	•		•			•				▲
Force majeur						•			•	▲
Change in taxes/laws									•	▲

BOT project whilst also stating that strong government support is a prerequisite for a winning BOT proposal. This fact tends to authenticate the developed frameworks, in which the elements are presented in a neutral light.

Concluding remarks

This paper presents an overview of how risk and success factors in BOOT projects have been addressed in the literature. A critical review of the literature reveals a lack of a comprehensive list that covers all areas of risks inherent in this type of project. Two broad frameworks for risk identification have been developed from a literature review and presented with the aim to include every possible aspect where risk may emanate from a BOOT project. The material presented in this paper reflects an introductory part of an on-going research program aiming towards better identification of risk and success factors and their interaction in international BOOT projects. Risk factor interaction is now under investigation by the authors. It is envisaged that once these interactions are critically examined, then a refined version of the proposed frameworks will be developed. Nevertheless, the two frameworks proposed herein can be considered a serious attempt to encompass available published material and present a step in the right direction for improving the quality of decision making on BOOT project investment.

References

Baba, K (1995) Principal differences of management in construction between U.S.A. and Japan. *Proceedings of the fifth East Asia-Pacific conference on structural engineering and construction – Building for the 21st century,* Gold Coast, Australia, July 25-27, pp. 907-912.

Chan, J K W, Chan, H C and So, F (1996) The management style of Japanese contractors in Hong Kong – The case study of Nishimatsu. *Asia Pacific Building and Construction Management Journal*, Vol. 2, No. 1, pp. 38-41.

David, A K and Fernando, P N (1995) The BOT option: Conflicts and compromises. *Energy Policy*, Vol. 23, No. 8, pp. 669-675.

Keong, C H, Tiong, R L K and Alum, J (1997) Conditions for successful privately initiated infrastructure projects. *Proceedings of the Institution of Civil Engineers, Civil Engineering,* Vol. 120, May, pp. 59-65.

Kerf, M, Gray, R D, Irwin, T, Levesque, C and Taylor, R R (1998) Concessions for infrastructure: A guide to their design and award. *World Bank Technical Paper No. 399, Finance, Private Sector, & Infrastructure Network,* The World Bank, Washington D.C.

Ma, T YF, Chan, A and Lam, P TI (1998) A study into the characteristics of BOT projects in Hong Kong (illustrated with the Tate's Cairn tunnel project). *The Sixth East Asia-Pacific Conference on Structural Engineering & Construction,* Taipei, Taiwan, January 14-16, pp. 1131-1136.

Sridharan, G (1997) Factors affecting the performance of international joint ventures – A research model. *Proceedings of the first international conference on construction industry development – Building the future together*, Singapore, December 9-11, pp. 84-91.

Tam, C M (1995) Features of power industries in Southeast Asia: Study of build-operate-transfer power projects in China. *International Journal of Project Management*, Vol. 13, No. 5, pp. 303-311.

Tiong, R L K (1990) BOT projects: Risks and securities. *Construction Management and Economics*, Vol. 8, pp. 315-328.

Tiong, R L K and Alum, J (1997) Distinctive winning elements in BOT tender. *Engineering, Construction and Architectural Management,* Vol. 4, No. 2, pp. 83-94.

Tiong, R L K, Yeo, K-T and McCarthy, S C (1992) Critical success factors in winning BOT contracts. *Journal of Construction Engineering and Management*, Vol. 118, No. 2, pp. 217-228.

Walker, C and Smith, A J (1995) *Privatized infrastructure: The build operate transfer approach*. Thomas Telford, London.

DEVELOPING A METHODOLOGICAL APPROACH TO THE IDENTIFICATION OF FACTORS CRITICAL TO SUCCESS IN PRIVATISED INFRASTRUCTURE PROJECTS IN THE UK

ROY MORLEDGE and **KEVIN OWEN**
Construction Procurement Research Unit, Department of Surveying, The Nottingham Trent University, Burton Street, Nottingham NG1 4BU, United Kingdom
roy.morledge@ntu.ac.uk, kevin.owen@ntu.ac.uk

Abstract

Infrastructure worldwide is increasingly being financed with private sector capital, generally referred to as Build-Operate-Transfer (BOT) agreements. In the UK, privatized infrastructure projects of this form come under the Private Finance Initiative (PFI). This paper outlines the beginnings of PFI and suggests that there is lack of knowledge about best practice. Consequently, there is a need to identify the criteria that must be satisfied if the aims of a PFI project are to be met and the individual processes involved in its procurement are to be successful. It is proposed that the Critical Success Factor methodology developed by Rockart (1982) can be adapted to suit this purpose. The authors review the concept of Critical Success Factors and revisit the corresponding methodology to assist in the establishment of a framework of factors considered vital to stakeholder and user success in PFI projects. A previous study identified some Critical Success Factors (CSFs) of PFI projects through literature review and questionnaire. However, since this pilot study the concept of critical success factors has been developed further to identify problems associated with the practical application of the theory in its current form. Concerns are raised regarding subjectivity, bias, failures in data processing, changing PEST environments, ambiguous definitions, and lack of adequate qualitative performance measures. It is submitted that factors may fall into one of two categories, labelled 'Failure Reduction Criteria (FRCs) and Critical Success Factors (CSFs).

Keywords: Critical success factor, failure reduction criteria, methodology, PFI, private finance initiative, privatized infrastructure.

Introduction

Infrastructure has traditionally been financed in the public sector. However, since the early 1970s there has been a return to private sector investments into infrastructure provision. In the UK private funds were first used this century during the mid-1980s in the provision of the Channel Tunnel. Since then a number of other infrastructure projects have been provided using private sector funding, and in 1992 a formal policy was announced by the government and termed the Private Finance Initiative (PFI). Under the PFI private consortia are invited to bid for a concession to build and operate infrastructure (and other privatized) projects in return for annual service payments from the public client. This involves using private sector finance in the form of debt and/or equity in varying proportions. Long-term expenditure and risk are borne by the private sector and thus the burden on public investment into building and maintaining the country's infrastructure and other public services is released. The CBI (1996), Private Finance Panel (1996), and the Construction Industry Council [CIC (1998)] have all indicated the importance of establishing criteria which contribute to success or failure, thus to ensure that only projects likely to meet their aims will proceed.

Genesis of privatized infrastructure

The Private Finance Initiative (PFI) was first introduced by the then Chancellor of the Exchequer, Norman Lamont, in his Autumn Statement of 1992. There is an assumption that the Conservative administration 'invented' the privatized infrastructure concept of the PFI, in 1992. This is purely fictitious as the PFI concept originated more than 200 years earlier, and privatized infrastructure concessions had existed since the middle of the 18th century. Even so, the current PFI concept is an adaptation of modern tried-and-tested methods (Build-Operate-Transfer, or BOT) to privatize infrastructure around the world.

Walker & Smith (1995) identified the first concession agreement for privatized infrastructure, granted to the Perier brothers in 1782, for water distribution in Paris, France. Following the success of this initial deal, concession agreements were introduced throughout the 1800s to fund infrastructure projects in Spain, Italy, Germany, Belgium and the UK including turnpike roads and railways.

However, new infrastructure was generally financed in the public sector from the tail end of the 19th century until the late 1970s. At this time a number of influences placed pressure on established methods of financing from fiscal resources and sovereign debt. It has been argued that the (re)introduction of private finance in infrastructure in the developed nations of Western Europe comes from two principal influences. First, 'the existing and limited additional infrastructure is unable to keep pace with growth' [Walker & Smith (1995)]. The Economist, 1993 (cited in Walker & Smith 1995), contended that the reason for congestion on roads in the UK was due to a lack of infrastructural development since 1970 that had failed to keep pace with the increasing use of cars. This supports similar arguments put forward by the Adam Smith Institute in 1992 [Hibbs & Roth (1992)] that although income from petrol and vehicle tax is continuously increasing the UK still faces 'an infrastructure crisis'. Similarly, the huge 'stock' of infrastructure built up since the Second World War from regenerating European nations is in dire need of upgrading and renewal. This has compounded the problems highlighted above. Secondly, 'the increasing longevity of the populations in the developed world has resulted in ballooning health care costs for these societies to shoulder' [Walker & Smith (1995)]. There is, argue Walker and Smith, a growing need for government to protect their nation for longer with a limited working population whom to tax.

Prior to 1977 all public capital expenditure in the UK was measured against the public sector borrowing requirement (PSBR), even when wholly financed from internal resources derived from user charges. From 1977, however, the focus for controlling the PSBR switched to using *external financing limits*, or EFLs.

Throughout the 1980s the UK Government continued with their privatization of the nationalised industries. This programme has been referred to as the 'most innovative policy success' since the Conservative Party took power in 1979, as it reduced the burden on the PSBR caused by the utility companies' requirements for investment [Plender (1996)]. Meanwhile, private finance was being used to fund infrastructure projects worldwide, under the acronym of BOT (Build-Operate-Transfer), and used increasingly among the Pacific Rim nations and the USA.

In 1989 John Major's speech to the Institute of Directors, when Chief Secretary to HM Treasury, drew attention to opinions that the Ryrie Rules were hampering the use of private finance in infrastructure. In the same address he abolished the Ryrie Rules as obsolete, and gave assurances that the *non-additionality* ruling would not occur on privately financed projects. The concluding remarks of the Atkins' Report [WSAtkins (1994)] on the European construction industry, highlighted the demand for infrastructure in Britain and the consequent necessity to reduce the PSBR to comply with Maastricht.

The decisive event leading to the foundation of the PFI was the crash in the foreign exchange markets on 16 September 1992, known as Black Wednesday. Reductions in Government capital spending programmes were inevitable. As a consequence, Prime Minister John Major, announced a *'strategy for growth'* on 20 October 1992, to offset these reductions by encouraging private sector involvement in large infrastructure projects. This *strategy* was made formal policy and announced by the Rt. Hon. Norman Lamont (when Chancellor of the Exchequer) on 12 November 1992, in his Autumn Statement to the Commons.

The private finance initiative - a concept

The principles associated with the Private Finance Initiative in the UK are derived from the Build-Operate-Transfer (BOT) form of procurement, used widely around the world. A Special-Purpose-Vehicle (SPV) or Project Company may be formed to bid for PFI projects, this would typically consist of contractors, a service operator, financiers, and other specialist partners as required. There are two fundamental characteristics of PFI:

1. There must be a genuine *transfer of risk* to the private sector. And this supports the overriding principle that:
2. A project must provide *Value-For-Money* to the taxpayer.

It is considered that the transfer of varying risks to the private sector, coupled with their greater efficiencies in management will outweigh the higher costs of private funding, resulting in greater value. This, in essence, is the ethos behind PFI.

Critical success factors: concept and methodology

From an original theory which Daniel presented in 1961 [Daniel (1961)], the concept of CSFs was developed by Rockart and the Sloan School of Management. The phrase,

'Critical Success Factors', was first used in the context of information systems and project management [Rockart (1982)]. Rockart's definition:

> 'Those few key areas of activity in which favourable results are absolutely necessary for a particular manager to reach his or her own goals...those limited number of areas where "things must go right".'

The Critical Success Factors (CSFs) methodology is a procedure that attempts to identify factors or areas vital to the success of the industry, organisation or the individual's work. Goals are identified from the organisation's strategies and objectives, and from these factors are determined which are critical to obtaining the identified goals.

The procedure begins by conducting interviews with senior management using the 'CSF interview process' [Bullen and Rockart (1981)]. Interviews start with the interviewer outlining the concept and methodology of CSFs, the interviewee then describes the company's mission and the role which they play in the company. Following a discussion of the interviewee's goals, CSFs are developed which are designed to best facilitate the interviewee in meeting their goals. General indications are then sought as to how the interviewee would prioritise the identified CSFs before attempts are made at determining suitable measures for each CSF. The collective sets of CSFs from all interviewees in the organisation are reviewed to check for areas that some interviews may have failed to cover, this collective set of factors is then analysed to identify the general areas considered as critical for success. In Rockart's studies, the final set of CSFs was used to develop the required information data bases.

Since then the concept of CSFs has been eagerly embraced by businesses and academics alike. Many sources can be found, especially on the internet, whereby large organisations have endeavoured to develop their own particular set of CSFs to be employed as part of their corporate strategy. Similarly, a number of academics have cited the CSF methodology in research.

Earlier research by the authors attempted to adapt the CSF approach for use in privatized infrastructure projects [Morledge & Owen (1998)]. The concept of CSFs has now been further defined to identify certain weaknesses associated with the practical application of Rockart's method. Six main areas of weakness have been identified;

1. Subjectivity
2. Bias
3. Human inability to process complex information
4. Change in relation to surrounding environments; time dependency
5. Imprecise definitions; generalisation
6. Qualitative performance measures;

each is considered below.

Subjectivity

Bullen and Rockart (1981) initially identified that the CSF concept required an interviewee to respond with 'a subjective judgment arrived at only after some thought'. This was confirmed by later musings on the concept; 'Any definition of critical success factors must be, by definition, subjective' [Rockart (1982)]; '...Many information requirements are 'soft' in nature, requiring subjective or expert opinion' [Henderson, Rockart & Sifonis (1984)]. Rockart and Bullen (1981) recognised this as a failing early on and suggested that the interviewer should try to be helpful, but not directive:

> "He should be careful...to not indicate a judgement on his part...[but] should be sensitive to the possibility...On the other hand, the interviewer should be sure that he has "stretched" the manager as far as possible in his thinking during the interview. Some [interviewees], however, do not understand the concept and need a great deal of prompting. This is a situation in which one must be extremely sensitive to the problem of 'leading the witness'. The interviewer must walk the narrow line of eliciting information without creating the answers".

Munro & Wheeler (1980), however, argue that a structured interview is required to remove the subjective nature of Rockart's method. This is discussed later.

Bias

Linked to subjectivity, Martin (1982) suggested that 'intensive participation by the interviewer' during interviews, may bring about the danger where the interviewer influences the outcome. This may certainly be worrying, especially when the interview process is described as involving "a lively exchange of ideas". However, Martin suggested that questionnaires may prove a more appropriate alternative to interviews when producing a "first cut" set of CSFs.

Rockart was questioned on this issue by Munro (1983) who commented that the CSF method was not scientific, and therefore the results obtained were "in danger of reflecting the interviewer's perception". Rockart's reply was that the research was of a descriptive nature, which was necessary and unavoidable in the development of the theory. Rockart did not, however, address the potential of interviewer bias and the question was again raised by Boynton & Zmud (1984). Subsequent to Munro's argument they put forward the theory that this threat could possibly be removed if skilled analysts carried out the interviews. However, this further reduces the potential to create a transferable methodology.

Human inability to process complex information

Davis on discussing the CSF concept, identified the potential for failure from "the executive's [poor] ability to respond with CSFs that are correct, complete and sufficient" [Davis (1979)]. He suggested that an interviewee, on listing critical factors, may name some irrelevant or incorrect factors, or respond incompletely. Davis classifies this as bias, stating: "...more recent events, or those easily remembered are more likely to be recalled than events which are less recent or less easily remembered". Boynton & Zmud supported this theory a few years later, stating that 'because of a human's general limited capacity to process much complex information...[CSFs identified would] represent the most pressing concerns and the more recent events' [Boynton & Zmud (1984)].

A further consideration which Martin (1982) put forward is that some critical factors may exist which the interviewee fails to mention. This may be purely by reason that the organisation's systems are perceived to contend with difficulties adequately and thus it is never perceived as a 'critical factor'. Selection of some factors may therefore occur because an improvement in that area is required, others *should* be selected, even though they are being adequately managed, because of their critical nature to the project.

The answer, Davis suggests, may be a model of the organisation that the interviewer can use in eliciting responses and with which to evaluate factors for relevance, correctness, and completeness. The CSF method may be useful if manipulated within the framework of a model and used alongside an awareness of the limitations on human ability to identify relevant factors [Davis (1979)].

Munro & Wheeler (1980) embraced Davis' arguments and suggested that the interview or discussion be structured by the presence of goals and objectives. As a consequence CSFs would be generated in response to stimuli, i.e. goals and objectives as opposed to relying solely on the manager's limited ability to process information. Munro & Wheeler also maintain Davis' argument that a model is required. 'Unfortunately a valid analytical model of a complex business unit is seldom available' and thus much depends upon the ability of the individual manager to articulate an analytical model of the business unit [Munro & Wheeler (1980)].

Change in relation to surrounding environments; time dependency

"CSFs are related to the specifics of a particular manager's situation" [Bullen & Rockart (1981)]. Therefore, it may be necessary to adjust or customize any generic CSFs identified to fashion them into factors which are more applicable to the particular industry, company, and the individual being interviewed. CSFs will differ from manager to manager due to the subjective nature of CSFs already discussed, but they may also vary in line with changes to the particular industry's environment, the company's position, or as particular problems or opportunities arise for a particular manager.

"As economic, political, social, or competitive conditions change, the industry CSFs will change. To understand the current set of CSFs for an industry, one must closely analyze all these factors, with particular attention to the changes that are taking place" [Bullen & Rockart (1981)]. This arguably supports the acknowledged theories that the construction industry exists within various and ever-changing PEST environments [Newcombe, Langford and Fellows (1990)]. CSFs are time dependent. "Thus, even if the appropriate factors are identified, events may alter the criticality of these factors." [Henderson, Rockart & Sifonis (1984)].

Imprecise definitions; generalisation

Rockart's method of eliciting CSFs fails to identify precise definitions of factors in preference of 'a short label or expression, as opposed to a narrative statement, which effectively communicates the area of activity' [Munro & Wheeler (1980)]. This imprecision coupled with the subjective nature of the interviews, as discussed earlier, perhaps questions the real value to the organisation of those final criteria identified as CSFs. Perhaps more worrying is that from past studies, when factors *have* been submitted with a distinct definition, they have been consolidated with other, broadly similar factors into general areas, even though respondents may have reported different aspects of the general area as the key to success. So says Martin (1982), 'we have grouped the specific success factors as stated by the managers into more general areas of concern.' This may serve to dilute the usefulness of the final set of factors. What is required is a method for identifying and employing precisely defined factors, rather than identifying the general areas of concern. This would require a more rigorous method to identify factors, possibly by employing triangulation processes using two or three methods to authenticate the precise definition of each factor. This is discussed later (theories on triangulation can be found in Fellows & Liu (1997) and Sarantakos (1994)).

Qualitative performance measures

Soft factors were mentioned earlier, but the distinction between *hard* and *soft* factors becomes all the more important once they have been identified, for the next step is to determine specific performance measures for each CSF. Hard factors can be measured by more rigorous measures using quantitative methods and hard data, e.g. financial information and statistics. The opposite, soft factors, may require the use of qualitative measures, where 'the use of indirect surrogate measures that infer progress toward an objective may be considered' [Munro & Wheeler (1980)]. This is not an entirely satisfactory answer, yet it is suggested that the interviewee with greater experience of the business will have a greater understanding of the business's directions and strategies, and therefore the surrogate measures suggested will be more valuable.

Factors critical to the success of PFI projects

Several sources of references have been found for PFI and since its inception numerous conferences and seminars have been held around the UK with many speakers discussing their respective roles and involvement. However, few sources, if any, have highlighted any conclusive patterns of success, and to date there has been few PFI projects which have been 'successfully transferred'. This may create future difficulties in identifying adequate sources of data, although it is contended that 'failed' projects should be studied as well as 'successful' ones[DF1]. Case studies and accounts of previous experience from individual projects [e.g. HM Prison Service & The Private Finance Panel (1996), Highways Agency & The Private Finance Panel (1997)] have presented some combinations of data but there is a lack of conclusive focus for use on future projects.

Due to the additional risk being carried in such projects by the private sector, as compared to traditional procurement routes, those factors critical for success need to be identified, and as far as possible, quantified. Furthermore, a number of public and private sector organisations, e.g. the CBI (1996), Private Finance Panel (1996), and the CIC (1998), have suggested that to progress the PFI common key success factors must be identified from experience of PFI projects in the past. Another consideration is that the current UK government supports the Initiative, and this has served to bolster the need for knowledge. Herein is the necessity to identify those factors that are peculiar to the successful realisation of these types of projects.

Reference was made earlier to a previous study [Morledge & Owen (1998)] in which the authors used the ideas of Critical Success Factors to identify fourteen CSFs applicable to PFI. Since, this earlier study has been revisited and questioned in light of the six problems of applying the CSF methodology as discussed above. The following problems, therefore, relate specifically to critical success factors of PFI.

1. In each organisation the most experienced person in terms of PFI was sought as a respondent for the questionnaire (and this provided the 'expert' opinion), however, from this the results are in danger of producing individuals' responses and not necessarily a company response; the response may also depend upon the individual's personal interest and involvement in PFI. The phrase 'Critical Success Factor' was not defined in the questionnaire, similarly the actual CSFs were never defined fully. This left a potential for difference of opinion and interpretation, although extensive editing and piloting of the questionnaire were carried out in an attempt to eradicate different interpretations. The results fail to provide a conclusive across-the-board opinion due to a poor response from operating firms. The response, therefore focused heavily on construction, finance and public sector organisations' replies.

2. Elements of researcher bias may have entered the earlier study by one or more of the following;

   • Original factors were derived from the authors' judgement of what may or may not constitute CSFs whilst reviewing the literature, much of which was drawn from construction related sources. Bias may also have been introduced via the next step where the identified factors were grouped using the authors' subjective opinions of which factors were similar. Therefore, only those factors that were seen to be repeated in the literature were selected, i.e. one-offs were discarded.

   • Although quite a random selection of organisations was targeted for questionnaires, only those companies identified from reading literature (and so companies known for their involvement in PFI projects) were used. Conclusively, the researchers failed to question those organisations not involved and so failed to discover why those companies were not involved and thus what may further be required for success.

   • Although respondents were given the opportunity to submit their own CSFs, only three did, and none of these new factors could be generalised and so were discarded from further consideration. The final CSFs therefore reflect the response to the authors' original submission of factors.

3. Most of the fourteen CSFs in the study reflected the problems and moods at the time the literature was written. Consequently, a number of those factors submitted are now out of date because of new legislation or improvement in processes and understanding. An interesting point to note is that a major review of PFI was carried out by the British government after the authors' original study had concluded, and a number of the authors' CSFs were reflected in the conclusions to the governmental review only a few months' later. Further evidence that a number of those factors were a result of the prevailing moods and problems.

4. From the literature review, similar factors were grouped accordingly, thus producing 34 general statements to encourage, it was thought, a clearer understanding by the respondents, and thus a useful return of data. However, as in Rockart's studies, this data is then based on general areas of concern, rather than specific factors, so creating problems of imprecise definitions in the data.

Improvements to the process

In order to strengthen the methodology first proposed by Rockart, the issues and weaknesses identified by a number of authors have been identified and some potential solutions proposed. The following is a basic summary of those points mentioned in the discussions above.

- On the point of subjectivity the researcher must be sensitive to the potential of leading the interviewee, this may be improved also with some structured, some unstructured interviews. From the experience of the pilot study, questionnaires are considered too dislocated and complex for eliciting such factors.
- It was suggested that using a skilled analyst would benefit the process, yet it is felt that the skill of the analyst must remain in the filed of contention, that is the particular industry being studied, else the methodology becomes very difficult to transfer.
- The use of a model was proposed to elicit initial responses and further to provide measures. However, as the authors' study relates more to a process, a new method of procurement, it is submitted that the model be based upon the PFI procurement process with measures for *soft* factors yet to be defined.
- Any study in the construction procurement field must acknowledge the ever-changing PEST environments as does this research.
- The use of triangulation to precisely define factors in order to limit the vagueness of using a general statement potentially misconstrued by interviewees.
- Adequate qualitative performance measures must be defined and this is thought to be possible via structured interviews and by way of an expert panel.

Furthermore, the research is to be triangulated using a wide and varied literature review, case studies and interviews, and an expert panel to validate factors and provide a weighting of importance to each factor (however, a robust method for this has not yet been agreed upon).

Critical success factors and 'Failure Reduction Criteria'

It is contended that factors identified as critical may fall into one of two categories; (a) necessary for success, but not critical, and (b) critical for success. Consequently, some factors, inherent in the project, may reduce the chance of failure but will not increase the chance of success. Other factors, if included, may increase the chance of success but if left out will increase the chance of failure. The former could be labelled as 'Failure Reduction Criteria' (FRCs), the latter as Critical Success Factors (CSFs). The fundamental question at this point, concerning FRCs/CSFs, is whether failure will always be the exact and pure opposite of success. Table 1 is an attempt to indicate how the identified factors could be divided into FRCs and CSFs in practice.

Table 1 A speculative view of how factors may be divided into
Failure Reduction Criteria and Critical Success Factors

Factor	FRCs	CSFs
Provide value-for-money		★
Risk Transfer:	★	★
Design & Construction		★
x% Return on investment	★	
x+y% Return on investment		★
Reimburse bidders	★	
Entrepreneur		★

(x and y are used purely as indicative as a potential measure to indicate importance of return ratios)

Conclusion

The commitment of the UK government to privatize infrastructure is persistent, yet because of the additional private sector exposure to risks inherent in these projects, it is vital to identify factors critical to success; and thus satisfying the two principle rules (transfer of risk and VFM). The original concept of critical success factors is considered to be robust, but the method of application which Rockart promoted contains a number of imperfections. These have been identified and discussed with reference to the authors' previous studies. Some improvements have been suggested, others weaknesses require further thought. The CSF concept has been developed and a distinction made between two potential types of 'success' factor. These are labelled Failure Reduction Criteria and Critical Success Factors. The continuing research programme is now seeking to develop a methodology that will better enable implementation and consequently improve predictability of success and assessment of risk.

References

Boynton & Zmud (1984) An assessment of Critical Success Factors. *Sloan Management Review*. Summer 1984, pp. 17-27.

Bullen & Rockart (1981) A primer on Critical Success Factors. Sloan Working Paper No. 1220-81. June 1981. Sloan School of Management. Massachusetts Institute of Technology.

CBI (1996) Private skills in public service - tuning the PFI. Confederation of British Industry.

CIC (1998) Constructors' Key Guide to PFI. Construction Industry Council. Thomas Telford. London.

Daniel (1961) Management Information Crisis. *Harvard Business Review*. September-October, pp. 111-121.

Davis (1979) Comments on the Critical Success Factors method for obtaining management information requirements in article by John F. Rockart, "Chief Executives Define Their Own Data Needs," Harvard Business Review. March-April, 1979" '. *MIS Quarterly*. September 1979, pp. 57-58.

Davis (1980) Letter to the Editor. *MIS Quarterly*. June 1980, pp. 69-70.

Fellows & Liu (1997) Research Methods for Construction. Blackwell Science. p. 20, 95.

Hancock (1995) The Private Finance Initiative. *Proceedings of ICE Municipal Engineer*. Vol. 109, December, pp. 278-283.

Henderson, Rockart & Sifonis (1984) A Planning Methodology For Integrating Management Support Systems. Sloan Working Paper No. 1591-84. September 1984, p 5. Sloan School of Management. Massachusetts Institute of Technology.

Hibbs & Roth (1992) Tomorrow's Way: Managing roads in a free society. Adam Smith Institute. London.

Highways Agency and the Private Finance Panel (1997) DBFO - Value in Roads: A case study on the first eight DBFO road contracts and their development. March 1997. Published jointly by the Highways Agency and the Private Finance Panel.

HM Prison Service & The Private Finance Panel (1996) Treasury Report on the Procurement of Custodial Services for the DCMF Prisons at Bridgend and Fazakerley. April 1996. Published jointly by HM Prison Service and the Private Finance Panel.

Martin (1982) Critical Success Factors of chief MIS/DP executives. *MIS Quarterly*. June 1982, pp. 1-9.

McAlpine (1995) Lord Alistair McAlpine: Premature Delivery. *Building Magazine*. 16 June 1995, p 32.

Morledge & Owen (1998) Critical Success Factors in PFI projects. *Proceedings of Fourteenth ARCOM Conference*. September 1998, Reading.

Munro & Wheeler (1980) Planning, Critical Success Factors, and management's information requirements. *MIS Quarterly*. December 1980, pp. 27-38.

Munro (1983) An opinion...comment on Critical Success Factors work. *MIS Quarterly*. September 1983, pp. 67-68.

NEDC (1981) Report of National Economic Development Council Working Party on nationalised industries' investment. NEDC. London.

Newcombe, Langford and Fellows (1990) Construction Management, Volume 1 organisation systems. Mitchell. London.

Plender (1996) Plender's perspective: When the silver is all sold off. *Estates Gazette.* June 15 1996, issue 9624 p 54.

Private Finance Panel (1996) Private Finance Initiative - Guidelines for smoothing the procurement process. April. Published jointly by HM Treasury and the Private Finance Panel.

Rockart (1982) The changing role of the information systems executive: A Critical Success Factors perspective. *Sloan Management Review.* Fall 1982, pp. 3-13.

Sarantakos (1993) Social Research. Macmillan. pp. 155-156.

Walker & Smith (1995) Privatized infrastructure: the BOT approach. Thomas Telford. London.

Willetts (1993) The opportunities for private funding in the NHS. Occasional Paper. Social Market Foundation. London.

WSAtkins (1994) Strategies for the European Construction Sector, a programme for change. A report compiled for the European Commission by consulting engineers, WSAtkins, for the EEC.

RISK MANAGEMENT OF BOT PROJECTS IN SOUTHEAST ASIAN COUNTRIES

C. M. TAM and ARTHUR W. T. LEUNG
City University of Hong Kong, 83 Tat Chee Avenue, Hong Kong.
BCTAM@cityu.edu.hk

Abstract

Over the past decades, the economic growth of some Southeast Asian countries has placed an unprecedented demand on their existing infrastructure. At the beginning of this decade, most governments started to realize that the inadequate infrastructure in this region would suffocate their economic growth. However, governments in these countries experienced strain on their financial spending; a large portion of which has been spent on social welfare for fulfilling the rising quality of life and the development of democracy in the region. The BOT arrangement in infrastructure development has suddenly become a solution for which most governments made a dash.

However, some unsuccessful BOT projects tell us that BOT is not a sure win business. It involves three different levels of risk: technical, financial, and political. Amongst which, technical is the most easy to manage; financial, although a bit harder, is still manageable; political is really the most difficult risk element to handle. The BOT franchisees must properly manage all these three risk elements to avoid burning their fingers in the developments.

This paper presents the risk elements in most BOT projects in the region and suggests ways to produce a win-win strategy in completing the projects.

Keywords: BOT, infrastructure development, risk management, win-win strategy.

Introduction

Traditionally in the Southeast Asian region, the public sector has financed the construction of the infrastructure. However, governments in these countries experienced strain on their financial spending; a large portion of which has been spent on social welfare for fulfilling the rising quality of life in the region. The BOT arrangement in infrastructure development, a private participation scheme without demanding public money, has suddenly become a solution for which most governments made a dash.

However, some unsuccessful BOT projects alert us that BOT is not a sure-win business. It involves three different levels of risk: technical, financial, and political. Amongst which, "technical" is comparatively the easiest to manage; "financial", although a bit harder, is still manageable; "political" is really the most difficult risk element to handle. To avoid burning their fingers in the development, the BOT franchisees must properly manage all these risk elements.

This paper uses a few BOT projects in the region to illustrate how "politics" can determine their success or failure and suggests ways to manage it.

BOT experience in Hong Kong

Before discussing the BOT examples, the explanation of the political systems and the corruption status of the Hong Kong government can illustrate a good contextual background. The political systems in Hong Kong have remained rather stable since its colonization until under the regime of the last British governor who started to introduce the Western democracy. Before that, there was hardly any trace of democracy affecting the decision making process of the government. After the return of sovereignty, the Basic Law has preempted the political structure of Hong Kong. Hence the Democracy Movement was rather slow and the interest of businessmen has been well protected by the system. In recent years, the pursuit of democracy has been growing vigorously. The progress of democracy in the territory, however, has to depend on the situation in China, which is in a very different stage of development.

As regards corruption, Hong Kong in the 1960s and early 1970s experienced massive economic and population growth, coupled with rapid urbanisation. At the same time, bribery and extortion became a way of life. Bribe-paying to public officials and secret commissions in the private sector were considered customary. In the police, the collection and distribution of bribes had become institutionalised.

Within the general growth of Hong Kong, a new citizenry, which was young, educated and concerned with public affairs, was emerging. An important event in 1973 provided the spark that ignited this new community's demands for action from the Government. Peter Godber, a police superintendent under investigation for corruption, fled to the United Kingdom.

The community was enraged. A Commission of Inquiry under a High Court Judge was set up in the same year to examine the situation of corruption in Hong Kong and make recommendations. A principal recommendation was to establish an independent, powerful agency to deal with widespread corruption. As a result, the Independent Commission Against Corruption (ICAC) was established in February 1974. It was independent of the police and the rest of the Civil Service; and its Commissioner was directly responsible to the Governor (now to the Chief Executive). Since then, the ICAC has been committed to fight corruption through effective law enforcement, education and prevention to help keep Hong Kong fair, just, stable and corruption-free.

a) Hong Kong Cross-Harbour Tunnel

The concept of privatization of infrastructure has been risen since the 18th century (Walker and Smith, 1995), but the evolved concept of BOT was only firstly introduced in 1970's. The first project adopting BOT was the Cross-Harbour Tunnel in Hong Kong. The tunnel was conceptualized in the 60's and built in the 70's. By then, the Hong Kong government lacked capital and technology required. Hence she decided to contract out the project on a BOT basis to a private developer. The concession period was 30 years. As this was the first project managed in the BOT mode and the government lacked confidence on letting the project 100% owned by the private sector, she has thus owned the project by around 20% in equity in order to exercise certain degree of control.

Since the concession period of 30 years included the construction time, the system had a built-in motivator for expedition. The tentative construction schedule of the tunnel was set at three years; however, it was completed in two years time. The estimated traffic flow was estimated to be 70,000 vehicles but turned out to be 124,000 vehicles per day. In coping with the unexpected increase in traffic flow, the tunnel management has implemented a number of innovations, including a traffic controlling system at the entrances to smoothen the traffic flow, the use of electronic toll collection systems, etc. In order to shorten time delay due to vehicle breakdown, the operation crews always help voluntarily changing flat tires.

The project provides a very high rate of return and can be classified as "very successful" in terms of return rates, speed of construction, charges, quality of services and management. The toll rates have not been increased since its operation in 1972 although the government has imposed a 100% tax to discourage road users to use the tunnel due to traffic congestion. The above demonstrates the superiority of BOT over government-owned projects.

b) Eastern Harbour Crossing

The tunnel was let on a BOT basis. The concession period is again 30 years and the construction was started in September 1986 and expected to complete in three and a half years time. Similar to other BOT projects in Hong Kong, the project was completed half year earlier and within budget.

As the government has acquired some experience in running BOT projects, its share of equity is reduced to 7.5% in this project. Alike other BOT projects by then, there was no agreement on the average rate of return. The toll rates and any toll increase in future need to be approved by the Executive Council by considering the average rate of return in the region. However, as the pace of democracy advances, the toll increase decisions were often hindered by the elected legislative council members whose prime concern is their constituents' interest. In fact, since the official opening of the tunnel in 1989, toll rise applications have been delayed till 1997. Fortunately, the BOT agreement contains an arbitration clause and the franchisee has successfully acquired a toll increase by going arbitration in 1998. This demonstrated the importance of an established and equitable legal system for BOT projects.

c) Western Harbour-Crossing

Construction of the tunnel was started in August 1993 and expected to finish in July 1997. Alike other BOT projects, the tunnel was completed earlier as at April 1997 and within budget. In this BOT project, a new contractual condition has been implemented and the government did not need to have any equity to monitor it operations.

In the former BOT arrangements, the toll rates, time for toll adjustment, and the degree of adjustments have to be agreed by the sponsors and the government. The agreement, then, needed to be endorsed by the Legislative Council. The rate of return should be the sole consideration factor in each toll rise negotiation. However, the Legislative Council would tend to protect the public. Hence, no agreement could be reached in most of the time. For instance, the toll rise negotiations of the Eastern Harbour Crossing and the Tate's Cairn Tunnel have been dragged for years. This arrangement puts the investors at risk.

The resumption of Hong Kong's sovereignty in 1997 has weakened the negotiation power of the government and forced the government to accept an alternative in toll rise arrangements. The new BOT agreement spells out the minimum and maximum annual income, the time periods of toll adjustment and the degree of toll rise. A government-managed toll stabilization fund will be set up to collect the excessive annual income which can be injected back to the project to delay or avoid toll rise. If the annual income falls between the maximum and minimum range, toll rise will be automated as scheduled. This forms an automatic toll adjustment mechanism to avoid political influence on business decisions. This mechanism has been applied to the two recent BOT projects including the Western Harbour Crossing and the Route 3 Country Park Section.

Negative BOT examples

Thailand has experienced an unprecedented economic growth in the past decade. From 1990 to 1995, an average of 8.5% growth in GDP was recorded (Asian Development Bank, 1996). Expansion has been driven mainly by exports and investments since the restructuring of the economy from an agricultural base to a relatively high level of industrialization. As a result, many "bottlenecks" have been created due to lagging behind in the provision of infrastructure and utilities to satisfy the demand fueled by the investment boom (Srisanit, 1994). This can be shown that traffic jams are legendary in Bangkok. Without any mass transit systems, private vehicles of all shapes and sizes pack the streets of this capital city. However, investments in roadwork ask for tremendous amount of money, which is unrealistic to rely on the government budget to fund. Hence, starting from 1986, many infrastructures in Thailand were developed by the private sector and most of them were in BOT contracts.

However, the political instability in this nation seems to be a major obstacle in the use of BOT. Ogunlana (1997) affirmed that the average length of the last seven Thai governments lasted for about 1 year. In addition, each Thai government was a fragile coalition of sometimes more than five political parties, each with its own agenda. A new government always wants to impress its constituents by seeing contracts signed by the previous government as targets for attack. This kind of political atmosphere has incurred extra risks to the BOT investors whose agreements usually span for 30 years.

In the past, BOT sponsors tended to focus, almost exclusively, on risks of commercial and technical aspects, which has been proven to be very dangerous for transportation projects in Thailand. As politics in developing countries is rather unstable. It is more common for the change of government. If the new government is unwilling or unable to meet the contractual obligations, the project promoters will fall into deep trouble.

From an interview with the Chairman of Hopewell Holdings Ltd., Sir Gordon Wu, he admitted that the greatest risk was the political stability and government supports were the most important factor to the success of BOT projects in developing countries.

The following describes the three projects that have fallen into trouble:

Bangkok elevated transport system (BETS)

The core of the plan was to construct a 60-km elevated rail system and a road through the heart of the capital. Hopewell (Thailand) Ltd. was selected to operate the project under a BOT contract, intending to collect revenues from toll income for 30 years as well as gaining the right to develop 900,000 square metres of land along the proposed route (South China Morning Post). This network would be built on State Railways of Thailand's (SRT) right-of-way land and divided into two main lines (South China Morning Post).

Although Stage 1 was supposed to be completed by the end of 1995, only a few piled foundations have been erected at the end of 1997. The project was ultimately terminated by the Thai Government.

Problems:

1.   **Land leasing cost**

     An interview with Sir Gordon Wu reveals that the major reason interrupting the project progress lied upon land acquisition. From Hong Kong Mass Transit Railway's experience, substantial profits of the project could be generated from property development along its route. The scheme will provide space for shops and offices along the transport network. However, when Hopewell negotiated with the SRT which provided land for the development, SRT demanded for 681 million Baht (US$26 million in 1993) for a plot of land which was near Bangsue in northern Bangkok, half of the amount was said to compensate SRT's loss.

2.   **Changes of government**

     After signing of the contract in 1990, Thailand has experienced several changes of government. They include a coup, two controversial elections and the ousting of a military junta (Levy, 1996). These have seriously hindered the progress of the project. From going contract till termination, there have been a total of eight different governments.

3.   **Changing construction scheme from elevated to underground**

     In mid-1993, the Thai government suddenly requested the whole or part of the project changing into an underground system, although the government made a promise to compensate for the extra costs (Levy, 1996). The move was made because the politicians wanted to please the public due to the growing public concern on the many elevated transport systems that would destroy the city's charm. The negotiation has wasted much time.

4. Tollway intersection

The Don Muang Tollway's north-south line runs parallel to the project and cuts into the project's space at some intersections, creating some conflicts between the two parties. However, the government seemed reluctant to do anything to help.

Second expressway system (SES)

Almost at the same time, a Japanese-Thai joint venture sponsored the development of the BOT expressway. The SES was planned in 1982 with the objective of complimenting the successful First Stage Expressway System (FES). With the completion of the FES in the 1980s, the government sought to accelerate the construction of the second stage of the transportation plan. The SES would connect to the FES and form a comprehensive road network for Bangkok residents. Apprehending shorter development periods of most BOT projects in the region, the government strongly favoured the fast track nature of BOT arrangements. Bangkok Expressway Company Ltd. (BECL) was selected among two tenderers in 1988.

Problems:

1. **Toll increment:** As agreed, ETA was required to raise the toll to 30 Baht for 4-wheel vehicles when the combined FES and SES started to operate. However, the Thai government in 1993 decided to lower the increase to 20 Baht due to the strong public objection. BECL regarded the failure to increase the toll to 30 Baht as a breach of obligation under the SES Agreement which has set a negative image to investors as the project was the first major BOT scheme in Thailand.

2. **Suspended loan:** BECL was found default on interest payments for loans already given. That might be due to the serious delay of the project. Further, the revenue sharing agreement between ETA and BECL was not clear. Hence the banks have suspended loans to BECL. BECL has then demanded an amendment to the contract since the toll was not set at 30 Baht. The banks preferred suspending loans until the amendment was made clear to all parties (Ogunlana, 1997; South China Morning Post).

3. **Right to operate:** ETA claimed that it had the sole right to operate the new expressway[5] and contended that there was no stipulation in the contract that the BECL had the right to operate the system for 30 years under the revenue-sharing formula. BECL, however, argued that since the project was based upon a standard BOT arrangement, they should have the right to operate the system.

4. **Toll sharing:** Under the BOT contract, ETA was required to acquire all land needed for the project and deliver them to BECL although without guaranteeing the timely delivery. As stipulated in the contract, revenue sharing would only start upon the physical completion of the project. However, there was a delay in delivering a section of land due to local residents' protests. Although the substantial parts of the project have been completed and used, the whole project cannot be physically completed and BECL was denied for toll sharing (Ogunlana, 1997; Levy, 1996)

5. **Cost escalation:** ETA was unable to deliver all the land required to BECL at the start. In fact, only 1% of its original hand-over target was achieved by 1 March 1990. This resulted in higher construction costs for the first phase of the project. Although the contract comprised a clause that additional costs due to delay not caused by BECL, such as late delivery of land, change in design, would be borne by ETA, the authority refused to pay (Ogunlana, 1997).

6. **Toll collection:** ETA argued that the project was a joint venture between the authority and BECL. As such, ETA could legitimately collect the toll revenue. Further, ETA insisted on supplying workers to man the booths but BECL must pay their wages.

When the negotiations on the above disputes broke down, BECL and ETA intended to assert their positions in court. Finally, under the strong influence by the host government, the leading venture partner of BECL, Kumagai Gumi, was forced to sell its shares to a local military contractor in March 1994 to settle the dispute (South China Morning Post).

Don Muang Tollway

Meanwhile a French-German-Thai consortium has fallen into trouble in another BOT tollway project: the Don Muang Tollway. The expressway was an elevated road link between the central part of Bangkok and the Don Muang International Airport. At the end of 1988, the Ministry of Transport and Communication, via the Department of Highways, called tender for the implementation of an elevated road. The tender was accompanied with a few guidelines for the design of the road and without other details. Bidders were required to finance the project on a non-recourse BOT scheme. The project was ultimately contracted to Don Muang Tollway Co. Ltd., comprising Dywidag (a German firm), Delta Construction (a Thai construction firm) and GMI (a French contracting company). The Tollway was granted a 25-years concession to build and operate the project.

Problems:

1. **Debt/equity ratio:** According to the original agreement, the developer would contribute 20% equity in foreign currency and 80% loans in Thai Baht. However, shortly after the concession agreement was concluded, the Thai government amended its rules for foreign investment which required foreign investors to provide both equity and loans in foreign currency. Might this rule be enacted earlier, investors would lose interest in the development as the revenue would be in Thai Baht and subject to exchange rate fluctuations. Finally, the shareholders had to increase the equity to 25% and secure additional privileges from Thai government in order to persuade lenders to proceed with the project (South China Morning Post).

2. **Competing project:** There was a consensus that no competing projects nearby would be approved by the government. However, soon after the award of the contract, Hopewell (Thailand) Corporation was granted the concession to build a combined tollway and train system less than 100 m from the project (South China Morning Post; and Levy, 1996).

3. **Removal of overpass:** To ensure the financial viability of the project, the Thai government agreed to dismantle two overpasses in Viphavadi-Rangsit (Ogunlana, 1997). The overpasses were directly under the elevated tollway at two busy intersections. It was necessary to demolish them to make road users to use the elevated tollway. By the time the tollway was being built, the Thai government had changed several times. The new government considered that dismantling the two overpasses would be regarded as unfair to road users. The problem remained unsolved for nearly two years after the tollway was opened (South China Morning Post).

As a result of the above, the tollway was reported losing 1 million Baht daily over two years and the developer could not meet the loan re-payment schedule.

Conclusions

From the above lessons, if the host governments and the investors intend to arrive at a win-win scenario, BOT projects must be operated under the following conditions:

1.   First, the project must offer a reasonable rate of return in order to attract private investments. The host government must guarantee a proper business environment, such as no competing projects nearby for a period stipulated in the contract.
2.   There must be a proper mechanism to fix and adjust toll rates which should be free of political influence in the decision making process.
3.   The projects need a reliable, committed and strong development consortium, which operation will not be affected by any short-term sufferings. In Hong Kong, most BOT franchisees are large, reputed and financially sound corporations.
4.   BOT projects need a strong and technical competent construction consortium, which can guarantee a timely delivery of the project. They have the know-how and resources to complete the project on time and according to the quality standards.

The above are important in achieving successful BOT projects. However, without an equitable and experienced government authority, BOT projects would fall into trouble. Hong Kong has gained her experience through years of managing BOT projects. The legal framework and contractual conditions have matured so that they protect both parties' interest. An equitable legal system can assure investors that any disputes can be resolved through litigation and make sure that both parties will respect the contract. Lastly but most importantly, an uncorrupted government is the key factor in the success of BOT.

References

Walker, C. and Smith, A. J. (1995) Privatized Infrastructure – The BOT approach. Thomas Telford, London.
Asian Development Bank Report, (1996) International Financing Review.
Srisanit, A (1994) "Some Issues in Signing Construction Contracts with the Government of Thailand", *International Construction Law Review*, 11, July, 344-352.
Ogunlana, S. O. (1997) "Build-Operate-Transfer Procurement Traps: Examples from Transportation Project in Thailand", *Proceedings of the CIB W92 Symposium on Procurement, May*

South China Morning Post (SCMP) articles
Levy, S M, (1996) Build, Operate, Transfer: Paving the way for tomorrow's infrastructure.
    John Wiley & Sons, Inc.

7. Culture

TOWARDS A CULTURE OF QUALITY IN THE UK CONSTRUCTION INDUSTRY

D. E. SEYMOUR
University of Birmingham, United Kingdom.

R. F. FELLOWS
Department of Architecture and Civil Engineering, University of Bath, Bath, United Kingdom. absrff@bath.ac.uk

Abstract

This paper is derived from initial, on-going research into cultural issues which impact on the provision of quality during the construction phase of projects. Current trends in seeking quality in construction projects are advancing in focus from Quality Assurance under ISO 9000 / BS 5750 to Total Quality Management. However, as Japanese industry demonstrates, provision of Quality requires continuous improvements grounded in culture and founded on practices conducive to such changes. The research investigates progress towards the development of a 'Culture of Quality' in construction firms. Formal questionnaires, interviews and participant observation are employed. The methods are, necessarily, intensive to facilitate comprehension of how beliefs, value structures and practices of personnel affect the form of and extent to which a Culture of Quality is developing. Case study projects provide live, longitudinal data for examination of the situations and elements of changes. Results (are intended to) demonstrate consequences of quality-oriented changes, to whom such consequences accrue and highlight areas and mechanisms for potential improvement.

Keywords: Behaviours, change agents, culture, performance, quality.

Background

In many countries, construction industries are becoming increasingly concerned with quality. The concern is prompted from both within and without construction based organisations - designers, constructors regulators and 'client functionaries'. There is also good evidence that the full implementation of Total Quality Management (TQM) increases competitiveness and customer satisfaction, reduces waste and improves the working lives of employees. Following Latham (1994), this research postulates that similar benefit may accrue to the construction industry and those it serves so long as it can break the vicious circle of mistrust, conflict and waste - the Culture of Confrontation - and replace it with a Culture of Quality.

Quality

To date, attention has tended to focus on Quality Assurance under the ISO 9000/BS 5750 provisions, despite acknowledgement by many (including the British Standards Institute - BSI) that QA does not, of itself, raise, provide and ensure good quality but does ensure realisation of the specification - i.e. provides confidence. Following the growing number of QA registrations, and the increasing requirement of such by clients, attention is progressing towards TQM. There are several consequences of this trend. First is the recognition of issues in the securing/providing of quality, notably concerning roles, responsibilities and behaviours of people - both individually and collectively - and the essential need to acknowledge such impacts and to take them into account in devising quality provisions. Second, quality is a strategic issue and that a culture conducive to its achievement must be fostered from the top through strong and supportive leadership.

Pirsig (1974) catalogues approaches to quality consequent upon the prolonged and extensive difficulties of definition. The typology of approaches consider quality as a presence of features of a product or service - including 'fitness for purpose'/functionality, appearance, maintainability, comfort - whilst others consider quality as being an absence of negative factors - including discomfort, ugliness. Especially in industries such as construction in which the supply process is complex and long term, attention is given to the quality of that process – procurement; this accords with a major focus of marketing theory – expressive performance in relationship marketing (e.g. Gronroos 1991).

Given increased maturity and sophistication of an industry, accepting that most construction is of relatively 'low' technical complexity, aspects of quality in setting targets for and in evaluating performance increase in importance. Such importance is manifested further in speculative activities and most pronounced during recessions when competition on the more tangible quantified variables of price and time is keenest.

Perspectives on quality may be aligned with major perspectives on management:
Theory X: quality must be demanded, specified and enforce.
Theory Y: the situation is what may inhibit the provision of quality.
Theory Z: involvement of the total workforce, by removal of fears etc. facilitates the realisation of quality.

The perspectives raise issues of how to secure quality. Indeed, acceptable quality may be regarded as a 'window' sufficient for the 'customer(s)' in the context of other performance targets – i.e. realising the specification as a minimum (given that setting the specification as the quality performance target has been carried out similarly and suitably) but constraining further provision to avoid affecting own performance requirements (notably immediate profitability) detrimentally.

Studies of quality on sites indicate that unacceptable quality is caused more by operatives lack of care than by lack of skills (Bentley 1981) (despite lamentations of de-skilling) but that managerial issues are, by far, the greatest impediments to achieving quality. The Building EDC (1987) noted that the overwhelming reason for poor quality was unclear or missing project information and continued 'contractual arrangements, as such, have little effect on the quality achieved but managerial structures have considerable influence'.

Bresnen et al (1990) support Bentley (1981) in that a good project working environment is important for achieving quality, notably a consultative approach to problem solving. The opposite resulted on sites exhibiting strong 'contractual behaviour'. "Quality achieved was far more the result of the total 'system' of individuals and tasks than it was of formal checking" (Bentley 1981). Thus, as the informal system is, generally, of much greater effect than the formal system (Higgin and Jessop, 1965; Higgin et al 1966), the contractual provisions provide a context/framework only for performance; one which is culturally grounded. Further, culture impacts through the informal systems' operation.

Seymour *et al* (1997) examined the cover of in-situ concrete in relation to that specified for a variety of structural components. Specifiers' practice involves statements of nominal cover which is subject to tolerance; the tolerance according to standards which have been derived from surveys of covers achieved. An inherent assumption is that such covers accord with a reasonably consistent pattern of distribution. However, in many cases, the covers prove to be unachievable due to the vast variations in site conditions for the operations involved. The result is problems of quality on site due to the use of global data in the design process without due regard to site conditions for operations and adequate (assurance) verification of what is reasonably achievable.

BS4778 (1987) defines Quality Assurance (QA) as, 'all those planned and systematic actions necessary to provide adequate confidence that a production service will satisfy the given requirements for quality.' Further definitions are provided in the various standards for quality, nb ISO 9000 series, covering quality planning, control etc etc and Total Quality Management (TQM). The emphasis on QA reflects lack of confidence/trust of/in the industry (Latham 1994) and hence, the perceived need to employ stringent control mechanisms to supplement (contractual) sanctions against 'failures'. Gunning and Leonard (1994) report that the reasons for companies seeking QA certification are, primarily, 'market advantage' and 'client requirements' – to a much greater extent than any desire to improve quality of project/service. Consequently, the benefits claimed are mainly market oriented, notably enhanced reputation and increased clientele.

Shadgrove (1994) assesses the advantages which accrued to Taylor Woodrow, a major UK based contractor, of using BS5750 to be that reporting lines are clearer, authority/ responsibility are defined, transfer of information is controlled, responsibility for supervision is clear and records are improved. Clearly, the advantages are a reflection of the organisations' adopting a visible management structure with known and followed procedures rather than a quality focus, per se!

Cole and Moyab (1995) note that a firm which implements best practice form others without an internal change of (paradigm) approach is doomed to be follower. This may occur due to impositions of requirements (e.g. QA certification) and/or through benchmarking practices. The situation of a follower is notoriously poor (Peters and Waterman, 1982). Thus, to gain from implementation of best practice, a cultural change is likely to be required – from the individual transaction, short term orientation of many businesses, notably those of Western market capitalism subject to the imperatives of meeting the liquidity requirements of investors, especially investment institutions (Hutton 1996), to the longer term orientation of continuous improvement firms (CIFs).

The successful implementations of quality provision have shown that the philosophies of, eg., Deming (1986) must inform the culture of all contributors to the value chain. The requirements of Deming's approach include involvement of the entire workforce, from a position of security, to aid continuous improvement of supply (product and process) in order to 'get it right first time, every time', coupled with continuity of advances in (quality) attributes sought by actual and potential customers - i.e. continuous improvement. Thus, checking is reduced, as is reworking, by (virtual) elimination of defects and, complementarily, encouraging Just-in-Time (JIT), cost-saving, inventory management and product/process enhancement. Deming's influence on Japanese management is acknowledged widely but its foundation lies in appreciation of the statistics of control. That provision of high quality requires an all-pervading commitment, necessitates the involvement of every member of the value chain at the corporate, project and individual levels; as manifest in exemplars such as Hewlett Packard, Toyota and Marks and Spencer.

Japanese structures - organisational and societal - lend themselves readily to adoption of the approach to quality provision advocated. The deep-seated involvement of each member of the value chain coupled with long term perspectives and security of employment plus the view of striking a bargain (e.g., Fellows, 1992) have provided fertile ground for rapid and major advances in outputs' quality. Today, with advanced technical sophistication and globalisation, emerging issues are non-price competition, customer involvement, broader perspectives of value, efficiency in employment of overheads' and the impacts of people (both demanders and suppliers; actual and potential).

Culture

Perspectives of what culture is range from the concept of how things are done to the definitions of culture (and, more particularly, organisational culture) -

'... patterns, explicit and implicit of and for behaviour acquired and transmitted by symbols, constituting the distinctive achievement of human groups, including their embodiment in artefacts; the essential core of culture consists of traditional (i.e. historically derived and selected) ideas and especially their attached values; culture systems may, on the one hand, be considered as products of action, on the other as conditioning elements of future action.' (Kroeber and Kluckhohn, 1952).

'... the collective programming of the mind which distinguishes the members of one category of people from another.' (Hofstede, 1980).

The notion of culture is one of groups - macro and micro (generating 'sub-cultures'). The differentiation of groups can result in 'managerial difficulties' - the issues concerning construction clients and their relations with the TMOs on projects (Cherns and Bryant, 1984). Thus, the nature of an individual involves consideration of character and personality (both shaped by and shaping culture(s)) whilst culture concerns collectivities.

Patterns of behaviour, artefacts and symbols may be observed and, in many cases, measured. Further manifestations of culture are heroes, icons and, most important, language.Even such 'hard' embodiments are interpretations of a culture and subject to interpretation by (an)other culture(s). Hence, manifestations of cultures are communication devices which are subject to considerations of effectiveness and efficiency of communications - notably, that of indexicality (Clegg, 1990) in which interpretation is acknowledged to be dependent upon socialisation; including education and training.

Generally, culture is acknowledged to be rooted in people's minds - their ideas, beliefs and values. Beliefs lie at the core and become hierarchically - ordered into a value structure which underpins behaviours, thereby creating the other manifestations of culture.

In researching national cultures, Hofstede (1980) concluded that appropriate dimensions were: Power Distance, Uncertainty Avoidance, Masculinity/Femininity and Collectivism/Individualism. A fifth dimension of Long-Termism/Short-Termism was added later, notably as a result of subsequent investigations in Asia (Hofstede, 1994).

Trompenaars (1993) advanced five value-orientational dimensions [Universalism - Particularism (rules-relationships), Collectivism - Individualism (group-individual), Neutral - Emotional (feelings expressed), Diffuse - Specific (degree of involvement) and Achievement - Aspiration (method of according status)] for examining cultures which he suggested, '... greatly influence our ways of doing business and managing as well as our responses in the face of oral dilemmas'.

The various sets of dimensions have much in common (perhaps, because many analyses have their roots in relatively similar cultures of developed, western countries). However, they do reinforce the requirement for determining the boundaries of cultural groups (communities).

People are believed to act in ways which they consider to be desirable - i.e. to give the results most satisfying to that person (the foundation of utility theory) - i.e. to yield most benefit according to the value structure. England (1975) suggested three primary orientations (from a Personal Values Questionnaire - PVQ) to predict behaviours [Pragmatic - behaviour accords with concepts believed to be <u>important</u> and <u>successful,</u> Moralistic - behaviour accords with concepts believed to be <u>important</u> and <u>right</u>, Affective - behaviour accords with concepts believed to be <u>important</u> and <u>pleasant</u>]. Other value orientations include Low context - High context communications (Hall, 1959) and Free Will - Collectivism (Kluchhohn & Strodtbeck (1961).

People are not free to act in ways which they perceive to be most beneficial. Constraints are imposed to yield a decision environment of bounded rationality (Simon, 1960) through the particular situation and norms of behaviour - both explicit (employer's dictates, professional institution rules of conduct, law) and implicit (moral codes etc.). Further, not only decision goals but many of the decision constraining factors may encourage opportunistic behaviour (seeking to take 'unfair' advantage) - the presence and potency of which is likely to be a function of the prevailing culture and, as such, will impact on all situations involving governance of relationships (e.g. via contracts) and/or negotiations. If management may be regarded as the human activity of making and implementing decisions concerning people; appreciation of culture is central to successful management.

The term 'culture' is used increasingly commonly; in many cases it is coupled with a desire to alter a situation or process resulting in attention to cultural changes (or change in/of culture) - almost implying a set of readily available cultures with an ease of rapid interchangeability. The majority of (active/interesting) management, if not all management, involves change; appreciation of both cultures' manifestations and foundations is powerful in determining appropriate changes towards the desired outcomes.

The nature of construction and the potential for effecting change are aptly précised by Seymour and Rooke (1995), 'The culture of the industry has been evolved by members in order to deal with its economic reality. Any attempt to change culture, which does not also tackle the problems which gave rise to it, stands little chance of success!'

The research

Given the importance attached to culture and its assumed impact on the quality of relations in the construction value chain and, therefore, ultimately on the quality of the products and services delivered, the key questions addressed by this research are:

1. Can organisational cultures be identified? 2. Can a 'Culture of Quality' be identified? 3. How does strategic orientation impact on a culture's development? 4. How can a 'Culture of Quality' be created and sustained? In addressing these questions, we note their similarity to the conventional 4-point framework identified by Wilkins and Patterson (1985) - the need for an organisation to be clear about: its overall business strategy, the character of the existing culture, the gaps in the existing culture and what is required to pursue strategic aims through a plan of action to close the gaps. This framework provides for a developmental strategy used, for example, by Dressel (1987) with respect to construction firms. However, in order to distinguish our own perspective from that of Wilkins and Patterson, and to indicate the novelty of this perspective, we note the difficulties in applying their approach and our methods for dealing with them.

First, there is a tendency to take a simplistic and 'objectivist' view of culture. Anthony (1994) has argued the difficulties of characterising a 'culture' at all, noting that quick and simple analyses mask a multitude of subtle processes that variously engage individuals over time. Eglin (1980) points to the dangers of using 'culture' as an analytical construct tending to gloss over actual conduct (including verbal behaviour) to which the term refers. He also highlights the inevitable involvement of the researcher as s/he engages with the members of a culture in order to understand and describe it (the problems of indexicality and reflexivity). Such problems undermine the otherwise useful concept of *networks* as applied to the study of culture (e.g. Rasmussen and Shove, 1996). Ethnographic description of networks of relationships and influence is vital to understanding cultural change provided the temptation to reify the concept is resisted.

Second, there is the difficulty of distinguishing culture as a *variable* separate from, notably, *strategy,* and *performance* or *outcome.* Thus, Weick (1985) argues that the concepts 'culture' and strategy' refer to exactly the same organisational phenomena. Similarly, Mintzberg (1987) notes how strategies are formulated - cultural processes which range from a craft-like response to the unknown to deliberate, rational appraisals.

Further, commentators like Schneider (1989) and Smircich (1983) have noted that even the manner in which 'rational' appraisals are carried out and acted on express wide cultural variation. Despite the Japanese phenomenon and the intuitive sense that there is a connection between culture and performance, research findings are highly ambiguous. Thus, some of Peters and Waterman's (1982) excellent companies were found (within two years) to be less than excellent, while there are many examples to challenge Deal and Kennedy's (1982) often cited thesis that *strong* cultures are high performers.

The third problem is that, even if the attributes of a culture are recognised, the problem of how to implement the necessary change identified in that culture remains. This is a contentious, not to say, emotive area. Certainly, there is much advice on how to manage cultural change, e.g. how to bring about the 'learning organisation' through such concepts as 'leverage' (Senge et al 1995). Frequently, advice centres on change agents and champions - charismatic role models who are capable of actively demonstrating the need for and the benefits of change (Deal and Kennedy 1982). Others note the role to be played by rewards - through procedures such as management appraisal (Kerr and Slocum, 1987, Randell et al 1990). Others are highly sceptical of the very attempt to change or manage culture and critical of the attempts to do so (e.g. Wilkinson and Willmott 1995). Nord (1985) and Burrack (1991) point to the ambiguities and inconsistencies in findings while Berg (1985) claims that we are so ignorant of the dynamics of culture that to propose being able to manage it is, simply, absurd - a view also expressed by Uttal (1983).

Aims

In order to address these problems and achieve an understanding of the relationship between culture and quality in construction, the aims of the research are:

1. To identify those features of an organisation which are understood by its members to constitute 'culture'.
2. To identify those features which underlie and lead to the provision of quality in the construction industry; i.e. a 'Culture of Quality'.
3. To identify the impact of strategic orientation on culture and quality.
4. To identify change strategies and methods that promote a 'Culture of Quality'.

Hypotheses

The following hypotheses provide the focus for the achievement of these aims.
1. A Culture of Quality can be identified and expressed in the form of a profile.
2. Achievement of high quality in the construction industry is dependent upon the presence of a 'Culture of Quality'.
3. Effecting cultural change is a function of specifiable strategic orientation(s).
4. The personal qualities of change agents who influence cultural change successfully are those of the charismatic leader.

Methodology

The research will be conducted in five contracting organisations using structured and unstructured questionnaires, observational and shadowing techniques. Questionnaires will be administered to clients of and suppliers and subcontractors to the five participating organisations. Ten specific *projects* will be selected and monitored longitudinally. These projects will be a further source of data on i) members' perception of the culture of the five focal organisations and ii) of the perceptions of clients', suppliers' and subcontractors' personnel regarding the quality of product/service delivered by the focal organisations.

1. Identifying organisational culture. The identification of organisational culture will be undertaken in three stages.

i)   Available frameworks, particularly those provided by Hofstede (1980) and Hall (1995), and the questionnaires associated with them, will be used.

ii)   A reputational approach will be used to establish the consistency of the findings produced in stage one - respondents will be asked to characterise the organisation to which s/he belongs and the organisations with which s/he has dealings by asking for descriptions of occasions and events that typify given organisations. The technique developed by Morgan (1994) of using exploratory metaphors will be used also. The theoretical assumption underlying this second stage is that it will be more revealing of culture as a lived and evolving experience. The benefit of the comparison will be that it will test the extent to which predetermined metaphors and characterisations, as provided by standardised frameworks, are found by respondents to be expressive of their experience. The comparison is intended to clarify the confusion to be found in management research in general between findings of fact which have clear empirical referents and findings which may be useful to practitioners; e.g. Argyris et al (1988) and Seymour (1996). The comparison will also address the problem, frequently noted in culture research, of distinguishing espoused and actual culture.

iii) An inventory of typifications, key terms, metaphors, etc., will be drawn up and numerical scores applied. A key measure is a consistency rating. Thus, to be able to talk meaningfully of an organisational culture, will require that there are consistent patterns amongst respondents with respect to a given organisation (Hypothesis 1). The approach will provide statistically valid characterisations or profiles. Results will have diagnostic uses and will be used for inter-organisational comparison also.

2. Identifying a Culture of Quality. The data generated will reveal the way a given organisational culture is perceived by those who have experience of it. In addition, it will provide profiles against which finite measures of performance and client satisfaction will be assessed. Some of the organisations selected for study are undergoing the process of transition from QA to TQM. Such organisations will provide the opportunity for using the methods described in stages (i-iii) above to monitor the nature and extent of cultural change.

3. Impact of Strategic Orientation. An organisation's strategic orientation is both an expression of its culture but, at the same time, is a formative influence on it. Despite the conceptual difficulties involved in distinguishing culture and strategy, our research to date suggests that strategic orientation is experienced frequently as a *constraining* influence on the *development* of a culture. Therefore, a major concern of the research is to establish ways in which to conceptualise 'strategic orientation' and to assess the ways in which it impacts on corporate culture and facilitates or impedes cultural change. In our recent research, two sets of distinction have been identified and are being fruitfully employed to elucidate this important issue. These are the distinctions between *occupational* and *economic control* which relate to styles of internal control and *integrative (relational)* and *distributive (contractual)* which concerns styles of

relating to and managing clients and the value chain. This may be expressed as a matrix.

4. Creating and sustaining a Culture of Quality. The literature on change management (e.g. Deal and Kennedy (1982), Kanter (1985), Buchanan and Boddy (1992)) emphasises the importance of change agents. Therefore, the research will attend particularly to individuals who are cast in this role, both those who are charged with the specific function to bring about change (e.g. quality managers) and those who are known to be championing some change initiative. Adopting this focus will, we believe, provide special insight into the dynamics of cultural change and address the issue of how change is brought about, through, for example, personal influence networks. McCabe at al (1996) have highlighted the crucial role of charismatic individuals in bringing about the transition from QA to TQM and the culture change associated with it. However, the findings also challenge much conventional wisdom concerning the strategies adopted by change agents, notably, that the agent must be adaptable (see also Seymour et al 1992).

Conclusions

The increasing emphasis on quality as a construction performance criterion has prompted a 'scramble' for QA certifications. However, QA provides confidence in specification realisation and has been adopted to secure market competitive advantages primarily, due to the short term perspectives which are common to enhance immediate profitability. Holistic quality provision relates to TQM and requires a long term perspective, securing commitment of all personnel, perhaps most influentially, top management and 'change agents'. Thus, cultural change is required including removal of fears – from managers regarding aspects of corporate survival (e.g. takeovers) and from operatives regarding employment and conditions.

Investigation is required to determine the current status of the industry and how an enhanced 'culture of quality' may be secured. In depth, research is essential to understand the complexity of human (cultural) factors involved which the instant study hopes to initiate by cases, benchmarking etc. to determine quality requirements, provision and appropriate control mechanisms, which, we believe, are founded in cultures.

References

Anthony, P. (1994) *Managing Culture,* Open University Press, Milton Keynes

Argyris, C., Putnam, R & Smith, D.M. (1985) *Action Science,* Jossey-Bass, San Francisco.

BSI: BS 4778 (1987) Glossary of Terms used in Quality Assurance. British Standards Institution.

BSI: BS 5750 (1987) Quality Systems. British Standards Institution.

Bentley, M.J.C. (1981) Quality Control on Building Sites, BRE Information Paper IP 28/81, Building Research Establishment.

Berg, P.O. (1985) Organisational Change as a Symbolic Transformation Process, in Frost, P.J., Moore, L.F., Louis, M.R., Lundberg, C.C. & Martin, J. (ed.) *Organisational Culture,* Sage, Beverly Hills.

Bresnen, M.J., Haslam, C.O. Beardsworth, A.D., Bryman, A.E. and Keil, E.T. (1990), Performance on Site and the Building Client, Occasional Paper 42, Chartered Institute of Building.

Buchanan, D.A. and Boddy, D. (1992) *The Expertise of the Change Agent,* Prentice-Hall, London.

Building Economic Development Committee (1987) The Achievement of Quality on Building Sites. HMSO.

Burack, E.H. (1991) Changing the Company Culture: The Role of Human Resource Development *Long Range Planning,* 24(1):88-95.

Cherns, A.B. and Bryant, D.T. (1984) Studying the Client's Role in Construction Management, *Construction Management & Economics,* 2:177-184.

Cole, W. E. and Moyab, J. W. (1995), The Economics of Total Quality Management: clashing paradigms in the global market. Blackwell.

Deal, T.E. & Kennedy, A.A. (1982) *Corporate Cultures: the Rites and Rituals of Corporate Life,* Addison-Wesley, Reading, Mass.

Deming W.E. (1986) Out *of the Crisis: Quality, Productivity and Competitive Position,* Cambridge University Press.

Dressel, G. (1987) Corporate Culture as a Basis for Organisation and Management, in Lansley, P.R. & Harlow, P.E. (ed.) *Managing Construction Worldwide, Vol 2,* E. & F.N. Spon, London.

Eglin, P. (1980) Culture as Method: Location as an Interactional Device, *Journal of Pragmatics,* 4:121-135.

England, G.W. (1975), *The Manager and His Values: An International Perspective,* Cambridge, Mass, Ballinger.

Fellows, R F. (1992) Karming Conflict, in Fenn P. & Gameson R (ed.), *Construction Conflict Management & Resolution,* E. & F.N. Spon, London. pp.122-127.

Gronroos, C. (1991) The Marketing Strategy Continuum: Towards a Marketing Concept for the 1990's, *Management Decision,* 29(1):7-13.

Gunning, J. G and Leonard, D. (1994) Total Quality Management: post certification views on EN 29000 and its development within the construction industry of Northern Ireland. Proceedings CIB Conference on Quality Management in Building and Construction (W88) Lillehammer.

Hall, E.T. (1959), *The Silent Language,* New York, Doubleday.

Hall, W. (1995) *Managing Cultures: Making Strategic Relationships Work,* Chichester, Wiley.

Higgin, G. and Jessop, N. (1965), *Communications in the Building Industry,* London, Tavistock.

Higgin, G., Jessop, N., Bryant, D.T., Luckman, J. and Stinger, J. (1966), *Interdependence and Uncertainty,* London, Tavistock.

Hofstede, G. (1980) *Culture 's Consequences: International Differences in Work-related Values,* Sage, Beverly Hills.

Hutton, W. (1991) The Student becomes the Master, *The Guardian,* 25 November, p.15.

Hutton, W. (1996) The State We're In, Vintage.

Kanter, RM. (1985) *The Change Masters: Corporate Entrepreneurs at Work,* Unwin London.

Kerr, J. & Slocum, J.W. (1987) Managing Corporate Culture through Reward Systems *Academy of Management Executive,* 1(2):99-108.

Kluchhohn, F. and Strodtbeck, F.L. (1961), *Variations in Value Orientations,* Evanston, Il, Row Peterson.

McCabe, S., Rooke, J.A. & Seymour, D.E. (1996) The Importance of Leadership in Change Initiatives, in Langford, D.A. & Retik, A. (ed.) *The Organisation and Management of Construction*, E. & F.N. Spon, Glasgow.

Mintzberg, H. (1987) Crafting Strategy, *Harvard Business Review*, July-August, 66-75.

Morgan, G. (1994) *Imaginization*, Sage, London.

Nord, W.R. (1985) Can Organizational Culture be Managed? A Synthesis, in Frost, P.J., Moore, L.F., Louis, M.R., Lundberg, C.C. & Martin, J. (ed.) *Organisational Culture*, Sage, Beverly Hills.

Latham, M. (1994) *Constructing the Team, HMSO*, London.

Peters, T.J. & Waterman, RH. (1982) *In Search of Excellence*, Harper Collins, New York.

Pirsig R. (1974) *Zen and the art of motorcycle maintenance: an inquiry into values*, London, Corgi.

Randell, G., Packard, P. & Slater, J. (1990) *Staff Appraisal*, Institute of Personnel Management, London.

Rasmussen, M. & Shove, E. (1996) *Concrete Conclusions: Report of a Pilot Study of Innovation and change*, Report to ESRC/EPSRC/IMI Award No GR/K67601.

Schneider, S.C. (1989) Strategy Formulation: The Impact of National Culture *Organisation Studies*, 10(2):149-168.

Senge, P.M., Roberts, C., Ross, R.B., Smith, B.J. & Kleiner, A. (1994) *The Fifth Discipline Fieldbook*, Nicholas Brealy Publishing, London.

Seymour, D.E., Hoare, D.J. & Ithau, L. (1992) Project Management Leadership Styles: Problems of Resolving the Continuity-Change Dilemma, *Proceedings of the Eleventh INTERNET*, Florence.

Seymour, D.E. (1996) Developing Theory in Lean Construction, in *Proceedings of Fourth Annual Conference of International Group on Lean Construction*, Birmingham.

Seymour D. E., Rooke J. A. (1995) The Culture of the Industry and Culture of Research. *Construction Management and Economics,* 13 (6), 511-23.

Shadgrove B. (1994) Constructive Teamwork, in BSI, *Quality in Action*, British Standards Institution.

Simon, H.A. (1960), *The New Science of Management Decision*, Harper and Row.

Smircich, L. (1983) Concepts of Culture and Organisational Analysis, *Administrative Science Quarterly*, 28:339-358.

Uttal, B. (1983) ;The Corporate Culture Vultures', *Fortune*, 17 October, 66-72.

Weick, K.E. (1985) The Significance of Corporate Culture', in Frost, P.J., Moore, L.F., Louis, M.R., Lundberg, C.C. & Martin, J. (ed.) *Organisational Culture*, Sage, Beverly Hills.

Wilkinson, A. & Wilmott, H. (1995) *Making Quality Critical*, Routledge, London.

THE IMPACT OF CULTURE ON PROJECT GOALS

A. M. M. LIU,
Department of Real Estate and Construction, University of Hong Kong,
Pokfulam, Hong Kong. ammliu@hkucc.hku.hk

R. F. FELLOWS,
Department of Architecture and Civil Engineering, University of Bath,
Bath, United Kingdom. absrff@bath.ac.uk

Abstract

Despite numerous reports, extensive debate and recognition of the importance of teamwork, the construction industry remains highly fragmented. Partnering and analogous attempts to overcome the negative consequences of individualism and opportunistic behaviours and to invoke advantages flowing from collaborative working have met with some success. Many fields of psychology and of management emphasise the importance of determining and working towards goals (as in the behaviour-performance-outcome B-P-O cycle) – which must be feasible, viable, appropriate and acceptable to all participants. Commitment to such goals is known to enhance project performance and participant satisfaction (and, hence, project success). However, the continuously shifting power structure of most construction project leads to changing goals and hierarchies, with their impositions on following participants, detracting from committed behaviours and hence, performance. By examining cultural factors pertinent to the development of project goals, compatible with participants' goals (with adequate inter-participant sensitivities and good communications), this paper demonstrates project performance and participant gains which will ensue.

Keywords: Behaviour, culture, goals, performance, success.

Introduction

Organisational analysis is evolving toward more complex, paradoxical and even contradictory modes of interpretation (Jelinek et al 1983). This is mainly underpinned by the fact that culture has many perspectives and can only be fully interpreted in a 'native-paradigm', i.e. culture as seen and understood by those immersed in it. Culture provides the platform on which values (and thus manifested behaviours) are formed to direct mass beliefs (affecting the setting of goals) and evolution of common standard norms (affecting perceived performance and its evaluation).

Instead of monochromatic thinking towards organisational analysis (based on one adopted set of perspectives), an interpretive framework more like a rainbow – a 'code of many colours' that tolerates alternative assumptions is advocated (Jelinek et al 1983). Such a rainbow is best explained as physicists' ideas of explaining light in various wavelengths. For organisational analysis, one needs to perceive and understand the complex nature of organisational phenomena, both micro and macro, organisational and individual, conservative and dynamic. Culture, as a root metaphor for organisation studies, may redirect the attention away from some of the commonly accepted foci of structure and technology to the other elements of shared understandings, norms and values. In conjunction with other approaches (e.g. the human behavioural school), culture may provide the tension that lead to new insight in organisational analysis.

The organisation: culture and goals

Organisations are often viewed as social systems equipped with socialisation processes, social norms and structures. As different societies presumably have different cultures, considerable research efforts have been expended in search of culture's influence on the structures and processes of organisations and on the attitudes, needs and motivators of managers (e.g. Haire et al 1966). Other studies focus on the cultural properties of the organisation itself (e.g. Schein 1985). Although culture has as many as 164 definitions (Kroeber and Kluckhohn 1952), the cultural dynamics model proposed by Hatch (1993) successfully encapsulates the processes of manifestation, realisation, symbolisation and interpretation to provide a framework within which to understand the dynamism of organisational cultures. The dynamism comes from the continual construction and reconstruction of culture as contexts for setting goals, taking action, making meaning, constructing images and forming identities.

An organisational goal is defined as "a desired state of affairs which the organisation attempts to realise" (Etzioni 1964:6). However, the desired state of affairs is many things to many people. Even with the three commonly stated goals of universities – teaching, research and public service – we are all too certain that these are too vague to serve even as a guide for organisational analysis or practice. Another example is the commonly stated project goals of "within budget, within time and meeting specification (quality) requirements" – successful completion of which (i.e. high performance by definition) does not necessarily mean producing the desired outcome for the participant (e.g. satisfaction – see Liu and Walker 1998).

Organisational goals can be approached from a variety of perspectives. Parsons (1960) points out that organisational goals are intimately intertwined with important and basic societal functions (hence, underpinned by culture), but this simple approach ignores the variations in goals and activities among organisations performing the same basic functions. Organisational goals by definition are creations of individuals, singly or collectively (as determined and steered by organisational culture). The determination of a goal for collective action becomes a standard by which the collective action is judged (Hall 1972), i.e. basis for performance evaluation. However, the collectively determined, commonly based goal seldom remains constant over time because new considerations imposed from without or within deflect the organisation from its original goal, not only changing the activities of the organisation, but also becoming part of the overall goal structure. As Hall (1972) points out, the important point is that the goal of any organisation is an abstraction distilled from the desires of members and pressures from the environment and internal system – so culture is fundamental in providing the values which dictate the desires. This approach is similar to that of Simon's (1964:2): "either we must explain organisational behaviour in terms of the goals of the individual members of the organisation, or we must postulate the existence of one or more organisational goals, over and above the goals of the individuals".

Goals are often treated as abstract. They must be converted to specific guides for the actual operations of an organisation. Perrow (1961) distinguishes between official (as put forth in the charter and public statements) and operative (through the actual operating policies of the organisation) goals. Where operative goals provide the specific content of official goals, they reflect choices among competing values (i.e. elements which form culture). Hence, operative goals become the standards by which the organisation's actions and performances are judged and around which decisions are made.

Unofficial operative goals, however, are tied more directly to group interests (Perrow 1961). Examples are: an interest in a major supplier may dictate the policies of a corporation executive (as allowed by the power structure in the organisation); the personal ambitions of a hospital administrator may lead to community alliances and activities which bind the organisation without enhancing its goal achievement.

Goal determination is vital. Members of organisations must know the system (including the underpinning culture and how to be part of it) if they are to operate within it or to change it. Organisations are often reluctant to allow the researcher access to records that show the formation of operative goals; they may tend to repeat the official goal as a form of rhetoric. However, as suggested by Hall (1972), the researcher can determine operative goals through the use of multiple methods of data collection from a variety of goal indicators, such as the deployment of personnel, growth patterns among departments, examination of available records etc. More importantly, since operative goals reflect what the major decision makers believe to be the critical areas and issues for the organisation, it follows that the operative goals will shift as internal and external conditions impinge upon the organisation.

Construction procurement

Temporary multi organisation

Emmerson (1962) identified an extensive array of problems which hampered performance in the UK building sector. Many of these problems are also apparent in other construction industries. These problems are caused by poor communications (Higgin and Jessop 1963) and other issues concerning the conduct of relationships between participants on building projects. Subsequent research reports have investigated the problems in greater detail (e.g. Cherns and Bryant, 1984). Despite identification of the problems and considerable effect in their investigation and attempts at providing solutions (including Latham 1994) the majority of them remain undiminished, if not enhanced.

A particular feature which has been identified consistently as a source of difficulty is that of the Temporary Multi Organisation (TMO) nature of building projects. The multi organisational facet has been enhanced through the expanding incidence of sub contracting whilst the temporary nature of the 'project organisation' has been reduced through more integrated and enduring procurement arrangement (turnkey, design and construct, BOOT, partnering etc.). Popularist expression has been given to focus project participants – through engendering a common identity focus and, for strategic rather then project duration relationships, to reduce the temporariness. Such 'partnering' arrangement have met with some success according to Bennett and Jayes (1995). However, that partnering represents a form of joint venturing and its operation under such a perspective is considered by Fellows (1997) in terms of holistic costs. In tuning governance structures to facilitate holistic economies (focus on transaction cost and production cost etc.), attention must be paid to behavioural factors – the essence of cultural analysis. Attention should be devoted to determining and communicating appropriate and acceptable project goals which should accord with performance criteria and parameters of the project and evolved to be compatible with goals of project participants. Only in such circumstances can holistic costs be minimised, taking account of cultural factors involved – notably those pertaining to bargains and relationships, including individual transaction versus long term perspectives.

The amalgamation of a number of organisations into a TMO fosters exponential expansion of organisational complexities. Thus, there are interactions and interdependencies between the TMO and individuals working in it filtered through groups and organisations (Gibson et al, 1982; Szilagyi and Wallace, 1983; Miner, 1988). Generally, individuals relate to organisations, and, hence, TMOs, as members of a group and are subject to particular cultural influencers (in the context of sentience). Such effects are reinforced by indexicality (meanings/ understandings being dependent upon socialisation, education, training etc) and rules/codes and norms of behaviour for the professionals and other groups.

The traditional fragmentation of the functions necessary for project realisation has fostered differentiation between construction management, structural engineering and the other professions/functions, thereby leading to degrees of cultural distinctions. However, integration of functions – multi-disciplinary design practices, design and construct organisations – plus attention to buildability/constructability and encouragement towards establishing better relations (organisational interface management) between participants (e.g. Latham 1994; partnering) seek to overcome cultural differences (and the effects of cultural similarities – e.g. profit maximising and opportunistic behaviours) detrimental to both project performance and participant satisfaction.

The determination of goals for a project is much more problematic than envisaged. A complication is that the power structure of a project is likely to alter over the phases of project realisation such that any hierarchy of goals will vary. Awareness of such a situation will encourage those who exercise power early in a project to put in place governance provisions (by fixity of design, procedure controls, contracts etc.) to 'concrete' the goals etc. which they have established, thereby restricting the inputs of subsequent participants. Commonly, such goal setting actions, even if executed with the best intentions, lead to performance parameters for the subsequent participants and, as they may influence how a project is procured too, may not lead to 'best' performance (of either product or process).

Shifting multigoal coalition

A particular issue for construction is not only the identification of participants' goals but their combination to yield project goals. The purpose or goal is the basis for organisational activities and most organisations have more than one goal. These multiple goals may be in conflict with one another, even then, they are still a basis for action (the Behaviour-Performance-Outcome cycle).

The action itself may or may not conflict with conflicting goals. The relative importance of the goals can be determined by the way the organisation allocates its resources to them. Since both external and internal pressures affect goals, along with the more rational process of goal setting, goals cannot be viewed as static. They change, sometimes dramatically, over time. These changes, it should be stressed, can occur because of decision making within the organisation. Goal alterations decided within the organisation are a consequence of the interactions of members who participate in the goal setting process. This can be done by an oligarchic elite or through democratic processes (but in very few organisations, if any, would a total democracy prevail).

Shifts in goals can also occur without a conscious decision on the part of organisation members – i.e. as a reaction to the external or internal pressures without a conscious reference to where the organisation is going. The TMO is portrayed as a shifting multigoal coalition dependant on the power structure prevailing at the time (Newcombe 1994).

According to Hall (1972), organisational goals change for three major reasons: (1) direct pressure from external forces leading to a deflection from the original goals, (2) pressure from internal sources leading the organisation to emphasise different activities than those originally intended, (3) changed environmental and technological demands leading the organisation to redefine its goal. Etzioni (1964) emphasised "goal displacement", i.e. a given set of goals may be altered drastically by changes in the power system of the organisation – new types of personnel and the development of new standards that supersede those of the past. Shifts in cultural values and their impact on the goals of organisations are obvious in the profit-making sector. For instance, while the official goal of profit making may remain, the operative goals shift as more energies are put into market research and as organisations redefine themselves as "young" organisations for the "now" generation – internal transformations have occurred to refocus the organisations' activities.

Donaldson (1985) asserts that priorities should be assigned to multiple goals based on the powers of the participants. He notes that managing organisational goals is a continuous process in which "competing and conflicting priorities must be balanced". Thus, the goal system is likely to be unstable due to constantly changing power structures. Satisficing (Simon, 1964) may be a necessary approach with multiple project goals (some of which may exhibit mutual exclusivity) however, overt satisficing may detract from higher levels of performance which bounded competitive maximising by participants could (otherwise) yield. Peters and Waterman (1982) found that effective organisations do not have internal debates over what is the culture of the organisation, its philosophy and its goals. Such factors are accepted (or people do not join or leave soon). Debates which such organisations do have concern the best mechanisms/processes by which the goals can be achieved.

Perrow's (1961) emphasis on the operative goals, however they are developed, and Simon's (1964) notion that goals place constraints on decision making both suggest that goals are relevant, even central, for organisational analysis. It does not matter what the source of operative goals might be, what does matter is that they come into the decision making and action processes of the organisation. The goal concept is vital in organisational analysis. The dynamics of goal setting and goal change do not alter the fact that goals still serve as guides for what happens in an organisation. If the concept of goals is not used, organisational behaviour becomes a random occurrence, subject to whatever pressures and forces exist at any point in time. Georgopoulos and Tannenbaum (1957) argue that measures of effectiveness must be based on organisational means and ends, rather than relying on externally derived criteria.

A consistent objective, congruent with projects and their participant organisations, is to achieve success through good or continuously improving performance to yield satisfaction, whether these be concerned with process, product or both; indeed, the inexorable and complex relationships between process and product – performance, success and participant satisfaction – are acknowledged and discussed widely. Achievements and their ratings are dependant upon goals. Diversity of goals, and hence, directed actions, generates disparate performances whilst congruence of goals yields focused performance.

It is postulated that goal setting for projects is a core issue such that until project goals are established appropriately and communicated, performances, success and satisfaction will remain impaired.

From goals to outcome

Porter et al (1975) asserts that goals must be defined to focus the attention of individuals and groups (and) to provide a source of legitimacy for decisions. In the decision process of searching for a satisfactory solution, the goals of the action, i.e. the constraints that must be satisfied by the solution, may play a guiding role in two ways. The goals may be used to (1) directly to synthesise proposed solutions and (2) to test the satisfactoriness of a proposed solution (i.e. evaluation).

In construction procurement, Liu (1995) discusses the behaviour – performance – outcome (BPO) cycle in the context of goal development, influence and (potential) conflicts relating to individual/groups, group/organisation(s) and organisation(s)/society. The B-P-O cycle in project procurement is triggered by project goals. Goal directed action is characterised by the utilisation of feedback information so as to keep actions directed towards the goal(s). The B-P-O cycle assumes that the basic conscious actions of the individual are the actions of choice, i.e. judgement and decision making. As explained in Liu and Walker (1998), the act (behaviour) creates projects as a result of carrying out the act (performance) within an input-transformation-output model. Projects (the results of the acts) are then perceived and evaluated either by the individual directly or by some other entity such as another person or an organisation. Behaviour in such a context is the result of the S-O-R (stimulus-organism-response) sequence, a fundamental concept in the study of behaviour, which is brought about as a result of the forces exerted by the environmental factors on the members of an organisation. The organisation's embers have to react by setting, adjusting or redefining their goals and actions.

As project participants exhibit goal seeking behaviour, they tend to behave in ways that result in goal accomplishment which gives rewards, i.e. they demonstrate motivated behaviour which is directly related to the desirability of the reward, the extent to which the reward satisfies personal motives (or personal goals), the belief that given behaviour will actually be rewarded, and the ability of the individual and his/her self-perceived performance capability to perform such behaviour successfully.

Locke et al (1988) identified four major determinants relating goals and outcomes (particularly concerning the personal consequences – 'valence' driven – of performance): (1) legitimate authority of the goal setter (2) peer and group pressures towards goal commitment (3) participants' expectations (that effort leads to performance) (4) incentives and rewards (to enhance goal commitment)

The first two factors emphasise establishment and communication of the goals whilst the third and fourth factors emphasise realisation. Legitimate authority to set goals can occur through governance mechanisms –such as being stipulated in a contract – or by command (rather than demand) of authority and respect from other participants; the latter situation is likely to invoke more peer and group pressures (due to the presence of respect), especially if consultation has occurred or/and the goal setter has (obviously) considered the needs of others. Hence, in determining goals, cultural factors concerning authority, individualism and uncertainty operate. In pursuance of goals, aspects of wealth and status apply to determine appropriate incentives/rewards to motivate desired behaviours which requires analysis of what the particular groups, from who the behaviours are required, value – notably the debate over whether wage/salary is a motivator or a hygiene factor!

Thus, given the stimulus – organism – response (SOR) perspective, modification of the BPO cycle to be preceded by goal determination is appropriate, thereby generating the GBPO cycle. Hence, goal determination is fundamental to project realisation, performance (in both process and product) and, consequently, success and satisfaction of participants. Role behaviour of an individual however depends on means-end premises as well as goal premises (Simon 1964). Thus particular professional training may provide an individual with specific techniques and knowledge for solving problems (engineering knowledge, legal knowledge etc.) which are then drawn upon to guide behaviour towards outcome. In this way, a project manager with an engineering background may find different problem solution from a project manager, in the same position, with a surveying background.

From culture to goals

Culture provides the framework for interpreting the interactive, ongoing, recreative aspects of organisations, beyond the merely rational or economic. Culture is defined (among numerous others) as "a system of meaning that accompany the myriad of behaviours, practices recognised as a distinct way of life" (Gregory 1983). Cultural meanings are "apparently shared", following Becker (1982) who suggests that people interact "as if" they shared culture. However, where culture is interpreted as the product-and-process of organisation's members' sense making through their ongoing interactions, power becomes an important aspect of culture (Riley 1983). The process by which the power structure is created – the process of structuralisation – is closely parallel to the process of culturation (e.g. described by Berger and Luckmann 1967). Power structure is created through images and the symbolic order, which, in turn, is shaped by the processes of domination and legitimisation. Structure expresses the hierarchy of the past, institutionalised in power arrangements and persist into the present by affecting people's behaviours. People's behaviours, so structured and constrained, recreate the structure that, in turn, guide thoughts and beliefs to create group values.

Hence, culture is seen as group values embedded in shared value-laden images or myths. An individual's behaviour in a given culture is guided by autocommunication (Broms and Gahmberg 1983), i.e. communication with the self – which is the meaning added in (or the self-cueing aspect of) qualitative communication. For instance, in strategic planning, the symbolic ideals embodied in the goals provides a pattern around which people can orient and gauge their performance against (via the B-P-O cycle). Here, the symbolic image (as represented in the goal) – the sense of purpose and progress, the continued self-cueing – is all that matters. By capturing the element of belief crucial to organisational action, goal setting and planning serve as self communication, to clarify the organisation's image of itself.

Communication (whether auto- or interpersonal communication) plays a central role in performance. Reluctance to communicate is reflected by Hofer and Schandel (1986) "... it is not always wise to communicate the company's plans completely or precisely to middle and lower management for various political and social reasons". Hopefully, such "strategic considerations" will not be too common or applied in other contexts. Communication reticence may be based on fear of competition – from other firms or people – or other fears. However cultures via ethics, lay down norms of acceptability, e.g. how site quantity surveyors report profitability (profile) of a project to the contractor's head office as construction progresses (feed-in early profits over later stages to achieve a smooth profit profile). Such provision of what it is assumed others want to see could have disastrous results. Other communication issues concern indexicality and, in a more extreme form, problems of good communications between high content and high context language/expression (e.g. English – Chinese) cultures. Without sensitivity to the receiver(s) and care in producing and transmitting the message it is likely that clarity will be poor or/and offence may be taken, especially if the unsuitability of messages is sustained. For either or both reasons, performance will suffer.

The shared view, the creation of a common image, and the importance of planning as cueing for common views and visions (via communication, whether auto- or interpersonal) is a perspective on goal setting significantly informed by the cultural metaphor and cultural processes (Broms and Gahmberg 1983).

Conclusion

Culture contributes an additional element in the existing paradigm, expanding the old, implicit models of machines or organisms to include the models of social process. The shift is not just from machine to organism but from organisation to organising. What is proposed is a dynamic and interactive model of organising as a process (B-P-O) that persists and changes over time (S-O-R). The nature of the process is conditioned by the nature of the human mind (which seeks to interpret) and the nature of the organisation as a human artefact of the sense-making process. Organisations and organising are too complex to be well explained by simple dichotomies (e.g. mechanistic v. organic organisations) or by monochromatic code of references. Culture provides a paradigm of multi colours (Jelinek et al 1983) allowing the impact of environmental dynamism over changing values, changing behaviour and changing perceptions (norms for evaluation) to be taken into account (see Hatch 1993).

Management involves change so appreciation of both cultures' manifestations and foundations is powerful in determining appropriate change processes towards the desired outcomes. Changes in peripheral behaviours are relatively unimportant to those from whom change is required and so, may be achieved quite readily; the more fundamental the change required, the more difficult and greater time necessary if it is to be achieved. Changes in attitudes are acknowledged to require considerable effort periods. Whilst changes in behaviours may be achieved rapidly, unless sustained through external means and subject to (increasing) acceptance, the change is likely to be temporary.

Thus the formal governance of construction projects (through standard forms of contract), is usually developed reactively to changes in law and industrial practices. Contracts which have been developed and introduced to promote (normative) changes have met with great resistance and hence, generally, have been abandoned. The nature of construction and the potential for effecting change are aptly précised by Rooke and Seymour (1995), "The culture of the industry has been evolved by members in order to deal with its economic reality. Any attempt to change culture, which does not also tackle the problems which gave rise to it, stands little chance of success!"

References

Becker H. S. (1982), Culture: a sociological view. *Yale Review.* 71, pp. 513-527.

Bennett J., Jayes S. (1995) *Trusting the Team,* Reading Construction Forum, Centre for Strategic Studies in Construction, University of Reading

Berger P. L., Luckmann T. (1967), *The social construction of reality.* Garden City, NY, Anchor Press.

Broms H., Gahmberg H. (1983), Communication to self in organisations and cultures. *Administrative Science Quarterly.* 28

Cherns A. B., Bryant D. T. (1984), Studying the client's role in construction management. *Construction Management and Economics.* Vol. 2, pp. 177-184

Donaldson G. (1985), Financial goals and strategic consequences *Harvard Business Review* 63(3), 57-66

Emmerson Report (1962), *A survey of problems before the construction industry.* London, HMSO

Etzioni A. (1964) *Modern organisations.* Prentice Hall, Englewood Cliffs, NJ

Fellows R. F. (1997) The Culture of Partnering. In Davidson C.H., Tarek A. Abdel Meguid (ed.) *Procurement - a key to innovation. CIB Proceeding.* CIB Publication no. 203. CIB. pp. 193-202

Georgopoulos B. S., Tannenbaum A. S. (1957) A study of organisational effectiveness. *American Sociological Review* 22, 5, pp. 534-540.

Gibson J. L., Ivancevich J. M., Donnelly J. H. (1982), *Organizations -- behavior, structure process* Plano : Business Publications

Gregory K. L. (1983) Native-view paradigms: multiple cultures and culture conflicts in organisations. *Administrative Science Quarterly,* 28, pp. 359-376

Haire M., Ghiselli E., Porter L. (1966), *Managerial thinking in an international study.* NY Wiley.

Hall R. H. (1972), *Organisations: structure and process.* Prentice-Hall, Englewood Cliffs, NJ

Hatch M. J. (1993), The dynamics of organisational culture. *Academy of Management Review.* 18, 4, pp. 657-693

Higgin G., Jessop N. (1963), *Communications in the Building Industry.* London. Tavistock Institute.

Hofer C. W., Schandel D. (1986), *Strategy Formulation: analytical concepts.* West Publishing Co.

Jelinek M., Smircich L., Hirsch P. (1983), Introduction: a code of many colors. *Administrative Science Quarterly,* 28, pp. 331-338

Kroeber A. L., Kluckhohn C. (1952), *Culture: a critical review of concepts.* NY Vintage.

Latham, Sir M. (1994), *Constructing the Team.* London. HMSO.

Liu A. M. M. (1995), *Evaluation of the Outcome of Construction Projects.* Unpublished. PhD Thesis. University of Hong Kong.

Liu A.M.M., Walker A. (1998), Evaluation of Project Outcomes. *Construction Management and Economics.* 16, 2, pp. 209-219

Locke E. A., Latham G. P., Erez M. (1988), The determinants of goal commitment. *Academy of Management Review*, 13, pp. 23-29

Miner J. B. (1988), *Organisational behavior. Performance and productivity.* NY Random House Business Division.

Newcombe R. (1994) Procurement Paths—a power paradigm, *Proceedings, CIB W-92 Symposium, East Meets West,* Hong Kong, Dec., pp. 243-250

Parsons T. (1960), *Structure and process in modern societies.* The Free Press, NY

Perrow C. (1961), The analysis of goals in complex organisations. *American Sociological Review.* 26, 6, Dec.

Peters L. H., Waterman R. H. (1982), *In search of excellence* NY Harper and Row

Porter L.W., Lawler E. E., Hackman J. R. (1975), *Behaviour in Organisations.* McGraw Hill

Riley P. (1983), A structurationist account of political cultures. *Administrative Science Quarterly,* 28, pp. 414-437

Rooke J. A., Seymour D. E. (1995), The NEC and the culture of the industry: some early findings regarding possible sources of resistance to change. *Engineering Construction and Architectural Management.* 2, 4, pp. 287-305.

Schein E. H. (1985), *Organisational Culture and Leadership.* San Francisco, Jossey Bass.

Simon H. A. (1964), On the concept of organisational goal. *Administrative Science Quarterly,* 9, 1, June, pp. 1 - 22

Szilagyi A .D., Wallace M. J. (1983), *Organizational behavior and performance* Glenview Scott Foresman

A LITERATURE REVIEW ON STUDIES IN CULTURE – A PLURALISTIC CONCEPT

S. BARTHORPE, R. DUNCAN and C. MILLER
School of the Built Environment, University of Glamorgan. Pontypridd. CF38 1AA
United Kingdom. sbarthor@glam.ac.uk

Abstract

The intention of this paper is to present an overview of the literature published on the subject of 'Culture' (in general, as well as with an eventual focus on the construction industry). The limitations of this brief paper preclude any comprehensive coverage of the vast amount of material published in this field. 'Culture' has become a pluralist concept and the interpretations, definitions and applications of "culture" are multifarious. The development and use of this evolving subject will be discussed under three headings:

- Historical and Sociological Cultural Perspectives
- Organisational and Corporate Cultural Perspectives
- Construction Industry Cultural Perspectives

Key Words : Cultural diversity, organisational culture, sociological culture, subcultures

Historical and sociological cultural perspectives

Introduction

The meaning of the term 'culture' has changed greatly over the last forty years. The word culture is no longer simply associated with the cultivation of soil and plants but has been redefined so that it describes the very fabric of society. Society is multi faceted and within each facet different 'cultures' are thought to exist. The 'culture' which exists within the construction industry may be very different to that which exists within other industries. The author's apologise for the omission of any material considered by others to be worthy of inclusion.

Culture - a definition

The use of the word culture has its roots in the ancient Latin word *cultura , 'cultivation' or 'tendering'*, its entrance into the English language had begun by the year 1430. The definitions of culture contained in pre-1960's dictionaries are likely to be restricted to the examples like those found in Webster's dictionary of the period:
1. The cultivation of soil.
2. The raising, improvement, or development of some plant, animal or product".

By the middle of the 20th Century, other connotations and definitions began to take precedence over the latin denoation. 'Culture' was later defined as, "the training, development, and refinement of mind, tastes, and manners" this definition is known today as 'high culture'.

However, the American Heritage English Dictionary provided a substantially different primary definition of culture, describing it as: "The totality of socially transmitted behaviour patterns, arts, beliefs, institutions, and all other products of human work and thought." The differences in the use of the word 'culture' have been heavily influenced by the academic fields of sociology and cultural anthropology. Over time, new uses for the word culture have replaced its older meanings, those associated with cultivation of the land and the production of crops (ref. Learning Commons 1998).

The modern definition of culture, as socially patterned human thought and behaviour was originally proposed by the nineteenth-century British anthropologist, Edward Tylor. In 1952 Alfred Kroeber and Clyde Kluckhom, American anthropologists, published a list of 160 different definitions of culture. Bodley (1994) prepared a simplified version of the table below:

Definitions of culture:

Topical:	Culture consists of everything on a list of topics, or categories, such as social organisation, religion or economy.
Historical:	Culture is social heritage, or tradition, that is passed on to future generations.
Behavioural:	Culture is shared, learned human behaviour, a way of life
Normative:	Culture is ideals, values, or rules of living
Functional:	Culture is the way humans solve problems of adapting to the environment or living together
Mental:	Culture is a complex of ideas, or learned habits, that inhibit impulses and distinguish people from animals
Structural:	Culture consists of patterned and interrelated ideas, symbols, or behaviours
Symbolic:	Culture is based on arbitrarily assigned meanings that are shared by a society

Bodley (1994) stated that culture involves at least three components: what people think, what they do, and the material products they produce. Thus, mental processes, beliefs, knowledge and values are aspects of culture. Culture also has several properties: it is shared, learned, symbolic, transmitted cross-generationally, adaptive and integrated. Bodley also stated that culture is learned, not biologically inherited, and involves arbitrarily assigned, symbolic meanings.

Without a shared culture, members of society would be unable to communicate and co-operate, and confusion and disorder would result (Haralambos et. al, 1990). In addition to the three components outlined by Bodley (1994), Giddens (1989) suggested that culture also has two essential qualities: firstly it is learned, secondly it is shared. Without it there would be no human society. Humans are human because they share with others a common culture, a culture which includes artefacts of both its living members and its ancestors, (Giddens et al, 1987). Culture consists of the values the members of a given group hold, the norms they follow and the material goods they create. Culture refers to the whole way of life of the members of a society. It includes how they dress, their marriage customs and family life, their patterns of work, religious ceremonies and leisure pursuits (Giddens, 1989).

Every culture, has three fundamental aspects: the technological, the sociological, and the ideological. The technological aspect of culture is concerned with materials, tools, techniques and, in contemporary society, machines. The sociological aspect of culture involves the relationships into which people enter, especially in work and in the family. The ideological aspect of culture comprises of beliefs, rituals, magical practices, art, ethics, religious practices and myths (Lewis, 1982).

Organizational and corporate cultural perspectives

Introduction

It could be argued that the concept of culture has been long on the agenda of many management theorists. The centrality of values in the management of organisations was emphasised from the 1930s through to the 1950s by distinguished and influential writers such as Chester Barnard and Peter Druker. The term management to these writers always meant more than simply decision making and the planning of procedures. The competent management of an enterprise implied that the motivation of people and effective strategic vision of the organisation's direction, were important to an organisation's productivity within its environment (Fincham, 1996).

"Culture" was introduced into the management mainstream by Peters and Waterman in their seminal work, ' In Search of Excellence' (1982). In the period before 1980 most of the issues relating to culture were not deemed as important in terms of organisational performance. In the past twenty years or so the concept of culture has become an important issue, that has been heightened with the realisation that, through technology and markets, organisations are becoming increasingly global (Kobrin, 1995, in Joynt, 1996, p.8).

Development of the concept of culture to organisations

Strategic management theory demands that the culture of organisations be held as an important element in the formulation of any strategy. The culture and values of an organisation are imperative in matching organisational resources to the environment. For an organisation to be effective, congruence must exist between the organisations values, resources and environment (Thompson 1995). Some of the key developments in the field of organisational culture have been outlined by Brown (1995, ch.1). Writers such as Hofstede (1980, 1991) Deal and Kennedy (1982) Peters and Waterman (1982), and Kanter (1983), placed culture within a wider context than just individual organisations and included industry and national dimensions.

Organisational culture

Culture within organisations is reflected in the way that people perform tasks, set objectives and administer the necessary resources to achieve objectives. Culture affects the way that people make decisions, think, feel and act in response to the opportunities and threats affecting the organisation (Thompson 1993). A popular method of explaining culture in simple terms is "the way things happen around here." Atkinson in (Mullins 1993) interprets culture as reflecting, the underlying assumptions about the way in which work is performed; what is 'acceptable and what is not acceptable'; and what behaviour and actions and encouraged and discouraged.

Determinants of culture

Hofstede (1980, in Fincham 1996) defined culture in its broadest term. His definition in terms of organisations is defined as "the collective programming of the mind which distinguishes the members of one category from another". In the studies undertaken by Hofstede (1980) of over 116,000 employees of IBM in 72 countries four dimensions were identified as being inherent within an organisations culture:

Power-Distance
Uncertainty-Avoidance
Individualism-Collectivism
Masculinity-Femininity

The IBM survey was dominated by Western ways of thinking and also excluded countries in East Asia. Bond (in The Chinese Culture Connection 1987), identified another cultural dimension which he called, 'Confucian Dynamism'. Hofstede (1991) has preferred to re-name this 'fifth dimension', Long-term / Short-term Orientation. Long-term Orientation is associated with East Asian countries, Short-term Orientation is associated mainly with Western countries. A strong correlation has been drawn between the long-term orientation and the economic growth of the East Asian countries in the past 25 years, (Hofstede & Bond 1988). Hofstede's third dimension of Individualism – Collectivism was further considered by Schein (1985) when he observed the differences between Western and Eastern cultures as exhibiting Low-Context (Individualistic) and High-Context (Collectivist) cultures respectively

Sub cultures are an important issue within the theory of culture, and it could be argued for the construction industry. It must be understood that if sub cultures exist within organisations due to the fact that culture does and will change over time then the firm has to manage groups of people with conflicting aims and objectives. Such conflict must be tackled and understood in the formation of strategic plans, as they can be either damaging or alternatively beneficial in terms of the future aims and objectives of the organisation.

Charles Handy (1976) offered four cultures that can be found within different organisations: Power or club, role, task and person. Handy showed that essentially, the type of culture inherent within organisations could be determined by the structure of the organisation. The structures identified are the Web, the Greek temple, the Net and the Cluster. Deal and Kennedy (1982) in their study of American companies found that to sustain competitive advantage organisations must actually believe in "something", and that these beliefs must permeate throughout the organisation. They identified five key elements that determine the culture within enterprises:

- The environment and key success factors
- The values of the organisation
- Heroes and role models
- Rites and rituals
- The cultural network: the communication system that determines how essential issues are diffused throughout the firm

The work of Pumpkin (1987) suggests that an organisation's culture is comprised of seven aspects which, dependant on the type of industry will be less or more significant to the individual firm. The seven aspects include; the extent to which the organisation is market oriented, the relationship between management and staff, the extent to which people are target oriented, attitudes toward innovation and costs and the disposition of the firm toward technological change.

Kanter (1989) found that for companies to be successful they must view innovation as a prerequisite to success. Kanter deals with issues such as the "culture of inferiority", a tendency to doubt their own ability to manage change and innovate. Interesting sets of cultural types put forward by Kanter are what she classes as cultures of "age versus youth". Companies in established industrial sectors that will not attempt to change, view change as unnecessary due to the belief that if the environment is ignored then the problems faced will finally depart. She argues that young companies, for example new hi-tech industries, tend to empower innovators resulting in young employees questioning existing patterns and norms creating "a culture of youth".

Construction industry - special cultural perspectives

Introduction

Despite its size and universality, the construction industry in the UK is not homogeneous, it is in effect a fragmented and hierarchical industry. Porter (1980) describes a fragmented industry as one in which no company has a significant market share. Or as Male (1991) suggests, there is no market leader able to significantly influence outcomes within the industry. A fragmented industry is typically comprised of a large number of privately owned, small and medium sized companies and a small number of large companies. Porter adds that a fragmented industry is populated by many competitors who are in a weak bargaining position with respect to both buyer and supplier groupings, and where profitability is marginal. These characteristics are common place in the UK construction industry. Ball (1988) indicates that construction is a hierarchical industry (designed by size of firm), where the many small companies tend to act as subcontractors to the large companies.

The employee profile of the construction industry

Historically, construction has been a casual industry due to its project-specific emphasis and the itinerant, peripatetic nature of its workforce. A report by NEDO (1970) states, 'casual employment promotes a casual response from employees. They do not identify themselves with the successful completion of projects and do not accept the conventions on punctuality, attendance and safety that apply in more settled work'. This casual, unsafe environment has its origins in the public work projects carried out in the 19th century. The extensive canal and rail infrastructures of Europe have largely been attributed to the phenomenal achievements of the British navigators or 'navvies'. The navvy culture has been well documented by Coleman (1965) who compares their harsh social and working environment with their remarkable work ethos. Coleman considers the navvy as the 'King of Labourers', exhibiting uncouth habits and manners yet revelling in perilous and physically exhausting work.

According to Agapiou, Price and McCaffer (1995), the traditional sources of construction labour have been predominantly young, white and male. Gale (1992b) suggests that exaggerated 'macho' behaviour among this profile acts as a kind of bona fide demonstration that legitimises a person's membership of the construction community. Women and ethnic minorities are under-represented in the industry. In the UK 1989, only 1.6 % of the CITB apprentice intake was female, and only 1.3 % was from ethnic minority backgrounds (IPRA, 1981). A study undertaken by the CITB in 1988 indicated possible reasons for this low level of interest as fear of harassment, inequality and the generally held negative image of the industry. They perceived the work as being 'dangerous', 'dirty' and not attracting any 'respect'.

Handy (1994), contends that organisations need talented women in their core jobs, not just for social equality reasons but because many women possess attitudes and attributes that new flat, flexible organisations need. Langford et. al. (1995) suggest that it would benefit the industry for a more female culture to be developed.

Root (1997) identifies and describes the cultural implications of the utilisation of standard forms of contracts and their effect on project performance. The existence of cultures and subcultures within the construction industry and their effect on the establishment of new roles and relationships as exemplified in the New Engineering Contract. Root's research makes five major assumptions that have been previously identified by Fellows et. al. (1994) concerning the shared understandings, expectations, differences and similarities that exist between subgroups, developing cultures and subcultures. The accommodative function of the industry's culture undergoes severe strain from the massive changes it experiences, which evidences itself in an exponential increase in litigation.

The confrontational nature of contracting

Latham's exhortation to the protagonists in the construction industry is to reduce adversarial relations and co-operate with each other. Gale (1992a) observed that, "… it seems that it is in the interests of those who have chosen to work in the industry to maintain the maleness of the culture, thus keeping conflict and crises as preferred aspects of everyday working life". He suggests that the construction industry is perceived as embodying :

- Crisis
- Conflict and
- Masculinity

He further observed that the construction industry's male culture must feminize if conflict is to be reduced. If male values include the propensity for conflict in human interaction Gale argues, then conflict becomes locked into the construction culture. An increase in the proportion of women into the industry is considered unlikely to change the entrenched 'macho' culture. The 'maleness' of the industry perpetuates adversarial, confrontational and often unsociable behaviour that discourages women and ethnic minorities. The harsh working conditions, insecure employment prospects and low esteem associated with construction also deter the better educated people from considering construction as a worthwhile career option. Gale also argues that educational establishments are gatekeepers to the construction industry's culture and it is here that changes to the philosophy and curriculum could bring about the 'feminization' of the construction culture.

The Construction Industry Board (CIB 1996) working group 8 report, *'Tomorrow's Team : Women and Men in Construction'* advocate that it is imperative to convert the current vicious circle of poor image, poor performance, poor delivery to a virtuous circle of improved delivery and better image, attracting the right people to continue the process.

Managing international cultural diversity

The main focus for global construction is now in the People's Republic of China (PRC) and it is therefore worthwhile focusing attention on recent research concerning this rapidly expanding part of the world. China's rapid development, combined with its recently introduced open-market policies have created a climate of co-operation and one which welcomes foreign management skills and technology. In Shanghai alone there are 20,000 construction sites and as many people employed there in construction as in the whole of the UK construction industry, (Li, 1997).

Chan (1997) draws the correlation between conflict and culture, stating that the cause of disputes is closely related to the culture of a society, which he claims is not a constant: it changes with society's social and economic environment. This phenomenon has been evidenced remarkably in the People's Republic of China. The potential for conflict is exacerbated when Western culture meets Eastern culture in joint-venture construction projects. Cultural misunderstandings are a major source of failure for joint ventures, with a 50% failure rate recorded in a study of 110 American / Asian joint ventures by Geringer (1991). Sweirczek (1994) argues that individualistic, low-context cultures, (i.e. Western countries) tend to be confrontational and direct. High-context, collectivist cultures, (i.e. East Asian countries) exhibit 'face-saving', indirect styles of conflict management. When these cultural values come into contact major difficulties emerge.

Hall & Jaggar (1997) suggest that there is an abundance of literature that addresses the impact of international cultural differences, citing (Klien & Ludin, 1992; Lucas, 1986; Baden-Powell, 1993; and Stallworthy & Kharbanda, 1985). However, Hall & Jaggar contend that little attention is paid to establishing procedures for mitigating the impact of culture on construction activities. If managing change is the issue of the nineties, then managing cultural diversity must arguably be the issue of the new millennium. Cultural preparedness or what Barnham & Oates (1998) call, *'intercultural competence'* will be essential skills for the management of those companies that have intentions of operating successfully in the global marketplace.

The subcultures of the construction industry

Greed (1997) describes the construction working environment as, 'a culture of intimidation' where there is an increase in emphasis upon subcontracting and a decline in direct labour. This trend is however reversing in the light of changes made to the taxation and status associated with the self-employed sector conditions contained in the 1998 UK budget. Greed observes that, 'the whole industry seems to work by each level putting pressure on the next one down'.

Seymour & Hill (1996) argue that a more robust conception of culture is necessary if the radical change that the industry is currently undergoing is to be properly understood and effectively responded to. They contend that the role of the first-line supervisor is considered as a crucial, prime articulator of the industry culture and therefore a key to changing it. Applebaum (1981), in his observations of construction workers in the USA noted that supervisors display a greater degree of consideration for subordinates that they perceive to be satisfied and responsive than those that are critical or dissatisfied. He contends that, 'in construction, actual authority is not granted by the formal organisational chart. It must be established in the courses of social interaction that is appropriate to the situation'.

Conclusion

This paper has provided an overview of literature concerning cultural issues within:

*   Historical and Sociological Cultural Perspectives
*   Organisational and Corporate Cultural Perspectives
*   Construction Industry-specific Cultural Perspectives

The development and evolution of the various definitions, interpretations and applications of 'culture' has been discussed with the construction industry specific perspectives receiving particular emphasis. Culture is an evolving, pluralistic concept, having many definitions and interpretations and there is an increasing tendency to use 'culture' in many diverse applications. In its simplest form, culture is what we are and what we do as a society. In essence, 'the way things happen round here'.

The modern definitions of culture, as socially patterned human thought and behaviour was originally proposed in the nineteenth century by Tylor. As the second millennium nears its completion, a plethora of definitions and interpretations have been applied to culture. The influence of culture appears to pervade most aspects of personal, organisational and national life. This influence is especially heightened as cross-cultural issues arise in an international context. Managing cultural diversity must therefore be considered as a major issue for the new millennium.

Businesses are increasingly aware of the importance and influence that organisational culture has on corporate performance. Management theory advocates that the culture of organisations is considered an essential element in the formulation of any corporate strategy. The culture of the UK construction industry has been significantly influenced by its traditional source of labour which have been predominantly young, white and male. This employee profile, combined with the often harsh working conditions have been translated into a culture embodying crisis, conflict, masculinity and embracing a casual approach to the working norms of punctuality, commitment and safety.

References

Agapiou, A., Price, A.D.F. & McCaffer, R. (1995) Planning future construction skill requirements: understanding labour resource issues in Construction Management and Economics (1995) 13, pp 149-161.

Applebaum, H.A. (1981) Royal Blue: The Culture of Construction Workers in Spindler,G. & Spindler L (eds.) Case Studies in Cultural Anthropology. Stanford University. Holt, Rinehart & Winston international.

Ball, M. (1988) Rebuilding Construction : Economic Change and the British Construction Industry. Routledge. London.

Banham, K. & Oates, D (1998) What is an International Manager ? in World Executive Management Digest. http://www.wed.asiansources.com

Bodley John (1994) An Anthropological Perspective. From Cultural Anthropology: Tribes, States and the GlobalSystem.

http://www.wsu.edu:8001/vcwsu/commons/topics/culture/culture-definitions/bodley-text.

Brown, A. (1995), Organisational Culture, Pitman Publishing.

Chan, E. H. W. (1997) Amicable dispute resolution in the People's Republic of China and its implications for foreign-related construction disputes, in Construction Management & Economics (1997) 15, pp 539-548

Chinese Culture Connection (a team of 24 researchers), 1987 'Chinese Values and the Search for Culture-free Dimensions of Culture' in Journal of Cross-Cultural Psychology, 18,2 (1987) pp 143-164.

CITB. (1988) Factors Affecting Recruitment for the Construction Industry. CITB. Bircham Newton. pp 4-5.

Coleman, T. (1965) The Railway Navvies. Pelican Books. London.

Construction Industry Board. (1996) Tomorrow's Team : Women and Men in Construction. (Working Group 8). London.

Deal, T and Kennedy, A. (1982) Corporate Cultures: The Rites and Rituals of Corporate Life, Addison Wesley.

Fellows, R.F., Hancock, M.R. and Seymour, D. (1994) Conflict Resulting from Cultural Differentiation : An Investigation of the New Engineering Contract in Fenn, P. (1994) Ed., The Effect of Culture on Project Performance.

Fincham, R (1996), Individuals and Organisations (2nd Edition), Oxford University Press.

Gale, A.W. (1992a) Proceedings of the First International Construction Conference. UMIST. Manchester, 1992. E & F. N. Spon. London.

Gale, A.W. (1992b) The Construction Industry's male culture must feminize if conflict is to be reduced: the role of education as a gatekeeper to male construction industry. in Fenn, P.& Gameson, R. (eds.) Construction Conflict Management and Resolution. E. & F. N. Spon. London.

Giddens, A. (1987) Introductory Sociology. Macmillan Press, London.

Giddens, A. (1989) Sociology. Polity Press, Cambridge.

Goldsmith, W and Clutterbuck, D The Winning Streak, Weidenfeld and Nicholson.

Greed, C. (1997) Cultural Change in Construction in the the Proceedings of the 13th ARCOM conference, Cambridge, 1997. pp 11-21.

Hall, M.A. & Jaggar, D.M. (1997) Should construction enterprises, working internationally, take account of cultural differences in culture ? in the Proceedings of the 13th ARCOM conference, Cambridge, 1997. pp 1-10

Handy, C. B (1985), Understanding Organisations (3rd Edition), Penguin.

Handy, C. (1994) The empty raincoat. Hutchinson. London.

Haralambos, M. Holborn, M. and Heald, R. (1991) Sociology Themes and Perspectives. Third Edition. Collins Educational, London.

Hoecklin, L (1994), Managing Cultural Differences: Strategies for Competitive Advantage. Addison Wesley Publishers.

Hofstede, G (1980), Culture's Consequences: International Differences in Work Related Values.Sage Publications.

Hofstede, G & Bond, M. (1988) 'The Confucius Connection: From Cultural Roots to Economic Growth' in Organisational Dynamics, 16,4 (Spring 1988) pp 4-21

Hofstede, G (1991) Cultures and Organisations : Software of the Mind. McGraw-Hill,

Joynt, P. and Warner, M. (eds.), (1996), Managing Across Cultures London: Thomson Business Press.

Kanter, R. M (1989), When Giants Learn to Dance, Routledge.

Mullins, L. J. (1993) Management and Organisational Behaviour Pitman Publishing.

Langford, D. Hancock, M.R. Fellows, R & Gale, A.W. (1995) Human Resources Management in Construction. Longman Scientific & Technical. Harlow. UK.

Latham, Sir M. (1994) Constructing the Team. Final Report of the Government/Industry Review of Procurement and contractual arrangements in the UKConstruction Industry, HMSO, London.

Latham, Sir M. (1994 b) Trust and Money. H.M.S.O. London.

Male, S. (1991) Strategic Management in Construction : Conceptual Foundations in Competitive Advantage in Construction. Male, S. & Stocks, R. (ed) (1991).Butterworth-Heinemann. Oxford.

Mullins, L. J. (1993) Management and Organisational Behaviour Pitman Publishing.

National Economic Development Office (1970) Large Industrial Sites. Report of the working party on large industrial sites. HMSO. London.

Peters, T. J. and Waterman, R.H. (1982). In Search of Excellence, Harper Row.

Porter, M. (1980) Competitive Strategy : Techniques for Analysing Industries and Competitors. The Free Press. New York.

Root, D. (1997) The impact of Culture on Construction Project Performance through the mechanism of Contract Utilisation – the case of the New Engineering Contract. University of Bath. UK.

Rowlinson, S. & Root, D. (1996) The impact of culture on Project Management: A Final Report (to be published).

Seymour, D. & Hill, C. (1996) The First-line Supervisor in Construction : A key to change? in the Proceedings of the ARCOM conference 1996 pp 655ff

Schein, E. (1985) Organisation Culture and Leadership. Jossey-Bass. USA.

Swierczek, F.W. (1994) Culture and conflict in joint ventures in Asia in International Journal of Project Management 1994. 12 (1) pp 39-47

Thompson J.L. (1993) Strategic Management (2nd Edition), Chapman and Hall.

CULTURAL DIFFERENCES BETWEEN DANISH AND BRITISH CONSTRUCTION PROFESSIONALS AND THEIR ATTITUDES TOWARDS THE EUROPEAN SERVICES PROCUREMENT DIRECTIVE

MICK HANCOCK
Danish Building Research Institute, Dr.Neergaardsvej 15, DK2970 - Hørsholm, Denmark.

Abstract

This paper concerns a study of cultural differences between Danish & UK Arkitekts/Architects, Civil Ingeniørs/Engineers and Bygningskonstruktørs/Building Surveyors and considers the implications for co-operation, competition and conflict under the centrally imposed Services Procurement Directive (92/50/EEC) of the European Union.

Analysis of data has been undertaken in three distinct areas:
1. A national comparison from the whole sample
2. A cross-national comparison of each of the professions
3. A cultural comparison of the professions within their own countries.

The study shows that cultural differences are much more significant between the professional groups within each country than between similar groups in each country or at a national level, although differences do occur within each category.

Despite cultural differences between the groups studied, opinions regarding European Procurement Directives were uniform across the whole sample and point to the need for amendments to these regulations and recommendations for wider ranging cultural studies within and between the construction industries of European Member States.

Keywords: Culture, Denmark, European Union, Procurement, UK.

Introduction

The purpose of the EU is to engender free trade, complete mobility of factors of production and the eventual harmonisation of fiscal and monetary policies among the member states. To achieve these ends, it has been necessary for the community to adopt a plethora of common and centrally imposed rules. The extent to which member states have difficulty in implementing these rules is likely to be a result of cultural differences. One such set of rules imposed upon the construction communities of Europe are the EU directives on procurement (particularly concerning tendering procedures for high cost projects and the provision of consultancy services).

Believing that culture constitutes a major variable affecting performance at both the strategic and project levels, a recent joint study, (by the University of Bath in the UK and the Danish Building Research Institute) identified significant cultures in the construction process in Denmark and the UK. and the ways in which these cultures affect the attitudes and behaviour of participants in the application, interpretation and negotiation of the EU Procurement Regulations.

EU procurement directives

Designed to encourage fair trade and an open market within the community the EU Procurement Directives require that all calls to tender with a value exceeding agreed thresholds are published in the Official Journal of the European Communities - Supplement (OJ S). All public works, supply and service contracts for members states of the EU are then subject to common legislation.

Apart from the procurement of goods (in all categories) and all purchases by Utilities authorities, both of which have been ignored for the purpose of this paper, threshold levels for contracts under Directive 92/50/EEC (applicable from 1 January 1998 to 31 December 1999) are set at:

- Contracts for supply of products and services by Central Government bodies (i.e. those subject to the WTP Government Procurement Agreement and in accordance with Schedule 1 of the Public Supply Contacts Regulations 1995) - 130.000 Special Drawing Rights. This equates to ECU 133.914 which in turn is equivalent to 988.777Dkr or GBP 104.435 (at May 1998 rates).

- Contracts for other public sector contracting authorities - 200.000 Special Drawing Rights. This equates to ECU 206.022 which in turn is equivalent to 1.521.195 Dkr or GBP 160.670 (at May 1998 rates)

Within these categories an alternative limit of ECU 200.000 (1.476.731 DKr or GBP 155.973) is set for a limited number of research and telecommunications services along with subsidised services contracts under regulation 25 of the Public Services Contracts Regulations 1993).

Larger contracts often comprise smaller individual components, the value of which may fall below the appropriate threshold levels, but avoiding the formal procedures through the process of disaggregation is prohibited. In cases where this cannot be avoided (e.g. where payment for ongoing services contracts are made in regular instalments), the regulations provide a method for calculating the threshold level.

Operation of the EU legal system & implementation of its Directives is complex. Practical problems arising from this complexity and difficulties in understanding combined with different mentalities and ways of working among Member States are likely to produce complications, misunderstandings and consequently provide a framework for conflict and unhappy working relationships. Given these points we might consider that, the competition promoted by the EU is not necessarily distributed fairly and may lead to a weakening of adherence to EU regulations.

Difficulties in defining and measuring culture

Despite its long historical consideration, an agreed definition of "culture" has been difficult to attain. Kroeber and Kluckhohn (1952) demonstrate this in their presentation of a list of definitions stretching from the 18th century. According to Jahoda (1995) the problem lies in confinement to the definition of the single word culture. He argues that this is misleading in that the impression is given that no general concept of the phenomenon underlies the modern usage of the term. Furthermore, Jahoda (1993) doubts whether a consensus definition is even possible, since culture is a social construct dealing with various and complex sets of phenomena and so its meaning will vary according to the user's purpose ie. what any of us accepts as a definition depends on what aspects of the world we are interested in.

In this study it is Hofstede's (1984) treatment of the term that has been adopted i.e. that culture is a kind of collective mental programming of people in a particular setting; a set of values, beliefs, attitudes and patterns of behaviour which are shared by a group of people in an environment. Hofstede admits that this is incomplete as a definition, but does represent what has been measurable in his work. Despite Hofstede's assertion as to the measurability of culture within his interpretation he actually suggests "differentiating" between the key factors within cultures which he offers as:

- power distance - the emotional distance between a boss and subordinate
- uncertainty avoidance - the desire for certainty about the future
- individualism
- masculinity / femininity

In fact, measurement of the component parts is rife with difficulties, particularly in attempting to identify cultural groups and boundaries. In a construction context this issue is clearly demonstrated at the project level. Here, cultural factors for participants in the temporary and shifting coalition of people who make up the team will be initially influenced by national, local and organisational determinants. The complexity is then increased as the project develops a culture of its own: a problem here is that the culture is not necessarily stable as frequent changes of personnel and relations between site and area / head office vary throughout its course. Accordingly, in this study, the subject has been considered at the national, sectoral and participant (consultants) levels.

The decision to base the research model on Hofstede's Value Survey Module (VSM) was taken despite a common perception of his work being that he has attempted to "measure the unmeasurable" (Root *et al* 1997: p.34) and that his attempts to measure cultures on a variety of indices are meaningless. Root *et al*'s observation demonstrates an understanding of the predominance in the construction management community of a mindset rooted in the rational philosophy of western science, where only that which can be measured according to the rules of objectivity and therefore provide "proof", has credence. Root et. al (1997: p.41) believe that the reliance on a single research paradigm is a "damning indictment on research carried out in this discipline" and accordingly (as a joint author of that paper) that criticism was addressed in this work. Hofstede's general model (a positivist approach) was then adapted to suit the particularities of this study and the results used in combination with and as a guide to interviews (an interactive approach).

Adaptation of Hofstede's Values Survey Module (VSM)

Hofstede's original VSM questionnaire comprised 33 questions and was applied across a wide ranging sample of employees within a single multi-national organisation. Given the nature of the sample to be tested in this project adjustments were made to a number of questions. Two questions, specific to traditional role and membership of professional bodies in the construction sector were included along with twelve questions concerning work related values specific to the construction sector, bringing the number of questions to 47 . For Hofstede, identification of national cultural characteristics was the main aim of his research. In this project however, such information constituted only one part of the overall objective which was to determine some practical significance of the characteristics to a specific situation i.e. the imposition of centrally imposed EU Procurement regulations. In order to relate the results of the VSM survey to the overall objective a third section of 16 questions was added to the questionnaire which investigated knowledge of and opinions about the EU Procurement Directives.

In all, the questionnaire consisted of 63 questions. These questions fall into four groups, each specific to their purpose i.e.,

- questions identifying ecological factors; age, sex, education etc.
- questions identifying work related values relating to Masculinity and Individualism
- questions tied to specific psychological indices/dimensions (Power Distance and Uncertainty Avoidance)
- questions relating to the EU Procurement regulations

The research format in both the UK and Denmark, was as follows:
- distribution of VSM questionnaire
- preliminary analysis of questionnaire data
- interviews with selected respondents

Difficulties encountered in adapting Hofstede's questionnaire

The original intention was to replicate Hofstede's study, but this proved impossible in respect to the Individualism and Masculinity factors due to the unavailability of detailed information regarding the form of factor analysis used in his work. Consequently the most commonly used form (Maximum Likelihood combined with Varimax Rotation) was applied to the questions that Hofstede had identified. An interesting result of this was that the analysis revealed only three (rather than four) cultural dimensions. The first (and strongest) of these was a fusion of Hofstede's Individualism and Masculinity dimensions: the other two were compatible with his Power Disatnce and Uncertainty Avoidance categories. Accordingly, it was necessary to prepare a single combined index which I have labelled the "Machismo" index. Whilst Hofstede's Individualism and Masculinity dimensions may not have been derived from the same form of Factor Analysis, the questions analysed were the same and the result is believed to be a result, not of the statistical analysis, but of the construction sector itself. Hofstede believes that a concept of "social-ego" is central to his Masculinity dimension and it is considered quite likely that this is so deep-seated in construction that it has become embedded within the Individualism factor. That construction is a sector dominated by masculine values and attitudes may not come as a surprise to anyone, but this particular finding may well go some way towards substantiating it.

A second difficulty in the study concerned the translation of the questionnaire. In order to ensure that the questions meant the same to the Danish respondents as they did to their British counterparts a double translation was used. Firstly, the questionnaire was drafted in English and then translated into Danish. The initial translation was then tested on a small group of Danes in the construction sector to test their understanding. It soon became clear that some of the questions, and /or phrases used simply did not translate. Having attended to discrepancies and difficulties identified by the small pilot group, the Danish version was then translated back into English by a member of the Danish construction sector with a first rate capability in the English language. The process outlined above was then repeated to erase any anomalies before being distributed to potential respondents.

Data collection

In total 366 questionnaires were distributed in Denmark and yielded a usable response numbering exactly 100. In the UK, 521 questionnaires were distributed resulting in a usable sample of 142. These were distributed as follows:

The Danish sample
Arkitekts - 44
Civilingeniørs - 45
Bygningskonstruktørs - 11

The UK sample
Architects - 52
Civil Engineers - 51
Building Surveyors - 39

Whilst the number of Arkitekter/Architects and Civilingeniører/Civil Engineers in the sample are both roughly equal and large enough to provide reasonably acceptable statistical results, the Bygningkonstruktører/Building Surveyor samples present some difficulties. With the UK sample of this latter group being more than 3 times as large as the Danish sample, it was inevitable that some inbalance in the results would occur if groups were pooled and their opinions then compared to other pooled groups. In the event, this has not been done, so that whilst the situation is not ideal, it is sustainable. A further problem exists in that the sample of Danish Bygningskonstruktører is very small (only 11 respondants). Ordinarily, this would prohibit their inclusion in any "proper" statistical testing, but in this case, where a main priority of the research was to establish a basic working methodolgy for such a study, they have been allowed to remain: the results for this group are however, treated with extreme caution and cannot be taken to be necessarily representative of that professional group.

The interviews

The research comprised a two part process beginning with the collection of "hard" data from returned and completed questionnaires. Analysis of the data thus gathered, enabled the identification of specific areas for investigation through the means of semi-structured personal interviews.

Interviewees were chosen at random from within each group being studied. A total of twelve interviews were conducted (six in each country) either in person or by telephone. The number of interviews conducted is smaller than would be necessary in order to draw secure conclusions: nevertheless, they do provide supporting evidence and a flavour of the opinions of the professional groups within each country.

Results

Hofstede's original work resulted in sets of indices which enable relative comparisons to be made between cultural dimensions of nations. In this research, Hofstede's method has been followed, but as explained earlier it has not been possible to fully replicate his study. As a result, the indices produced in this study cannot be directly compared with those produced by Hofstede.

Indices for Machismo (MAC), Uncertainty Avoidance (UAI) & Power Distance (PDI)

Following use of the method of Factor Analysis known as "Most Likelihood / Varimax Rotation" new indices were calculated thus:

Table 1 - Indices for Machismo (MAC), Uncertainty Avoidance (UAI) & Power Distance (PDI)

DENMARK	MAC	UAI	PDI	UK	MAC	UAI	PDI
Arkitekter	59	50	41	Architects	51	52	45
Civilingeniører	56	43	45	Engineers	48	45	49
Bygnings - konstruktører	55	52	48	Building Surveyors	47	53	52

Note: The indices are calculated on a scale of 1-100. As a result, when indices are referred to as "low" the reader may infer that the score is below 50, whilst "high" denotes an index in excess of 50.

Data analysis constituted a combination of approaches ranging from the complexity of Factor Analysis & the compiling of indices to simple statistical testing (identifying significant differences by means of the Chi Square test) and analysis of individual questions through interview. It is not possible in a paper of this length to report in detail on the results and findings, but the following summarises the main points.

Firstly, the results very clearly demonstrate is that the strongest identification of cultural differences did not occur at the national level (as it did in a similar comparison of The UK and Hong Kong - Rowlinson & Root, 1996), or between comparable professional groups in both countries. By far, the greatest number of cultural differences were identified between the professional groups within their own countries.

The professions and Machismo

Statistical analysis showed a significant difference between countries, but not between occupations. The fact that the Individualism and Masculinity dimensions are inextricably combined in this study can be interpreted as being a direct result of the nature of the construction industry from which the sample populations were drawn. It is suspected that the overwhelming male dominance that exists in the industry, has meant that respondents replies to questions have been so consistent that the Masculinity component has not been identified separately, but has simply been subsumed within the Individualism dimension. This has occurred even in the Danish sample, where some 11% of the sample were women. This may lead us to conclude that women in the Danish construction industry accept and play by the men's rules at work; there is no reason to believe that it is any different in the UK. It is also perhaps a sign of just how masculine the industry is, that despite the growing masculination of women, the industry is still not feminine enough to attract greater numbers of that sex.

The professions and uncertainty avoidance

The desire for mental security about the future underpins the notion of Uncertainty Avoidance and is measured through attitudes towards rules, stability of employment and levels of stress. Between the two countries, virtually no difference was found.

The professions and Power Distance

The scores for Power distance represent the perceptions of respondents in respect to the degree of equality between boss and subordinate in the workplace and are calculated from data concerning management style, decision making methods and fear of disagreeing with the boss. A low score indicates more equality , a preference for consultative management and low fear of disagreement with superiors.

Arkitekts and Civilingeniørs in both countries score low, and this may be a result of their working practices and the structure of the professions. For the most part firms representing these groups are small and cannot afford to employ professional managers. Management is carried out by senior members who continue to practise as designers alongside their subordinates. A high Power Distance in these conditions would tend to hinder the smooth running of such offices.

Bygningskonstruktørs / Building Surveyors score highly and this may well be a reflect a combination of the same factors in reverse with attitudes inculcated during their education. This group see themselves as professional "all-rounders". It is a matter of some pride that their education and work requires them to have quite specialised knowledge in a wide range of construction related areas, including management and economics, which the Arkitekts/Architects and Civilingeniørs /Engineers don't have to the same degree. In the work situation, Bygningskonstruktørs / Building Surveyors spend less time working in teams (which necessitate a low PDI) than is the case with Architects and managers tend to be more removed from daily operational activities.

Inter-professional comparisons (in both Denmark and the UK)

In both of the inter - professional comparisons the results were largely similar (in both countries) with stereotypical images being confirmed along with the accompanying prejudices and beliefs held by members of one profession about the others. However, the images and opinions held about members of a group are frequently different from those held by them.

The following significant differences between the groups were demonstrated:

- Arkitekts/Architects and Bygningskonstruktørs / Building Surveyors place a higher degree of importance on having good working conditions than Civilingeniørs / Engineers.

- Arkitekts seem to have more confidence in themselves than the other groups, but rather less confidence / trust in other people.

- Bygningskonstruktørs / Building Surveyors are subject to higher levels of anxiety at work than either Arkitekts/Architects or Ingeniørs / Engineers. Whilst this may lead this particular group into more instances of conflict, the supporting evidence would seem to show that the cause is not a matter of aggression on the part of the profession, but rather one of uncertainty brought about by working in a very wide-ranging discipline.

- Bygningskonstruktørs / Building Surveyors are less likely to break rules and behave unconventionally at work than either of the other professions.

The causes for these differences are not all clearly determined or clear-cut, but there do appear to be three main features, that all parties agree, have a significant influence on the professions:

1. Each profession is greatly affected by the education and training they receive.
2. The breadth of the professional discipline i.e. whether wide or narrow and .....
3. The extent to which teamwork is required in fulfilling one's role.

The European Union procurement directives

Uncertainty of future workload

Generally, both British and Danish Arkitekts show little belief in the idea that the Directives will reduce uncertainty over future workload, although a significantly higher proportion of the British sample are hopeful in this respect. Negotiation rather than competitive tendering, and the establishment of longer term relationships through strategies such as partnering are very common approaches in Denmark and so it is perhaps not suprising that the consultants interviewed expressed such a strongly negative view of the EUPD in respect of workload certainty.

The EU belief is that the Directives provide all with an equal chance of gaining work, but during periods of economic recession it is not unusual for tenderers to bid at below cost in order to win the work. The overall result is an increased level of conflict within the industry. This then begs the question as to who is likely to be the main beneficiary? None of those interviewed appeared thought that consultants would benefit from the introduction of the Directives.

The EUPD and value for money

At a conceptual and theoretical level, the term Value for Money (VfM) presents numerous difficulties in definition. Nevertheless VfM is a commonly used term (in construction and other fields) and despite their cultural differences, all of those interviewed held the same notion of VfM: each in their own way suggesting that getting value for money means ensuring that benefits are balanced with costs. With regard to whether or not the EUPD will help Clients to achieve VfM there were some minor differences of opinion, but in general the view seems to be that enforced competition does not hold long term benefits for Clients, because you don't get what you don't pay for.

The prospects for transnational co-operation under the EUPD

Civilingeniørs/Engineers and Bygningskonstruktør/Building Surveyors differ in their ideas about transnational co-operation resulting from the EUPD. The samples were somewhat divided here with British Engineers demonstrating a much more optimistic view than the Danes. This result was reversed among the Building Surveying / Bygningskonstrucktør groups, but the admitted lack of knowledge of the EUPD by these groups combined with a lower likelihood of being involved in such projects renders this finding less than certain. Overall there was a vague sense that transnational co-operation is OK provided that national interests are given priority. This is in line with the general view of the EU held in both Denmark and the UK; two member states concerned to avoid complete immersion in a Federal Europe.

Working within the EUPD regulations

Those interviewed generally expressed a preference for a set of guidelines or framework rather than a set of rigid rules. One interviewee suggested that something like the Building

Regulations, where the objectives must be achieved, but the method is open to interpretation, would be preferable. This view is not surprising given that they are all employed on the basis that they are people educated, trained and experienced in problem solving and decision-making. This finding is also in line with the identification (in this study) of the Machismo index which combines the individualism associated with professional workers and the masculinity associated with the construction industry.

Whilst the EU clearly believes that these directives will be of benefit to both Clients (achieving value for money) and to consultants (providing equal opportunities for work) the clear majority of consultants do not hold this view.

Concluding remarks

This study has attempted to determine the relative degrees of homogeneity and heterogeneity between compared groups at three levels i.e. national, industry and professional participant.

In brief, the findings show that cultural differences are much more significant between the professional groups within each country than between similar groups in each country or at a national level, although differences do occur within each category.

In respect to the European Procurement Directives, there tended to be a general agreement amongst all participants regardless of their cultural grouping. It was not however established whether the views expressed were common as a result of some unexamined culture of professional attitudes in general, or out of a general aversion to change. In the main, it was agreed that the European Procurement Directives were:

- unlikely to help improve uncertainty of levels of future workload
- unlikely to help Clients achieve "value for money"
- unlikely to improve opportunities for transnational co-operation (although British Engineers showed some small signs of optimism here)
- unlikely to create more opportunities for small firms to win work.
- too rigid as a set of rules, and that a set of simple guidelines would be preferable.

The foregoing, demonstrates that cultural differences are a reality even within the narrow focus of this project which has only been concerned with 3 sub-groups of professional workers within the industry. Given a rough commonality of education, training and professional values amongst those surveyed, it is suspected that a study that included a wider variety of respondents e.g. clients, contractors and siteworkers would have yielded rather less subtle differences than have been uncovered here.

References

Hofstede Geert (1984) *Culture's Consequences: International Differences in Work Related Values,* Sage, London

Jahoda G, (1995), *The Ancestry of a Model*, in Culture & Psychology Vol.I, 1, 11-24

Kroeber A.L. & Kluckhohn C., (1952) Culture: A critical review of concepts and definitions, in *Papers of the Peabody Museum of American Archaeology & Ethnology*, 47 (1), Harvard University Press, Cambridge, MA

Root D.S., Fellows R.F. & Hancock M.R (1997) Quantitative versus Qualitative or Positivism and Interactionism - A Reflection of Ideology in the Current Methodological Debate, *Jounal of Construction Procurement*, Vol.3 No.2, pp 34 - 44, 1997

Rowlinson Steve & Root David (1996) *The Impact of Culture on Project Management*, unpublished report to The British Council, Hong Kong

Williams R. (1983) *Keywords*, Flamingo, London

FUTURE CHALLENGES IN CONSTRUCTION MANAGEMENT: CREATING CULTURAL CHANGE THROUGH EDUCATION

DENNIS LENARD
Faculty of Design Architecture and Building, University of Technology, Sydney
P.O. Box 123 Broadway NSW 2007 Australia. Dennis.Lenard@uts.edu.au

Abstract

The next wave of industrial development will necessarily incorporate the adoption and utilisation of innovative technological processes and developments, and see the emergence of highly focussed organisations capable of exploiting transient and niche markets. This environment demands a responsive and dynamic construction industry with the diversity to cope with and initiate change, that is capable of employing a range of approaches to the procurement and delivery of construction projects. This will necessitate a much stronger emphasis on innovation than ever before and this must start in the Construction Management Schools. Education is the key and fresh approaches in education need to be developed. The suggestions made in this paper are based on an extensive study of the Australian Construction Industry and the following issues are discussed in this paper, in terms of how the education process can create cultural change and foster innovation in the construction industry.

Keywords: Construction, culture, education, innovation.

Introduction

The importance of innovation has long been recognised by the manufacturing sector, where it is regarded as a tool for maintaining competitive advantage, process improvement and cost efficiency. However, the construction industry has rarely shared the same vision nor, until recently, has it fostered research on the nature, generation, scope and application of technological or organisational innovation. Certainly, there have been many technological advances in the construction procurement process since the Second World War but these have largely been developed in response to the increasing level of litigation.

The suggestions made in this paper are based on an extensive study of the Australian Construction Industry (Lenard 1996). The first part of this investigation was to identify the essential characteristics of an innovative organisation; what factors best promoted innovation and the adoption of innovative ideas; and how this approach could be achieved in the construction industry.

The first stage of the investigation involved an initial field study of six Australian construction companies, followed by an extensive survey, comparing the outlook, organisational structure, culture and management personnel of a number of manufacturing companies well known for their innovative approach, with companies selected at random from the construction industry. The results indicated there are significant differences in the approaches and priorities of the two sectors, and identified a number of key areas where the construction industry needs to improve its organisational structure and management style if innovation is to become a key objective.

Manufacturers placed great emphasis on strategic planning, research and development, and investment in new technology. Considerable effort went into extending their market share, improving products and monitoring competitors. The organisational structure was generally non-bureaucratic, yet processes and projects were monitored, regulated and reviewed on a regular basis. Management personnel were drawn from a diverse range of backgrounds and employed an equally diverse range of management styles. They were highly motivated and promoted a culture of innovation as a way of improving the organisation's competitive edge. Importantly, innovative ideas were encouraged and rewarded.

The construction industry focused on project management rather than strategic planning and as a result tended towards a reactive rather than pro-active stance. While construction organisations perceived themselves as 'risk takers', innovation generally occurred outside the industry and was only adopted once a product or technique had been tried and tested elsewhere. This approach was in direct response to the low capital intensity and poor operational profit margins evident throughout the industry. Cost efficiency rather than innovation was seen as the way to maintain competitive edge and on the whole, the emphasis was on maintaining rather than extending market share as cost restricted efforts to seek out new markets.

Construction organisations were highly structured with a clear chain of command and a cost-efficiency driven management style. Whilst this ensured correct procedures were followed, it often inhibited innovation and restricted the learning process. The industry demanded and attracted management personnel predominantly from construction/civil engineering backgrounds. Although this provided a solid knowledge base in the construction discipline, it also created a mono-culture that may not provide the impetus for deviation from established practice.

There was also a significant difference in the attitudes and perceptions of management personnel from the two industry sectors. Personnel from the manufacturing sector generally identified with and supported the organisation for which they worked and perceived this support was reciprocated and rewarded.

The perceptions of construction industry participants were less positive. On the whole, management and senior management personnel regarded their positions as transient and, in the long term, saw themselves making several career moves. This attitude permeated through the organisation and has created a predominantly static organisational structure, with less incentive to explore new technology or deviate from established practice.

The construction industry is subject to a number of historical, industrial and market forces that perpetuate the existing culture and management style, and inhibit the extent to which the industry can initiate change. Cost efficiency and quality can only provide incremental improvement. Innovation, with its potential for large scale improvement, is the key to competitive advantage and the concept must be promoted in the education process.

Additional drivers for a revitalised curricula promoting cultural change

Impact of Information Technology

The development of the "information technology" enables the exchange of massive volumes of numerical, textual and graphical information at high speed and at relatively low cost. This allows a new form of connectivity through the electronic media. Construction data can now be stored in electronic form and be distributed to students quickly. The technology helps to increase students' access to data and research material; and scholars' ability to exchange research results. Interaction with the lecturers and fellow students is facilitated directly through e-mail, and information and teaching materials can be transmitted as attachments. The University of Technology, Sydney now has a number of courses offered in construction and based on this mode of delivery throughout South East Asia.

Explosion of new knowledge

New knowledge in construction dictates the necessity for a flexible and responsive education process. Students need to have up-to-date knowledge and world's best practice solutions need to be high on the agenda.. Curricula needs to be regularly reviewed so as to keep pace with the development. Professors need to take a leading role to examine critically the appropriateness of the contents of teaching programs to check whether they respond to the needs of the industry

The concept of "through-life" learning needs to be established at University and the concept is presented below.

Construction organisations are endeavouring to develop a learning environment. Initiatives in this area included the adoption of workplace reform, the use of smaller, highly skilled and flexible site teams, and the creation of partnering arrangements between client and contractor. However, a reluctance to invest time and resources into new technology remains.

If the industry is to develop a learning environment (as shown in figure 1) it must accept that initial use experimentation, development, refinement and continued iteration through each of these steps usually creates frustration.

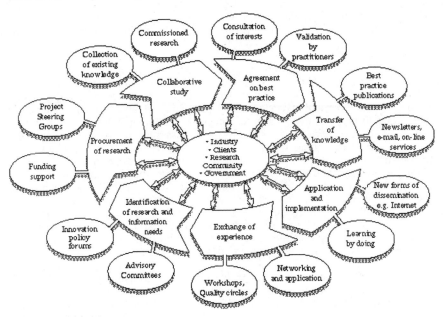

Figure 1: The organisational learning circle

Impatient and production-oriented construction supervisors may find this phase especially difficult. However, if time allows, practising the operation prior to actual use on a permanent construction can decrease risk (and anguish). Again, this is also a matter of investment and cost-benefit analysis. Without a sufficiently large and detailed frame of reference to assess such costs, it may well be impossible for project managers to determine the risks and benefits associated with experimentation, prototyping, simulation and other forms of learning.

The industry should also capitalise on knowledge gained on individual projects. At present, there is little transfer of knowledge gained on one project to another. This is largely because projects tend to operate autonomously, therefore conventional organisational processes are generally unsuccessful in transferring information. Informal communication within the organisation and parties to the construction process can assist knowledge transfer, but this is often unreliable and important opportunities may be overlooked. Establishing specific mechanisms for the transfer of information therefore is a high priority. Senior managers can organise meetings to appraise organisational performance (and failure) on a particular project. However, to select a particular innovation or the most appropriate means of transfer requires analysis of the technology itself; its likely application on future projects; existing means of coordination between projects; and its comparative cost advantage. This can only be achieved through the establishment of an industry-wide database and expert knowledge in the area of costing.

Advanced technology

Advanced technology represents a wide variety of modern technologies including computing, telecommunications, artificial intelligence, fuzzy logic, database and expert system technologies that improve the efficiency and effectiveness of manufacturing processes. The prime motivation for installing such technology is to improve an organisation's competitiveness. The adoption of advanced technology can bestow not only operational benefits but also marketing and strategic benefits, including increased market share, improved responsiveness to changes in the market place, the ability to offer a continuous stream of customised products, faster product innovation and improvement of the company's image.

Each technology offers its own basket of operational, strategic and marketing capabilities. By judiciously combining technologies, management should, theoretically, be able to obtain desired levels of system performance. Design and engineering technologies (such as CAD and CAPP) are particularly beneficial to companies involved in the manufacture of products prone to rapid design and processing changes, or in industries with highly differentiated products. Fabricating/machining and assembly technologies offer the opportunity to reduce set-up times which in turn allows for faster changeovers as well as improvements in the consistency and quality of manufactured products. Automated material handling systems allow for reductions in in-process inventory and for improved safety. Communication and control technologies have been credited with improving processing and procedural efficiencies and assisting in the integration of factory operations. The adoption of just-in-time and manufacturing resources planning offer such benefits as stable production rates and balanced capacity. Table 2.3 presents a partial list of advanced manufacturing technologies, their capabilities, and the competitive opportunities offered as a result of these capabilities.

The emphasis on innovation is crucial here because sheer technology alone does not necessarily generate industrial and social advances. The internet is a perfect example of a case in point. From its development in the late 1960s to early 1980s it remained largely the province of technocrats and researchers; only with the widespread adoption of PCs, modems and bulletin board services that the internet spread into the public domain. While internet technology had to exist, is was equally as important that end-users incorporated the technology into their work environments and lifestyles. That is to say, the culture that evolved around the internet was as innovative as the technology itself, and ultimately proved to be the most important factor in its growth.

The construction market is becoming increasingly competitive and organisations must achieve a level of sophistication not previously envisaged. New technologies are developing to allow significant performance improvement in construction. Further, it has been argued that innovation is fostered when supporting technologies (such as those listed

Table 1 Competitive opportunities from advanced manufacturing technologies (see glossary for definitions)

Technology	Capability	Competitive Opportunities
CAD/CAD CAM/CAPP	Reduction in design to production lead time Better engineering analysis Better design	Reduced new product lead time Faster response to customer requests Better products Faster response to customer needs
CAM/FMS ROBOTS/MRP	Reduce work in progress Higher productivity	Reduced costs
CAM/FMS/ MRP	Reduction in manufacturing lead time.	Reduced delivery times Faster product innovations
ALL	Error reduction Higher quality Consistency Economies of scale Parts variety reduction	Higher quality products Reduced costs Higher product variety at the same cost Same product variety at reduced cost.

Source: Adapted from Voss 1988.

above) are available and easily adapted to encourage modification and process improvement. Computer databases, robotics and automated process control are examples of fast-developing technologies that offer promise.

It is therefore argued that embracing this technological innovation is crucial if the construction industry is to regain competitiveness; and, being primarily focused on delivering services, the industry is ideally suited to take advantage of such technology. Educationalists have a role to play in promoting the use of the technologies.

Future needs in construction management education

The following discussion offers a different approach to construction management education with a key focus of creating cultural change to foster innovation in the construction industry.

Education for the enterprise of the future

Information technology provides the means whereby organisations that have remained fundamentally unchanged for decades, and arguably for centuries, can be transformed. The theory is that the new structure is possible when each member understands the team vision; has the competencies required; has the trust of others; and, importantly, has access to the information and tools required for functioning and collaborating within the team in a broader context. While critics may argue that that such radical change is not possible in the construction industry, the increasing number of information based enterprises indicates that that this scenario is achievable.

The structure of the new enterprise is shifting from a multi-layered hierarchy to flatter networks, or relatively autonomous organisations. The responsive, entrepreneurial business team is becoming a key organisational entity rather than the traditional department locked into a traditional organisation chart. The concept of the organisation is being expanded to include links with external business partners, suppliers and customers. The resource focus is shifting from capital to human and information resources. Rather than remain static and stable, the enterprise must be dynamic and constantly changing. The professional (not the manager) is emerging as the central player, often working in multi-disciplinary teams that cut across traditional organisational boundaries. Interpersonal commitment, rather than traditional reward and punishment mechanisms, are the desired basis for organisational cohesion and stability.

The new project team is self-managed. Individuals are empowered to act, and do so responsibly and creatively. Freed from bureaucratic control, they take initiatives and even risks to get closer to customers and work more productively. They are motivated by one another to achieve team goals rather than to satisfy superiors. With common interests that are immediate and clear, cooperation flourishes.

In the construction enterprise of the future, the project is a working-learning environment, where individuals develop strong specialised expertise and broader competencies, not just specific skills. The notion of learning job skills that require periodic updating is replaced with the notion of life-long learning. Income is tied to level of competence and accomplishments rather than to position in the hierarchy. Information technology has enabled a reduction of the middle layers of management and separate departments housing specialised information and knowledge are not longer required.

As well as construction, the construction enterprise of the future will supply management and organisational expertise and provide knowledge based services. The new style curriculum should look like the following:

- research and development on world best practice solutions;
- feasibility;
- marketing;
- design;
- information systems;
- cost time and quality systems;
- work breakdown structures;
- facility and asset management;
- construction technology solutions including advanced manufacturing technology;
- components and systems;
- financial, legal and logistical functions;
- information on operative productivity and processes.

Concluding comments on aspects of education and integration

Fundamental to the new approach to education is the concept of enterprise integration. The Japanese *keiretsu* (the association of a number of industries centred on a bank) has proved to be particularly effective in reducing time to market and creating long-term competitiveness. Variants of the *keiretsu* are emerging in North America and Europe, with companies such as Ford and IBM acquiring equity positions in suppliers, and participating in various external organisations. Underpinning to the success of such partnerships is a rapid and free information exchange and major business transformations where entire processes (both production and management) are integrated. Paper-based systems, bureaucratic approval processes, labour intensive clerical activities, project development phases, production cycles, and multi-layered decision-making processes are being replaced by source data capture, integrated transaction processing, electronic commerce, real-time systems, on-line decision support, document management systems, and expert systems. These tools emerge as mechanisms to drive change and create a climate for innovations. Such management concepts must replace the outdated organisational management curricula.

In recent years, there has been a steady move from national to international competition, with emphasis on the expansion of free trade zones and the removal of access barriers.

This has presented a number of challenges and opportunities for the Construction industry, on the one hand the opportunity to capitalise on markets opening up in Asia (for example China and Korea), on the other, the local industry now has to compete with aggressive competitors) not only for overseas markets, but in terms of its market share within local markets. To remain competitive, construction companies must be able to adapt to rapidly changing market conditions, competitive threats and customer demands.

To this end, a number of construction organisations have established alliances with other key players in both similar and disparate markets. These partnerships (which involve enterprises of all sizes) involve the creation of research and development consortia, joint ventures and cross-licensing arrangements. Such relationships enable organisations to develop comprehensive approaches to markets; jointly fund large efforts in their common interests; respond quickly to new or ephemeral opportunities; create new markets; share information; combine as interest groups or lobbies; and rapidly expand geographically.

It is often assumed that the global market place is the domain of larger enterprises, as competing in this environment requires substantial investment in communication networks, databases and high level telecommunication. However, strategic partnerships (such as those outlined above) can lower the risks substantially, allowing smaller entities to specialise in areas where they have competitive advantage and then augmenting their services with intelligent systems provided by partners. This desegregation of services is in sharp contrast with the vertical integration evident in large multi-national construction companies. This flexibility must now be emphasised and encouraged in the education process.

Education needs to lead the change. Traditional outdated subjects such as residential construction, contract administration, simple cost estimating etc should dissappear and be replaced with advanced technolgy construction solutions, world's best practice, data access and retrieval systems, and professional construction services.

References

Lenard D., and Bowen-James A., 1996a, Innovation: The Key to Competitive Advantage, *Research Report* no. 9, Construction Industry Institute, Adelaide, Australia

Voss, C A 1988, 'Success and failure in advanced manufacturing technology', *International Journal of Technology Management*, 3(1), pp. 285-297.

Glossary of advanced technology

Automatic Guided Vehicle System (AGVS)
> Vehicles equipped with automatic guidance devices programmed to follow a path that interfaces with work stations for automated or manual loading and unloading of materials, tools, parts, or products.

Automatic Storage and Retrieval System (AS/RS)
> Computer controlled equipment providing for the automatic handling and storage of materials, parts, subassemblies, or finished products.

Cellular Manufacturing (CM)
> A type of factory layout in which machines are grouped into what is referred to as a cell. Groupings are determined by the part families that require similar processing. Cells can consist of one or more machines but each cell must be arranged to handle all the operations necessary for a group (family) of similar parts.

Computer-aided Design (CAD) and/or Computer-aided Engineering (CAE)
> Use of computers for drawing and designing parts or products and for analysis and testing of designed parts or products.

Computer-aided Design/Computer-aided Manufacturing (CAD/CAM)
> Use of computers to help plan the manufacturing process. CAPP is usually considered to be the link between CAD and CAM.

Computer-integrated Manufacturing (CIM)
> A system for linking a broad range of manufacturing and other organisational activities through the interconnection of computer systems in the plant. A CIM system may integrate engineering design, flexible manufacturing systems and production planning and control, or simply connect two or more FMS through a host computer.

Distributed/Direct Numerically Controlled Machine (DSC)
> A single machine controlled by program instructions which are stored in a separate computer attached to the machine through a communications channel.

Flexible Assembly System (FAS)
> A flexible manufacturing system dedicated to assembly. An FAS can include automatic insertion machines, robots, automated inspection systems and computer control systems.

Flexible Manufacturing System (FMS)
> Two or more machines with automated material handling capabilities controlled by computers or programmable controllers, capable of multiple path acceptance or raw materials and multiple path delivery of finished product. this system may be linked in series or parallel.

Group Technology (GT)
> The process of identifying items that have similarities in design characteristics or manufacturing characteristics and grouping them into part families.

Just-In-Time (JIT)
> An approach that emphasises continual efforts to remove waste and inefficiency from the production process through small lot sizes, high quality and teamwork.

Manufacturing Resources Planning (MRP II)
> The integration of MRP with production control, purchasing, marketing, finance and other organisational resources.

Material Requirements Planning (MRP)
> A computer-based information system designed to handle ordering and scheduling of dependent-demand inventories (eg. raw-materials, parts and subassemblies). A production plan for a specific number of finished products is translated into requirements for component parts using lead time information to determine when and how much to order.

Materials Working Laser
> Laser technology used for welding, cutting, treating, scribing, and marking.

Numerically Controlled/Computer Numerically Controlled Machine (NC/CNC)
> A single machine either numerically controlled (NC) or computer numerically controlled (CNC) with or without material handling capabilities. NC machines are controlled by numerical commands punched on paper or plastic mylar tape while CNC machines re controlled electronically through a computer residing in the machine.

Pick and Place Robot
> A simple robot, with one, two, or three degrees of freedom, which transfers items from place to place by means of point-to-point moves. Little or no trajectory control is available.

Robot
> A reprogrammable multi-functional manipulator designed to move materials, parts, tools or specialist devices through variable programmed motions for the performance of a variety of tasks.

8.   Information and Decision Systems

THE DEVELOPMENT OF A RICH INFORMATION MODEL USED TO IDENTIFY THE FACTORS CONSTRAINING DESIGN DEVELOPMENT

M. A. MIOR AZAM, A. D. ROSS, C. J. FORTUNE and **D. M.JAGGAR**
School of the Built Environment, Liverpool John Moores Universtity, Clarence Street, Liverpool, L3 5UG. D.JAGGAR@livjm.ac.uk

Abstract

This paper aims to report upon ongoing research at Liverpool John Moores University investigating decision making within design teams in design and build organisations. The aim of the research is to develop an information model which reflects the constraints that exist in the teams involved in design development, consider the impact procurement arrangements have upon team learning and culture. The paper will report upon a literature search in the area of design development, consider how contractors have responded to the design and build procurement route in the United Kingdom and suggests some of the factors that impact upon decision making by the design team. Some preliminary findings will be reported upon and a research approach is identified for the next stage of data collection.

Keywords: Design decisions making, group communication, information modelling, procurement.

Introduction

Since the publication of the Latham Report (1994), much work has focused on improving the management of the construction process. One of the areas within construction management being given attention, partly due to the increased uptake of Information Technology (IT), is the field of construction information management. Construction as a process, relies heavily on information being communicated effectively between the parties to ensure the successful management of projects. The problems of effective communication are primarily due to the multi-party, fragmented nature of the building industry, particularly between the parties involved in the design and production stages. To redress the situation, the industry, has invested resources in ongoing initiatives to identify and find new ways (and technology) where integration can be made possible. This can be seen in the introduction of new procurement arrangements, management techniques and highly sophisticated information technology.

New and alternative procurement arrangements have been seen as having the potential to achieve better productivity by improving the level of integration between building design and production (Akintoye 1995). Within the context of information management what was hoped from the application of these alternative procurement routes was that information both in content and meaning, would be communicated more effectively. This is particularly the case when design and construction personnel work towards a common object under the Design and Build procurement route. Within the Design and Build route, parties managed under one administration, it was anticipated that information on production cost and buildability (i.e. information essential to economic design) would be incorporated. However current research has established this was not the case (Ross and Fortune, 1997). From a sample of 92 leading design and build contractors in the UK it was established that the mahjority use designers who are "external" to their organisation. The type of project undertaken was found to have little influence on this practice.

Type of Project	Internal	External
Utilities	12	88
Industrial Facilities	10	90
Admin. and commercial	6	94
Health and Welfare	10	90
Residential	5	95

Figure 1: Location of designer by type of Project, generally(% response)

The survey also captured data on the criteria used by contracting organisations in selecting designers which indicated that the production of an economic design was the most important criteria for the selection of an external designer. The following were also identified in descending order of importance, past experience of design team, good record in design and build projects, quality of design, familiarity with project type and potential reciprocity for future projects.

Economic design	88
Quality	67
Durability/maintainability	50
Aesthetic	33
Buildability	33
Standardisation	23

Figure 2: Contractors criteria for selection of designer

The overall results of this earlier work have shown that there is some evidence to support the common perception of design and build as a procurement option susceptible to poor quality product in that, as well as the location and selection of the designer, there seems to be little use of sophisticated budgeting techniques and more significantly evidence of low usage levels of life cycle and value related techniques as tools.

The survey also considered information flows and sixty nine percent of contractors pass to their designer at conceptual design stage, an outline budget and sixty one percent of these respondents considered that this information flow was important for design development. The format of this was generally cost per m2..The results also indicated that contractors were influential in determining the quality of the design by passing outline specification information to the designer. Seventy three percent of the respondents indicated that they felt that this was also an important information flow.

These results and follow up intervierws indicated that contracting organisations in the UK were tending to rely on more traditional forms of estimation. The results have shown that information used by the estimator in estimating the cost of resources tends to rely upon historic project cost data. This is shown in figure 10 below.

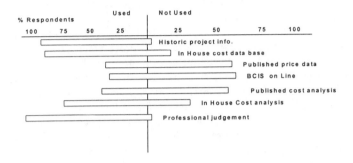

Figure 3 Use of cost information for estimate development

These results accord with the work of Green showing that estimators are not taking a radical approach to the development of estimates relying on traditional techniques i.e. Bills of Quantities. They tend to use a form that they are familiar with and that their sub-contractors are familiar with for possible reasons of risk management. The results support the proposition that this conservatism could be due to the availability of cost data in a format that cannot be easily translated to be applied to alternative forms of measurement.

Procurement systems and the communication problem

The process of building consists of several diverse and fragmented sub-processes. Effective management of the construction process depends on the systemic treatment of its constituent sub-processes. To function as a whole, it is necessary for the sub-systems (sub-processes) to interact in some way, i.e. there must be some form of communication between them. Each subsystem can be said to receive **inputs,** which stimulate further activity to produce **outputs,** passing either to other subsystems, or to the environment. Many of these messages are concerned with control, defined by (Checkland, 1981) as the means by which a whole entity retains its identity and/or performance under changing circumstances. The issue of communication within the Construction Industry has been an active area of research ever since the publication of the Banwell Report in 1964. That report and subsequent others, published, attributes the problems faced by the industry to the separation between the design and production sequence of the construction process. This conclusion was arrived at by comparison to the minimal problems faced by those in the field of manufacturing.

The success of any construction activity rests on its ability to manage information (Austen and Heale, 1984). The high dependence of the construction process on information puts pressure on the industry to improve its flow of information and communication capabilities. The structure upon how information is managed and organised is formalised by the form of procurement strategy employed. Among the main objectives of building procurement systems is to facilitate the delivery of projects, which will fulfil clients' objectives with respect to time, cost and quality. A major obstacle to improving communications, however, is created by the traditional form of contract, where most construction projects are currently being tendered. This procurement route has tended to nurture the already competitive and adversarial culture prevalent in construction. Teams and working groups that operate within such antagonistic cultures tend to encourage defensive routines that hinder effective communication (Argyris, 1990). To improve the situation, various approaches have been adopted by the industry to redress the issue at hand. Among them is the introduction of alternative forms of procurement, one of them Design and Build. They were introduced to bring together the fragmented parties and processes together. What Design and Build, in particular offered, was to bring together the relevant parties under the management of a single contractor to co-ordinate the various design, planning and construction activities. This was with a view to managing the critical interface between design and production.

Reflecting on the conclusions highlighted by the Latham Report (1994) and research conducted into the use and utility of Design and Build it can be seen that little progress has been made to improve the situation. As identified above , it organisations were found to be inherently conservative in their approach to non traditional procurement arrangements. Several possible explanations may be inferred, firstly, it may be due to the reluctance of the industry to fully embrace the new "culture" initiated by the alternative system or secondly, that the industry is still undergoing a substantial learning curve (AMEC 1998).

A way forward to redressing the situation was observed with the introduction of the Value Management (VM). Studies into VM practice have indicated that success in attaining the stated client's objectives lie in the proper evaluation of strategic decisions at the front-end incubation period of construction (Mohsini, 1996). It is within this front-end stage that decisions with respect to alternatives in design solutions, in relation to, cost and buildability may consciously be appraised. This is important as considerations made and decisions taken during the design stage have significant impact on informing the subsequent construction stages. In practice however, the benefits of such a procedure are constrained by demands imposed by clients (Allen Building, 1998). Due to pressures of time, designers rarely have the opportunity to consider alternative designs. Clients seem less inclined to invest resources at this front-end stage.

Figure 3 above indicates that the use of value management by contracting organisation is not used extensively. This background indicating a lcak of involvement in design decision making by the contractors supported the case for a study into how building designs were developed within the design and build procurement route and how the designs were being informed by the participating parties.

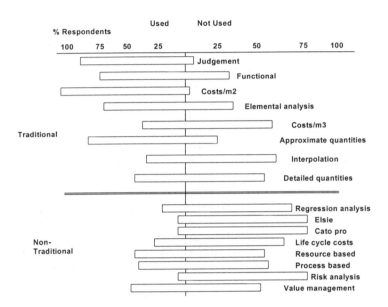

Figure 4 Incidence of techniques in use (% respondents)

Informed design decisions

Basic to the design process is the decision-making process. The thrust of any managerial process relies on the process of decision-making and its communication (Mead, 1994). Communication is essential within the decision-making process to ensure not only that shared meaning is constructed in the organisation but that the solution has considered all angles of the problem. Furthermore, design teams are increasingly multi-disciplinary, they approach the design situation with pre-existing patterns of work activities, specialised work languages, different expectations and perceptions of quality and success, and different organisational constraints and priorities. Therefore it is important for construction-related decisions to be properly informed, integrating the various perspectives and differences, leading towards a satisfactory solution. Failure to do so may result in decisions made that have a negative impact on other members' work and on the built product as a whole. It is acknowledged however that most decisions are based on incomplete knowledge (Harris, 1996). This is dictated frequently by pragmatic considerations of time, cost and availability. Nevertheless, research has indicated that most construction-related decisions, typically those pertaining to design and estimating were not comprehensively informed, due to an over reliance on the use of self-informing strategies employed by designers (Mackinder and Marvin, 1982, Newland and Powell, 1995). For such decisions to be effectively informed this requires the collaborative participation of all parties, contributing actively to the design process. However, a major obstacle to making this possible, exists within how designs are presently developed.

Sonnenwald (1996) in her review of design research literature, concluded that studies into design methods and how designers carry out design have traditionally focused on design tasks and design management whereas studies into communication among design team participants have not been explicitly discussed. This she assigns to the prevalent view, that it is the responsibility of design professionals to know what is best when it comes to design. This view is also shared by (Reich et. al, 1996), in their research on user involvement. They pointed out that active user involvement is often considered not useful and in most cases it is avoided. Such a viewpoint, to a degree, emerged as a result of isolating architects from the process of building, introducing into the scene the professional designer (Fowles, 1984). Professionalism was further reinforced by the establishment of professional bodies and in the way professionals were educated and trained (Schön, 1983). This separation between the design and construction processes, coupled by the multi-party nature of the industry has been a major source of uncertainty within construction projects. This uncertainty and the ambiguity stemming from it, is a by-product of individual parties' perception and interpretation of data. In practice this is observed, as contractors are required to "interpret" design drawings into physical structures. This, more often than not, exposes the project to variations from the intended design purpose. Without providing essential communication channels, problems of miscommunication and abortive work are anticipated. What intensifies the problem is the variety of perception that the individual participants contribute in (and to) the construction design activity (Reich et. al, 1996).

In the context of communication and decision-making, several studies have been carried out to further enhance the organisational potential within Design and Build. Recent work conducted has sought to study the flow of information within these processes (both between and within the design and construction domains). This is with a view to model the processes involved and hence formulate strategies for intervention. Work recently conducted by Fisher et. al (1997) presented several case studies of contracting organisations that adopt a variety of procurement routes. The study looked into the problem of information transfer during the development of design stage and had modelled the flow of information. The data collected was then used to develop conceptual models for use in the development of Object-Oriented database systems. Others such as Baldwin et. al (1997) modelled the flow of information involved at the briefing stage within Design and Build organisations, to enable the timely provision of strategic cost advice prior to design development. The primary focus of the research quoted above was centred on the co-ordination of processes. A similar line of reasoning is used in research into Construction IT application. The employment of modelling techniques, database technology and process engineering, among others, sought to establish a normative model of the industry, whereupon its information structures and requirements can be identified. Lera et. al (1984) are of the view that these approaches tend to be underpinned by two basic premises. First, that if information is well presented and has relevant content and if it is stored accessibly, then designers will use it. Second, if designers use such information, they will produce "better" buildings. This approach to managing complexity in construction very much follows the problem solving, optimisation model of design, expressed by Simon (1969). This paradigm has driven research towards searching for a more systematic method at reducing complexity by emphasising the need for the structuring of information and formalising procedures (Holt et. al, 1985).

The focus of these studies consider the management of information. Research on communication conducted by Bowen (1995) highlighted the more personal (human) element involved in the use of information. He reinforced, in the context of provision of cost information, the need for shared mutual understanding of information by the parties involved. This reduces the likelihood of misrepresentation and misinterpretation so critical to ambiguity. This is achieved, he asserted, by facilitating the process of interpersonal communication through an idealised framework. An observation on the model proposed by Bowen is that its feedback loops to the communication process was limited to what is termed single-loop learning as opposed to double-loop learning (Argyris, 1988). Within the design process, this implies that parties' input to the design solution is merely communicated to the architect and feedback is relayed back as a procedure to ensure that shared meaning has been attained. This, as in the case of Bowen's work into cost planning, merely directs the design process, so as to make certain that it keeps within the budget limits. It does not however, encourage the participants to rigorously question assumptions and to creatively formulate more effective alternatives based on informed discussion. Double-loop learning involves surfacing and challenging deep-rooted assumptions and norms of an organisation [or process] that have previously been inaccessible, either because they were unknown or known but undiscussible. Many attempts have been made to model the complex, iterative nature of design (Lawson, 1997). Such attempts were deemed necessary to provide an understanding into the information needs of the designer, enabling its associated decision-making process to be informed. But due to its creative, almost ad-hoc manner of development, such models have proven to be of little use or insight. However, the need for an understanding of the design process is still essential. Design, being positioned at the front-end of the construction process has a strategic influence on the effective management of subsequent stages (Mohsini, 1996). Design according to Schön (1983) unlike that proposed by Simon (1969), above, is more likened to that of a learning process. Holt et. al (1995), where they describe design as a creative endeavour, which involves an imaginative leap beyond what already exists, shares this perspective. This imaginative leap, they assert, is not spontaneous but the result of a physical development of the gnostic mechanisms of the brain; design is a learning process. Hence any attempt at improving the design decision-making capabilities of the design team should be focused at facilitating the learning process of the design team. Learning, according to Espejo et. al (1997), may be defined as an increase in the individuals or organisation's capability for effective action. This implies individual team members whose actions will be based on common understanding. Team learning is possible through the process of shared mental models. Capturing (and enhancing) individual mental models alone is not sufficient to achieve organisational learning. There needs to be a way to get beyond the fragmented learning of individuals and spread the learning throughout the organisation. The spirit of the learning should be one of active experimentation and inquiry where everyone participates in surfacing and testing each other's mental models (Kim, 1993). There needs to be what Schön terms as *reflection-in-action*. Among the barriers to learning is the practice of defensive routines (Argyris, 1990), where individuals are unwilling to undertake dialogue, so critical to exchange shared mental models. It is within this process of dialogue that double-loop learning (mentioned above) take place.

Research approach

Past research as quoted above, in information management and communication in construction has informed us of the extent of the communication problem in the industry. These problems are characterised by the structural fragmentation of the industry, minimal appreciation of industry processes, lack of co-ordination and conservative attitude to

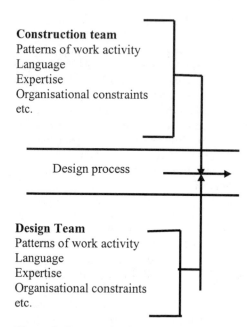

Construction team
Patterns of work activity
Language
Expertise
Organisational constraints
etc.

Design process

Design Team
Patterns of work activity
Language
Expertise
Organisational constraints
etc.

Figure 5: Communication between constructors and designers

change. This particular research extends on previous work by investigating the extent of information use within the development of conceptual design and how the interpersonal character of communication may be extended to that of design groups within the Design and Build system. The study, instead of looking only at information (and its flows), or the interaction between designer and information, seeks to focus on the interaction between the members of the design team within the collaborative process of designing. From this, the study may then establish the main factors that may facilitate (or impede) informed decisions.

One of the limitations of past research was that of the methodology employed. Methodologies adopted to gather information concerning the communication processes have thus far been unable to capture the problem in sufficient depth. In order to uncover and understand the dynamics involved in arriving at design decisions, a qualitative methodology in determining the human input and an information modelling methodology to determine the outputs is proposed Among the considerations for such a choice was the nature of the research question that attempts to understand how design decisions were made and informed, and subsequently how they may be improved. The focus here is on developing a context-based, process oriented description and explanation, rather than an objective, static description expressed strictly in terms of causality. It does not aim to produce information suitable for conventional statistical analysis and hypothesis testing, but rather adopts an interpretative epistemology that seeks to understand the complex social process within design practice. This is achieved by taking the point of view of individual practitioners, their values and practices, as the focus of the research. This includes how they view or perceive the issue under study, the factors involved, views on the practice of others and how the situation, in their view, may be improved.

The research strategy considered for this study involves describing and analysing several in-depth case studies to develop rich insights of the design development process, at the concept stage, and how decisions pertaining to them are informed by the relevant parties. This approach therefore fits within the interpretative case study approach to Information Systems (IS) research advocated by Walsham (1995). The case description will be based on a field study carried out on several typical projects involving large Design and Build contractors undertaking design development at the conceptual stage. It is worth noting here the sampling strategy adopted. Purposive sampling (Patton 1990) is adopted, where samples (subjects) for the research are selected based on its potential and ability for informing the research question. Bias responses are addressed within the analysis procedures. Design and Build contractors are selected as the arrangement most representative of an integrated construction team. Within this arrangement, members responsible for the design, planning and constructing of the building are put together under one management. Projects as opposed to Design and Build contracting organisations are selected as units of analysis. This was due to the existence of a more coherent taxonomy with regards to project types. The industry has found it difficult to establish an acceptable taxonomy in classifying contracting organisations, small, medium or large. In the past, project turnover has been used, but this has proven to be unreliable. However project types may be easily categorised as, commercial, process, housing or retail. Activities within the tender stage are selected for as indicated by previous studies (quoted above) decisions taken at this stage have a significant impact on the management of subsequent stages.

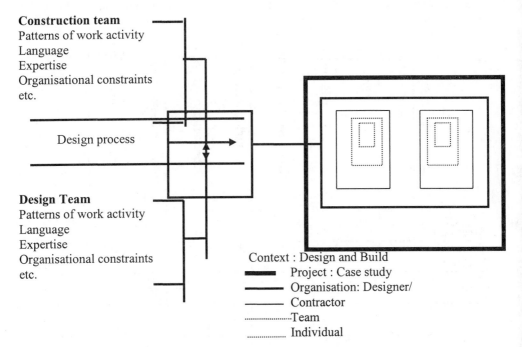

Construction team
Patterns of work activity
Language
Expertise
Organisational constraints
etc.

Design process

Design Team
Patterns of work activity
Language
Expertise
Organisational constraints
etc.

Context : Design and Build
Project : Case study
Organisation: Designer/
Contractor
Team
Individual

Figure 6: Context for research investigation.

The difficulty in this area of research is to try to capture data that is rich enough to provide some insight into the learning behaviour of groups and individuals link this with the design processes and provide a level of abstraction that can be taken to another case study to validate the model developed.

The methodology adopted encompasses two broad areas, one of information modelling to capture the process of design and the structure and type of information passing between the parties. A Create, read, update, delete matrix is to be developed from information gathered during the early design stages. This will allow the research team to map the process of design which can be used in conjunction with the qualitative phase of the work. The use of the CRUD matrix has been adopted as it provides a level of abstraction that captures some of the complexity of the information flowing between the parties and identifies both inputs and outputs to the design process. It has been favoured over the other methods of process modelling such as data flows and activity hierachies because it is easily understood by practitioners and consequently will allow the model to be confirmed as the design develops. It also allows for manipulation, by clustering the key areas of creation, updating, reading and deletion the model can reflect some of the dynamics of the design process.

The CRUD matrices are to be read in conjunction with the qualitative data that is to be collected through unstructured interviews. As identified above these unstructured interviews aim to gather some of the richer background to the decisions taken by the project actors. The data is to be analysed using a qualitative analysis tool such as Nudist to attempt to establish a set of common factors that influence decisions taken during the design phase.

Activity	Outline design	Employers requirements	Site survey	Outline design programme	Estimating programme	Subcontractor schedule	subcontracor design enquiry	Internal design resource prog	Notes
Meeting 1		R	R	C	U	U	C		
Design team mtg.				R	R			C	

Figure 7 CRUD matrix indicating key areas involved in design development

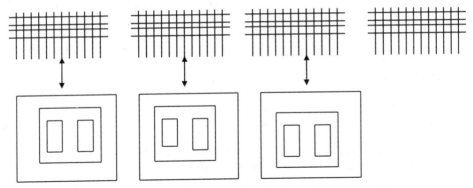

Figure 8. Comparison of the qualitative analysis and clustered CRUD matrices

The analysis of the data collected will allow the influences on the teams to be identified and categorised. The data is to be used to develop a model that will be at an appropriate level of abstraction, it is then aimed to validate this model by testing it on further case studies.

Conclusion

This paper has reported on earlier research into the design development processes between contractors and designers in design and build organisations in the UK. The problkesm of location of designer, information flows, pressure, conflicing culture between organisationn involved in team design development and the reliance on traditional techniques of estimate development have been highlighted as areas that warrant further study.

It has suggested a resrach methodology for data gathering that is both qualitative in identifying the cultural influences on organisations, teams and individuals be used in parallel with a methodology that has the potential to reflect the dynamics of the early design development.

References

Akintoye, A. and Kennedy, S. (1995), Quantity Surveyors Role in Design and Build Procurement Method, *COBRA Conference '95, Royal Institute of Charted Surveyors.*

Allen Building (1998), Private conversation.

AMEC (1998), Private conversation.

Argyris, C. (1988), Problems in producing usable knowledge for implementing liberating alternatives in *Decision Making - Descriptive, normative and prescriptive interactions*, Bell, D. E., Raiffa, H. and Tverskey, A. (Eds.), Cambridge University Press, Cambridge.

Argyris, C. (1990), *Overcoming Organisational Defenses - Facilitating Organizational Learning*, Allyn and Bacon, Boston.

Austen, A.D. and Heale, R. H. (1984), Managing Construction Projects: A Guide to Processes and Procedures, International Labour Organisation, Geneva, Switzerland.

Baldwin, A. N., Austin, S. A. and Pendlebury, M. C. (1997), The Interface of Early Design and Cost Advice in the Building Design Process, Proceedings of the 13th ARCOM Conference, Kings College, Cambridge, 395-404.

Bowen, P. A. (1995), A Communication-Based Analysis of the Theory of Price Planning and Price Control, *Research Paper Series, Vol. 1, No. 2, Royal Institute of Charted Surveyors, London.*

Checkland, P. B. (1981), Systems Thinking, Systems Practice, John Wiley and Son, Chichester.

Espejo, R., Schuhmann, W., Schwaninger, M. and Bilello, U. (1997), *Organizational Transformation and Learning*, John Wiley & Sons, Chichester.

Fisher, N., Barlow, R., Garnett, N., Finch, E. and Newcombe, R. (1997), *Project Modelling in Construction*, Thomas Telford Publishing, London.

Fowles, (1984), Design-build projects in architectural education, *Design Studies*, Vol. 5, No. 1, 7-14.

Green, S.D. (1989), Tendering; Optimisation and rationality, *Construction, Management and Economics, 7, pp 53-63.*

Harris, S. (1996), *Human Communication and Information Systems*, 4th edition, NCC Services Ltd., Manchester, UK.

Holt, J. E., Radcliffe, D. F. and Schoorl, D. (1985), Design or problem solving - a critical choice for the engineering profession, *Design Studies*, Vol. 6, No. 2, 107-110.

Management Review, Fall 1993, 37-50.

Latham, M. (1994), *Constructing the Team: Final Report of the Government/Industry Review of the Procurement and Contractual Arrangements in the UK Construction Industry*, HMSO, London.

Lawson, B. (1997), *How Designers Think*, 3rd. Edition, Architectural Press, Oxford, England.

Lera, S., Cooper, I. and Powell, J. A. (1984), Information and Designers, *Design Studies*, Vol. 5, No. 2, 113-120.

Mackinder, M. and Marvin, H. (1982), *Design Decision Making in Architectural Practice*, Institution of Advanced Architectural Studies, University of York, Research Paper 19, York.

Mead, R. (1994), *Cross Cultural Management Communication?*

Mohsini, R. A. (1996), Strategic Design: Front End Incubation of Buildings, Proceedings of CIB W-92 International Symposium on *'Procurement Systems: North Meets South'*, University of Natal, Durban, South Africa, 382-397.

Newland and Powell, J. A. (1995), An integrating interface to data, *Integrated Construction Information*, Brandon, P. and Martin, B. (eds.), E & FN Spon, London.

Patton, M. Q. (1990), QualitativeEvaluation and Research Methods, 2nd Edition, Sage, Thousand Oaks.

Reich, Y., Konda, S. L., Monarch, I. A., Levy, S. N. and Subrahmaniam, E. (1996), Varieties and issues of participation and design, *Design Studies*, Vol. 17, No. 2, 165-180.

Ross, A.D. and Fortune, C. J. (1997), *Production based design cost modelling in design and build organisations*, RICS Education Trust Research Report, RICS, London.

Schön, D. A. (1983), *The Reflective Practitioner - How Professionals Think in Action*, Basic Books Inc., New York.

Simon, H. A. (1969), *The Sciences of the Artificial*, The M.I.T. Press, Cambridge, Massachusetts.

Sonnenwald, D. H. (1996), Communication roles that support collaboration during the design process, *Design Studies*, Vol. 17, No. 3, 277-301.

Walsham, G. (1995), Interpretive case studies in IS research : nature and method, *European Journal of Information Systems*, 4 (2), 74-81.

Yin, R. K. (1994), *Case Study Research: Design and Methods*, 2nd. Edition, Sage, Newbury Park, California.

THE PERFORMANCE OF A SYSTEMATIC RISK MANAGEMENT PROCESS FOR PRE-TENDER COST ESTIMATE FOR PUBLIC WORKS IN HONG KONG

STEPHEN MAK
Department of Building and Real Estate, The Hong Kong Polytechnic University, Kowloon, Hong Kong. bssmak@polyu.edu.hk

DAVID PICKEN
School of Architecture and Building, Deakin University, Geelong, Victoria 3217, Australia. davidp@deakin.edu.au

Abstract

Risk has long been recognized in the construction industry, as can be reflected from the inclusion of a contingency item in professional quantity surveyors' project cost estimate. The contingency sum, usually expressed as a percentage markup on the base estimate, is used to allow for the unexpected outcome. The use of risk premium money is regarded as standard practice in construction. The practice of a presenting project cost estimate in a deterministic figure comprising a base estimate and a contingency has been adopted in the construction industry for a long time. This is often required as the client needs to have a figure for budgeting and financing purposes. The Hong Kong Government has implemented a technique called Estimating Using Risk Analysis (ERA) since 1993 to estimate and calculate an amount of money to allow for uncertainties associated with a project. ERA produces similar base plus contingency cost estimates. The usefulness of this technique lies in its ability to help predict the uncertainties or the unexpected expenses that might arise during the course of a project. The more contingency money is allowed for a project, the more funds are tied up until the project is completed before the excess fund can be released for other use. In public finance excessive funds being tied up in contingencies lead to mis-allocation of resources and criticisms. This paper compares the performance of project cost estimates against final account valuations between projects that adopted ERA and those that did not. Two sets of data were collected from the government in 2 consecutive years. The first set of data indicated a significant improvement in project cost estimate with the ERA; whereas the second set did not show any improvement. This paper will try to explain why these contrasting results could happen. Recommendations will be given to improve performance.

Keywords: Estimating using risk analysis, contingency

Introduction

Risk has long been recognised in the construction industry. Risk can be reflected by the inclusion of a contingency item in professional quantity surveyors' project cost estimates at pre-tender stage. The contingency sum, usually expressed as a percentage markup on the base estimate, is used in an attempt to allow for the unexpected. At different stages of project development, there are different types of risk. At the feasibility and inception stages, for example, the client might not have decided exactly on the floor area that is required, or an amount of additional floor area over that in a basic scheme may be in abeyance. Such matters will represent an uncertainty from the estimator's point of view. The depth of piling for foundations is another typical example of uncertainty. Some of the uncertainties will be eliminated or clarified as the planning of the project develops towards detailed design stage when, for example, the client has decided the floor area required. Some uncertainties will be carried forward to tender stage. The use of risk premium money is regarded as standard practice in construction (Raftery, 1994). The practice of presenting project cost estimates as a deterministic figure comprising a base estimate and the addition of a single contingency amount (usually as a percentage addition) has been adopted in the construction industry for a long time for budgeting purposes. The contingency will usually be transferred to the provisional and prime cost sums in the tender document (or Bills of Quantities). In an attempt to deal with the determination of contingencies in a more analytical way the Hong Kong Government implemented a technique called Estimating Using Risk Analysis (ERA) in 1993. ERA produces similar base plus contingency cost estimates at pre-tender stage. For building projects, which usually use the government's fixed quantities contract, the magnitude of the final account variations (comprising additions and omissions) can be compared with the contingencies included in estimates. This comparison can be used to assess the accuracy of the allowance made for the contingencies at the planning stages. This paper compares the contingency estimates and final account variations of public works projects by analysing data sets of pre-1993 (non-ERA) and post-1993 (ERA) projects.

Contingencies

Traditionally cost estimates are point estimates. That is, single value estimates based on the most likely values of the cost elements. These point estimates may or may not accurately indicate the possible value of the estimate, and they certainly do not indicate the possible range of values an estimate may assume (Toakley, 1995). When estimating, the most common method of allowing for uncertainty is to add a percentage figure to the most likely estimate of final cost of the known works. The amount added is usually called a contingency (Thompson & Perry, 1992).

Contingencies are often allowed in cost estimates. The objective of contingency allocation is to ensure that the estimated project cost is realistic and sufficient to contain any cost incurred by risks and uncertainties. There are a number of methods in risk analysis practice which are used to deal with uncertainty. For budgeting purposes, however, rather than indicate a range of figures there is a need to present an estimate as a fixed amount so that the client can arrange for the financing of the project. The estimate is also important in reaching a decision on whether to proceed with a project, given that the client has only a fixed budget.

There is a tendency for estimators to include an inflated buffer in the contingency estimate. Raftery (1994) has identified personal bias and differences in personal risk attitude. Kahnaman and Tversky (1972) referred to this as 'conservatism'. Moreover, due to the effect of negative sanctions, that is, imposing a penalty for an underestimate (where tender bids are above the pre-tender estimate), but no reward/penalty for an overestimate, an over-exaggerated contingency is not uncommon in many project estimates. For public works projects, this leads to misallocation of resources as more than sufficient funds are locked up in projects.

Estimating using risk analysis

To alleviate these usually over-exaggerated contingency estimates, the Hong Kong Government has introduced the technique of Estimating using Risk Analysis (ERA) in all public works projects. Details of ERA can be found in United Kingdom Government publications (HM Treasury, 1993). A similar technique is referred to by Raftery (1994) using the terminology MERA (Multiple Estimating using Risk Analysis). The following section provides an overview of the ERA technique.

ERA is used to estimate the contingency of a project by identifying and costing risk events associated with a project. The starting point for the ERA process is a base estimate which is an estimate of the known scope and is *risk free*. The contingencies as determined by the ERA process are added to the base estimate. Firstly, risks are identified by the project team. They are then categorised into either (i) Fixed or (ii) Variable. For each risk event, an average risk allowance and a maximum risk allowance are calculated. The relationship between risk category and risk allowance is shown in Table 1.

Table 1. Relationship between risk allowance and risk category

	Average Risk Allowance	**Maximum Risk Allowance**
Fixed Risk	Probability x Maximum Cost	Maximum Cost
Variable Risk	Estimated Separately	Estimated Separately
Assumption	50% chance of being exceeded	10% chance of being exceeded

Fixed risk events are those which either happen in total or not at all. If the event happens, the maximum cost will be incurred; if not, then no cost will be incurred. An example is the need for an additional access road. At the early stages of a project, the client might be uncertain as to whether an additional access road will be required, rendering this as a fixed risk item. The scope of the road, should it be required, can be known and used to determine the maximum risk allowance. The uncertainty is whether the road will be needed or not. The maximum risk allowance is the cost of constructing this access. The average risk allowance is the probability of the client requiring it multiplied by the maximum risk allowance.

Variable risk events are those events which will occur but the extent to which they will occur is uncertain. The cost incurred will therefore be uncertain and variable. An example is the depth of piles required to be driven. The maximum risk allowance is estimated by the project team members based on past experience or records. This means the most expensive type of piling being required at the maximum length. The Hong Kong Government's ERA method requires an assumption to be made that there is only a 10% chance that the actual cost incurred will exceed this allowance. The average risk allowance is estimated with an assumption of a 50% chance of being exceeded. There can be mathematical relationship between the average and maximum risk allowance but it is also legitimate for these two allowances to be estimated separately.

The rationale for using a 50% chance of being exceeded in the average risk allowance is that it is unusual for all identified risks, that is the worst case, to occur. The swings and roundabouts effect of the totality of the risk events identified should be able to cover the most likely costs incurred.

Having identified all risk events and calculated their average and maximum risk allowances, the summation of all events' average risk allowance will become the *contingency* of the project concerned. Figure 1 shows a typical ERA worksheet.

ERA Calculation

(1) Risk	(2) Type	(3) Probability (Fixed Risks Only)	(4) Average Risk Allowance $	(5) Max. Risk Allowance $
Design Development	V		8,400,000	12,600,000
Additional Space	F	.70	11,760,000	16,800,000
Site Conditions	V		525,000	1,000,000
Market Conditions	V		4,000,000	8,500,000
A/C Cooling Source	V		250,000	1,250,000
Access Road	F	.50	250,000	500,000
Additional Client Requirements	V		1,680,000	4,200,000
Contract Variations	V		8,400,000	12,600,000
Project Co-ordination	V		500,000	1,500,000
Contract Period	F	.60	1,000,000	1,750,000
			Σ $36,765,000	

Base Estimate   = $168,000,000
Average Risk Estimate = Base Estimate + *Total* Average Risk Allowance
                      = $204,765,000 (21.88% on base)

Figure 1. An example of a ERA worksheet

Advantages of ERA

The ERA process is usually carried out several times during the pre-tender period for any one project. As the project develops, some events which were originally identified as uncertain will be clarified and will be either deleted from the list of risk events or included in the base estimate (as a certainty). For example, the requirement for the secondary access road may have been confirmed. The remaining uncertainties will form the final contingency allowance. Figure 2 shows the relationship between the base estimate and contingency estimate at different stages of a project. A distinct advantage of ERA lies in its ability to retain the traditional method of presenting a project cost estimate in the form of a base estimate plus a contingency. It imposes a discipline from the outset to systematically identify, cost and consider the likely significance of any risks associated with a project. It also aids financial control in having risk and uncertainty costs identified before action is taken to determine precise requirements. Further, the itemised and substantiated contingency forms the basis on which to evaluate the impact of risk and uncertainties upon completion of the project, thereby providing useful data for use in investment appraisal. In short, it is a mechanism for accountability for public money. Rigorously done, it will reduce the usually conservative and excessive percentage add-on contingency and lead to a better allocation of resources.

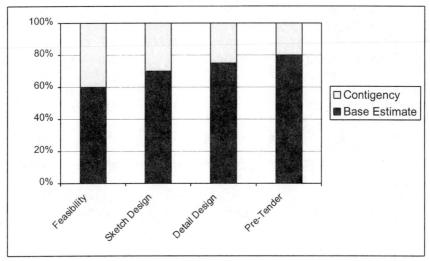

Figure 2. Relationship between Base Estimate and Contingency Estimates at different stages of project development

Empirical data

Two sets of completed project summaries were received from the government for 2 consecutive years (1996 and 1997) which detailed the contract sum, original contingency allowance, amount of contract variations (as additions and omissions), final account amount and start date of building projects, as shown in Table 2. Of these projects, some of them have used the ERA process for determining contingencies and some were done by the traditional method. As these are all building projects using the government's standard forms of contract it is possible to compare the contingency allowance against variations.

A new variable, DEVI, defined as the ratio between the amount of contingency and the amount of variation at final account, was used to reflect the accuracy of the original contingency estimate, having included all the additions and omissions in the final account.

In most building contracts, a fixed contract sum is used whereby a contractor is paid according to a sum shown in the contract documents. Only if there are variation orders will the contractor be paid accordingly. Therefore, the summary of variations should, by and large, reflect how accurately the estimator has calculated the uncertainties at pre-tender stage. If the value of DEVI is 1 then there is a perfect match between contingency and variation and the contingency is capable of covering uncertainties. If the value is higher than 1 then there is surplus in the contingency fund and vice versa. A negative DEVI value means that the amount of omissions is larger than additions, that is, a net reduction in the final contract sum.

Table 2. Summary Raw Data

	Year 1		Year 2	
	non-ERA	ERA	Non-ERA	ERA
Number of Projects	72	19	315	46
Contract Sum (min)	0.79M	0.99M	0.24M	0.99M
Contract Sum (max)	1331.01M	86.13M	1331.01M	208.48M
Contingency (min)	0.07M	0.08M	0.015M	0.08M
Contingency (max)	110.00M	6.00M	110.00M	38.00M
Contingency/Contract Sum (min)	0.67%	4.04%	0.67%	4.02%
Contingency/Contract Sum (max)	137.40%	13.57%	137.40%	18.23%
Final Account Variation (min)	-1.74M	0.04M	-3.49M	0.04M
Final Account Variation (max)	85.98M	6.39M	85.98M	27.39M
DEVI (min)	-1.55	0.55	-68.00	0.00
DEVI (max)	38.60	5.58	150.00	16.00

Analysis

It was hypothesised that the ERA process should produce better project estimates by making more realistic allowances for contingencies. Better project estimates means both the mean and standard deviation of DEVI should be consistently smaller for projects with ERA than those without ERA. Table 3 shows the summary of the various statistics.

First year

Table 3 shows that the standard deviations for ERA and non-ERA projects are 1.416 and 5.917 respectively. The standard deviation for non-ERA projects is surprisingly high as this is contrary to the law of large numbers due to its sample size compared with ERA projects. This can be attributable to the fact that a fixed or blind percentage was added to the base estimate to arrive at the contingency amount for non-ERA projects.

An analysis of the F-statistic, which is the square of the ratio between the two standard deviations, shows that the F value is 17.47 with a corresponding probability of less than

0.0001 (extremely low). This means the hypothesis that the variances of the two populations are equal is rejected. It can be concluded that there is a significant difference between the variances of the two populations. This also means that the variability of contingency allowance for non-ERA projects was much higher, and the contingency allowance for ERA projects was more consistent.

Table 3. Summary of Statistics (1ˢᵗ year)

Contingency Type	DEVI Mean	s.d.	Critical Values
non-ERA (N=72)	2.9597	5.917	
ERA (N=19)	1.6422	1.416	
F-statistic		17.47	2.7335 (1%, 71,18 df)
t-statistic	1.71		1.6647 (5%, 89 df) 1-tail

Table 3. also shows the mean values of DEVI of the 2 groups of projects. It can be seen that the mean for ERA projects is 1.6422 whereas that of non-ERA projects is 2.9597. This means that an average of 64% more funds have been set aside for uncertainties in ERA projects and an average of 195% more for non-ERA projects. It is not surprising that both methods resulted in excess funding reserved for uncertainties as estimators are usually conservative (Raftery, 1994; Kahnaman and Tversky, 1972). However, the discrepancy of more than three times (64% versus 195%) the contingency allowance in non-ERA projects has resulted in severe misallocation of resources. When contingencies are exaggerated in this way, some projects under consideration in a given phase of a works program have to be foregone or deferred due to insufficient funds.

It was hypothesised that the ERA process should lead to a smaller DEVI value than the non-ERA process. A 1-tail *t*-test was used to test whether the means of the two groups are different, with the assumption of a different variance (supported by the *F*-statistic). The resultant *t*-value is 1.71 with 89 degrees of freedom. This is larger than the critical value for *t* at the 5% significance level with 89 degrees of freedom. The hypothesis that the means of DEVI for the two groups are equal is rejected. It can be concluded that the ERA group had a much smaller DEVI value than the non-ERA group of projects.

The *F*-statistic supports the hypothesis that the contingency allowance was more variable for non-ERA projects and more consistent for ERA projects. The *t*-statistic supports the hypothesis that the contingency allowance was much smaller for ERA projects.

Second year

Table 4 shows the summary of statistical analysis from the second year's set of data. Using the same methodology of analysis, it was found that, from the second year's data, the standard deviations between the non-ERA and ERA group were 117.5483 and 6.9734 respectively. The *F*-statistic shows a significant difference between the 2 groups at 1% significance level.

However, the difference in DEVI between the non-ERA and ERA projects was not significant at all. Indeed, the DEVA for non-ERA was 2.7744 whereas that of ERA was only slightly smaller at 2.4476. This means that a very high percentage of funds were reserved in both groups for the uncertainties; and that the ERA process did not produce the expected reduction in contingency allowances.

Both sets of data demonstrated that estimators were highly conservative and therefore supported earlier researchers' findings (Raftery, 1994; Kahnaman and Tversky, 1972). Reasons for this will be discussed more fully in the next section after analysing users' feedback.

Table 4 Summary of Statistics (2nd year)

Contingency Type	DEVI Mean	s.d.	Critical Values
non-ERA (N=315)	2.7744	117.5483	
ERA (N=46)	2.4476	6.9734	
F-statistic		8.1324	1.7834 (1%, 314, 45 df)
t-statistic	0.4511		1.6501 (5%, 289 df) 1-tail

Users' feedback

Beyond the statistical analysis of contingency funds, variations and contract sums described above, it was also considered useful to examine the views of the estimators who were required to implement ERA. A questionnaire was developed for this. In the questionnaire survey of estimators who have used ERA, the estimators were asked to rank in a scale of 1 to 5 the importance of 3 factors of using ERA (they ranked the factors independently instead of ranking the relative importance of the factors). It was revealed that ERA was applied to all types of government building projects such as government offices, fire stations and airport projects. ERA was used on an average of four times throughout the pre-tender stage. A Severity Index was used to measure the perceived importance of various factors in motivating the respondents to use ERA. In particular, three areas were identified; namely, ERA being a policy requirement, ERA helping to identify accountability and ERA's potential for accuracy in determining contingency allowances. Table 5 shows the summary of the findings.

It can be seen that the respondents, who were government officials, saw policy as the most important factor for using ERA. Achieving accuracy in estimating was ranked least important among the three attributes. It was not surprising to obtain these results as one of the reasons for introducing ERA was accountability at management level due to severe under-spending of budgets in the early 1990s.

The respondents also expressed difficulties in implementing ERA due to a lack of data and records for them to determine both the probability of occurrence of risk items and the cost associated with these items. These would make estimating less realistic. They expected more feedback to be made available from past projects.

Table 5 Summary of Feedback

Rank	Policy Requirement	Accountability	Accuracy
1 (most important)	15	2	1
2	2	9	7
3	3	5	6
4	1	3	4
5 (least important)	0	2	3
No. of Respondents	**21**	**21**	**21**
Severity Index	86.90	57.14	48.81

Discussion

ERA was made a mandatory process for all public works projects since 1993. The comparison of its accuracy in estimating contingency allowances can only be done when the project was completed and final accounts settled. The first batch of data for 19 ERA projects was available in 1996. Because of the small number of projects involved, the results looked excellent. A substantial reduction in contingency allowance was achieved. The second batch of data for 46 ERA projects, however, did not show any improvement in contingency allowance estimates. Several reasons contributed to this. First, the 19 projects from the first year (1996) were all small scale projects. Uncertainties involved in these projects were fewer. Second, feedbacks from the estimators indicated that ERA was used simply because it was a policy. Accountability ranked second whereas improvement in accuracy in estimated was not regarded important. It would not be surprising that estimators would gradually treat ERA as another way of substantiating contingency without considering the ultimate objective of improving estimating accuracy. Third, having said that, estimators did express concerns for the lack of historical data from completed projects to help with identifying and estimating financial impacts of risks. When ERA has become a standard practice and estimators were faced with more complex projects, they lacked information to fulfil the requirements of ERA. Last but not least, sampling theories stipulates that the standard error of the sampling distribution decreases as the number of sample increases. Apparently, with 19 samples in the first year, the chance of error is much higher than the second year's data with 46 samples. Nonetheless, since no adjustment has been made on the second set of data, chances are that a few projects with abnormal figures, which can be the result of significant change in scope of works, will distort the whole analysis. As can be seen from Table 2, the value of DEVI ranged from 0 to 16 for ERA and –68 to 150 for non-ERA projects. Perhaps if some of the extreme cases were taken out from the data set, the results of the analysis would have been different. After all, more data should be collected and clustered in order to reveal patterns of *good practice*. International benching-marking can also be established.

Conclusion and recommendations

ERA is a simple to use method for implementing the use of risk analysis in estimating for construction projects. The way it handles risk analysis is consistent with the traditional practice of estimators at pre-tender stage to produce a deterministic figure in advising clients of likely bid levels. Empirical findings from a government department showed that the use of the ERA approach has improved the overall estimating accuracy in determining contingency amounts. The samples in this study revealed significant improvement in project estimates as both the mean and standard deviation of ERA projects are consistently smaller than those of non-ERA projects. Further research should be carried out to increase the sample size, especially of ERA projects. Clustering of projects into different sizes may also be useful to explore the effects of project size (complexity) and type on uncertainty estimates. Feedback from practitioners indicates that cost data of past projects should be made accessible to estimators. Such data would provide invaluable information to help make more accurate forecasts about uncertainties. The potential importance of information technology can be considered in this respect.

References:

HM Treasury (1993) *Managing risk and contingency for works projects.* Guidance Note No. 41, Central Unit on Procurement, HM Treasury (UK).

Kahneman D. and Tversky, A. (1972) Subjective probability: A judgment of representativeness. *Cognitive Psychology*, 3, 430-454.

Mak, S. & Raftery, J. (1992) Risk attitude and systematic bias in estimating and forecasting. *Construction Management and Economics.* 10, 303-320.

Raftery, J. (1994) *Risk Analysis in Project Management.* London: E&FN Spon.

Thompson, A. and Perry, J.G. (1992) *Engineering Construction Risks: A Guide to Project Risk Analysis and Risk Management.* Thomas Telford.

Toakley, (1995) Risk management applications - a review. *Australian Institute of Building Papers.* 6, 77-85.

HARMONIZING CONSTRUCTION PLANNING, MONITORING, AND CONTROL: AN INFORMATION INTEGRATION APPROACH

A.S. KAZI and C. CHAROENNGAM
School of Civil Engineering, Asian Institute of Technology, P.O. Box 4, Klong Luang, Pathumthani, 12120, Thailand. scc68494@ait.ac.th

Abstract

Researchers have time and time again emphasized the importance of decision making, information management and the incorporation of information technology (IT) in the operations of an organization. This emphasis holds a special importance in the construction industry as value added project delivery is basically a consequence of the degree of proper incorporation of these factors during the planning, monitoring and control of construction processes. To facilitate the incorporation of all these factors, different approaches have been identified and related proposals presented. Success has however not been easy to come by as even in today's software and hardware dominated world, decision making, information management and usage of IT in construction operations is not truly up to the challenge.

This paper presents the preliminary findings of an ongoing study on the evaluation of different project control parameters and their appropriate integration for providing an integrated decision making environment facilitated by an efficient information management strategy and through the incorporation of IT. A review of the problems instigating the study is presented and a structured methodology for research conduction highlighted. Cross-functional integration of project control functions and associated parameters is illustrated and further steps towards the development of an integrated system proposed.

Keywords: control, decision making, information management, integration, monitoring, planning

Introduction

The competitive nature of the construction industry in today's work environment is instigating construction organizations to compete on the basis of value added project delivery in terms of costs, schedules and resources. Since managers at three distinct hierarchical levels make decisions the nature, detail, and form of information required by these management levels may significantly vary. It becomes a necessity hence, to ensure the availability of timely, pertinent, and accurate information to them in order to facilitate a better decision making environment.

The construction industry constitutes a complex, open and dynamic work environment instigated by the simultaneous execution of independent, interdependent and dependent activities. Construction process dynamism leads to uncertainty [Saeed and Brookes

(1996)] in project control function delivery as a consequence of either an insufficiency of project information, or an overload of it. The presence of different parties with different goals, interests, and project control function authorities infect the implementation of information systems which in turn only adds to project uncertainty and risk [Kazi and Charoenngam (1996), Saeed and Brookes (1996)].

Different information systems have been implemented to coordinate the cost, schedule, resource procurement, quality adherence and overall performance of construction projects to facilitate better decision-making. These systems have however not fully delivered as per expectations. This may be due to several reasons. The systems usually incorporate the assistance of a multitude of minor systems to provide an information database for the organization as a whole to serve the organization's information requirements for decision making. What some systems may fail to recognize is that the minor systems may not efficiently share data and information amongst themselves during data and information input and processing stages. Another area of concern during system implementation is that rarely are multiple user needs appropriately satisfied.

Major problems

Three major problems pertaining to information management delivery through IT enablement were identified. These major problems were, poor decision making and information management, lack of integrated solutions and automation in planning and control, and, the under utilization of information technology.

Poor decision making and information management

Poor decision making may be identified as being congruent with information mis-management. Yates, and Rahbar [1990] mentioned this issue at length. According to Shapira et al. [1994], "Effective planning and its essence, decision making, is the key to the success of a project. Oddly enough, not much systematic research has been devoted to the decision making process." Other researchers [Tenah (1986), Abudayyeh and Rasdorf (1991)] also reported similar findings.

Lack of integrated solutions and automation in planning and control

The need for integrated solutions and automation in project planning, monitoring and control has been a persistent need. This need was implicitly outlined by Teicholz, and Fischer [1994]. Integrated construction planning and control systems, though required and asked for are yet to truly demonstrate their potentiality and effectiveness in the construction industry. Tatum and Fawcett [1986] elaborated on the required system. According to an investigation made by Kazi, and Charoenngam [1996], "Most adopted construction management information system environments have failed upon implementation due to several reasons. The reasons for failure may be attributed to the lack of system integration and user-system adaptability." Thus, though making an attempt to satisfy the need for integrated construction management information systems mentioned by Tenah [1984], most currently available systems fail due to poor system integration and user-system adaptability.

Under utilization of information technology

Despite numerous studies voicing paradigmatic changes in terms of value added delivery of business objectives, there has been significant resistance to proper utilization of information technology. In his classic study on Human-Centered Information Management, Davenport [1994] pointed out that most technological innovations usually, "disregard how people in organizations go about acquiring, sharing, and making use of information. In short, they glorify information technology and ignore human psychology." Despite the current stress on the, "use of the microcomputer for sorting, computing, and sorting data...has not been without the introduction of difficulties. the problems include input inefficiencies (e.g. multiple input of the same data) and output problems (e.g. extensive tabular listings of data)" [Rowings (1991)].

Research objectives

For appropriate consideration of the problems instigating this study and a search for an amicable solution that may help resolve these problems, seven key research objectives were identified;

- Analysis of current planning and control functions, processes, and tools
- Schematic identification of the extraction, flow and transmission of data and information within the planning and control processes
- Development of a framework for the integration of all planning and control functions and processes at different levels of abstraction; corporate, project, and functional
- Analysis and development of different planning and control function modules; cost planning and control, resource planning and control, and management reporting
- Integration of different planning and control function modules
- Prototype integrated planning and control information system development
- Identification of potentialities of and recommendations on methodologies of integrated planning and control information system implementation.

Scope and focus

The demands imposed by the research objectives of this study were limited in scope to cover prototype development, project management level focus in a matrix type organization, project life cycle stages, and project control function integration. Project life cycle stages were further restricted to cover detailed design and construction stages. In a similar fashion, project control function was restricted to cover four main project control functions, schedule planning and control, cost planning and control, resource planning and control, and management reporting.

Alongside the investigation and development of a prototype integrated system environment, the focus of this study was broadened to look into several key information related areas, information value and sensitivity, integration, filtering, and communications.

Methodology

A structured methodology is being employed for this ongoing study. Using system development principles, a modular methodology comprising of seven distinct and yet interrelated modules was identified for research purposes as illustrated in Figure 1.

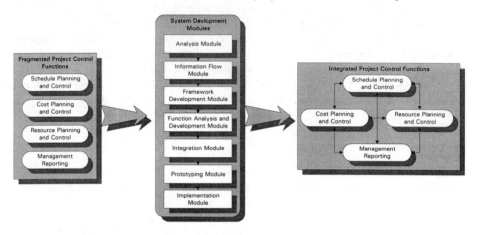

Figure 1. Research Methodology

Information

The WWWebster dictionary [1996] defines information as, "the communication or reception of knowledge or intelligence." At the same time, differentiation between information and data must be sought, as data is the "information output by a sensing device or organ that includes both useful and irrelevant or redundant information and must be processed to be meaningful," [WWWebster (1996)]. There is a clear indication here that during normal transactions and processes, both meaningful and useless information may be generated.

Information in the construction industry

In the construction industry, information is generally scattered with information coming in from different sources that may be contradictory. There is a need therefore, to ensure the validity of this information. A possible mechanism is through integration at different levels of abstraction.

Information filtering

As a project progresses through its different life cycle stages, more and more function and project specific data and information is generated. This data and information needs to be appropriately managed, though this becomes difficult when there is an overload of the same. An overload may lead to information pollution (Figure 2) that may render unmanageable project control functions and associated decision-making needs.

Information integration

Information in construction as derived from different project control functions, is generally scattered. Each control function for example may have a different interpretation of an entity. For example, the work breakdown structure (WBS) being used for cost planning control purposes may be different from that used for schedule planning and control. Furthermore upon integration with the organization breakdown structure (OBS), each of the said work breakdown structures may designate responsibility to different persons. This would only lead to more confusion and complexity.

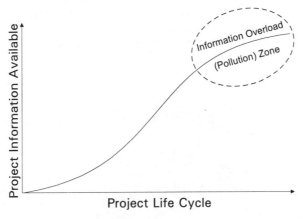

Figure 2. Increase in Availability of Project In with Project Progression

The induced complexity and confusion mentioned may be significantly tuned out through a proper coding structure provided that a consistency of work breakdown structures is maintained. However, to ascertain better planning and control for each project control function, the level of detail of the work breakdown structure may be varied.

Information complexity and inconsistency is not only a consequence of a poor-coding structure. Each project control function has its own set of inputs, processes, and outputs. Generally in construction, each project control function is planned, monitored, and controlled separately. This at times leads to confusion as there are certain input, process, and output parameters that are common to different project control functions. Therefore, mapping, and thence integration of these common parameters will not only lead to the elimination of the need for multiple inputs, but will also add more consistency to the generated information.

Information Communications

Proper information filtering and integration is only one step towards improving decision-making by different managerial levels. The type of information and its associated level of detail generally vary from one managerial level to the next. From a construction planning and control perspective, data is present at a field level, which is then processed to information for middle management and then further rationalized to constitute knowledge for executive managers. Furthermore, it should be noted that the domain of decision making too varies from one managerial level to the next (Figure 3). The prime objective of communications from a management perspective is the provision of information to all managerial levels in the desired format and at the desired level of detail in a timely, pertinent, and accurate manner. A properly integrated information environment should hence be able to generate and disseminate information to all managerial levels in an appropriate manner. Flow in such an environment, will generally be from field to middle and middle to executive management in terms of information, and executive to middle and middle to field in terms of decisions made and directions set.

Building on the information concepts outlined in the preceding sections and paragraphs, it may be said that for a proper information management set-up, information filtering, integration, and communication systems must exist in unison. While these systems do not have to be information technology based, the use of information technology, or any other similar technology can however once appropriately implemented, allow for quicker system operation and minimization of manual error.

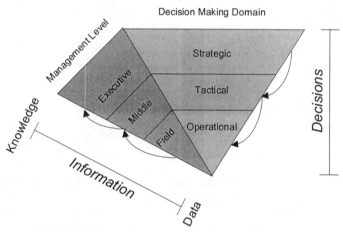

Figure 3. Management Information and Decision Making Domains

Information Value and Sensitivity

The value of information may be measured in terms of its relevance to the managers using it and its consequent level of impact on decisions made. Proper information may furthermore be used as an effective tool for risk management purposes. Here the issue of information sensitivity is worthy of exploration.

During all project stages, for the purpose of risk measurement and aversion, information is a very critical factor. Risk management principles strongly emphasize that risk should be transferred to that party that can handle it best. This analogy may be extended to construction information also.

Sensitive issues are generally in the realm of head offices, whereas other less sensitive issues may be dealt with at site levels. Thereby, some issues are centralized and others decentralized. Extending this thought to construction information for purposes of better project management through risk control, certain pieces of information may be centralized whereas others decentralized.

Minato and Ashley [1998] classified risks into two broad categories, corporate and project. Since some risks may be generic to different projects, these risks may be dealt with at a corporate level in the form of a group than individually at each project level. Unique project risks on the other hand may be handled at respective project levels.

Based on the aspects mentioned in the preceding paragraphs, and application of these concepts to construction information management, a different insight into construction information may be made. Firstly, this concept may be used as a vehicle for classifying project control function parameters and to serve as an instrument for project control function information integration, and second, based on the degree of associated sensitivity/risk during implementation, project control function processes may be appropriately classified and designated to relevant management levels.

Cross-functional approach to information integration in construction

One of the prime objectives of this study was to identify, study, and then devise a mechanism through which information pertaining to planning, monitoring, and control in construction could be efficiently integrated. A bottom-up approach was initially used to study the process and identify possible integration areas. Conceptualizing an integrated environment as a starting point, this environment was systematically decomposed to finer and finer elements. After decomposition to a significant degree, some key project control function parameters were identified that could be used for integration purposes. From here a top down approach was used to conceive the initial integrated environment.

Decomposition and parameter identification

Assuming an integrated environment as identified for the scope of this study each project control functions was systematically decomposed into a collection of modules (Table 1). Each of these modules was then decomposed to finer details yielding a collection of input and output parameters, and tools and techniques (Figure 4).

Table 1. Project Control Functions and Associated Modules

Project Control Function	Module 1	Module 2	Module 3	Module 4	Module 5
Schedule Planning & Control	Activity Definition	Activity Sequencing	Duration Estimating	Schedule Development	Schedule Control
Cost Planning & Control	Resource Planning	Cost Estimating	Cost Budgeting	Cost control	
Resource Planning & Control	Resource Planning	Resource Estimating	Resource Budgeting	Resource Control	
Management Reporting	Communications Planning	Information Distribution	Performance Reporting	Administrative Closure	

Parametric mapping and classification

Parametric mapping was used as a tool for eventual planning, monitoring and control information integration. First of all, parameters within each control function module (such as Activity Definition with Schedule Planning and Control), were mapped on to each other in terms of information flows. Later, this concept was extended to all parameters within a

given project control function such as, cost planning and control. Finally, this mapping was done at a system level for all project control functions being studied.

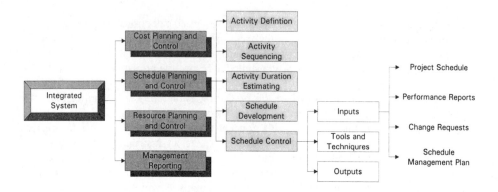

Figure 4 System Analysis and Functional Decomposition

The analysis of the parametric mapping at different levels of abstraction and detail indicated that certain project control function parameters were generic. Furthermore, upon a more micro level investigation, it was observed that generic parameters were of two forms, those that were generic with one project control function, and those that were generic across more than one project control function.

Taking the generic nature of the project control function parameters as a possible classification criteria, the parameters were classified as follows.

Local Parameter: A parameter, which was unique to a given project, control function module.
Functional Parameter: A parameter, which was generic across more than one project control function module of a given project control function.
Global Parameter: A parameter, which was generic across more than one project control function.

A total of one hundred and twenty one project control function parameters were considered during the analysis. After the analysis and subsequent parameter classification, thirty-three local, thirteen functional, and fourteen global parameters were identified. Hence, after integration and removal of redundant parameters, a total of sixty project control function parameters would be required for data and information integration purposes. A summary of the analysis is given in Table 2, and the integrated environment based on local, functional, and global parameters presented for illustration purposes in Figure 5.

Table 2. Summary of Parametric Mapping and Classification Analysis

Project Control Functions	Parameters Considered	Parameters			Parameters Required	Parameters Eliminated
		Local	Functional	Global		
Schedule Planning and Control	47	15	4	13	32	15
Cost Planning and Control	27	5	3	12	20	7
Resource Planning and Control	27	4	3	12	19	8
Management Reporting	20	9	3	5	17	3
Integrated Environment	121	33	13	14	60	61

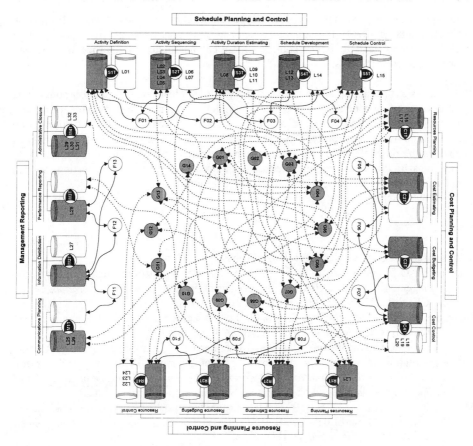

Figure 5. Integrated Environment based on Local (L), Functional (F), and Global (G) Parameters.

Conclusions

Information management for efficient decision making during construction planning, monitoring and control is a key entity that may contribute towards better project performance and delivery. Researchers have hinted however that several obstacles do exist however. These have been attributed to poor decision making and information management, the lack of integrated solutions and automation in planning and control, and due to the under utilization of information technology.

The ongoing research presented in this paper attempts to observe the problems restraining proper information management and information integration in construction. A micro-level analysis coupled with parameter and process classifications and mappings helped explore the possibility of the cross-functional integration of project control functions. It was observed that efficient data and information integration may be done through project control function parameters and in the process, conflicting and redundant parameters eliminated.

Furthermore, an exploration of the type and nature of project control function parameters and processes through theoretical analysis and feedback from construction practitioners may lead to better control of critical parameters and processes, at both corporate and project levels.

The study is still ongoing, and a detailed analysis of project control function parameters and processes being conducted. Once this is complete, data and information blueprints and maps will be generated to provide a better overview of possible data and information integration domains. A prototype integrated system environment will then be developed to observe the rationale and applicability of an integrated environment. Finally, based on the findings of the study and the feedback acquired through prototype testing, system implementation domains and techniques will be explored.

References

WWWebster Dictionary, Merriam (1996)- Webster, New York.

Abudayyeh, O Y and Rasdorf, W J (1991) Design of construction industry information management systems. *Journal of Construction Engineering and Management,* ASCE, Vol. 117, No. 4, pp. 698-715.

Davenport, T H (1994) Saving IT's soul: human centered information management. *Harvard Business Review,* March-April 1994, pp. 119-131.

Kazi, A S and Charoenngam, C (1996) The need for information management in construction: a strategy for improved project planning, monitoring and control. *Proceedings of the International Conference on Urban Engineering in Asian Cities in the 21st Century,* Bangkok, Thailand, November 20-23, pp. D.226-231.

Minato, T and Ashley, D B (1998) Data-driven analysis of "corporate risk" using historical cost control data. *Journal of Construction Engineering and Management,* ASCE, Vol. 124, No.1, pp. 42-47.

Rowings, J E (Jr.) (1991) Project-control systems opportunities. *Journal of Construction Engineering and Management,* ASCE, Vol. 117, No.4, pp. 691-697.

Saeed, K and Brookes, K (1996) Contract design for profitability in macro-engineering projects. *System Dynamics Review,* Vol. 12, No.3, pp. 235-246.

Shapira, A et al. (1994) Anatomy of decision making in project planning teams. *International Journal of Project Management,* Vol. 12, No.3, pp. 172-182.

Tatum, C B and Fawcett, R P (1986) Organizational alternatives for large projects. *Journal of Construction Engineering and Management,* ASCE, Vol. 112, No.2, pp. 49-61.

Teicholz, P and Fischer, M (1994) Strategy for computer integrated construction technology. *Journal of Construction Engineering and Management,* ASCE, Vol. 120, No.1, pp. 117-131.

Tenah, K A (1984) Management information organization and routing. *Journal of Construction Engineering and Management,* ASCE, Vol. 110, No.1, pp. 101-118.

Tenah, K A (1986) Construction personnel role and information needs. *Journal of Construction Engineering and Management,* ASCE, Vol. 112, No.1, pp. 33-48.

Yates, J K and Rahbar, F F (1990) Executive summary status report. *1990 AACE Transactions,* D-3, pp. D.3.1-D.3.9.

COMPUTER-BASED SIMULATION FOR FOSTERING HARMONY IN CONSTRUCTION

MOHAN R. MANAVAZHI
School of Civil Engineering, Asian Institute of Technology, P.O. Box 4, Klong Luang, Pathumthani 12120, Thailand. manav@ait.ac.th

Abstract

Construction projects are complex undertakings and are often executed in an environment that provides a fertile ground for the development of adversarial relationships amongst project participants. Such relationships have a negative effect on productivity and economy in construction. Although it is difficult in practice, the key to achieving economy in construction lies in developing and sustaining a certain degree of harmony amongst the project participants. This paper investigates the possibility of developing harmony through project integration that is facilitated by the development and use of a computer-based product-centric simulation model that encompasses all phases of a project.

Keywords: Simulation, project integration, simulation model, product-centric model

Introduction

In many parts of the world, the construction industry represents a significant part of a nation's economy. It plays a central role in development through the construction, maintenance and repair of infrastructure facilities, housing, commercial complexes, office buildings and industrial plants. Thus, a high level of construction activity is one of the primary indicators of economic growth. It follows then, that the achievement of economy in construction has far-reaching consequences on national and global economies.

The conceptualisation, planning, design and construction of a facility involve major commitments of time and money in addition to material, equipment and manpower resources. Efficient utilisation of these resources in the various operations required to construct the facility is key to achieving economy and thereby realising profit for the project participants.

Harmony in construction

The key to achieving economy in construction lies in the establishment of a certain degree of harmony between the various participants in the project. However, the establishment of harmony in a construction project can prove to be an elusive goal. Construction projects are often plagued by adversarial relationships (Oglesby *et al.*, 1989). Such relationships can exist between the following:

1. Project participants
2. Management and labour
3. Crafts
4. Design/Quality assurance staff and line management
5. Line management and support function staff
6. Contractor and subcontractor
7. Client and contractor
8. Project participant and third-party

The list is by no means exhaustive and many other forms of adversarial relationships can exist in construction projects.

Although several mechanisms like project partnering (Nicholson, 1992) aimed at developing harmonious relationships between the various contractual entities in a project are being tried, the basic environment in which a construction project is executed still falls within a broad spectrum that could range from intense but fair competition at best to "cut-throatism" at worst.

Notwithstanding the efforts directed at replacing human labour with equipment in construction, the industry still remains a predominantly labour-intensive one. Strategic and tactical decision-making is still carried out by humans. Thus the human component forms the backbone of both creative and productive effort in a construction project. However, this component is often the source of a number of problems that impede progress of work in a project. At the individual level, a typical construction project brings together people with varying skills, cultural backgrounds, intellectual attainments, behaviour patterns, values, financial conditions, emotional characteristics and motivations. At the organisational level, one finds project entities with varying organisational cultures, value systems, reputations, economic power and political clout. Yet this brew of variegated individual and organisational traits must somehow be channelled to execute a project in a manner that will help achieve the project objectives of cost, schedule, quality and safety. To add to the chaos is the fact that there is a great deal of competition both at the individual and organisational levels. Thus the construction project environment is a fertile ground for the development of adversarial relationships. These thoughts have been echoed by Hellard (1992) who noted that "the organisation of the construction industry today has a built-in recipe for conflict."

Thus, there is ample scope for adversarial relationships to develop in construction projects and once these develop the parties involved shift their focus and energies from meeting project objectives to "winning the dispute."

Fragmentation in construction

According to Brandon and Betts (1995), one of the main causes for poor productivity in the construction industry is fragmentation. They note that fragmentation "refers to the fact that the different participants in the design and construction of buildings have been trained, and gained experience through different educational and professional paths, are employed in distinct organisations have distinct value systems and methods of working, and have derived separate information management systems."

Fragmentation has the adverse effect of raising barriers that hinder co-ordination among the various organisations and individuals involved in a project. This results in the sacrifice of the larger good of the project for individual and/or organisational gains. If carried to the extreme, fragmentation can result in these individuals and organisations working at cross-purposes to each other and to project objectives. This can in turn result in poor productivity, quality and the loss of harmony and profit in the project. If it is accepted that such a state of affairs should not exist in the construction industry, which typically represents about 10% of a nation's economy (Brandon and Betts, 1995), then the tendency to drift towards fragmentation in construction projects must not only be arrested but also reversed. This is one of the primary reasons for construction integration enjoying the attention that it receives from researchers and practitioners all over the world today.

Role of simulation in project integration

Betts *et al.*, (1995) provide a generic definition for integration and define it as "the sharing of something by somebody using some approach for some purpose." The term integration, therefore, can have different interpretations to different individuals and organisations and in different contexts and environments. The authors have also listed several technologies that can facilitate integration namely, object-oriented systems (OOS), hypertext, data exchange standards, virtual reality and multimedia. This paper takes a look at the role that simulation can play to facilitate project integration through the development of a unified, product-centric model of the project.

Simulation could be defined as the representation and manipulation of a real-world system, normally in a medium outside of its natural environment, with a view to studying the behaviour of the system in response to external stimuli. The representation of the system used in a simulation is called a simulation model. Vincent and Kirkpatrick (1995) advocate the development and use of an "effective" model for conducting (i) financial and life cycle simulations, and (ii) use, construction, operation and maintenance simulations for the purpose of predicting and avoiding problems. Such a model should of necessity be predominantly product-centric. Manavazhi (1998) developed a configuration-based simulation modelling methodology using the object-oriented programming paradigm for simulating the construction of structures with specific configurations. The methodology is based on the progressive decomposition of the structure into its components and then integrating the representations of these components in a computer through inter-nodal communication links to facilitate simulation of the construction of the structure. Although this methodology was used to develop a predominantly product-centric model for simulating construction operations, it could be enhanced to include other phases of a project.

The use of simulation to facilitate project integration is based on the following core ideas:

(a) Extending the capabilities of the simulation model of the structure that is to be built to include other phases of the construction project namely feasibility studies, design, commissioning, use and maintenance. Although such a model will not be as comprehensive as the one envisaged by Vincent and Kirkpatrick (1995), it covers almost all of the project life cycle.

(b) Use of simulation to play out various scenarios involving technological options, organisational groupings and schedules that will help the project team come up with the best alternative or alternatives under the circumstances.

(c) Use of simulation-facilitated problem solving as a means of breaking down communication barriers and fostering a feeling of unity and satisfaction among individuals and organisations.

Model generation and refinement

As a first step in the model generation process, a preliminary simulation model is created. The preliminary model is based on the information available within the project team at the point in time in the feasibility study phase that the development of the simulation model is undertaken. The preliminary simulation model will contain information or will have access to information necessary for conducting simulations that will help evaluate conceptual alternatives for the project from the technical and economic perspectives. Output generated from these simulation runs can be selectively stored for future use at the discretion of the user of the simulation model. Additional information available as part of activities of the project team in the design phase will be utilised to enhance the preliminary simulation model. Simulation runs at this stage will be used for evaluating various design alternatives based on predictions generated by the model with regard to satisfying given design specifications. In the construction phase, the model could help the project team take decisions on matters such as the types and quantities of resources that will be required for a particular operation. It could also be used to predict the duration of an operation or of the whole project based on a detailed analysis of the operations involved in the project. In the start-up phase simulations could be used to warn the project team of potential "teething" problems that could occur in systems and subsystems within the project. Finally an "as-built" simulation model could be handed over to the owner or maintenance staff. The "as-built" model could be used for pointing out problems and the need for preventive maintenance within the completed facility. It could also be used for carrying simulations of repair scenarios after a problem has occurred in a system or subsystem within the facility.

It is important to note that the "as-built" simulation model has to be developed by a series of successive refinements carried out to the preliminary simulation model. These refinements are based on suggestions received from within the project, knowledge available with the implementation team and historical information from other projects. To enable ready access to such knowledge and information the simulation model must have access to an information base that can consist of information stored in databases, hypertext documents, spreadsheets, local area networks and the Internet. In large and complex projects the simulation model may also turn out to be large, complex and unwieldy. In such cases multiple simulation models may have to be developed with each model being linked to others for accessing information. The model creation and refinement activities must be carried out by a team representative of the various entities that are connected with the execution of the project. The composition of this team could change with time and with the particular phase that the project is passing through at any point in time in the life of the project. Similarly the end-users of the simulation model can also change. For example, in the design phase the users could be architects, structural designers and services designers while in the use and maintenance phase, the end users could be maintenance staff. Feedback and suggestions for improvement should be welcomed from anyone involved in the project irrespective of their background or role in the project. The feedback and suggestions are placed before a model development group consisting of project management expert(s) drawn from both within the project and outside. Ideally they should represent both software engineering and project expertise particularly expertise that is relevant to the current phase of the project. The project management expert(s) act as technical adviser(s) to the software professionals. The model development group will decide on whether to refine the model based on the suggestion or reject the suggestion. Refinements of the simulation model can continue to take place till the end of the useful life of the facility or structure. A graphical representation of a framework for achieving project integration using simulation is shown in Fig. 1. An organisational scheme for the model development process is shown in Fig. 2.

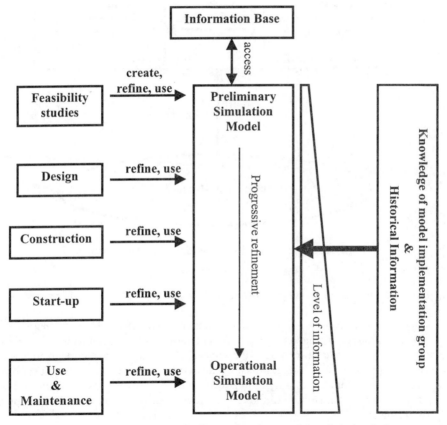

Figure 1. Framework for achieving project integration using simulation.

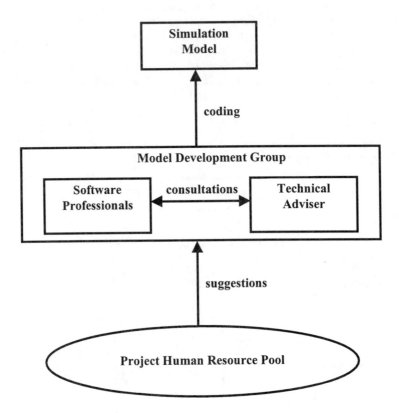

Figure 2. Organisational scheme for model development

The use of the simulation model promotes project integration in the following ways:

a) A feeling of unity and sense of satisfaction can be achieved among participants through a process of conducting repeated and participative problem-solving exercises facilitated by the simulation model. Participants in such problem-solving exercises can see the effects of their suggestions within a very short time frame. Therefore a relatively large number of suggestions can be tried out inexpensively and within a short time when compared to similar exercises conducted without the aid of computer-based simulation.

b) The simulation model can also serve as a medium for the sharing and transfer of information between project phases and between project participants. It can serve as a focal point of group activity directed at constructive problem solving. Such activity in any particular phase can have the active participation of project personnel from temporally adjacent phases. For example, in the design phase, inputs on a particular problem can be solicited from the contractor and/or subcontractors hired to construct a facility according to design specifications. This could result in a more workable and economic design. Similarly, inputs can also be obtained from the owner's representative or the architect-engineer providing guidance and the rationale for certain upstream decisions taken in the feasibility study phase. Interactions between groups from different phases could also help determine the validity of the simulation model and the need for updating it.

Limitations

1. The effort and costs associated with developing a simulation model of this magnitude may be justifiable only in very large and complex projects.
2. The use of simulation for routine analysis, planning and problem solving in construction must precede its use as a medium for project integration. This would require large-scale acceptance of simulation by practitioners in the construction industry.

Conclusion

Although the use of simulation in the construction industry is still in its infancy, the spin-offs from such use in developing project-wide integration and fostering harmony among project participants could result in tangible benefits in improving productivity and economy in construction. A feeling of unity and a sense of satisfaction can be achieved among project participants through a process of participative problem-solving facilitated by a simulation model. Furthermore, a larger number of suggestions could be tried out inexpensively and in a small time frame. The simulation model can also serve as a medium for the transfer of information between project phases and between project participants resulting in the realisation of higher productivity, greater profit and a less stressful working environment.

References

Betts, M., Fischer, M.A. and Koskela, L. (1995) The purpose and definition of integration. In *Integrated Construction Information,* Brandon, P. and Betts, M. (eds.), E & FN SPON, Chapman & Hall, London.

Brandon, P. and Betts, M. (1995) The field of integrated construction information: An Editorial Overview. In *Integrated Construction Information,* Brandon, P. and Betts, M. (eds.), E & FN SPON, Chapman & Hall, London.

Hellard, R.B. (1992) Construction conflict - management and resolution. In *Construction Conflict Management and Resolution.* Fenn, P and Gameson, R. (eds.), E & FN SPON, Chapman & Hall, London.

Hendrickson, C. and Au, T. (1989) *Project Management for Construction,* Prentice Hall, N.J. USA.

Manavazhi, M.R. (1998) A configuration-based modelling methodology for the automated generation of simulation models in construction, Ph.D. Thesis, University of Alberta, Edmonton, Canada.

Nicholson, M.P. (1992) Peace, love and harmony. In *Construction Conflict Management and Resolution.* Fenn, P and Gameson, R. (eds.), E & FN SPON, Chapman & Hall, London.

Oglesby, C.H., Parker, H.W. and Howell, G.A. (1989) *Productivity Improvement in Construction.* McGraw-Hill, New York.

Vincent, S. and Kirkpatrick, S.W. (1995) Integrating different views of integration. In *Integrated Construction Information,* Brandon, P. and Betts, M. (eds.), E & FN SPON, Chapman & Hall, London.

DECISION SUPPORT SYSTEM FOR PHASE COORDINATION ON FAST-TRACK PROJECTS

P.L.S. RAVIKANTH and KOSHY VARGHESE
Department of Civil Engineering, Indian Institute of Technology, Madras.
koshy@civil.iitm.ernet.in

J.T. O'CONNOR,
Department of Civil Engineering, The University of Texas at Austin.

Abstract

Executing projects on fast-track basis poses numerous management challenges. Meeting these challenges are critical for the harmonious and profitable completion of a project. A survey of owners, construction managers & contractors by the Construction Industry Institute has reveled that one of the key challenges faced is to ensure that flow of information and deliverables from engineering and procurement to construction is synchronized. As the designer, vendor and contractor view the project differently, and face different internal constraints, it has been found that the sequence of executing the work activities is different in each of these phases. Further, the inherent uncertainty in the fast-track concept also disrupts planned sequences.

This paper presents a detailed definition of the sequencing problem in the context of industrial construction. It is proposed that an information system representing the deliverables at each stage and their interdependence would assist in making rational decisions on the appropriate sequence at each stage. This system can be extended to model uncertainty and generate optimal sequences to satisfice design, procurement and construction requirements. The overall architecture of the system, identifying the components and the functions of each component is also presented.

Keywords: Decision support systems, engineering-procurement sequence, sequence coordination

Introduction

To meet the increasing demand to deliver projects within a short time and minimise financing costs many mega-projects are executed on a fast-track basis. On fast-track projects the construction begins when design is between 55-70 % complete. This poses numerous management challenges.

A study carried out by the Construction Industry Institute has reveled that one of the key challenges faced is to ensure that there is close coordination between the Engineering-Procurement-Construction activities [O'Connor & Liao (1996)]. As the designer, vendor and contractor view the projects differently, it has been found that the natural sequence of executing the work activities are different in each of these phases. For example, the contractor looks at the project physically and would prefer to sequence the construction based primarily on physical constraints. However, to the designer would not view the project physically but as systems & sub-systems and would proceed to design in a logical manner - governed by design constraints. While to the vendor/fabricator the best sequence of delivering the components would be based on manufacturing/fabrication constraints.

As a result of the differing ideal sequences in each phase and inherent uncertainty in the fast-track concept, it has been found that on numerous projects items to be installed as per the construction schedule items might be still undergoing design or fabrication. In contrast, items scheduled later in the construction phase might have been delivered and will be taking up storage space at the site. Thus there is a need for a tool which can assist in the coordination of deliverable flows within and between the phases.

The objective of this work are to: (i) understand and document the sequencing and coordination problems faced in the engineering-procurement and construction of process facilities. (ii) Develop a framework to classify the problems (iii) Design and develop a decision support tool to facilitate solving the sequence and coordination problems.

This paper is organised into five sections. The following section presents the information flow through different stages for a heavy industrial project. Next, the obstacles which arise in this information flow are presented, and in the context of these obstacles, the sequence problem is defined. The fourth section, reviews previous studies which are relevant to this problem, and proposes a solution to the problem. Finally, the conclusions of the study are discussed.

Stages of information flow

The first step taken was to define the information flow in the different stages & sub-stages of a project. This was done through a series of interviews with industry experts. Figure-1 shows the information flow between the various stages considered in the scope of this study. An analysis of this information flow was then done to determine and classify the possible sequence problems, which can arise.

In the preliminary stage of the project, broad issues such as deciding on the products and the capacity of plant are considered. These decisions are made by the client (with assistance from specialty consultants if required). Based on these decisions, the licensor who provides the chemical process required for production is selected. The basic details of the chemical process are communicated in the form of a document called the Process Flow Diagram (P.F.D). The P.F.D contains the details about various units to be constructed and the flow parameters across these units.

Based on the process requirements specified in the P.F.D and the site characteristics, the location of the different production units in plant is determined by the client/consultant. In addition, the preliminary layout of areas within each unit is also determined based on the major equipment requirements within the unit. The numbering of unit/areas is conventionally done in the direction of the flow of the fluid. A typical layout is shown in Figure-2.

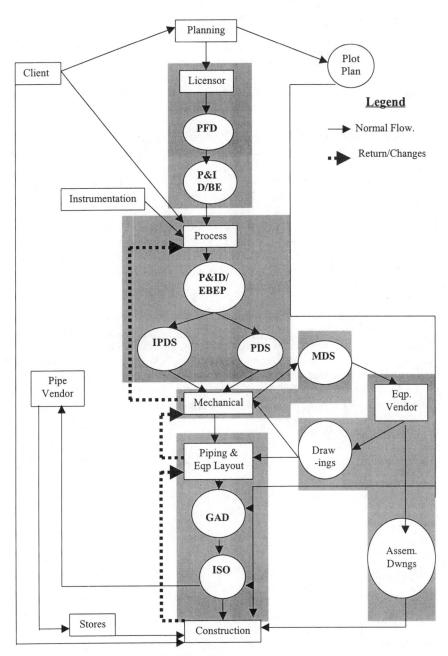

Figure 1 Information flow diagram

The Piping and Instrumentation Diagrams (P&IDs) are critical documents as they are used as a reference throughout the project. There are two stages of P&ID development The first stage is called Basic Engineering Package (B.E.P) and contains information about major equipment, connectivity of piping and all piping related details such as line sizes, material type, pressures, temperatures, flow velocity etc. In order to develop the P&ID the process licensor mathematically simulates the flow in various units to derive many of the parameters specified in the P&IDs. The P&IDs are released unit by unit in the direction of process flow. For each unit there will be approximately 25 to 30 P&IDs. The instrumentation component of the P&ID is not completed at this stage as it needs detailed survey report of the site.

The process team receives the P&ID/B.E.P from licensor and makes minor modifications to accommodate all the client specific requirements such as additional by-products or further treatment of effluents. Additional details are included in the process package with the help of instrumentation personnel. This results in an improved form of the P&ID/B.E.P referred to as Extended Basic Engineering Package (E.B.E.P).

In the next step, specification sheets called Process Data Sheet (P.D.S) and Instrumentation Process Data Sheet (I.P.D.S) are prepared. The P.D.S contains information about equipment list, line list, material specification etc. The I.P.D.S contains the details about control valves and other instrumentation components. These documents form the input for mechanical design.

Based on the process requirements specified in the Process Data Sheets, the Mechanical group does the preliminary design for the components such as pressure vessels, compressors etc. Based on the preliminary design, the exact specifications of the equipment are specified in the Mechanical Data Sheet (M.D.S). The MDS forms the basis for the vendors to carry out detailed design. The vendors do the detailed design and submit the resulting drawings to the Mechanical group for approvals. Fabrication of the equipment begins only after the approval from engineering is received.

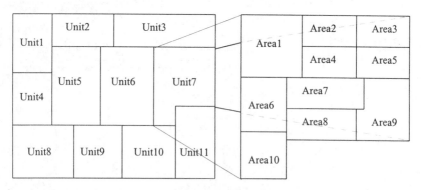

Figure 2.a Site layout   Figure 2.b Areas within unit

The approved design drawings, the P&ID/EBEP and the overall site layout is required by Piping & Equipment Layout group to finalise the detailed layout of the equipment and route of pipes. The General Arrangement Drawings (G.A.D) contain details of the layout of various areas at different elevations based on equipment and pipe rack heights in the area. Depending on the complexity of the route or fittings on individual pipe lines within an area, isometric drawings showing the details of the route and special fittings are prepared. These isometrics are required by the vendors to fabricate the pipes and fittings.

After fabrication, the vendors deliver the pipe fittings and equipment with relevant assembly drawings to the construction site. Based on the sequence of construction these components are erected utilising relevant resources. After the erection of all the equipment and piping, detailed tests will be done for each unit before the start-up commences.

Problems in flow sequence

Although the broad procedure for the information flow between the various stages is well-structured, it does not ensure that the construction team will receive the necessary information and components in the required sequence. At each stage there are delays in specific documents or components, which affects the sequence of flow. The delays can be generally categorised as unforced delays and forced delays.

Unforced delays are caused due to errors and omissions in the specification sheets and mismatch between component specifications and available standards. In addition, the lack of awareness within a group on the downstream priorities can also cause delays. If a document contains errors or omissions, the document will be sent back to the originating group for correction. Similarly, if the specifications for a component cannot be met by the vendor, it will be sent back to engineering for redesign. If errors are not detected early and are allowed to propagate to construction, the resulting impact can be significant. Unforced delays are common in practice and numerous instances have been documented by the Construction Industry Institute. [O'Connor and Goucha (1996)]

Forced delays are caused mainly due to non-availability of data, changes in scope and delays in approval. If data is not available to complete a particular document, a "hold" is placed on the document. This affects the progress of all related downstream documents. Holds are common at early stages of the design process where there are more chances for unavailability of data. As the P&ID development process is one of the early design stages it is common to have numerous holds on P&IDs. As there are interrelationships between P&IDs & other documents/drawings these holds will cause design delays in other stages also. If a designer releases the hold by making an guesstimate, and freezes the drawing too early, it may necessitate redesign/rework if the guesses made are outside the tolerance limits.

Scope changes are initiated by the client due to a change in project requirements or operating considerations. Such changes can have a ripple effect through the entire project especially if it requires the modification of upstream documents such as P&IDs. Case studies relating to the impact of changes and holds on industrial projects have been documented by the Construction Industry Institute [O'Connor & Liao (1996)].

In addition to delays, the outputs from a sequence based on design logic will differ from construction sequence. Thus, even if the impact of delays are not considered, and the natural (optimal) sequence of design is followed, the inputs into construction will not arrive at the desired sequence. It might be possible to change the sequence of some design activities through conservative assumptions, when this is not possible the expected completion dates should be communicated to the construction team so that appropriate action can be taken. It is also likely that as the project progresses the construction sequence can change, in reaction to this change the design team should verify if they could reschedule to accommodate this change without violating design logic.

As there are numerous dependencies and iterative relationships between design activities, which cannot be, modeled using conventional planning tools thus only an abstract level of planning is done. In the abstract plan, relationships between design documents are not enumerated and communicated to the designers. Further, no resource planning and duration estimation of design of activities is carried out. Thus, little control can be established to ensure the design and fabrication outputs are sequenced to satisfy the construction sequence. A tool, which can assist in managing the design phase and its coordination with procurement and construction phases, would certainly ensure harmony between the various groups and enhance the profitability of the project.

Solution approach

Past work

Past work related to this problem has focused on developing management tools as well as computerised tools to support the decision making process. Studies conducted by the Construction Industry Institute have focused on (i) enhancing the piping and instrumentation diagram development process [O'Connor & Liao (1996)], (ii) improving the design-fabrication process [O'Connor & Goucha (1996)] and (iii) managing uncertainty in the piping function [Howell & Ballard (1996)]. All of these studies were carried out with extensive input from industry and resulted in management tools to improve the levels of current practice.

A study focusing on the interface between piping fabricator and construction was modeled using computer simulation [Tommelien (1997)]. The study focused on the strategies to enable construction to progress continuously by selecting work areas for which all resources for installation were available.

In manufacturing, the need to alter production sequences in response to external factors is common. The process of developing an alternate plan is called reactive scheduling [Zweben & Fox (1994)]. In the current problem this is analogous to re-sequencing design to meet changing construction sequence. Computer based tools have been developed to optimally re-sequence production plans. As uncertainty modeling is critical to the effectiveness of these tools various approaches such as fuzzy logic [Schmidt (1994)] have been used to model uncertainty.

The use of techniques such as Design Structure Matrix have been proposed and tested for optimal sequencing of design tasks [Huovila et. al.(1995)]. This study focused primarily on building design. Another study focusing on optimising building design sequence and improving the overall design management function proposed the use of a management tool called the last planner [Koskela et. al (1997)]. The study found the application of the tool resulted in better design effectiveness.

Proposed solution

The sequencing problem discussed in the previous section can be modeled at two levels. The first level would be deterministic, here it is assumed that there are no delays. The second level would be probabilistic, in this case the uncertainties caused by the delays and changes would also be modeled. The first step towards developing the tool is to design an information system to represent relevant data efficiently. Further, in each of these levels decision support features should be provided to facilitate the predictive and reactive use of the tool.

The information system basically consists of a database, which represents the design documents and defines the relationship between various design documents. The interdependencies and intra-dependencies between various drawings, specifications, plant units and areas are represented through the entity-relationship diagram in Figure-3. Other data which need to be represented includes:

1. Current status of the drawing/document and stage of fabrication of equipment & pipe spool
2. Relationships between drawings/documents and the areas in the plant
3. Relationship between plant areas and sequence of construction
4. Duration for developing design documents
5. Ideal design sequence
6. Penalties for changing sequence

As there is a need to represent and integrate graphical information along with the text information, it is proposed that a geographic information system environment would be ideal for implementation.

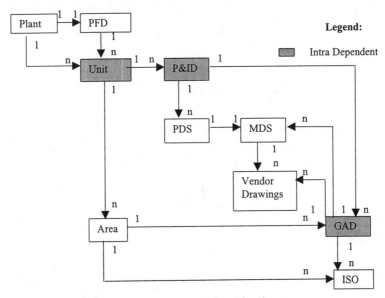

Figure 3 Entity relationship diagram

Based on the data represented, this system can be queried to provide information such as: conflicts in sequence, current status of design/fabrication, history of any drawing or document, impact of pre-determined delays on construction sequence. In addition, the system can be used to carry out what if analysis, thus enabling the systematic development of a non-conflicting sequence.

Automating the sequence generation can further enhance this basic information system. This will permit the planner to develop near-optimal sequences in a short time thus permitting the design sequence to be fine-tuned for any change in construction sequence. Depending on the size of the sequencing problem, the automation technique can be based on conventional optimization techniques or combinatorial methods. Work on evaluating the suitability of different techniques is being carried out.

The elements of uncertainty, which are inherent in the system are: (i) variability in the duration required to complete a design stage or fabricate a component, (ii) the probability of errors in the document and stage of detection, and (iii) changes in design due to scope changes. Only the first two types will be represented in the current system. It is proposed to model the information flow process and these uncertainties using a simulation platform. Using this module of the system the decision-maker will be able to test the impact of alternate management strategies on the overall project schedule.

The solution concepts presented above are being implemented in a software environment as shown in figure-4. The key component – the information system is being implemented using a Geographic Information System - Autocad Map, linked to a database - Access. This will provide an integrated environment to link the graphical and non-graphical information and manage the design documents. Based on the technique selected for automated sequence generation, the module can be implemented within and appropriate software environment and will be linked to the database. A simulation environment such as AweSim will be utilised for modeling the information flow process with delays. A customised Visual Basic interface will be provided to allow user-friendly access to the system functions.

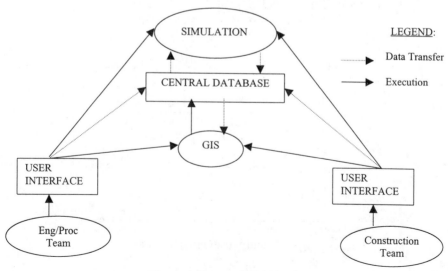

Figure 4 System Architecture

Conclusions

This paper presented the details of a coordination problem between the engineering-procurement and construction activities in a fast-track industrial construction project. It specifically focused on the sequencing issue in the design phase as design is a complex activity and case studies have shown that (i) design being an upstream activity has a tremendous impact on the project and (ii) there is little planning done in the design phase.

Based on the nature of the defined problem and capabilities of conventional tools it was concluded that: A tool which can assist in sequencing and coordinating engineering, procurement and construction activity would ensure harmony between the various groups and enhance the profitability of a fast-track project.

The paper also discussed the specific functions of the tool and identified components of such a tool. It was concluded that an information system would be the core component of the tool. Extension of this information system to provide decision support functions and alternate strategies to represent and model uncertainty was also discussed. Architecture identifying possible software platforms for implementing system components was also presented.

In addition to implementing and testing the proposed system, future work will focus on the investigation of concurrent engineering and reactive scheduling concepts for alternate solution strategies.

References

Howell G.A (1996), Ballard G. (1996). "Managing Uncertainty In The Piping Functions". Report to Report 47-13 to Construction Industry Institute, Austin, TX.

Huovila, P., Koskela, L., Lautanala, M., Pietiläinen, K., Tanhuanpää (1995), . "Use of the Design Structure Matrix in Construction" International Conference on Lean Construction, Albuquerque, October 16 – 17.

Lauri Koskela, Glenn Ballard, Tanhuanpää Veli-Pekka (1997). "Towards Lean Design Management" International Conference on Lean Construction, Gold Coast, Australia.

Michael E.Leesley (1982). "Computer-Aided Process Plant Design", Gulf Publishing Company, Houston, Texas.

O'Connor J.T, Hatem Youssef Goucha (1996). " Improving Industrial Piping Through Vendor Data And Packaged Units Processes". Report 47-11 to Construction Industry Institute, Austin, TX.

O'Connor J.T, Shih-jen Liao (1996). "Enhancement Of The Piping And Instrumentation Diagram Development Process". Report 47-12 to Construction Industry Institute, Austin, TX.

Schmidt Gunter (1994). "How To Apply Fuzzy Logic To Reactive Production Scheduling". Knowledge based reactive scheduling, Elsevier Science B.V (North Holland).

Tommelein I.D (1997). "Discrete-Event Simulation Of Lean Construction Processes" International Conference on Lean Construction, Gold Coast, Australia.

Zweben M, Fox, M. (1994), "Intelligent Scheduling". Morgan Kaufmann publishers Inc.

DEVELOPMENT OF A DECISION SUPPORT SYSTEM FOR COST ESCALATION IN THE HEAVY ENGINEERING INDUSTRY

WILLIAM BATES and NASHWAN N. DAWOOD

School of Science and Technology, University of Teesside, Borough Road, Middlesbrough, TS1 3BA, England. n.n.dawood@tees.ac.uk

Abstract

The lack of structured cost escalation models for construction activities in the heavy engineering industry has led to the development of this paper and the knowledge based system introduced. Research by the authors looks into the development of a methodology for improving cost management for construction works in the heavy engineering industry, through the addition of a cost escalation factor. This paper, as part of the research, reviews and discusses a knowledge-based system for applying a cost escalation factor. The cost escalation factor is made up of market variation, a risk element and a component for bias. A knowledge elicitation strategy was utilising to obtain the required knowledge for the system. The strategy included questionnaires, interviews and workshops. The deliverables came in the form of influences and their affect of project cost escalation. From these deliverables the concept of a decision support system (DSS) for applying cost escalation to base estimates was established.

Keywords: Decision, support, knowledge, outturn cost

Introduction

The overall research looks into the development of a structured cost escalation model for improved estimating management and control in the heavy engineering industry. Cost escalation, in this context, is defined as the percentage increase between the initial base estimate and the final achieved cost. As part of this research, the axiomatic goal of this paper is to introduce and discuss a decision support system based on knowledge provided by expert personnel from the heavy engineering industry [Bates (1997)] and a theoretical model derived using this knowledge [Bates (1998)].

More often than not, an arbitrary figure (say 10%) is added for cost escalation. This is naive and without explanation. Cost escalation, in the field of project management tends to be an addition to project costs based on an estimator's 'gut feeling'. Although this simple method produces an addition or 'fat' component to the base estimate it does not supply any explanation of how the figure was formulated, even if it was accurate. Therefore, as part of a change in thinking the authors will break up cost escalation into three key elements:

- Market variation dealt with by cost indices [Bates (1996)]

- Risk/contingency dealt with by a risk engineering strategy [Bates and Dawood (1998)]

- Bias dealt with as part of a company addition through the monitoring of previous projects.

These components are introduced subsequently. The aim of decomposing the cost escalation is to make it more comprehensible. These are several reasons for developing a structured cost escalation decision support system with three components. It is based on the fact that:

1. The accuracy of published cost index predictions are highly questionable [Fitzgerald and Akintoye (1995)]. These authors showed that even naive models perform better than the subjective predictions made by the Building Cost Information Service (part of the Royal Institute of Chartered Surveyors)

2. Risk Allocation strategies tend to vary from the extremely simplistic (checklist) to extremely complex probabilistic analysis [Russell and Ranasinghe (1992)]

3. Bias has never really been looked at part of the cost escalation phase. However, if there is a general continual under or overestimation, then the bias factor becomes extremely important. It can be closely related to the area of bidding when contractors are monitoring their bids in comparison with other firms.

Firstly, a breakdown of outturn cost including all of the components and how they can be brought together is provided. Secondly, the authors will introduce the business processes in the heavy engineering industry. This will highlight where our decision support system is to fit. As the objective of the research is to develop a model and prototype system for developing outturn cost from an initial base estimate, within the heavy engineering industry, prototype, components developed by the authors are introduced. Finally, a typical consultation with the introduced system is shown through screen captures, highlighting its simplicity and ease of use.

Outturn cost breakdown

The authors propose that outturn cost is made up of four components: Initial Estimate, Mark-up (accounting for Bias), Index Variation and Risk. This links in with the statement made by a prominent cost engineer in the heavy engineering industry where he states that every project has two elements, an achieved cost and an attached story. The story refers to the component additions suggested by the authors. These come together to form the outturn cost as in figure 1.

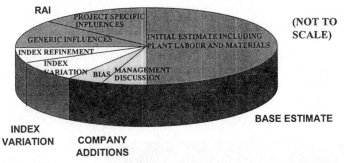

Figure 1 Outturn cost breakdown

Each component requires a software title or analysis mechanism to prepare the factor for that element. However, they should all be brought together under one user interface. This will be highlighted later, although, the position of the proposed decision support system within the estimating process is outlined next.

Where will the DSS fit?

This section works the reader through the business processes, work processes and estimating process involved in the development of a project for a client and where exactly the Decision Support system will fit. At the highest level there are client needs and these must be met (Figure 2)

Business processes

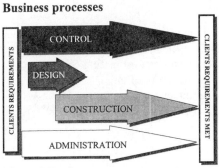

Figure 2 Business processes

To meet the needs of a client, in the heavy construction industry, there are four main business processes, which include:

1. Control

2. Design

3. Construction

4. Administration

Work process

The business processes can be broken down further into work processes which are represented in Figure 3. The authors' work fits into the control aspect of the business processes and into the estimating work process.

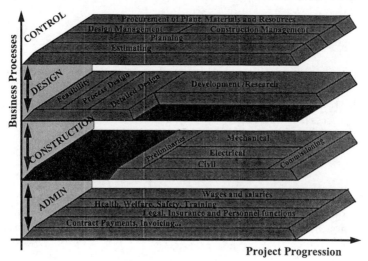

Figure 3 Work processes

The next level of process breakdown is that of estimating process and where exactly our DSS applies. Figure 4 depicts the flow of work in the estimating process (excluding any of the other elements involved in the control aspect). The normal estimating course requires past data to help prepare the estimate, either in the form of published data or past experience. Harnessing this data or knowledge is an extremely worthwhile task. The estimating process model (Figure 4) shows estimates being made continuously in the project life, determining such issues as:

- Feasibility
- Sanctioning
- Control
- Final analysis
- Estimating Process

By including analysis mechanisms as part of the overall Decision Support System shown in Figure 4, the decision making process is simplified, based on the opinion of a group of experts as opposed to a single expert, thus improving the estimate production speed and accuracy. The information and figures presented on estimating process show where the authors' research is located.

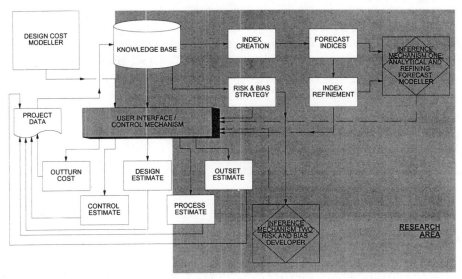

Figure 4 The estimating process

Figure 4 shows two inference/analysis mechanisms. The first is an analytical forecast modeller which provides predictions for market variation through cost indices. The second of the inference mechanisms uses risks and influences based on knowledge elicited from expert personnel [Bates (1997)]. Then, a user interface, is used to develop an overall cost escalation factor using the information from the two mechanisms. This is detailed next.

System map

The user interface, developed by the authors to control the two analysis tools, is know as TAROT (Total Automated Response for Outturn cost Tally). A system map (Figure 5) demonstrates the components of the system and their interaction with other external components. The user interface shown in Figure 4 not only interacts with the user but controls all of the other elements involved in the estimating process. Each of the elements described in Figure 5 are reproduced in more detail subsequently, starting with the TAROT user interface/controller element.

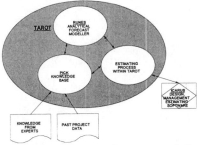

Figure 5 System map for cost escalation modelling software

Total Automated Response for Outturn Cost Tally (TAROT) in Detail

Expertise for the system came from many areas. A high level interaction diagram is shown (Figure 6) depicting the flow of knowledge and data within the TAROT set-up.

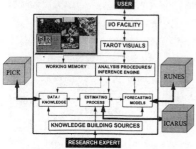

Figure 6 TAROT data movement model

TAROT, although being a new software, has similar capabilities to those software's in existence, utilising common procedures found in most software packages with elements suited only to it. Therefore, to emulate common processes (cutting and pasting) in the form of a process model would be needless. Nevertheless, the most important elements, namely escalation and prediction are simulated in Figure 7.

Apart from the normal functionality of the software, the analysis of data falls into the paths highlighted in Figure 7. The options include forecasts of indices short and long term and the escalation of a base estimate to form an outturn cost estimate with reasoning attached. The decision support comes in the form of:

- Project plans
- Cash flows
- Additions made for risks
- Index variation additions
- Mark-up additions (Bias)

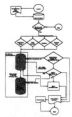

Figure 7 Flow of information through TAROT

Index forecasting element

Indices are both analytically and subjectively forecast. This is highlighted in [Bates (1996)]. First of all a solution that can provide analytical forecasts of past indices will need development. The author decided in his wisdom that, a model complementing the whole system needed development. There were several reasons for this decision and these are highlighted subsequently. The ideal opportunity for the development came in the form of Visual Basic for Applications (VBA) with Excel. Reasons for using this approach include:

- Simplicity
- Most consultancies and engineers are already familiar with excel
- It is the most common spreadsheet package
- Already in-built visuals and mathematical functionality
- Automatic upgrade with Excel
- Capable of being developed to work with all Mircrosoft Office products including access and project.

The name for the new forecast modeller is Runes© for Excel. "Runes" is currently being developed further to improve its capabilities. Packages such as @Risk for Excel and other similar models for Excel means that software is optimised on the cost engineers desktop computer. The interaction between other software titles is also fairly simple with the use of Dynamic Data Exchange (DDE) and Object Linking and Embedding (OLE). The redevelopment capabilities of this software model are endless. To illustrate how Runes for Excel works, a data flow model is shown in Figure 8. However, for a fuller description of the mechanism refer to [Bates (1998)].

The Runes for Excel data process model (Figure 8) shows the procedures involved in making a forecast using Runes for Excel. Initially, the user is asked to provide an index series (either stored or manually input). Following on from this, the user is asked for a model to represent the historical data and provide a forecast. The user has ten possibilities (although this list can be added to at any time):

1	Moving Average	6.	Classical Time Series Decomposition (Linear)
2	Simple Exponential Smoothing	7.	Classical Time Series Decomposition (Quadratic
3	Browns One Parameter Smoothing	8.	Two Weight Linear Adaptive Filtering
4	Holt's Two Parameter Smoothing	9.	Simplified Box Jenkins
5	Winter's Three Parameter Smoothing	10.	Automatic Choice Based On Modelled Past Data

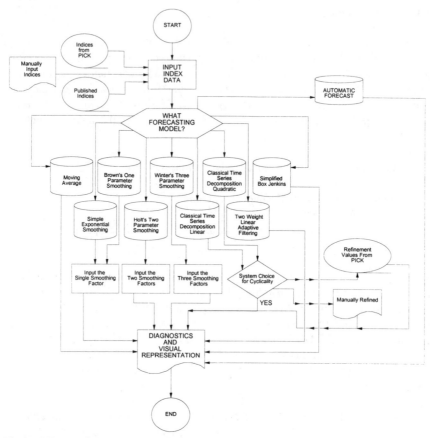

Figure 8 Runes data process model

After having supplied the required model, output diagnostics and a visual representation of the data are supplied. Further options include the refinement of the forecast manually or through supplied knowledge.

Knowledge base element

The knowledge base elements form the backbone to the overall system. It is called upon to develop factors for Risk Allocation, Mark-up (Bias) and Index predictions based on the expert knowledge supplied through the knowledge elicitation strategy [Bates (1997)]. There are no existing knowledge bases that store this specific information, this was the main reason behind the development of PICK (Process Industry Cost Escalation Knowledge base). Again, simplicity was the main objective. Therefore, Microsoft Access was used to store the relevant knowledge in the form of factors and additions. Furthermore, the Access Basic language allowed for inference and processing. The linking capabilities were also on a par with those of other Microsoft products allowing for ease of manipulation and understanding on the part of new users. Figure 9 shows the structure of the knowledge base. The user can simply view the data, however, the expert can use the system as well as update the knowledge base periodically.

The entire knowledge base can be updated without the use of any other software. The system regularly selects the indices and forecasts them using Runes for Excel and then stores the predicted indices for use within the TAROT control software. For fuller detail refer to [Bates (1998)].

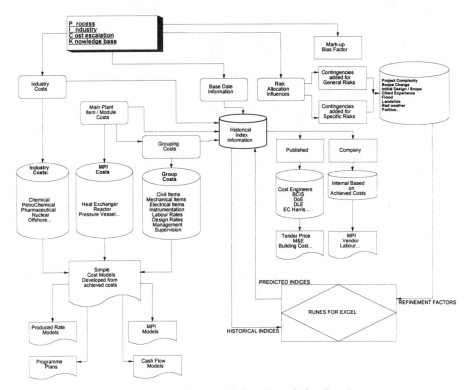

Figure 9 PICK (Process Industry Cost escalation Knowledge Base)

Summary

This paper has introduced a decision support system for predicting cost escalation in the heavy engineering industry. It utilises knowledge from processes undertaken by the authors previously [Bates (1997)]. The aim of the decision support system is not to give a definitive answer to the problem of cost escalation, but suggest possible answers based on the information fed to the system at the time, rather like the expert. The ultimate decision still lies with the user, although his estimates are now better informed.

This system has already been agreed a useful tool by personnel within the heavy engineering industry. The authors have shown where they think they system would be best utilised. However, it's optimum usage, as the experts see it, is in the boardroom in the very early stages of a project, discussing what-if scenarios.

References

Bates, W., and Dawood, N., N., Development Of A Cost Indices Forecasting Methodology For The Heavy Civil Engineering Industry, Proceedings of COBRA 96 University Of The West of England, Bristol, 1996.

Bates, W., and Dawood, N., N., Development Of A Risk Allocation Strategy For Construction In The Process Industry, Proceedings of ARCOM, University of Reading, 1998

Bates, W., Internal Report On The Estimating And Forecasting Practices In The UK, University Of Teesside, Middlesbrough, 1997.

Bates, W., Internal Report On The Development Of A Decision Support System For Cost Escalation In The Heavy Engineering Industry, University Of Teesside, Middlesbrough, 1998.

Russell, A., D., and Ranasinghe, M., Analytical Approach For Economic Risk Quantification Of Large Engineering Projects, Construction Management And Economics, 10, pp277-301, 1992.

Fitzgerald, E., and Akintoye, A., The Accuracy And Optimal Linear Correction Of UK Construction Tender Price Index Forecasts, Construction Management And Economics, 13, pp493-500, 1995

DEVELOPMENT OF A FORECASTING METHODOLOGY FOR COST INDICES IN THE GREEK CONSTRUCTION INDUSTRY

NASHWAN. DAWOOD

School of Science and Technology, University of Teesside, Middlesbrough, TS1 3BA
UK. n.n.dawood@tees.ac.uk

Abstract

Forecasting is the process of estimating or predicting the future. The principle aim is to provide reliable estimates of future business that will assist management in making accurate decisions and sound plan for the future. Essentially, there are only two forms of approach to forecasting: Subjective, which incorporates the judgement of 'experts' (persons who work closely with the industry or product they are set to forecast). Objective, which apply a scientific process to the analysis of previous data using a statistical approach to create 'a fit' to the historic data then applying the 'model' to predict future occurrences

Forecasting can be considered to perform most effectively as a two-stage process with trends extracted from historical data and subjectively analysed by an expert predictor. There have been several attempts to see whether analytical mapping can indeed outperform the judgement of a so-called expert and vice versa. However, numerous publications have shown that combination methods are undoubtedly the most comprehensible. The prime objective of this paper is to introduce a forecasting methodology, which has been developed during the course of this research. The methodology combines analytical process of forecasting with judgmental input to tune and adjust forecasting figures. The methodology is being applied to cost indices in the Greek Construction Industry. It is concluded that such methodology can improve forecasts and provide managers with a vital tool to analyse historical data. The methodology is encapsulated in a form of a computer program.

Keywords: Cost indices, forecasting, Greek construction industry

Introduction

The principle aim of accurate cost forecast is to provide reliable estimates of future business that will assist management in making accurate decisions and sound plan for the future. Essentially, there are only two forms of approach to forecasting: Subjective, which incorporates the judgement of 'experts' (persons who work closely with the industry or product they are set to forecast). Objective, which apply a scientific process to the analysis of previous data using a statistical approach to create 'a fit' to the historic data then applying the 'model' to predict future occurrences.

Forecasting performs most effectively as a two stage process with trends extracted from historical data and subjectively analysed by an expert estimator (Bates, W., Dawood, N. N. (1996)). There have been several attempts to see whether analytical mapping can indeed outperform the judgement of an expert and vice versa (Edmundson, et al. (1988)). However, numerous publications have shown that combination methods are undoubtedly the most comprehensible (Bunn, .D, Wright, G. (1991), Dawood, N. N., Neale, R. H. (1993), Pereira, B. D., et al. (1989), Bates, W., Dawood, N. N. (1996)). This paper will introduce a methodology to illustrate the concept of combining subjective methods and objective techniques in order to produce a reliable cost forecast. Objective approaches tend to use historical data and/or economic indicators to predict the future. This might be not enough to produce a reliable forecast as current market information and intelligence, and expert opinion is not taken into consideration.

The authors' are not striving to say that this is the answer to all our forecasting problems but the most understandable. Estimators in the construction industry tend not to be statisticians and like the analytical approaches to be kept simple, so they can apply some of their knowledge of the market to the prediction.

Trends and projections tend to aid the estimator with his prediction as a visual understanding of the past can help users develop the future. To make the process of forecasting simpler, indices can be developed from antecedent data and predicted using a variety of models. Explanations of these modelling techniques appear in several publications (Bates, W., Dawood, N. N. (1996)) and are outlined in a subsequent paragraph.

To prevent loss or erroneousness, the knowledge mentioned should be harnessed in an expert system. Forecasting time series information has been made simple with the development of expert forecasting software such as Forecast Pro and generally by the world of IT (Woodward, S., et al. (1994)). Complex statistical methods are carried out by the processor without the intervention of the user. This may all seem a little 'black box'. Nonetheless, as the estimator is only interested in a visual or tabular representation of the underlying trends and not the analysis that goes into producing these trends, this only aids the outcome.

Expertise is inherent with an experienced estimator as market feel. However, this feel is difficult to grasp and to put into a logical format. Nevertheless, if it was to be controlled and verified within the construction industry it could be an extremely useful tool in:

1. Speeding up the experts response to the client
2. Accelerating the training of junior estimators
3. Providing a means for encapsulating expert knowledge from experienced personnel.

All of these provide confidence for both the estimator with his figures and the client with an improved and guaranteed estimate and possible reasons for error.

The difficulty in producing reasoned information from a computer processor has gone with the simplicity of knowledge based shells. The only doubtfulness is extracting the information from the expert. In the case of the mathematical analysis, codes and formulas are used with confidence. However, in the case of market feel, there are no formulas but reasoning on different sets of circumstances at that time. The remainder of this paper discusses and introduces forecasting techniques and a methodology for integrating the subjective and objective styles. The subjective refinement utilised as part of the joint approach is also reviewed and detailed in other publications by the author. This joint methodology has been applied to actual cost indices data from the Greek Construction Industry.

Forecasting techniques

As aforementioned there are several techniques available for forecasting time series data. The approaches tend to fall into two overall categories, either analytical or subjective. In this section the author will be reviewing the most common of the techniques available in the business management field.

Review of the subjective techniques

Subjective techniques are those based on the judgement of 'experts'. There are several different forms of subjective forecast and they are reviewed in a proceeding paragraph. The techniques are based on the knowledge of the persons working closely with the element they are forecasting. The predictions are usually fairly accurate and should not be considered unusable because there is little or no mathematical background to the theories. In fact a lot of the companies in the UK only use judgement in their predictions (Bates, W. (1997)) as they found statistical methods to be far more cumbersome, expensive and no more accurate (Sparkes, J. R., McHugh, A. K. (1984)).

Normally the forecasts are produced by people who are directly involved with the item they are forecasting, and have a sixth sense about how they are going to move in the future. The familiar techniques can be found in any judgmental forecasting text (Wolfe, C., Flores, B. (1990)) and are summarised below. Table 1 (shows the judgmental techniques):

Review of the objective techniques

As the prime intent of this paper is to produce a joint approach to forecasting in the Greek Construction Industry, it is best to review the mathematically based methods also. The statistical approaches are the backbone to the conjunctive method with the subjective evaluation being the refining element. The judgmental methods are normally applied to incorporate expert judgement to the analytically forecast data. In later publications a more detailed structure for applying the subjective refinement will be introduced. Nonetheless, as we are trying to produce a statistically sound model a prose of the most common analytical methods is surveyed in table 2.

Table 1: Subjective forecasting techniques

SUBJECTIVE METHOD	DESCRIPTION
Sales Force Estimates	Sales force estimating is the method where the salesperson is required to produce an estimate of future movements in the market himself.
The Survey Method	This method is extremely simple to carry out with very little expertise required. Basically it asks customers their opinions.
The Average Judgement Method	Based on the judgement of personnel concerned with the business
Delphi Method	Based on a questionnaire approach given to a panel of experts, were the answers from the first round of the questionnaire are applied to develop the next.
Identification Methods	Based on identifying who the experts are.
Graph And Table Methods	Based on eyeball judgements of previous data, either by graph or by tables of results

Table 2: Objective forecasting techniques

OBJECTIVE METHOD	EQUATIONS	DESCRIPTION
Moving Average Models	$$\hat{Y}_{t+1} = \frac{(\sum\limits_{i=0}^{N-1} Y_{t-i})}{N}$$	This approach is ideal for finding seasonal effects, by smoothing the peaks in the actual data
Smoothing Models	$$\hat{Y}_{t+1} = \alpha Y_t + (1-\alpha)\hat{Y}_t$$	This group of models is based on the equation (left) that can be used to smooth the actual series, the trend and the seasonality depending on which model.
Decomposition Models	$Y = T * S * C * I$	This model breaks up the data into workable components, which are easier to forecast, however requires some expert intervention
ARIMA Models / Box Jenkins	$z_t = f_1 z_{t-1} + ... + f_p z_{t-p} + q_t - t_1 q_{t-1} - ... - t_q q_{t-q}$	This is a very complex set of models which are made up of auto-regressive components and moving average components based on the generic equation(left)
Causal Models	$Y_i = b_0 + b_1 X_i + b_2 X_{2i} + b_3 X_{3i} + ... + b_k X_{ki}$	These models are moving away from the projectionist type models by considering the factors that influence the overall factors movement
Learning Models	$$\hat{Y}_{t+1} = \sum_{i=1}^{N} w_i x_{t-i+1}$$ $$Y = \sum_{i=1}^{k} \alpha_i f(\sum_{j=1}^{n} \gamma_j X_j)$$	These models fall into two categories where the weights applied to the actual series for learning or as weights to the influencing factors. (Neural Networks)

In conclusions, the simple forecasting methods might perform better than complicated methods and almost all objective techniques use the past to predict the future, see Dawood, 1993. In this paper, an attempt to use historical information and knowledge about market behaviour to forecast the future has been made. The next section introduces and discusses the joint approach methodology.

Joint approach methodology

What the authors plan to show here is a systematic methodology for applying subjective intervention within a statistical procedure. The model is made up of an objective semi-complex method which allows the use of judgmental intervention, so as to refine and influence the outcome. The author will be looking in great detail at its procedure and a simple application to the Greek Construction Industry.

The semi-complex approach utilised is the Classical Time Series Decomposition (Proportionate Model). Decomposition refers to the breaking down of the data into workable components, then aggregating the prediction of each component to form an overall prognosis using Equation 1 Classical and can be carried out by following the steps discussed (vide infra).

$$Y = T * S * C * I$$
Equation 1 Classical Time Series Decomposition
Y = actual series; T = trend; S = seasonality; C = cyclicality; I = randomness

The trend may be extracted by using least squares regression.
The seasonality produces seasonal variations about a unity value and also reduces some of the randomness of the series. The stages required to achieve the values are shown next.

1. Find the moving average of the data (using a suitable seasonal period such as twelve months)
2. Divide the actual historical indices by the moving average values for the same period
3. Find the average seasonal factor for each period over the whole term
4. Normalise the seasonal factors (by multiplying them by the reciprocal of the mean of the factors calculated in 3.)

The cyclicality, which represents the ups and downs in the economy and the general movement of the business cycle (still including some randomness) is calculated as follows:

Cyclicality = Moving average ÷ trend
Equation 2 Cyclicality

Figure 1 shows the main components of the decomposition method applied to a cost index in the Greek Industry.

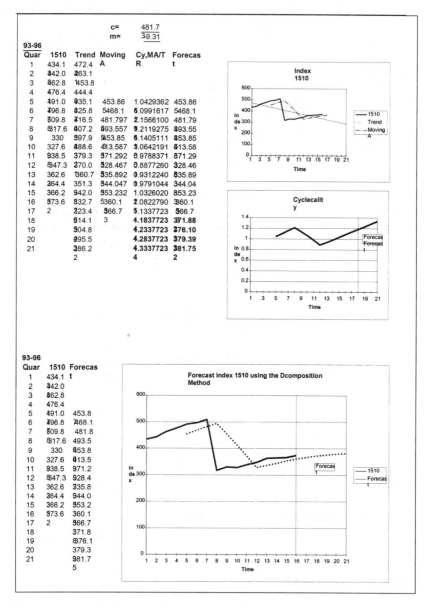

Figure 1. Main component of the decomposition method

Forming the forecast

Once the seasonal factor is identified, the trend is calculated and the cyclical factors are estimated, the forecast is the aggregating the components as shown in equation 3.

Forecast (t)= Trend (t) * seasonality (t)* Cyclicality (t)
Equation 3 Forecast

An element of judgement is applied to the forecast of the cyclicality as shown in the following section.

Subjective intervention

The subjective intervention is applied by forecasting the cyclicality that can have up to 40% on the forecast. Two different approaches have been used, the first is to use the Delphi and Identification methods (mentioned earlier in this paper) to forecast the cyclicality to be incorporated in the overall forecast (equation 3), the second is to use weighted average risk factors. The latter uses aggregation of influences to predict cyclicality as shown in equation 4.

$$\text{Cyclicality} = 1 + \left[\sum_{K=1}^{M} \phi_K \times E[MR]_K \right]$$

Equation 4 Cylclicality Movement

ϕ_K Represents the weightings of each of the factors

$E[MR]_K$ Represents the expected addition value for each factor found through knowledge elicitation.

The component inside the squared braces represents the variation about the unity value for cyclicality as an addition or subtraction. Furthermore, until these factors have been investigated a detailed evaluation such as this cannot be carried out.

The paper is focused on using the Delphi and Identification method to forecast cyclicality. In this method the experts (Engineers and Consultants) were asked to prepare a forecast of the cyclicality for 12 different cost indices and asked to specify their confidence in the results. This is used later to assign weighting for each expert. The outcome of this process is a weighted average of the cyclicality of each index. The following section discusses the case study used in this paper to explain the methodology.

Forecasting in the Greek construction industry

The historical data that is used in this paper is cost values of activities (cost of laying square meter of sub-base for major roads) and published by the Greek Ministry of Economy. The indices are fairly sensitive to economical factors such as inflation, changes in prices of materials (like petrol, cement, tyres), changes in employment fees as well as changes in taxation or interest rate. The indices are used by Greek Cost Engineers to estimate projects and by some contractors to calculate accurately the discount price in order to enter a tender submission process.

Figures 2,3 and 4 show the analysis of one of the cost indices.

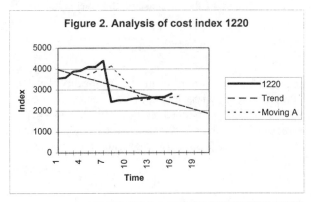

Figure 2. Analysis of cost index 1220

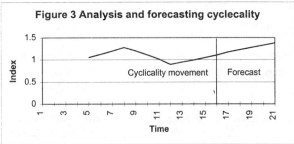

Figure 3 Analysis and forecasting cyclecality

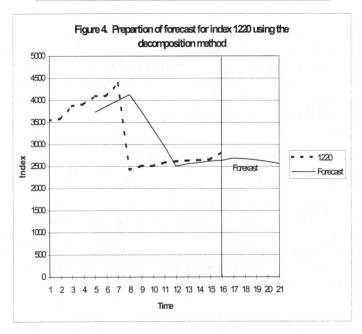

Figure 4. Prepartion of forecast for index 1220 using the decomposition method

Conclusions and future work

From visual inspection of the graphs and table above it is obvious that accuracy is not directly related to complexity with regard to forecasting, with our joint approach comparing well with some of the complex methods but with greater understanding. Further, combination forecasts leave the necessary room for improvement and show that a little judgement can improve the accuracy significantly. The expert opinion alone tended to underestimate the situation especially the seasonal jump in the data.

The heavy engineering industry is going through cultural changes in contract set-up and execution strategies. Alliances and Incentives are the buzz words. Initiatives like CRINE and ACTIVE hope to achieve cost reductions on project cost, related to the budget. However, if the budget is not accurate then cost reductions cannot be measured as there is no benchmark. This is where indices are most useful. Therefore, this area of change shows the need for accurate forecasting and model development.

As part of the overall research further aspects of the field of cost engineering will be looked into in an attempt to develop an index and a forecasting strategy for the heavy engineering industry. The indices and forecasting approach will form part of the makeup for an outturn cost development from an initial estimate.

All of the methods outlined in this paper will be harnessed as part of an expert system known as TAROT which will utilise a statistical forecasting package known as Runes for Excel and MSACCESS databases to formulate the forecasts mentioned. Further publications on the development of outturn cost figures, expert systems and risk allocation will be produced in the coming months.

References

Bates, W (1996). Internal report on statistical forecasting methods and their application to the Heavy Civil Engineering, University of Teesside, Middlesbrough.

Bates, W (1997a). Internal Report on the estimating and forecasting practices in the UK, University of Teesside, Middlesbrough

Bates, W (1997b). A Risk Allocation Strategy For Construction Activities In The Process Industry, University of Teesside, Middlesbrough

Box, G.E.P., and Jenkins, G.M., (1970). Time series analysis forecasting and control, Holden Day, San Francisco

Bunn, .D, Wright, G. (1991) Interaction of judgmental and statistical forecasting methods: issues and analysis, *Management Science*, 37,5, pp501-518

Dawood, N. N., Neale, R. H. (1993) Forecasting the sales of precast concrete building products, *Construction Management and Economics*, 11, pp81-98

Edmundson, B., Lawrence, M., O'Connor, M., (1988) The use of non-time series information in sales forecasting: A case study, *Journal of Forecasting*, 7, pp201-211

Fitzgerald, E., Akintoye, A., (1995) The accuracy and optimal linear correction of UK construction tender price index forecasts, *Construction Management and Economics*, 13, pp493-500

Fortune, C., Lees, M., Early cost advice for clients: The use of judgement by practitioners, *13th International Cost Engineering Congress* CE 7

Lycett, J. Artificial Intelligence and expert systems Internal Report University of Teesside

Moore, C., J., and Miles, J., C., (1991).Knowledge Elicitation Using More Than One Expert To Cover The Same Domain, *Artificial Intelligence Review*, 5, 4, pp255-271

Nkado, R. N., (1995) Construction time influencing factors: The contractor's perspective, *Construction Management and Economics*, 13, pp81-89

Pereira, B. D., et al., (1989) Experience In Combining Subjective And Quantitative Forecasts, *Journal Of Forecasting*, Vol. 8, 3, pp343-348

Sparkes, J. R., McHugh, A. K., (1984) Awareness and use of forecasting techniques in British industry, *Journal of Forecasting*, 3, pp37-42,

Wolfe, C., Flores, B., (1990) Judgmental adjustment of earnings forecasts, *Journal of Forecasting*, 9, 4, pp389-405

Woodward, S., et al., (1994) Cost Management: The Software Solution, *Construction Papers 31, Ascot CIOB* pp14

IDENTIFYING BEST PRACTICES IN CONTRACTOR SELECTION FOR ENHANCED HARMONY AND PROFIT

E. PALANEESWARAN, MOHAN M. KUMARASWAMY, AND P.W.M. TAM
Department of Civil Engineering, The University of Hong Kong, Hong Kong.
palanees@hkusua.hku.hk

Abstract

Contracts are fundamental to the management of all engineering projects as they are used to procure direct workers, consultants, contractors, sub-contractors, materials, plant/machinery and other services. Various procedures are followed for selecting consultants, vendors, suppliers and contractors. Of these, the process of selection of construction contractors has far reaching impacts on the success level of any construction project, given the larger proportion of the work done by the contractor(s) and the expenditure incurred on them. A survey confirmed that there are many different practices followed around the globe for the procurement of contractors. The survey also revealed that a number of innovative initiatives have recently emerged to guide and improve procurement decisions, following the disillusion with some previous practices which for example focused only on the lowest cost criterion. The findings that are presented in this paper demonstrate the need for more structured and comprehensive approaches that are based on guidelines that incorporate all relevant selection criteria.

Keywords: Best practices, contractor selection, procurement

Introduction

Construction contractors form a vital part of any construction team and the right choice of contractors is extremely crucial to the success of the project. Adversities and complexities are common in construction projects, hence the inescapable risks of contractor failure. As pointed out by Holt et al. (1995), construction is normally 'purchased' before it is 'manufactured' and the client has to repose confidence in one selected contractor to satisfactorily execute the construction contract. Moreover, all contractors have varying capacities to perform in different environments. Private sector clients enjoy relative freedom and their procurement procedures can be quite flexible to suit their own objectives. However, public sector clients are bound by legislation and regulations that dictate the methods and practices of procurement, arising from the needs for public accountability and social responsibility. While various procedures are followed to procure contractors, Holt et al. (1993) have found that subjective approaches in these areas do not necessarily serve the best interests of the construction owner. Furthermore, it is not common practice for procurers to share experiences and data in this field even in the public sector (Hatush and Skitmore, 1997). Latham (1994) reported a lack of communication between the procurers, a feature that is symptomatic of the construction industry as a whole. This paper discusses some aspects of contractor selection procedures practiced in Hong Kong, USA, Australia and Singapore and highlights some interesting innovative approaches that are practiced by various public clients.

Traditional practice versus current practice

The traditional contractor selection procedure is mostly based on opting for the lowest tendered price. The inherent simplicity and fairness of the low bid method favors its choice. The straightforward selection of the lowest bidder is based on the assumption that the specification used clearly defines the product that will be delivered by the contractor. Furthermore, it assumes that all contractors are the same and it is therefore best to choose the lowest cost alternative. In general, the public clients prefer to award contracts to the lowest responsive bidder. The Maryland Department of Transportation in USA defines a responsive bid as a bid submitted in response to an 'Invitation for Bids' that conforms in all material respects to the requirements contained in the 'Invitation for Bids.'

Some clients follow a postqualification procedure, which permits any contractor to obtain project plans and specifications and prepare an estimate. In the postqualification process, the clients receive bids from contractors and then evaluate the bidders' qualifications. If the lowest bidder is unqualified, then the next lowest bidder is considered and so on. On the other hand, prequalification is the process of screening candidates prior to issuing the complete project plans and specifications that contractors need to prepare a bid. Though the owners have to expend more effort and resources at the initial stages, prequalification saves owners substantial costs and efforts at later times in the life span of a construction project (Russell, 1996). It also saves time and money wasted by unqualified bidders in preparing tenders. Prior to 1988, the US Postal Service (USPS) used postqualification. As the Department observed various problems and risks, USPS introduced prequalification procedures from 1988. Now, prequalification is practiced widely in the construction industry. Some clients follow prequalification procedure on a project-by-project basis, while others perform it on a periodical basis (for example annually), the later being more commonly referred to as 'registration' where lists of registered contractors are maintained and drawn upon when needed to prepare shortlists of prospective tenderers. Project-specific "prequalification" of eligibility is useful for projects of unusual scope, complexity, value, technology requirements, quality levels or time constraints, or with special forms of funding or contract (Kumaraswamy, 1996). The relative merits and demerits of annual prequalification (registration) and project-by-project prequalification are enumerated in Russell (1996). Because the complexities and risks are on the increase in present day construction, clients prefer to opt for low but not necessarily the lowest bidders who are responsive, capable and responsible.

Best practice principles

The choice of opting for the lowest bid is not always the best route to meeting the client's goals such as time, cost and quality standards. When selecting a 'best bidder' for public clients, the twin objectives of best value for money and maintaining open and fair competition plays a vital part. In general, procurements by public clients are based on the following key principles:

1. *Public accountability*
   As public clients are utilizing taxpayer's money for construction projects, clients are accountable to the public.

2. *Value for money*

In order to achieve the best value for money, the tender evaluation should consider not only price competitiveness, but also compliance with users' requirements, reliability of performance, qualitative superiority, and whole-life costs.

3. *Transparency*

In order to encourage contractors to submit responsive and competitive bids, all necessary information should be provided in the tender documents and all procurement procedures and practices followed by public clients should be clear and transparent to facilitate better understanding amongst contractors.

4. *Open and fair competition*

All bidders should be treated on an equal footing and it should be ensured that all bidders are given the same information to prepare their bids. There should be no undue discrimination.

The best practice principles referred to in the Code of Practice for the South Australian Building and Construction Industry are as follows:

- conduct tendering honestly and fairly and refrain from seeking or submitting tenders without a firm intention to proceed
- seek to constrain the costs of bidding
- apply the same conditions of tendering for each tenderer and avoid any practice that gives one party an improper advantage over another
- produce tender documents that clearly specify the principal's requirements and evaluation criteria
- preserve the confidentiality of all tender information nominated as confidential during the tendering process, other than public opening of tenders and disclosure of tender prices which are acceptable
- comply with all statutory obligations, including trade practices and consumer affairs legislation
- refrain from practices such as collusion on tenders
- be prepared to attest to their probity by statutory declaration, in particular on issues concerning collusive practices and conflicts of interest
- recognize that tenderers retain their right to intellectual property, unless otherwise provided in the contract
- do not conduct post-tender negotiations solely on price. Neither clients nor contractors shall seek to trade off different tenderers' prices against others in an attempt to seek lower prices

Best practices and innovative approaches

The foregoing examples of good international and national practice need to be compared with other practices, with a view to distilling 'best practice'. This section of the paper is based on a recent study of some contractor selection practices followed by various public clients.

Most of the federal states of US prequalify capable contractors on an annual basis and assign various ratings such as maximum capacity rating or aggregate rating, work class ratings or project rating, performance factor, ability factor, current ratio factor, adjusted annual worth, etc. Under these procedures, contractors are classified according to the type of work and the amount of work on which they may bid. Each contractor is classified for one of more of the types of work categories requested by that contractor and is rated in accordance with his financial ability, adequacy of plant and equipment, organizational composition, record of construction and any other factors that are deemed pertinent. Contractors are assigned a classification, designating the types and dollar values of work upon which they are eligible to bid. The maximum capacity rating is the total value of uncompleted prime contract work a contractor is permitted to have under contract at any time. Some federal clients consider the contractors' performance reports as a measure of responsibility. They prefer to award contracts to the lowest, responsive, responsible bidder. For instance, the Division of Design and Construction of the State of Missouri, USA consider the terms responsiveness and responsibility as follows:

Responsiveness

The bidder may be deemed non-responsive and ineligible for award recommendation should the proposal be found to contain any of the following irregularities:

1. Illegible firm name or bid amount
2. Proposal form with no signature or invalid signature
3. No bid security provided or improper security
4. Bid bond not fully executed or without current power of attorney attached
5. Failure to properly list subcontractors for each category of work identified on the proposal form
6. Failure to acknowledge addenda on proposal form
7. Numerical and written amounts on proposal inconsistent
8. Incomplete or insufficient information provided on contractor qualification sheet
9. Failure to submit, if project size requires, the specified MBE/WBE qualification requirement
10. Bidder has altered the proposal form or otherwise qualified the condition of bid

Responsibility

If a bidder is found to have previous failures of performance in quality, timely completion or administration of the contract, the client may recommend the bidder's rejection. The responsibility of the bidder is evaluated based on performance records and reports.

1. If the state's Architects/Contractors/Engineers (ACE) computer database contains a satisfactory performance rating, the bidder may be deemed responsible. If the rating indicates a low or questionable previous performance, the specialist may contact references to determine responsibility in performing previous contracts.
2. If the ACE system does not list the bidder under evaluation, the specialist may contact listed references to determine performance history on similar construction contracts.
3. If the bidder's financial condition is unclear or questionable, the bonding company supplying the bid bond may be contacted for additional or detailed information.

The Queensland Government in Australia has developed 'Pre Qualification Criteria' (PQC), by which contractors are assessed against prescribed criteria including technical capacity, management approach, people involvement and business relations with commitment to continuous improvement. Accordingly, contractors are rated at one of four levels ranging from a base level (1) to world's best practice level (4). A review of contractor's financial capacity is also undertaken and contractors will generally be limited to allowable maximum project values in relation to their net tangible assets. In order to match projects to contractors, the Project Risk Level (Level 1,2,3 or 4) will be determined and to be eligible to tender on a project, contractors must have a PQC level not less than the Project Risk Level. PQC has been developed with the aim to assist in the reduction of the cost of tendering and to ensure that Queenlanders receive value for money by ensuring a match between the size and complexity of projects and the abilities of suitable contractors.

In 1997, the Government of South Australia developed a prequalification process and associated prequalification requirements, which will assess a contractor's capacity and capability to satisfactorily contract with the government, and complete government building and construction projects. The contractors are evaluated by the assessment panel based on the contractors' particulars submitted in three forms (general form, financial form and contractor performance reports). Benchmarking criteria for the contractor prequalification evaluation parameters - such as technical capability, financial capacity, quality assurance, skill formation, industry initiatives, human resources management and occupational health safety and welfare - have been established for 1997, 1998 and 1999 separately with increasing levels of standards. The following section highlights some innovative approaches followed by some client organizations.

Contractor performance

Most clients include the contractor past performance data as one of the evaluation criteria for the contractor prequalification and bid evaluation. Most of the federal states in USA have well-structured performance monitoring systems and the contractors' performance is rated and recorded systematically. The contractors' past performance records will have significant impact in the contractor prequalification and bid selection. Also, the performance inputs will increase the maximum capacity rating, so that the better performing contractors will increase their contracting opportunities. For instance, the West Virginia Department of Highways determines the contractor's prequalification rating using the following formula:

$$R = P (A+I+L+E), \text{ in which,}$$
R = Prequalification Rating
P = Performance factor based on contractor's past performance record in West Virginia
A = Net current assets (working capital)
I = Cash surrender value of Life Insurance
L = Line of Credit statements limited to fifty percent (50%) of net current assets
E = Unencumbered book value of highway and /or bridge equipment in good operating condition

Different client departments follow different procedures and different rating methods. The "Construction Quality Assessment System (CONQUAS) in Singapore assigns tendering price advantages based on the CONQUAS premium list up to a maximum of 5%, for contractors who have achieved higher CONQUAS scores in previous jobs. Similarly, preferential tendering opportunities are provided for better performing contractors in the Hong Kong Housing Authority contracts. The Housing Authority follows a performance assessment scoring system (PASS) which comprise an "output assessment" component and an "input assessment" component. The output and input scores are then merged together to produce composite scores, which form the basis for the comparative score leagues. The project league is categorized into three bands with two benchmark lines as Composite Target Quality Score (CTQS) and Composite Lower Score Threshold (CLST). The highest tendering opportunities for upcoming projects are then given to those contractors who fall in the upper band of this league (above CTQS) whereas those contractors in the lower band (below CLST) will not be invited to tender in the next quarter.

Transparency

Some of the federal states lead by setting examples in following one of the best practice procurement principles – 'Transparency'. For instance, Washington State Department of Transportation, USA has an exhaustive prime contractor performance rating procedure, which has a numerical rating and a narrative rating section. The contractors are given a copy of the performance report and the contractors are aware of their right to appeal. It also serves the purpose of verifying that the report has been reviewed for the purposes of assuring objectivity in its preparation and for elimination of the influences of personalities. On similar lines, Holt et al. (1996) emphasize that constructive feedback to unsuccessful tenderers would encourage firms to address their identified areas of weakness. As the contractors become more aware of their strengths and weaknesses they can improve their performance levels to obtain more contracting opportunities and remain competitive. This will be beneficial to the industry as a whole.

Value of time factor

Although 'time is money', it is often ignored by many. Normally, clients decide on the time of completion of a project and invite bids within the specified time frame. There are some instances in which a bidder specifies less construction time than the client's estimation. In general, a bonus/penalty clause in the contract provisions will take care of such situations. Some clients practice an innovative approach for these situations by introducing a 'value of time' factor in the tender invitations. The Florida Department of Transportation, Florida, USA follows one such innovative approach. The dollar value per day as established by the department is included in the request for proposal package. The bid evaluation is performed with the bid adjustment for the value of time. This adjustment is used for selection purposes only and will not affect the department's liquidated damages schedule or constitute an incentive/disincentive to the contract.

Hong Kong perspective of contractor selection

Hong Kong has achieved tremendous economic growth during the last few decades. The construction industry in Hong Kong includes many private sector clients, as well as large group of Government and quasi-Government clients. Government clients include various departments under the Works Bureau. The quasi-Government clients such as the Hong Kong Housing Authority (HKHA), Mass Transit Railway Corporation (MTRC), Kowloon-Canton Railway Corporation (KCRC), Airport Authority (AA), etc. also handle large amounts of construction projects, but they follow different procurement guidelines.

For public projects administered by the Works Departments, only contractors on the approved lists can tender for contracts. They are classified into five categories (buildings, port works, roads and drainage, site formation, and waterworks) according to their relevant expertise and managed by the relevant Works Departments. The lists of approved contractors are in three groups based on their capacity. There are two status levels such as 'probationary' and 'confirmed' in each group. The confirmation of probation depends on the satisfactory completion of works with good performance records. The promotion of contractors to a higher group, with initial admission on probation, depends on satisfying requirements of financial criteria, appropriate technical and management capabilities and continuous satisfactory completion of contracts under the present group. Every Works Department maintains separate approved lists of contractors and the relevant Department manages each category of contractors.

Methods of tendering

1) *Selective tendering.* Notices are published every Friday for general civil engineering and building works in the Government Gazette and also in parallel press releases, inviting contractors in the appropriate categories and groups of 'Approved Lists'. For contracts with nature of work comprising more than one category, invitation from contractors of either or both appropriate groups are made.

2) *Prequalified tendering.* Prequalified tendering is followed in some special circumstances such as a contract that is time critical or one that requires particularly high levels of skills and proven reliability. The selection of pre-qualified contractors is carried out in two stages. The first stage is primarily intended to screen out those applicants who are so patently ineligible or unsuitable so as to make further assessment of their applications unnecessary. The second stage assessment comprises a detailed technical evaluation of all applicants who have fulfilled the requirements of the first stage screening.

Criteria for contractor prequalification

The project team of the procuring department determines criteria for selection for both the stages. The criteria for first stage assessment are based on determination of whether an applicant can satisfy the basic requirements stipulated in the gazette notification or advertisement. The items of assessment require only simple "Yes/No" evaluations involving no detailed checking and include the following where applicable:

- Whether the applicant is confirmed within the relevant category or group of works stipulated in the Gazette or advertisement;
- In case of joint ventures, whether each of the joint venture partners satisfies the criteria stipulated in the Gazette or advertisement;
- Whether adequate financial resources are available. Applicants will be considered as meeting the financial requirements for pre-qualification, provided they submit an appropriate undertaking to make good any shortfall, if awarded the contract;
- Whether special plant, equipment, workshops etc., as stipulated for the works will be made available to the contractor.

The criteria for second stage assessment contain all technical aspects of the checking process. The criteria for this stage include the following where applicable:

- Experience in Hong Kong Government contracts in the past 5 years

- Experience on other local contracts and overseas contracts in the past 5 years
- Experience of the types or size of buildings etc., or forms of construction involved in the project in the past 5 years
- Experience, availability and organization of staff.
- Availability and details of special plant, equipment, workshops, etc.
- Proposals for undertaking the project will be demonstrated by the submission of a preliminary method statement, quality assurance plan and contractor's safety policy.
- Contractors' safety record, including accident records and records of convictions for safety violations and suspensions arising from regulating actions relating to safety.
- Details of financial resources
- A brief history of all claims, litigation and arbitration proceedings in connection with Government building and engineering contracts in which the applicant has been involved in the capacity of the contractor over the last five years.
- Past environmental performance, including convictions under any of the pollution control ordinances and breaches of environmental protection clauses in any contracts in which the applicant has been involved.
- Record of convictions related to construction offences such as employment of illegal immigrants, site safety, etc.

Tender evaluation

The procuring department is responsible for evaluating the tenders to determine whether they meet the conditions and specifications laid down in the tender document. Normally, the department will recommend acceptance of a tender, which fully complies with the tender conditions and specifications and is the lowest in tender sum. Where pre-determined factors other than price are included in the tender assessment, the recommended tender is the one attains the highest combined technical and price score. The scoring system and weightings for various factors considered in tender evaluation are determined by the procuring department on a case by case basis and the procedure varies for different projects. The procuring departments will then submit their recommendations in the form of a tender report to the relevant tender board for approval. Based on the tender value, either the Central Tender Board (tender value exceeding HK$ 15 millions) or the Public Works Tender Board (tender value not exceeding HK$15 millions) will consider and decide on the acceptance of tenders.

Contractor performance assessment

Quarterly assessment of contractors performance on a three point scale is performed for assessing the following ten factors: workmanship, progress, site safety, environmental pollution control, organization, general obligation, industry awareness, resources, design (design & build contracts only), and attendance to emergency (term contacts). Warning letters will be issued in cases of adverse reports on performance, and the non-performing contractors will be called for interviews. If the performance of a contractor is continuously adverse, disciplinary actions such as suspension, downgrading or removal from the list will be taken in accordance with the Works Branch Technical Circular 13/96.

Conclusions

This study of different approaches to construction contractor selection unveils various practices and innovative initiatives adopted by different clients. Best practice principles such as 'public accountability', 'value for money', 'transparency', and 'open and fair competition'; and innovative approaches such as integrating contractor performance assessment with contractor selection and incorporating 'value of time' factor in tender invitations are identified in this study. The study also reveals the significance of factors such as 'responsiveness', 'responsibility' and 'competency'. The findings will be useful for framing the basis and formulating guidelines for following best practices in contractor selection by public clients in different countries. Even though the traditional low bid system continues to be the norm, there is an increasing trend among public clients to move away from mere low bid selection towards a multiple criteria approach, while still aiming at commonly pursued general objectives. Contractor performance is monitored and recorded by many clients for managing of ongoing contracts. Some clients such as the Works Bureau in Hong Kong use the performance reports only for controlling the contractor performance and for penalizing 'bad performance'. As there is no incentive for 'good performance' in that approach, 'better performing' contractors may not derive any motivation and may therefore tend to lower their output standards, which may be just sufficient to match the clients' requirements.

However, some clients in USA, Australia, Singapore and the Hong Kong Housing Authority offer morale boosters for 'better performing contractors' by way of providing some 'bonuses' such as preferences in contractor prequalification and bid evaluation, tender price adjustments, preferential tendering opportunities, increasing the maximum capacity rating. 'Transparency', another innovative concept/practice discussed in this paper, will form the base for a 'level playing field' among clients and contractors. Also, the contractors can identify their areas of weaknesses needing improvement before winning more work opportunities. Clients can enhance project success by helping to minimize contractor failures. The 'value of time' factor discussed in this paper will pave the way for challenging designs, especially in 'Design and Build' contracts. Also, this approach may prove to be handy where alternative bids are allowed.

Benchmarking of the best practices and innovative approaches discussed in this paper may be undertaken not only for the sake of evaluation and comparisons, but also for achieving improvements, better harmony and higher productivity/ profits. The new dimensions in the foregoing approaches such as by incorporating value of time factor, performance specifications and competitive negotiation require a more systematic rational approach to reduce the elements of subjectivity. Although, the elements of subjectivity can not be completely eliminated, developing computerized decision support systems incorporating relevant knowledge bases can minimize them. The ongoing research study aims at developing such a knowledge based client advisory decision support system for contractor selection. The envisaged decision support systems should facilitate more efficient and effective contractor selection, thereby enhancing the profits of each participant, and promoting harmony between them.

References

Arditi, D, Khisty, C J and Yasamis F (1997) Incentive/Disincentive provisions in highway contracts. Journal of Construction Engineering and Management, Vol.123, No.3, pp. 302-309.

Chinyio, E A, Olomolaiye, P O, Kometa, S T and Harris, F C (1998) A needs-based methodology for classifying construction clients and selecting contractors. Construction Management and Economics, Vol.16, No.1, pp. 91-98.

Construction Industry Development Agency (CIDA) (1994) The Australian construction industry prequalification criteria for contractors and subcontractors.

Construction Industry Development Board (CIDB) (1995) Higher premium for certified contractors. Construction focus. 7(3), 2

Corcorn, J and McLean, F (1998) The selection of management consultants – How are governments dealing with this difficult decision? An exploratory study. International Journal of Public Sector Management, Vol.11, No.1, pp. 37-54.

Finance Bureau, Government Secretariat, Hong Kong (1997) A guide to procurement by the Government of Hong Kong Special Administrative Region.

Government of South Australia, Australia (1995) Code of Practice for the South Australian Building and Construction Industry.

Government of South Australia, Australia (1997) Prequalification – Contractor categories 1,2,3 and 4. Application Kit.

Hatush, Z and Skitmore, M (1997) Criteria for contractor selection. Construction Management and Economics, Vol. 15, No.1, pp. 19-38.

Hatush, Z and Skitmore, M (1997) Evaluating contractor prequalification data: selection criteria and project success factors. Construction Management and Economics, Vol.15, pp. 129-147.

Hatush, Z and Skitmore, M (1997) Assessment and evaluation of contractor data against client goals using PERT approach. Construction Management and Economics, Vol.15, pp. 327-340.

Herbsman, Z J (1995) A + B bidding method – hidden success story for highway construction. Journal of Construction Engineering and Management, Vol.121, No.4, pp. 430-437.

Holt, G D, Olomolaiye, P O and Harris, F C (1994) Evaluating performance potential in the selection of construction contractors. Engineering, Construction and Architectural Management, Vol.1, No.1, pp. 29-50.

Holt, G D, Olomolaiye, P O and Harris, F C (1996) Tendering procedures, contractual arrangements and Latham: the contractors' view. Engineering, Construction and Architectural Management, Vol.3, No.1,2, pp. 97-115.

Kumaraswamy, M M (1996) Contractor evaluation and selection – a Hong Kong perspective. Building and Environment Journal, Elsevier Science, Vol.31, No.3, pp.273-282.

Kumaraswamy, M M and Dissanayaka, S M (1996) Performance – oriented building procurement systems. Building Technology and Management, Malaysia, Vol.22, pp. 17-26.

Latham, M (1994) Constructing the Team, HMSO, London.

McGeorge, D and Palmer, A (1997) Construction management – new directions, Blackwell Science Limited, Oxford, UK.

Queensland Government, Australia (1998) A contractor's guide to prequalification – Competing for Government building work.

Queensland Government, Australia (1998) Queensland Government PQC – Project risk levels for contractor selection: Do-it-yourself guide – Draft.

Rankin, J H, Champion, S L and Waugh L M (1996) Contractor selection: qualification and bid evaluation. Canadian Journal of Civil Engineering, Vol.23, pp.117-123.

Russell, J S (1996) Constructor prequalification – choosing the best constructor and avoiding constructor failure, ASCE Press, New York, NY.

Russell, J S, Hancher, D E and Skibniewski, M J (1992) Contractor prequalification data for construction owners. Construction Management and Economics, Vol.10, pp.117-135.

OPTIMIZING PROFIT AND SELECTING SUCCESSFUL CONTRACTORS

AMARJIT SINGH
University of Hawaii at Manoa, Department of Civil Engineering; 2540 Dole Street, Honolulu, Hawaii 96822 ,USA. singh@wiliki.eng.hawaii.edu

MAX MAHER SHOURA
Carter & Burgess, Honolulu; 1001 Bishop St., Pauahi Tower, Ste. 850, Honolulu, Hawaii 96813, USA. shoura@hawaii.edu

Abstract

Finding a delicate balance between a low bid and a reasonable profit is a scientific problem that is usually an anxious management ordeal in estimating departments. Owing to the inexactness of estimating, we always see a range of bids submitted for any project by the different contractors.

In this study, 32 bids were collected from Bills of Quantities for building projects. A statistical definition of success contributed to the development of Operating Characteristic Curves to quantify bidding success. These were combined with cost underrun profiles to formulate an overall profitability index. It was consistently found that successful and profitable contractors had an overall profitability value, OP_{sp}, between 0.203 and 0.297. Obviously these contractors, bidding in secret, and using different methods and techniques, are nevertheless exhibiting consistent OP_{sp} values. This speaks something about the art of bidding. By knowing the history of the market and range of costs, it is possible to screen bids and fine-tune them for a contractor to become more competitive without compromising profitability. By the same token, clients can screen contractors to discover who is bidding with OP_{sp} values in the optimal range, thereby minimizing chances of unscrupulous behavior by contractors during project execution. This allows clients to adopt alternate bid selection criteria in lieu of making the award to the lowest or second-lowest bidder.

Keywords: Alternate bid selection criteria, operating characteristics curve, overall profitability value, profit optimization, the art of bidding

Introduction and background

Winning a bid and making a profit on a project is a scientific balancing act. However, this act or formula is influenced by market factors that are statistical in nature. Estimators, commonly and arithmetically, calculate direct costs, overheads, and profit in a simple deterministic method. This paper capitalizes on the statistical nature of bids and market costs to propose a bidding strategy. This technique analyzes bid prices distribution at the elemental work level of grouped cost packages; then, an optimal bid is determined maximizing the overall bid success and project profitability.

Earlier work, applying statistics in a bid selection model, was undertaken by Fuerst (1977) who used random variables for identifying probabilities of winning. Carr (1977), and Clough and Sears (1994) used utility theory to study risks and strategies in competitive bidding. Park (1979) did competitors   bid comparison to calculate a probability of underbidding, and then finding an optimal markup that would generate an optimal expected profit. The unique method introduced by the authors here, combines optimizing the probabilities of success and minimizing the probability of cost overrun. The application of this method is enhanced by the advent of modern computers that provide the capability of managing customized databases.

Research methodology

The introduced technique uses the following methodology and steps as outlined in Table 1. The database used to analyze and apply this technique was obtained from unit prices of BOQ   s submitted in the market of French Polynesia airport projects.

Arrangement of data and cost groups

A data set was collected on airport projects with prices ranging from 32 million Polynesian French Francs (PFF) (approximately US$ 336,842) to 170 million PFF (US$ 1,789,473). Raw data was in the conventional Bill of Quantities (BOQ), as adopted by European system of quantity surveying. A data set, of 32 bid items, was rearranged according to the following work groups suitable for airport runway construction:

Mobilization and Start up	Rough Earthwork	Grade & Base Work
Topping and Surfacing	Concrete & Joints	Drainage Work
Soil Finishing & Special Works		

Other grouping methods, depending on size and type of projects, can be validated for other target applications. The authors have tested a variant of this grouping on building projects [Singh and Shoura (1998)]. To set up the study of cost group bids, a bid breakdown is necessary. In every case, the bid winner is considered to be the drive behind the price of a cost element. A projection of a lowest winning bid can be made from a probabilistic study of market data.

Table 1. Sequence and Steps of the Process:

Sequence	Steps
- Data Acquisition	Collecting of recent and reliable market data on work packages (officially submitted)
- Database Development	Grouping of work packages of similar nature into major cost group unit prices
- Profit Information	Generating an expected profit range model used by active contractors in a considered market.
- Data Simulation of Cost	Recreation of database by generating a Monte Carlo random number for
Groups	cost group unit prices
- Operating Characteristic sampling	Finding a probability of winning, P_s, done using acceptance
Curves (OCC)	method of quality control
- Underrun profiles a frequency	Finding a probability of making a profit, P_u, done by making
	histogram of grouped units
- Bid Optimizing & and Op_{sp} Positioning Process	Balancing a winning bid and a profit project -- obtaining P_{sp}

Analysis

The analysis is undertaken in the following steps:

A specific quantity unit is selected to represent each cost group from the most influential work item in each work group. Next, the prices of all work items in a cost group are totaled, and the total is divided by the selected quantity to produce a group unit price. Once this is done for the major cost groups, we have a modified database consisting of a *representative quantity* and a *representative unit price*. These representative parameters allow researchers to handle a relatively fragmented BOQ in a meaningful way.

To allow more statistical objectivity into the analysis, a Monte Carlo simulation is done generating 10,000 (infinite) data points. Simulated frequency distributions are used in all subsequent calculations for probability.

The cost group by successful contractors (ones awarded the projects) establishes a lower and an upper limit to the success range. These signify the lowest and highest prices, respectively, bid by successful contractors on specific cost groups. Figure 1 shows (*i*) the upper and lower limits of successful bids (the success range), and (*ii*) a distribution of all bidders on Surfacing Work. These limits are different for each cost group. On occasion, a lowest bidder was the successful bidder, while on other occasions, the highest bidder for a particular cost group turned out to be awarded the project. The prevalent theme here is to bring out the fact that successful bidders can bid high on some cost groups and still be awarded the project. This bidding range, or fluctuation, is what we feel makes it necessary to look at bidding from a statistical variation and sampling perspectives. These success limits shall be used in developing the Operating Characteristic curves through use of Acceptance Sampling to determine specific probabilities of winning a bid, P_s. It is significant here to realize and understand that the successful bidders bid within the success range. This is a truism. We infer that bids lower than the lower success limit would result in unprofitability, while bids higher than the highest success limit would result in reducing chances of winning a bid. Thus, contractors need to focus on trying to keep their bids within the success range for each cost group.

The authors conducted candid interviews with highway & airport contractors to obtain information on profit margins, often held as closely guarded information. The interviews resulted in a number of factors affecting profit margins. This can simplify the complexity of profit analysis for the case of runways civil works. The researchers found that contractors apply profit percentages between 5% and 15% during the bidding stage. The actual profit after work is over, however, was between 3 and 13%. All contractors agreed that they never make an actual profit of 15%. They all agreed that even on the toughest jobs, they made profits of at least 3%. Hence 3% and 13% were selected as "upper profit point" and "lower profit point" respectively. These limits shall be used to develop overrun profiles, and consequently in calculating a probability of being profitable, P_u.

Obtaining an estimate for profit is important to allow an estimate to be determined for the probability of making a profit. This has been called here the probability of profitability, P_u. Bidding too low reduces the profit percentage and might result in a loss on that particular work group. Bidding too high would obviously reduce one s chances of winning the bid.

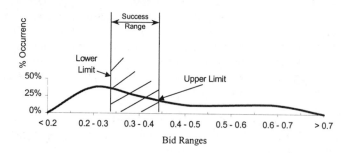

Figure 1: Percent Occurrence Distribution and the Success Range of Surfacing Work.

! To conduct the Probability of Success, P_s, one needs to design a hypothesis before the quality control technique is conducted. The authors set up the following hypothesis (Figure 1).
"A bid is successful when represented by the area under the curve in the *success range*."

The results of the probability -- and hypothesis -- tests give us the Operating Characteristic curve (OCC) of a process (Figure 2a). OCC represents the probability of being a successful bidder in context to the particular cost group. The acceptance probabilities calculated thus are the Probabilities of Success, P_s, of the bidding model. Statistically, there is a probability of being successful even if the bid value is outside the success range. The researchers believe that this is very realistic considering the risks and context of construction planning.

!   Probabilistically, any bid lower than the lower limit of cost group unit price will register a 100% chance of cost overrun. Similarly any bid higher than the upper limit will register a 100% chance of cost underrun. Intermediate points for cost groups will register intermediate cost overrun/underrun probabilities in correspondence to the shape of the frequency histogram (Figure 2b).   The Probability of Underrun, P_u, is given by:

$$Pu = 1 - P_o \tag{1}$$

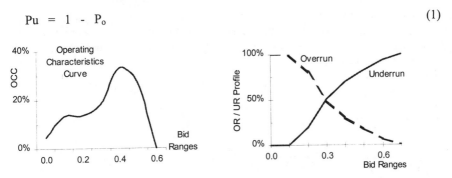

Figure 2a: OCC Graph for SurfacingFigure   2b: Overrun and Underrun Profiles.

Overall success & profitability model

The next step is to utilize the two previously introduced concepts to obtain a profitability index for each work package. This calculation produces an optimal relative index of success and profitability, P_{sp}, for a particular work group, and is obtained by taking the product of P_s and P_u, thus:

$$P_{sp} = P_s * P_u \tag{2}$$

This states that the *profitability* reduces if the chances of underrun are low, and increases if the chances of winning the bid -- being successful -- are high. Thus an optimal trade-off is designed between two opposing parameters of the construction cost estimate. This completes the bid competitiveness analysis model. It is noted that P_{sp} is not highest when P_s is highest, or when P_u is highest. Obviously, P_{sp} is an optimization between P_s and P_u. Plots of P_{sp} for some cost groups are shown in Figure 3.

From P_{sp} values for individual cost groups we develop an index for the bid as a whole, OP_{sp}. An important step to be considered is the impact of each cost group on the whole bid. In a game of risk, the prices that vary or fluctuate the most have a greater impact on the entire outcome than the prices that are relatively fixed. The determining measure used here, for fluctuation, is the standard deviation of unit prices bid on runway projects. Other measures can be used as weight factors, as long as they are scientifically justified. This will yield an Overall Probability of Success and Profitability, OP_{sp} for the whole bid.

The two tasks that have to be ultimately achieved through the optimization process are:

i. Cost group bids are placed within success ranges (or as close to it with a strategic probability).
ii. Cost group bids have highest chances of cost underrun (or least chance of overrun).

In usual cases, a high OP_p indicates that a contractor included a profit margin high enough to increase P_u s, and a probability of success, P_s, within the success range. However, attempts to increase P_u beyond certain limits would push P_s out of the success range. This note indicates that a balance must be achieved to win the bid by using this formula and give the contractor an edge in bidding.

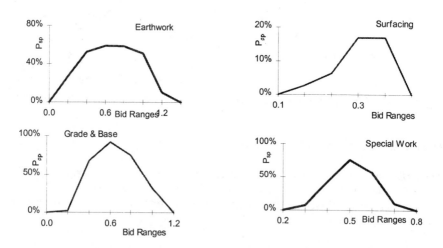

Figure 3: Profitability Index Profiles of Some Cost Groups.

Sample process

The following steps outline the sequence of applying the profitability method:

1. Prepare a cost estimate for the project, then arrange it in 'cost groups'.
2. Select a representative quantity and compute a representative unit price for each cost group.

3. Select an operating range for company profitability. This is a range that the management chooses to operate within. It is usually based on the following:
    a. The company's condition such as work load, success rate, etc.
    b. The market condition such as economy, jobs available, etc.
    In general, the operating range should be between 50% and 75% of peak profitability. It is best selected around the 60% to 65% range. The narrower the band selected, the easier the job of the optimization would be; this is the same as saying that *the company s bidding strategy is well defined.*
4. Using these newly set bid percentages, readjust the bids and obtain a new total based on the new bid values. Then compute the adjusted bid percentages to enter the graph for each work package and obtain a profitability value, P_{sp}. The weighted cumulative of all P_{sp}'s is the OP_{sp} for this 'bid position'.
5. Fine tune the overall new bid position by readjusting bids of major groups to optimize for a high OP_{sp}.

It is noted here that many methods can be devised for the use of profitability graphs; the above listed method is considered aggressive (quick to target an optimum) in its application to optimize the bid positioning. This method was tested on a real set of data, and was successful in achieving profitability [Singh and Shoura (1996)].

Research findings

The following findings were made:

1. On all bids analyzed, OP_{sp} values range from a low of 0.103 to a high of 0.438.
2. Successful bidders had OP_{sp} range closely banded, 0.203 to 0.297, with a low standard deviation, = 0.04.
3. 71% of unsuccessful bidders have OP_{sp} s outside this range of success and profitability, = 0.08
4. Another 14.5% of unsuccessful bids were very close to the successful OP_{sp} upper limit; that would *tend* to increase the percentage of unsuccessful bidders outside the range of success and profitability to 86%. The authors feel this speaks volumes in determining probabilities of success and profitability. Owners can also potentially use this to determine whether a bidder has greater chances for being relied upon to produce a balanced estimate. In principle, reasonable estimates reduce chances of claims and disputes.
5. Average OP_{sp} of successful bidders is 0.245; for unsuccessful bidders it is much higher, standing at 0.311.
6. No relation exists between the bid volumes or the number of bidders on a project and OP_{sp}.
7. Contractors winning bids had been in business for fifteen years or more inferring that these companies have grasped the art of doing their bidding right.
8. On one occasion, an owner decided to go with a bid s alternate indicating problems with initial bid s documents and technical feasibility. The analysis proved that OP_{sp} can assist owners in evaluating bids.

Conclusions

If a contractor bids with an OP_{sp} outside 0.203 to 0.297, the success range, we can say with a confidence of 100% that the contractor will not be the successful bidder. If the contractor has bid between 0.203 and 0.297 then we can say with a confidence of 71 to 86% that the bid will be successful. However, if the average OP_{sp} bidding practice of the contractor deviates considerably from 0.245, the doubts for success increase.

Sixty percent (60%) of projects have clear cut decisions concerning a successful and profitable bidder versus the non-successful bidders. Owners can use this data for weighing appropriate bidders for award. A contractor bidding outside this range may not display adequate professional competence, or else displays some bidding deficiency that is of potential harm to an owner causing claims, dispute, litigation, or contractor bankruptcy. Owners should be informed of this fact to ensure optimal service and prevent *bad business*. Bidders practicing OP_{sp} s within the successful and profitable range have made reasonable profits as well. Maintaining their presence in the construction market and the practice of OP_{sp} factor confirm this; they are obviously doing something right.

Further, the contractors must maintain a good database with continuously updated information. The application of this method may improve construction bidding and help industry by eliminating non-profitable bidding and the risks associated with it. Contractors, owners, public agencies, all can benefit from employing profitability index methods whether in bidding or in selection for award. Clients can use this method to select contractors who perform within stated OP_{sp} values, so that reliable contractors are hired.

References

Fuerst, M (1977) Theory of Competitive Bidding. *Journal of the Construction Division*, ASCE, March.

Carr, R I (1977) Paying the Price for Construction Risk. *Journal of the Construction Division*, March.

Clough, R H and Sears, G A (1994) Construction Contracting. John Wiley and Sons, NY.

Park, W R, (1979) Construction Bidding for Profit. John Wiley and Sons, Inc., NY.

Singh, A and Shoura, M M (1998) Optimization of Bidder Profitability and Contractor Selection, *Cost Engineering*, June.

Singh, A and Shoura, M M (1996) Optimizing Bidding Success and Profitability in Airport Runway Construction in French Polynesia. *Construction Management & Economics*, November.

9. Change, Technology and Value Management

NECESSARY CHANGES IN CONSTRUCTION SYSTEM TO COPE WITH AND UTILIZE RECENT ECONOMIC CRISIS IN ASIA

KEIZO BABA
Department of Infrastructure Systems Engineering, Kochi University of Technology, Tosayamada, Kochi Prefecture,782-8502, Japan. baba@infra.kochi-tech.ac.jp

Abstract

The recent economic recession in Asia has made a great impact on the execution of construction projects in the region. Many construction projects have either been stopped or canceled. Tremendous efforts have been made to cope with these adverse conditions.

The result of these efforts shows the need for some fundamental changes in the construction systems, which were in operation before this economic crisis, for projects to survive. However, these required changes could be almost of the same nature as the "inevitable changes" which will be required for sustainable construction in the next century.

These inevitable changes could be a new paradigm that will meet the requirements of the next century. Therefore, changes in the construction systems to cope with the recent adverse economic conditions in Asia should eventually result in the utilization of these economic conditions to establish a new paradigm of construction systems for the next century in Asia.

Keywords: Changes, construction systems, economic recession, sustainable construction

Introduction

Recently, many countries in Asia have been suffering from economic crisis, which began in Thailand in the middle of 1997. Certainly, this situation will slow down, to some extent, the speed of development of many countries in Asia. Actually, there are so many construction projects, which have either been totally stopped or left in under-construction stages in many major cities in Asia now. Also, many projects have been canceled due to financial problems.

These phenomena show that by the unpredicted and inevitable tremendous changes in economic conditions, the feasibility of these projects, which were carried out at the planning stages of the projects have been completely negated, and there is no economic justification for them anymore. Also, these big waves may cause another worse economic condition to form a vicious circle. Before the outbreak of this economic crisis, Asia was considered the most prospective fast developing region in the world towards the next century.

Nevertheless, the present adverse economic condition does not imply that the world center of development from the end of this century to the beginning of next one has shifted to any other place than Asia. On the contrary, the situation could be viewed from a different perspective. That is, the fact that the financial crisis of Thailand greatly affected the world economy shows the increasing importance of the Thai economy in the world. Towards 21st century, Asia will become the dominant region of the world economy again after this recession.

However, in order to cope with this recession, remarkable countermeasures should be introduced in every aspect of the economy. In the construction industry, all projects should be checked to reduce the required total cost of execution of projects so as to recover its feasibility. This action requires new procedures of execution of construction projects to cope with this economic recession.

After recovery from this recession, as a well-known sociologist pointed out before, "what is happening in Asia is the most important in the world today. Nothing else can come close, not only for Asians, but also for the entire planet. The modernization of Asia will forever reshape the world as we move towards the next millennium." He also predicted that this modernization will be done not as Westernization but rather Asianization [Naisbitt (1996)]. Therefore, the development in Asia by Asianization will have some impact on the whole world in many aspects in the next century.

The philosophy of the people who are involved in the development of Asia is very important as a key factor in determining the effect of development in the region. Among them, the most important issue for future of the world should be the relationship between the development in Asia and its influence on the world environment, and of the saving resources.

Before discussing this relationship, the concept of development and its mechanism should be clarified. The definition and procedure in the previous development should be analyzed to make adjustment for sustainable development in the future. In the 20th century, many wars were fought in the world. Through experience of these wars, many people especially in Asia have come to realize that the economic development of their countries is more important than the military power. However, this concept should be checked from point of view of the world environment and the saving resources.

The coming century should be a challenging era for everyone as to whether new philosophy will be established for the maintenance of the environment level of the world in the level as it has been.

Rapid deterioration of condition of construction market in Asia

Interruption and suspension of construction projects

The economic summer storm 1997 which started in Thailand and immediately spread over the entire region, is reaching not only across the Pacific, but has also become one of the biggest issues that is often discussed in international political conferences today.

After a long flourishing period, a very strong setback suddenly came upon construction market in the middle of 1997. As stated before, this slack began with the monetary crisis in Thailand due to rapid devaluation of the Thai Baht in the middle of 1997. In the downtown of Bangkok, there were many buildings, which were under construction, but construction works were stopped due to financial reason. Also in the suburb of Bangkok, there is a new town developed for the housing of new middle class people who had enjoyed the upgrading living standard by previous economic prosperity in this region. There are many projects by foreign investors of almost same nature, which provide so big number of rooms of apartment and office that the relationship between demand and supply has become off balance to make sharp decline of the price of real estate.

Growth rate of economy

One of the important indexes which shows the actual state of the economy, is growth the rate of economy. Table 1 shows sharp drop in the growth rate of the economies of major countries of Asia in 1998 [Diamond (1998)].

Table 1. Growth rate of economy of major countries in Asia

Country	Average Growth 1985-94	Estimated Growth in 1998	Estimated Year of Recovery in Economy to the Level of More Than 5% Growth Rate
Thailand	8.6	-5.0	2001
Malaysia	5.6	1.7	2001
Indonesia	6.0	-8.5	2004
Philippines	1.7	2.9	2000
Singapore	6.1	2.9	2000
Korea	7.8	-1.5	2001
Taiwan	6.4	5.9	Already in economic growth rate of 5%
Hong Kong	5.3	1.8	2001
China	7.8	7.5	Already in economic growth rate of 5%

This growth rate of the economy is a key factor indicating the condition of the construction market. Figure 1 shows the relationship between the growth rate of the economy and the gross national product per capita. In this figure, a dotted line shows the trace of the development of Japan. On this trace line of the very rapid development period from point X to point Y, the construction market in Japan has been very busy and there has been remarkable amount of construction projects [Economic Review (1998)].

Economic forecast; revival and cold period of construction market

There are many forecasts on economies of the countries in Asia as to when this recession will end. This information is of great interest to most people in Asia, whether they are working in the industries or not. The third column in Table 1 shows the recovery year of economies of major countries in Asia estimated by one of the major research institutes of economy in Japan [Diamond (1998)].

There will be many discussions on this forecast. Therefore, the numbers of years showing the year of the economic recovery only suggests some idea of the length of the period of the recent economic recession in every country. It also sheds some light on the fact that there is no objection on the opinion that the economies of the countries of Asia will sooner or later revive and become almost as active as it was before, to have ordinary active trades in industries.

GNP per Capita   All data 1995

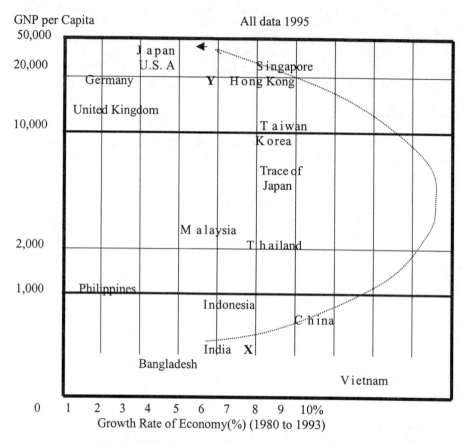

Figure 1. Growth Rate of Economy & GNP per Capita

Therefore, the expected cold period of the construction market which may continue for several years could be considered an appropriate time to evaluate the methods and procedures of the execution of projects under very strict economic conditions, and to establish new and suitable systems to cope with such adverse conditions. These new systems may be useful even when the construction market becomes busy again. Also, these new systems may be adopted to meet the new requirements of the next century. That will be the requirement from the cost reduction and sustainability of construction.

Change of basic condition for choosing construction systems

Major governing economic conditions to choose construction system

Shrinking construction market. During the recession, the economic activities in industries are slowed down to provide less business opportunities than before. This slow down causes the shrinking of the construction market, thus requiring a strict analysis of feasibility of a project in the planning stage.

Low ROI. Due to the recession in the economy, every activity in business is slowed down. ROI is kept at a low level in every industry.

Unemployment rate. By the slow down of business activities as a result of the economic crisis in the countries in Asia, so many people have lost their jobs. Therefore, unemployment rate is going to increase quite rapidly. Accordingly, wage rate of labor especially that of common works in construction is going to decrease.

Foreign currency exchange rate. Currency exchange rates among major foreign countries, and the exchange rates in the countries in Asia have been fluctuating so remarkably that this changes has resulted in the change of the local prices of imported materials and equipment.

Required new construction system to cope with economic recession

Previous and new construction systems applied for executing works of many projects

Before this recession affected the countries in Asia, the major concern of the people in the region was how to increase their GNP as quickly as possible so as to catch up with the developed countries. To meet this target, a construction system had been established and adapted to most of projects as a common and routine execution system. This system is very similar to the system used in the developed countries with slight adjustments to meet the local conditions. However, this system is only good in an economy with high growth rate. Therefore, once the growth rate of the economy declined, this system ceased to be the basis of a suitable system for the execution of projects after the recent recession.

The existing system of execution of the projects that were started before the recession should be reviewed and adjusted to the new conditions. Similarly, all the planning of new projects should be checked by feasibility analyses on new economic conditions from every aspect. Above all, the construction system to be used for execution of the projects should conform to important requirements to be strictly checked to reduce the construction cost.

Major components to be considered for reviewing previous construction system to establish new one.

Construction period. The construction period of previous projects before the current recession was usually rather too short than it should be. The main purpose of the previous projects is to increase GNP of the developing countries in Asia where everybody was very anxious to get out of poverty by developing their countries within a short period of time. However, with this recession, time factor for development is not so important as it was before, because the pace of business in every industry has slowed down. Therefore, instead of making construction period short, more effort should be put into other important requirements such as cost reduction and quality of construction.

Materials. Before this recession hit the countries in Asia, imported materials for construction were rather easy to use in projects. However, due to the shortage of foreign currencies and rapid change of currency exchange rates in almost all countries in Asia, the prices of imported materials has become quite high compared with local materials. Therefore, effort should be placed on using more local materials as practically possible, to reduce the total construction cost of projects.

Construction equipment. Many countries in Asia import construction equipment from Japan, U.S.A. or European countries. Therefore, costs increased remarkably since the economic slump. Also, the cost of spare parts jumped up so high that construction equipment could not be used as before. Fuel and other materials, which are products of crude oil, are also imported in many non-oil producing countries in Asia. Therefore, increased costs of operation and maintenance has made it difficult to use construction equipment as before.

Method and procedure. Methods and procedures to be used for execution of construction works should be selected mainly from three different viewpoints, namely quality, cost and time. These items are so much influenced by economic conditions that sometimes changes of these items directly affect the feasibility of projects.

New construction system based on new economic condition

Since the influence of recent recession in the economy is quite big, every system to be employed for execution of the construction works of projects should be changed. The previous system was almost same as those employed for the construction projects in developed countries for many years. However, these systems were not checked from the viewpoint of the environment. The new construction system to be used after recession is almost on the same line of saving resources and consideration for the environment. It results in effective use of local materials, skilled labor and less use of oil and other resources even with rather construction period.

This change will have some possibility of developing a new system which is the most optimum for the execution of construction works in the coming new era in Asia. This new system should not only be good for reducing construction cost to meet new requirements of the recession, but it is also required to be sustainable for the new century and for future generations.

Paradigm shift required for sustainable development in Asia

Previous mechanism of development

The meaning and measurement of development. Nowadays, it is a common thinking in Asian developing countries that a major strategy for poor countries to escape from poverty is economic development or economic growth. This concept is supported by Japan's experience of rapid modernization in the previous half century. During the Second World War, almost all of Japan's manufacturing base was destroyed by air raids of U.S. Air Force. As a result, approximately half of the workforce of Japan worked in the agricultural sector. At that time, Japanese average income was very small and ranked below that of many developing countries [Nafziger, (1997)]. However, much has changed since then.

Also, the group of NIEs (South Korea, Taiwan, Hong Kong and Singapore) are showing the most successful economies in the world. These are the typical example of the Asian miracle pointed out by the World Bank report in 1993. The keywords for the progress of these countries are "development" and "growth of the economy". [World Bank, (1995)]. However, these two terms are not identical in the strict sense of the terms. Economic development is linked with economic growth accompanied by some changes in output distribution and economic structure of the country.

In addition, measurement of economic development is quite important to know the actual state of the country. For this purpose, two indices are usually used. One index shows the rate of development and the other shows the level of development. The rate of development indicates developing speed, and it is explained by using growth rate of economy, which is computed in the form of annual change of GNP per capita after adjustment by price deflator.

On the other hand, the level of development is measured by using the figure of annual GNP per capita, which is usually used for classification of countries. One classification used by the World Bank divides all countries into three categories by using GNP per capita. According to this classification in 1993, three categories were "low-income" countries (less than $700), "middle-income" countries ($700 to 12,000), and "high-income" countries (more than $12,000). [World Bank, (1995)]. It is quite natural for developing countries to put in their best to pull up their rank in the classification by income level.

Principle and concept of development. It is quite a remarkable that there are three countries in Asia among high-income economies, Japan, Singapore and Hong Kong. Singapore and Hong Kong surpass or almost surpass the United Kingdom, which was their model country. Thus, the growth rate of the economy of NICs is quite high. This means that development rate of the NICs is quite high. [World Bank, (1995)]. These facts generated some expectation in other developing countries in Asia - that for the development of a country, industrialization is the most effective way, and that the principal concept of the development of a country cannot be discussed without industrialization. Many people in Asia believe that development of a country can be achieved by economic modernization and industrialization.

Previous economic modernization in eastern the world. To begin with, it is necessary to know the origin of the economic growth. What did make sustained economic growth? In the history of the Western world, capitalism produced it. Capitalism is the dominant economic system in Western world since the breakup of feudalism from the fifteenth to the eighteenth centuries. In principle, capitalism is the economic system of relation between private owners and workers. Pieces of land, mines, factories, and other forms of capital as means of production are privately held, and workers who were free but without any capital sell their labor to employers [Nafziger (1997)].

Capitalism, as an engine for rapid economic growth, spread beyond Europe to the outposts of Western civilization, and it is most suitable system for industrialization. Capitalism provided economic growth only in Western countries with little exception. In Eastern countries, there have been some barriers to capitalism not only in the traditional societies but also in the effect of colonization by Western countries. It is clear that most Asian countries are lacking the strong capitalists and the effective bureaucratic and political leadership, which are essential for rapid economic modernization [Nafziger (1997)].

Japan was devastated economically during the Second World War. In the late 1940s and beginning of 1950s, technical and economic assistance from U.S.A., like the Marshal Plan, made it possible to reorganize international trade system and financial system as well as provide the basis of creating economic miracle. It is obvious that the remarkable rapid growth of Japan occurred because technical knowledge and human capital were intact, even in the ruins of other capitals due to war [Nafziger (1997)].

The fastest developing countries in the world now are the countries called "Asian Tigers". They are also referred to as "Niches" as mentioned before, and they are South Korea, Taiwan, Singapore and Hong Kong. Taiwan and South Korea are countries, which have experienced quite high rate of growth through the years at the end of this century. The models of development of these two countries are almost the same as that of Japan.

State of the World

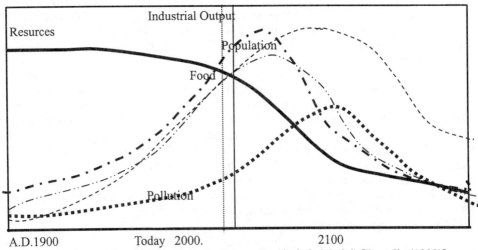

Figure 2. Estimated State of World in the Future (Pessimistic Model) [Donella (1992)].

Like Japan, their governments systematically intervened to invest in the construction of infrastructure for further development of the countries, and the promotion export of their products, and education for better human capital under political and economic stability. Also, these two countries like Japan had a high quality of their own management backed up by ancient Chinese philosophy of Confucianism [Naisbitt, 1996]

Effective execution of infrastructure projects: The World Bank report indicates that infrastructure has a high potential payoff in terms of economic growth [World Bank, (1994)]. However, infrastructure alone does not guarantee growth of the economy. There are many infrastructures, which have smaller return on investment. The disparity in return on investment may be due to the differences in the efficiency of investment across countries [Rongviryapanich, (1997)]. In the developing countries, poor maintenance results in inefficient performance of infrastructure. Therefore, efficiency in the life cycle of the project will be dominant.

The influence of development on global environment: World Bank report shows that energy consumption per capita equivalent to kilograms of oil will increase proportionally to GNP per capita [World Bank, 1995]. This means that the rapid development of countries in Asia causes big amount of consumption of oil resulting in emission of green gases as discussed in the Kyoto conference.

Conclusion

There is a well known theory of relationship between economic index and environment index by Kuznets [Kuznets, 1955]. In this old theory, the environment will improve after the income level has reached a certain level. Application of this theory may be limited to environmental problems in the country but not for the global level, but some people have denied this theory itself [Matuoka, 1997].

The importance of saving resources for future generation: Some pessimistic mode shown in Figure 2 suggests that the society is to run out of the non-renewable resources.

From the middle of 1997, a quite strong recession of economy spread out all over Asia. Many projects were interrupted and suspended due to the serious changes of economic conditions. However, after several years of tremendous effort to cope with this situation, there will be revival of the economy, which will bring more projects as before. Therefore, sometimes at the beginning of next century, considerable amount of infrastructure construction projects will be carried out again in the region of Asia. These construction projects will so much affect the world environment that some counter measures would be needed. However, the new construction system established to cope with adverse economic condition will be almost same as the construction system required for sustainable construction in the future.

For this purpose, the system and methods of "construction management" should be reviewed to put the concept of environment and saving of resources into its theory as the most important principle. Also, professionalism of civil engineers should be revised in order to put the concept of the environment and saving of resources into its fundamental philosophy.

References

Naisbitt, J. (1996) *Megatrends Asia,* Nicholas Brealey Publishing Limited, London
Diamond, Japan (1998) The Diamond Weekly 7/25, Japan
Far Eastern Economic Review (1998) Asia 1998,Year Book, A Review of The Events of 1997, Hong Kong.
Nafziger, E. W. (1997) *The Economics of Developing Countries, 3rd Ed.,* Prentice-hall, Inc. New Jersey
World Bank (1995) *World Development Report 1995,*Oxford University Press
World Bank (1994) *World Development Report 1994,* Oxford University Press
Rongviryapanich, T. and Shibayama, T. (1997) Infrastructure Development Stages: Comparative Study of Thailand and Japan's Experiences--*Proceed. of the Conference of the Japan Society for International 8th Annual Development*, Urawa, Japan
Kuznets, S (1955) Economic Growth and Income Inequality, *the American Economic Review,*45(1) pp1-27
Donella, M et al.,(1992) Beyond The Limits; Confronting Global Collapse, Envisioning a Sustainable Future (Post Mills. VT Chelsea Green, 1992) p.133

RESIDENTIAL CONSTRUCTION IN COUNTRIES IN THE PROCESS OF TRANSITION - THE QUESTION OF PROFIT

I. MARINIC and S. VUKOVIC
University of Novi Sad, Faculty of Technical Sciences, Institute of Industrialized Building, Trg Dositeja Obradovica 6, 21000 Novi Sad, Yugoslavia.
vukovics@uns.ns.ac.yu

Abstract

The transition from a centralized planned economy to a market economy is marked by changes in the economic relationship between the subjects of economic activities, business goals, organizational structures and the structure of production. Instead of adapting the activities of economic subjects to the requirements of a central plan, those activities are continually being adapted to the requirements of the market, according to the criteria for maximizing profit.

General changes in the economic system have also made possible a corresponding change in construction firms according to the demands of the market. Based on the results of market research and the research of solvent demand for residential buildings and units, the technical/technological and organizational structures of construction firms are undergoing changes.

The paper will analyze the changes accomplished in the organizational structure of construction firms, technical/technological changes in the construction of residential buildings and the production of residential units as well as the degree of success of the adaptation to the demand for apartments in Yugoslav conditions.

Keywords: Profit, residential construction, transition, technological change.

Introduction

A significant part of the national product has always been invested in residential construction due to the fact that it falls into the category of basic human existential requirements, such as food, clothes, footwear, etc. When productive forces were at a lower level of development, housing standards were also lower, providing only elementary living conditions and shelter from bad weather. The growth of these forces brought about higher housing standards, whose use value also contained certain hedonistic elements. On a higher level of development, apart from satisfying one's basic existential needs, an apartment also fulfills one's needs for a comfortable living environment which provides and allows the development of conditions for harmonious family relationships and the development of one's personality. Therefore, there is no single set scheme for satisfying housing requirements, in other words a set housing standard. A housing standard is the product of the totality of economic and social development, that is to say, it depends on the objective material circumstances of satisfying housing requirements and also on the development of those requirements, the willingness of the individual and the society to set aside a certain part of their earnings for the purpose of providing a certain level of housing.

Residential construction in countries with a centralized planned economy prior to their transition

In former European socialist countries with a centrally planned economy, satisfying the population's housing requirements was based on the realization of planned goals in residential construction and the planned distribution of the finished housing fund to the users.

The money for financing residential construction was obtained by fiscal or parafiscal methods from the profit of firms and paid into government funds for residential construction. The planned use of these funds was handled by a specially appointed government agency.

Table 1. The number of apartments constructed per 1000 inhabitants

Country	Year	No. of apartments constructed per 1000 inhabitants
1	2	3
Developed Western-European Countries		
- Austria	1977	6.0
	1984	5.0
- Belgium	1980	5.0
	1988	3.6
- Denmark	1980	6.0
	1988	4.9
- Finland	1980	10.0
	1989	3.9
- Norway	1981	8.0
	1988	7.2
Eastern-European Countries (presently in transition)		
- Bulgaria	1981	6.0
	1987	6.9
- Czechoslovakia	1981	6.0
	1989	5.6
- Hungary	1980	8.3
- German Democratic Republic	1980	10.0
	1988	13.2
- USSR	1980	8.0
	1987	0.3
- Yugoslavia	1981	7.0
	1989	1.2

Source of data: The Statistical Yearbook of Yugoslavia for the years 1985 and 1991, published by the Federal Statistical Office, Belgrade.

The planned distribution of the existing housing fund was based on economic and social criteria. The economic criteria were based on the economic contribution of the individual (which most often meant the individual's position in the social hierarchy or social reproduction) and the social criteria were based on the individual's domestic situation (the number of members in the household, material status, etc.).

These countries used to set aside 6-7% of their national products for residential construction. In the period of economic prosperity, 6-10 apartments were built per 1,000 inhabitants every year, which corresponded to the number of apartments built in developed Western European countries. However, depending on their area, equipment and other relevant indicators, the value of residential construction was in fact lower.

Residential construction in countries in the process of transition

The abandonment of the central planned economy in these countries led to the decomposition of institutions which were responsible for the functioning of the economic system and also to the creation of new ones which were supposed to provide a more efficient and dynamic growth in conditions of free enterprise. However, many elements from the old system survived and the creation of new institutions and relationships was very slow. These circumstances led to radical changes in relations in the economic system, but they didn't lead to significant increases in the system's efficiency. All relevant economic indicators in the period between 1990 and 1995 show the deterioration of the economy of the countries in transition. The market as a general regulator of relations in an economic system in its rudimentary form did not allow the optimal allocation of resources, which affected the economy. The instability of the political systems in these countries lead to an increased economic uncertainty and, instead of an increase in the level of cooperation with the economies of developed countries, many countries in transition displayed a decrease in economic cooperation with these countries.

While the relative volume of residential construction in most countries of the European Union, as well as other countries of the European Community and Non-European developed countries, has continually been increasing, residential construction in Eastern and Central-European Countries has continually been decreasing (Diagrams 1 and 2).

In all Eastern and Central-European countries in the process of transition, as well as in countries which were formerly part of the USSR, both the absolute and relative volumes of residential construction were lower in 1995 than in 1990. Considering the high level of interdependence between residential construction and other sectors of the economy, the fall in the volume of residential construction affected the volume of the entire national economy.

In most of these countries the population's savings did not represent a relevant factor in the dynamics of economic growth. Instead, the financial potential of government funds, budgets, firms, etc. were the important factors. In the absence of control over the use of these funds in the period of transition, they were drained into other forms of expenditure and gradually lost the role of regulator of economic activities. This resulted in a decreased investment into the satisfaction of the needs, especially housing needs, of the dominant social groups in the population (the middle class).

Diagram 1. The number of apartments constructed per 1000 inhabitants in Eastern and Central-European countries in the process of transition in 1990 and 1995.

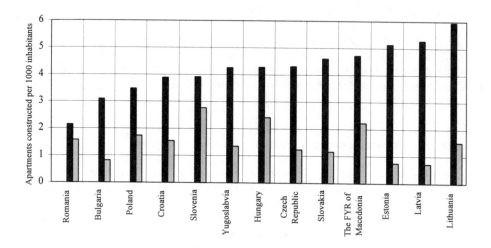

Source of data: Trends in Europe and North America 1996/97, The Statistical Yearbook of the Economic Commission for Europe, United Nations, New York and Geneva, 1997.

The establishment of political and economic institutions in order to realize basic aims of development on democratic bases forms a foundation for the transition of the economic and political systems of certain countries towards a more efficient economic and social development. Material shortage, the absence of elementary living conditions, such as a roof over one's head and other physical necessities, cannot be the aim of a society in general. A long period of decline in the economic activities of a society represents a Source of data: Trends in Europe and North America 1996/97, The Statistical Yearbook of the Economic Commission for Europe, United Nations, New York and Geneva, 1997.

Diagram 2. The number of apartments constructed per 1000 inhabitants in the countries of the former USSR in 1990 and 1995.

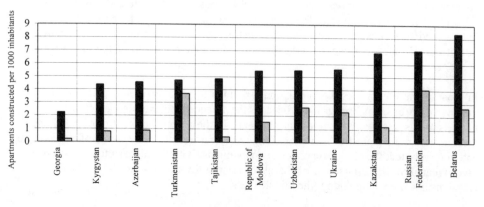

serious obstacle to modern development and possesses elements of instability of social communities. That is why there is a necessity for activities on a larger scale which would speed up the processes of transformation in countries in transition towards modern societies, the activation of economic resources on the basis of common interests. Profit as motivation for business is a significant economic instrument for allocating resources, but it is not enough to ensure the realization of a wide variety of economic and social goals. "In the free world, individuals, groups and nations are free to make decisions any way they choose, but what they cannot avoid is paying the price for their choices." (Prokopijevic, 1996) It is obvious that the paths and methods of transition in most countries are not the best choice and that empirical data indicates that the price of transition is high, both from the economic and social aspects. Therefore it is essential to channel joint efforts towards finding a solution which would provide an adequate economic growth in the future in order to achieve future economic and social goals.

Residential construction in Yugoslavia from the end of World War II to the beginning of the transition

In different stages of Yugoslavia's development after WW II there were different forms of redistribution of profit whose purpose was to regulate the volume of residential construction in accordance with the aim of satisfying housing and other requirements of the population. The government's decisions to raise the standard of housing also meant a certain redistribution of income for the purpose of residential construction, at the expense of other economic development goals, whereas insisting on other development goals meant the reduction of the volume of residential construction.

From the end of World War II to the beginning of the 90's, the dominant portion of residential construction in big towns was public property. However, on the whole, private residential construction made up the majority of the total residential construction.

The entire organization of institutions responsible for residential construction was subordinated to public residential construction. The concentration of funds for residential construction and the centralization of their management allowed the realization of large investment ventures in big towns. As a result of scientific research new construction systems are introduced along with the industrial production of construction elements. Large construction firms are established in big cities. They are equipped with heavy mechanization and have a large number of employees for the production of large series of residential units. Beneficial economic effects were being achieved in conditions of continual influx of funds for residential construction and its increased volume. The technical progress of that period was made evident in the lowering of construction costs and increased production output.

Financing residential construction in the public sector was made possible by administrative measures, by obliging firms to set aside part of their profits for residential construction. However, the use of the residential fund was not based on economic criteria, so that the rent which was collected did not provide the necessary means for the reproduction of the housing fund. Instead, it was almost exclusively financed by fresh funds from firms.

Private residential construction was treated as marginal throughout the entire post World War II period, even though it represented the bulk of residential construction. Private residential construction in big towns was incorporated into the construction of large buildings which, along with apartments in public property, contained privately owned apartments as well. Housing cooperatives assumed the role of representative of private interests in the construction of residential buildings.

Changes in the economic and political systems in the late 80's and early 90's were declaratively aimed at the affirmation of entrepreneurial economy, free market and the development different forms of capital ownership. In the process of transforming the economic and political systems, numerous limitations on capital handling were abolished. In that sense earlier obligations of firms to allocate a part of their profit for solving workers' housing problems also ceased. The previously established public housing fund was mostly purchased by the tenants at subsidized prices.

After abandoning the administrative system for forming and using funds for residential construction, no appropriate market institutions which would establish a consistent system for financing residential construction were established. As a consequence of that as well as of all the problems of economic reproduction, the volume of residential construction sank under the level of simple reproduction.

Residential construction in the Federal Republic of Yugoslavia during the period of transition

The beginning of transition in the Yugoslav economy is associated with the significant changes in the political and economic system in the late 80's. However, the transition is taking place under very complex circumstances which saw the disintegration of the former

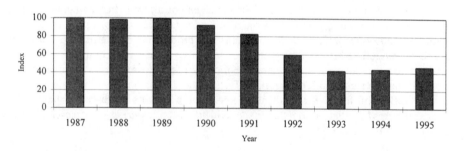

Diagram 3. The index of the national product on the territory of the Federal Republic of Yugoslavia between 1987 and 1995 (1987=100)

Source of data: Statistical Yearbook of Yugoslavia 1997, Federal Statistical Office, Belgrade, 1997.

Socialist Federal Republic of Yugoslavia and the formation of the Federal Republic of Yugoslavia. The entire period is characterized by stagnation and decline in economic activities on the territory of the Federal Republic of Yugoslavia.

The period of transition in the Yugoslav economy coincides with the drastic fall of the national product. The new ownership structure did not lead to a more rational use of capital either in the economy in general or in its individual segments. The volume of residential construction fell below the level of simple reproduction of the housing fund thus leading to a decrease in the national wealth. The increase in the participation of private residential construction is taking place alongside of a significant decrease in the volume of residential construction in the public sector.

Diagram 4. The number of apartments in the public and private sectors in the Federal Republic of Yugoslavia between 1983 and 1995.

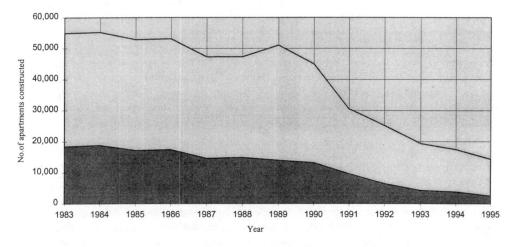

Source of data: Statistical Yearbook of Yugoslavia 1997, Federal Statistical Office, Belgrade, 1997.

There have not been any significant changes in the structure of residential construction according to the number of rooms. The reduction in demand caused a reduction in the volume of residential construction.

In the field of construction, design and constructive material industry, there are 6715 registered firms in Yugoslavia, out of which 556 are public, shareholders' or mixed capital firms and 6169 are private. The above mentioned 556 firms employ 177,000 workers and the private ones 6,800.

The average age of those employed in the construction industry is 50, while younger people turn to professions which offer more perspective, better working conditions and salaries. The salaries in the construction industry are evidently low. One can ask, "How can high quality skilled be provided labor under present economic circumstances?"

It is also interesting to mention one peculiarity of the Yugoslav housing market - the sale of the existing housing fund on the basis of the tenancy right represents an overwhelming competition to construction firms. These apartments are significantly cheaper than those newly built, so that potential buyers who have cash frequently opt for them because they are cheaper and do not fall behind the new ones in quality.

The appearance of a large number of small construction firms means, above all, the construction of residential buildings with one or two entrances, up to four stories high (plus the ground floor), with no lifts, or family houses with the exclusive use of classic construction materials as well as manual labor.

The trend of constructing multi-story residential buildings is significantly declining, whereas the trend of constructing family houses is rising. The bearer of that trend is a narrow circle of nouveaux riches. This is further aided by relatively low construction material prices and very cheap labor. According to what has been said so far, there exists a process where the stress falls on poorly paid labor and a method of building which does not require the contractor to be very well equipped.

In the public sector, the available mechanization is used at half capacity and lower, while a high percentage of the employees are on involuntary leave due to lack of work. A paradox appeared: on the one hand there is all the unused mechanization, on the other there is a number of newly established small firms which lack mechanization who insist on pushing the construction industry towards classical construction with the primary use of manual labor.

The disappearance of the so-called "middle class" in Yugoslavia follows the shift in the structure of constructed apartments. Smaller two-room, one-room and bachelor flats with areas between 25 and $50m^2$ are very much in demand as well as large flats - above $80m^2$. The stratification in the society is also reflected in the structure of demand for new apartments. The market ruthlessly shows the relationships of social stratification. However, the logic of buying smaller flats is connected to the possibility of their renting because the price of hiring an apartment is often higher than the average salary or pension.

The informal pressure of the state on larger construction firms to do business abroad for the purpose of earning the much-needed hard currency is evident. However, it is much more difficult for Yugoslav firms to do business abroad, especially in the case of construction firms and also because of the unacceptance of guarantees by Yugoslav banks and because foreign banks are not interested in financing Yugoslav firms.

The technology of residential construction in Yugoslavia

The Yugoslav construction industry has achieved admirable business results in the past decades both in Yugoslavia and abroad. Today in times of crisis and a very narrow construction market, the running of construction firms is inefficient, processes are poorly organized and productivity is low. In such conditions the construction industry, and residential construction as well, are unable to follow international trends. However, the return to traditional construction, in the desire to offer employment to workers at all costs, is not going to lead to more efficient business. Development ought to be directed towards the production of construction and other materials, as well as products of known specifications possessing standard quality, which will adhere to European standards. Until the construction industry reaches its full intensity again, especially in the area of residential construction, there will be many transitional stages and the existing systems, semi-prefabricated and prefabricated, must be constantly improved and adapted to the possibilities and demands of the market. By determining the direction of its trends, the Yugoslav construction industry must preserve that which it already possesses in order to be able to rejoin international trends at some point.

The question of profit

The business activities of Yugoslav firms were motivated by making profit even before the transition. Unlike firms in countries with a central planned economy where business activities were subordinated to the requirements of the central plan, Yugoslav firms possessed a high degree of independence in business and so they had a tendency towards achieving maximum profits, which were at the end of the year split between those who participated in its making (for the firm's funds, workers' salaries and government income taxes). The co-ordination of all the participants' interests was realised through the distribution of profit.

In the period when the Yugoslav economy is undergoing transition, making profit is the basic motivation for business activities in firms, where the method of its distribution has been significantly changed. In other words, the distribution of profit is based on the ownership of capital, so that workers, not having capital of their own, are mostly excluded from the distribution of profit. Despite the fact that the owners of capital are strongly motivated to improve business activities and thus increase profit, new conflicts arise between the owners of capital, workers and the state, which diminishes the importance of profit and the possibility of its growth.

Previously presented data indicate a decline in the efficiency of the Yugoslav economy in general and of production within the construction industry in the period of transition. Changes which took place in the economic system in the form of general circumstances for the business activities of firms did not contribute to the improvement of business efficiency. Profit as the principal motivator of technical and technological changes and improvements in business activities did not become a sufficiently powerful tool, considering the fact that the majority of those who participate in making profit are excluded from its distribution. The elimination of those contradictions is an important condition for creating more harmonious relationships in the country's economic development.

Conclusions

The period of transition in the Yugoslav economy is taking place in conditions of reduced economic possibilities, due to the effects of numerous factors on the business activities of economic subjects and the decline of the national product. The force of those factors has significantly surpassed the positive effects of profit as a goal of economic activities. As a result of that, contradictions in the economic development have grown stronger and they have to be solved by using the numerous instruments of market economy along with the active intervention of the governments in countries in the period of transition, on their way to a free and open society with economic and other freedoms.

References

Marinic, I. and Vukovic, S. (1995) Cost estimation of outfitting land and the construction of apartments in Novi Sad. Proceedings of the International Conference, Newcastle, UK.

Marinic, I. (1995) Residential construction in Yugoslavia - the present state and perspectives. DaNS, No. 13-14, Novi Sad, Yugoslavia. (in Serbian)

Prokopijevic, M.(1996) Without a Reasonable Alternative. Ekonomika, No. 2-96, p. 96, Belgrade, Yugoslavia. (in Serbian)

Vukovic, S. and Marinic, I. (1996) The development and requirements of residential construction in Yugoslavia. Proceedings of the International Conference, Architecture - urbanism at the turn of the millenium, Vol. 1, Belgrade, Yugoslavia.

Vukovic, S. and Marinic, I. (1997) Planning and monitoring construction production - Yugoslav experiences. Proceedings of the International Symposium, Construction Technology - Construction Management '97, monograph, Subotica, Yugoslavia.

Vukovic, S., Marinic, I. and Milidragovic, G. (1998) Legalizing illegal residential buildings – a solution or chaos. The proceedings of the CIBWorld Building Congress '98, Gavle, Sweden, (Poster presentation in Symposium C)

MODELLING CONSTRUCTION WASTE PERFORMANCE – AN ARISING PROCUREMENT ISSUE

P. FORSYTHE and P.K. MARSDEN

School of the Built Environment, University of N.S.W., Sydney, 2052, Australia,
P.Marsden@unsw.edu.au

Abstract

Environmental issues are beginning to play an important role in construction contracts. Within this framework, construction waste is a growing area of concern. No longer is it possible to dump waste at the community's expense. The paper looks at the way that construction industry clients are responding to the need for better performance in this area. In the cases discussed there is a need for contractors to demonstrate how they will address waste performance by means of submitting a waste management plan as part of the tender assessment criteria. This requirement identifies the need for an analytical approach for characterising waste cost and environmental performance. There is also a need for an on going basis for auditing the achievement of performance. A cost analysis model is proposed and trialed on two residential projects. The approach implements the use waste performance standards which have been derived from extensive field studies. The cost planning approach allows the standards to be applied in a way that is relevant to the cost profile of the specific projects. The method is considered to be of interest to those involved in management oriented approaches to contracting and procurement because it identifies where waste minimisation strategies will be most useful in meeting perceived requirements.

Keywords: Construction performance, construction procurement, construction waste, cost planning

Introduction

Construction waste is increasingly presented as a performance criterion affecting clients, project managers and lower tier contractors in the supply chain. In this regard, there are two key areas of performance likely to be considered when procuring construction services. First, the need to comply with environmental and sustainability requirements [Kumaraswamy and Dissanayaka (1996), Hill and Bowen (1997), New South Wales Government Construction Policy Steering Committee, (1998) pp. 29] and second, the need to reduce non-value adding production costs [Koskela (1994), Hammer and Champy (1993)]. These two areas are related in so far as increased environmental regulation tends to cause greater expense in traditional approaches to dealing with on-site waste. This potentially places pressure on project participants to not only respond to environmental regulations but also to reconsider if avoidance in waste can realistically lead to improved process efficiency.

The paper identifies how environmental pressures are influencing construction sector clients in their procurement of construction services. Where this is an issue tenderers are typically required to submit a formal waste management plan as part of the tender assessment criteria. As a response to this, the paper identifies an approach to developing a waste management plan based on the use of expected waste margins applied in a cost planning framework. In this context value management principles can be applied to strategically address expensive waste items. An extension of this approach shows which cost centres will also have an environmental impact in terms of waste being disposed of at landfill. The methodology aims to provide a basis for construction managers and lower tier service providers to transparently justify their strategic decision making relating to waste performance.

Why environmental waste performance is becoming important

There is a strong community push to make the construction industry accountable for the waste it creates [Combined Regional Organisations of Councils, (1996)]. Studies in the US, Netherlands and Australia suggest that the construction and demolition industry is responsible for 20%-30% of all solid waste entering landfill [Bossink & Brouwer (1996), du Blet (1995), Johnson & Minks (1995)]. Germany has a slightly lower rate of 19% that has been achieved through taxing new materials more highly than recycled materials [Wilson (1996)]. New trends in environmental policy seek to identify "waste creators" then make them accountable for all associated costs by taking a "cradle to the grave" approach in terms of costs and responsibility. This means that construction projects are likely to be seen as the "creators" of waste. As such, construction industry clients and providers will become inherently responsible. An argument in favour of this approach is that until now, landfill has been a cheap form of construction waste management. If this perception changes to a user pays principle, then the full life cycle value assigned to wasted materials "from the cradle to grave" may cost significantly more than the current situation.

Waste cost performance

Waste cost performance varies depending on whether waste can be avoided or if lower grade alternatives are the only realistic option. For instance, if avoidance cannot be achieved then the next best option is to re-use or re-cycle the materials. Unfortunately, this involves significant devaluation by virtue of extra costs incurred in re-working the materials which makes them relatively expensive compared to new raw materials. If this intermediate step is not economical then materials have lost all residual value and must be disposed. Traditionally in Australian construction projects, there has been relatively little attempt to manage waste minimisation. Therefore, the lack of waste avoidance and the under-development of re-use and recylcing markets means that designed-in-waste will lose full value and will then incur the added cost of disposing of those materials.

Fortunately, this situation is changing in Australia as demonstrated by McDonald and Smithers (1998) who examined an Australian Police and Court Complex project where waste minimisation strategies were trialed. The results of this trial revealed substantial savings when compared with a similar project where there was no formal waste minimisation strategy used.

Public authority clients and waste performance

The N.S.W. Department of Public Works is the largest procurer of built assets in N.S.W. As a government department it has a responsibility to lead by example and comply with the NSW Environmental Protection Authority's 60% waste reduction target for solid construction waste disposed of as landfill - as documented in the Waste Minimisation and Management Act. [New South Wales Parliament, (1995)]. This is formally addressed as a performance criterion in the recently released New South Wales government white paper, Construct (1998) which identifies the intention to formally integrate waste performance in tender processes on government construction contracts. It will require tenderers to submit a waste management plan to identify how they will implement and monitor waste performance [New South Wales Government Construction Policy Steering Committee, (1998) pp. 79].

From a slightly different perspective, the Sydney Organising Committee for the Olympic Games (SOCOG) have an environmental imperative arising from commitments made when tendering for the 2000 Olympic games projects. There is a commitment to use recyclable and recycled building products in the construction of new facilities. In an operational sense, the Olympic Co-ordination Authority (OCA) has responded to this need by creating an Environmental Tender Specification [Olympic Coordination Authority, (1996)]. This forms part of the OCA policy when entering into Construction Management, Consultancy, Construction and Services contracts as before, a key feature of the standard tender specification is the need for waste management plans.

Waste cost analysis

A project cost profile, similar to that used by quantity surveyors when preparing an elemental cost plan, provides a useful technique for identifying and quantifying waste in new construction projects. Such a cost profile could provide the essential data needed to develop a strategy to reduce or optimise waste levels. From a financial perspective it provides the opportunity to look at large contributors to waste costs and to examine the entire cost of waste relative to the overall cost profile of a project. This allows waste to be seen as an independent cost centre and on this basis the merit of reducing waste can be assessed in a comparative way. This is the basis of the waste analysis cost model described hereinafter.

Waste Research

In order to construct the waste analysis cost model it was necessary to ascertain appropriate waste margins. The necessary data was obtained from a series of integrated studies as outlined below. These studies aimed to quantify and qualify the occurrence of waste on residential scale projects.

Study 1: Quantification and qualification of trade waste. This study focused on detached housing and medium density housing sites. Only trades which were identified as key contributors to the waste stream were investigated including: concrete floor and footing construction, roof tiling, plasterboard fixing, brickwork, wall and floor tiling and frame carpentry. Sample sizes ranged from 6 – 30 dwellings for each trade. Most samples were over 15 projects. Waste was determined by subtracting insitu quantities (taken-off from drawings or site measurement) from ordered materials (taken from delivery and order documents).

Study 2: Quantification of the removal of waste from site (i.e. Landfill). Waste from 10 brick veneer project homes in Sydney's north western suburbs were assessed and analysed by weighing and quantifying materials taken to landfill.

Study 3: Qualification of waste (source evaluation). This study identified detailed causes of why waste occurred on site in keeping with previous methods developed by Skoyles (1976). These causes were traced back to initial origins such as construction design, quantity take-off, materials manufacture, materials supply, contracting practises, site management and work practises and customer preferences. This was undertaken by way of focus groups containing a mixed representation of the above participants. In addition, approximately 20 site based interviews were conducted with tradespeople for each trade.

Expected waste margins

From the research, minimum and maximum waste margins were determined as shown in Table 1. It was found that large ranges were attributable to a variety of situational causes. The qualitative interviews and cross tabulation of the data were used to identify causal variables. An example of variables is shown for concreting in Appendix A. It is proposed that an expected waste margin can be estimated by interpreting how the relevant causes are likely to affect the waste margins shown in Table 1.

Table 1. Waste expressed as extra proportion to insitu requirements

Material	Minimum waste margin	Maximum waste margin
Concrete raft slab	6.0%	22.0%
Brickwork	3.0%	10.0%
Timber framing	4.2%	6.0%
Roof tiling	2.5%	8.0%
Plasterboard wall linings	5.0%	10.0%
Wall and floor tiling	4.5%	15.0%

Operationalisation of the waste cost model

The previous research identified a number of different materials used in construction that generate on-site wastage due to the manner in which these materials are stored, handled and fixed. Typically, these materials are delivered as standard manufactured units and require cutting and fabrication on site as distinct from other materials that are prefabricated off-site and only require site fixing eg. doors, windows, roof trusses, kitchen cupboards, sanitary fittings, etc. In order to provide a clear understanding of the cost of on-site material waste the cost of these materials was applied in a summarised cost plan format to two residential projects.

The first project, the HIA House, has a total net cost of $58,380. It is a single storey detached brick veneer house constructed with reinforced concrete floor slab and timber framed pitched roof covered with concrete roof tiles. The Housing Industry Authority provided cost information. The cost of the selected materials, as noted in Table 1, are expressed as a percentage of the total net cost of the building in Table 2.

Table 2. HIA House						
		Cost Performance Information				
	Cost	% of total cost	Min. waste	Max. Waste	Min. Waste	Max. Waste
	$	%	%	%	$	$
Selected Materials: (Supply only)						
In-situ concrete	2,765	4.74%	6.0%	22.0%	165.90	608.30
Bricks	2,287	3.92%	3.0%	10.0%	68.61	228.70
Timber framing (excluding trusses)	1,942	3.33%	4.2%	6.0%	81.56	116.52
Fix out timber	713	1.22%	4.2%	6.0%	29.95	42.78
Concrete roof tiles	1,982	3.39%	2.5%	8.0%	49.55	158.56
Plaster linings	2,224	3.81%	5.0%	10.0%	111.20	222.40
Floor and Wall Tiling	242	0.41%	4.5%	15.0%	10.89	36.30
Low cost items fabricated on site	984	1.69%	5.0%	10.0%	49.20	98.40
Sub Total:	13,139	22.51%				
Waste removal and disposal	650	1.11%			400.00	750.00
Labour, Plant and Prefabricated Materials not included above:	44,591	76.38%				
Total Net Cost:	58,380	100.00%				
Total Cost of Waste including removal and disposal:					966.86	2,261.96
% of Total Net Cost:					1.66%	3.87%

Research into waste margins, refer Table 1, indicate that the expected waste margin for the selected materials is not constant but will vary depending upon many project specific factors. As such, Table 2 demonstrates the range in cost by portraying minimum waste values as "good waste management practice" and maximum values as "poor waste management practice". The difference between the two indicates the potential for waste savings. In this case, the total cost range of the selected materials including removal and disposal was $966.86 to $2,261.96, or 1.66% to 3.87% of the total net cost of the building. This equates to a potential saving of $1,295.10 or 2.21% between "good waste management practice" and "poor waste management practice".

The second project, Villa Units, has a total net cost of $389,083. This project is a single storey multi-unit brick veneer residential development comprising four two bedroom villa units with external works including landscaping, roadworks and paving as shown in Table 3. The building is of similar construction to the HIA house described above. The cost of the selected materials and waste was determined in the same manner as the first project.

Table3. Villa Units

	Cost performance						Environmental performance	
	Cost $	% of total cost %	Min. Waste %	Max. Waste %	Min. Waste $	Max. Waste $	Min. Landfill kg	Max. Landfill kg
Selected Materials: (Supply only)								
In-situ concrete:								
Floor slabs and beams	12,700	3.26%	6.0%	22.0%	762.00	2794.00	0	0
Driveways	5,044	1.30%	5.0%	7.0%	252.20	353.08	0	0
Paving	909	0.23%	5.5%	7.5%	50.00	68.18	0	0
Bricks	13,125	3.37%	3.0%	10.0%	393.75	1312.50	2,475	8,250
Timber framing (excluding trusses)	8,100	2.08%	4.2%	6.0%	340.20	486.00	206	295
Fix out timber	2,722	0.70%	4.2%	6.0%	114.32	163.32	46	65
Concrete roof tiles	8,489	2.18%	2.5%	8.0%	212.23	679.12	698	2,232
Plaster linings	8,196	2.11%	5.0%	10.0%	409.80	819.60	420	841
Floor and Wall Tiling	2,990	0.77%	4.5%	15.0%	134.55	448.50	70	234
Low cost items fabricated on site	6,679	1.72%	5.0%	10.0%	333.95	667.90	113	225
Sub Total:	68,954	17.72%					4,028	12,142
Waste removal and disposal	2,069	0 53%			1800.00	2,400.00	132,000	132,000
Labour, Plant, and Materials not Included above:	318,060	81.75%						
Total:	389,083	100.00%					136,028	144,142
Total Cost of Waste including removal and disposal:					4,802.99	10,192.20		
% of Total Net Cost:					1.23%	2.62%		

The total cost range of the selected waste materials including removal and disposal for the Villa Units was $4,802.99 to $10,192.20, or 1.23% to 2.62% of the total net cost of the building, refer Table 3. This equates to a potential saving of $5,389.21 or 1.39% between "good waste management practice" and "poor waste management practice".

One reason for the lower percentage values in this case when compared with the HIA House results, is due to the difference in the elemental cost profile between the two projects. For example, the external works is 5.05% of the total project cost for the HIA House and 15.39% for the Villa Units. Naturally, there will be some variation in the nature and extent of various cost categories between different housing projects due to different specifications, site conditions and external works. These differences will have some effect on the relative percentage of total cost of the selected materials.

Apart from the materials dealt with in the model, the building process generates other waste materials that are not physically part of the structure. For instance, excavation and site vegetation are highly variable factors which cannot be addressed in the predictable setting used previously for trade waste. This is apparent in the "equal cut and fill" excavation method used in the HIA house compared to the $88m^3$ of spoil to be removed on the Villa Units project. Another item excluded from the trade oriented approach to waste quantification is material packaging e.g. cardboard, plastic wrap, and strapping. At this stage, there is insufficient data available to determine the nature and extent of packaging waste though it is apparent that this is a key contributor to the volume rather than the mass of landfill.

An extension to the model

The Villa Units project contains an extension to the model by way of including information relating to environmental performance. It is notable that all waste costs do not necessarily transfer to being disposed of at land fill. For instance, over excavated trenches and inaccurate formwork will contribute to the need for extra concrete over and above that shown on the engineer's design drawings. This is obviously waste retained on site and therefore will not contribute to waste disposed of at landfill.

As such, the extended model allows waste minimisation strategies to developed in a way that allows a synthesised view of environmental and cost performance needs. Even so, it is apparent that while the model allows an informed view of waste performance information it is only a basis for justifying the implementation of waste management strategies. It is apparent that such strategies will come at an implementation cost which requires further extension of the model. Even so, the waste margin benchmarks taken from the research provide indicators of good and poor practice and therefore allow proposed strategies to be evaluated within a comparative framework.

Strategies to address the key areas derived from the plan can take on a wide variety of forms. For instance, the usual method for procuring brickwork involves the head contractor supplying bricks and a separate contract for labour. There is express incentive in this situation for the labour to avoid waste. The primary concern of the labour is to maximise productivity. As such, there is an excessive amount of breakage mainly due to poor brick cutting practises. In another instance, there is incentive to waste bricks if the labour is paid on the quantity delivered rather than laid. Strategy development is under way in a variety of areas to allow the above basis to waste minimisation planning to have a valid effect.

Closing discussion and future research

The paper has identified how construction waste is gradually becoming a more important performance priority for clients. As such, there is likely to be the need for a corresponding response from lower tier providers. This will especially be the case given proposed increases in disposal costs and exclusion of certain materials from landfill sites. Also the current practice of burying waste on site will become more difficult as being an accepted site practice. Though this is unlikely to seriously affect procurement systems in the short term, it does mean there will soon be another item on the client's shopping list which must be resolved at the design, tender and construction stages. Initially, community environmental concerns are likely to drive the impact of waste management on procurement needs. In the future, the impact of this may improve the potential for re-thinking work processes along the lines of lean construction [Koskela (1994)].

The minimum and maximum waste figures used in this paper aim to act as comparison standards. These figures were obtained from site observations at residential sites where there was no formal attempt to reduce or minimise waste on site. The results reflect typical site practice. On this basis, total project cost reductions in the range of 1.39% to 2.21% appear possible by the achievement of minimum traditional waste levels. Although it is recognised, that some of these savings may be absorbed by additional costs required to adopt good waste management standards. However, it must also be considered that the introduction of greater levels of prefabrication and improved design that aims to minimise site fabrication waste would result in lower waste margins than those observed in the foregoing field studies. Therefore, there is a potential to achieve lower minimum waste rates than the traditional performance rates used in the tables.

In reality, "expected waste" margins on a project will normally fall somewhere in between the minimum and maximum range. Even so, the key issue is that the normal limits of trade waste combined with the implementation of the causal list, refer Appendix A, create a means for checking that "expected waste margins" and any strategies relating to achievement of these margins, is realistic.

Most of the discussion about implementation of the model has been with reference to the waste minimisation plans being submitted as part of the tender process. Even so, it is also apparent that the proposed model can provide a basis for waste auditing to ensure that original assumptions were valid and commitments to waste reduction are upheld.

The application of project cost planning principles to create a "waste performance model" appears to allow many situational variables to be taken into account. However, it is apparent that the method requires a degree of openness about the cost structure of a project. Fortunately, this openness is only essential on large material costs. These materials are often competitively available and as such approximate costs are readily available which tends to reduce the need for commercial confidentiality. The availability of similar cost plans from past jobs also reduces the need for confidentiality. Even so, the more transparency in the procurement systems, the more applicable the model is likely to be.

On going research will focus on the identification of waste minimisation strategies and how effectively these strategies meet cost and environmental performance requirements.

References

Bossink, B.A.G. and Brouwers, H.J.H. (1996) Construction Waste: Quantification and Source Evaluation. *Journal of Construction Management and Engineering* Vol. 122, No.1, pp. 55-60

Combined Regional Organisations of Councils, (1996) *Waste Not – A Model Development Control Plan and Local Approvals Policy,* Combined Regional Organisations of Councils, Sydney.

du Blet, R. (1995) Dodging the Dump. *Housing New South Wales*, Housing Industry Association, February/March, Sydney, pp. 96-99.

Hammer, M. and Champy, J. (1993) Re-engineering the Corporation: a Manifesto for Business revolution. Nicholas Brealy Publishing.

Hill, R.C. and Bowen, P. (1997) Sustainable construction: principles and a framework for attainment. *Construction Management and Economics*, Vol 15, No 3, pp. 223-239.

Johnston, H. and Mincks, W. (1995) Cost-Effective Waste Minimisation for Construction Managers. *Cost Engineering*, Vol. 37, No. 1 pp. 31-39

Koskela L., (1994) Lean Construction, *National Construction Management Conference*, Sydney, Australia pp. 205-217.

Kumaraswamy, M. and Dissanayaka, S.M. (1996), Performance – Oriented Building Procurement Systems, *Building Technology and Management,* Vol. 22 pp. 17- 28.

McDonald B. and Smithers, M. (1998) Implementing a waste management plan during the construction phase of a project: a case study. *Construction Management and Economics*, Vol. 16, No 1, pp. 71-78.

New South Wales Government Construction Policy Steering Committee, (1998) *Construct New south Wales – White Paper*, New South Wales Government Construction Policy Steering Committee

New South Wales Parliament, (1995) *Waste Minimisation and Management Act 1995 (NSW)*, Statute, New South Wales Parliament, Sydney

Olympic Coordination Authority, (1996) *Environmental Tender Specifications for Olympic Games Projects and Development at Homebush Bay*, Olympic Coordination Authority, Sydney

Skoyles, E.R. (1976) Materials Wastage – a misuse of resources, *Building Research and Practise*, July/August, pp. 232-242.

Wilson, D. (1996) Stick or Carrot? The Use of Policy Measures to Move Waste Management Up the Hierarchy. *Waste Management and Research*, Vol 14, pp. 385-398.

Appendix A. Causes of Concrete Waste in Typical Rank Order of Importance

Causes of waste	Characteristic of waste variables	Waste source
1. Inaccurate trench formwork.	The less accurate the excavation, the larger the proportional waste. Contributors include • Wrong bucket size, • Earth breakout when removing bucket/auger (worse in dry soils) • Poor depth control where stepping footings • Erosion or collapse of trenches (rain, traffic, non-cohesive soils) • Poorly formed "fill" at beam edges	• Work practise • Site management • Contracts administration
2. Inaccurate edge board formwork.	The greater the length of edge form and the greater the height of edge form, the greater the likelihood of high waste. Contributors include: • The use of plastic vapour barrier as a side restraint • Deep edge beam forms • Pour attention to height control of edge boards	• Work practise • Site management • Contracts administration
3. Depth control.	Waste from the contributors below tends to remain proportionally constant with increases in floor area (e.g. if there was 5% waste on a 100m2 slab then the waste would still be 5% on a 150m2 slab – excluding mistakes and other variables). • Settlement in the ground/fill under the slab causing an unforeseen increase in slab depth • Deflection in suspended slab soffits • Variances associated with normal tolerances in level (typically +/-10mm max.) • Incorrect or omitted use of levelling equipment during pour	• Work practise • Site management • Contracts administration
4. Waste in pump line and subsequent wash out	The smaller the pour the greater the impact. Specific factors include • Hose length • Hopper size	• Work practise • Site management • Equipment technology
5. Concrete escape from formwork	Occurs on an isolated basis. Highest risk is where: • Vapour barrier rips when used as an edge restraint • Concrete escapes under hollow void formers in waffle raft construction.	• Work practise

Notes:
1. Despite the rank order, each cause will vary between projects.
2. The "Characteristics of waste variables" column identifies issues which may cause variation. This should be used as a guide when determining "Ideal", "Average" or "Upper range" waste margins.
3. The first three items are the key areas of concern.

THE IMPACT OF THE DECLINE OF THE ASIAN ECONOMY ON AUSTRALIAN CONTRACTORS: DEVELOPMENT, PROFILE AND STRATEGIES OF AUSTRALIAN CONSTRUCTION IN ASIA

G. R. NAJJAR, M. Y. MAK and M. C. JEFFERIES
Department of Building, The University of Newcastle, Callaghan, NSW, 2308, Australia.
bdgn@cc.newcastle.edu.au

Abstract

The growth of most Asian economies during the 1990's resulted in a steady increase in the number of foreign contractors seeking to capture some of the rapid growth in construction activity. A number of Australian contractors are currently involved, at various stages, in several major construction projects throughout South East Asia. The contracts having been tendered for and negotiated prior to the rapid decline experienced by most Asian Economies in 1997.

The problem facing many foreign contractors, as a result of Asian currency devaluations is yet to be fully recognised. Although many contracts are negotiated in US$ there are still significant impacts on these contractors in terms of payment of subcontractors and suppliers as well as strategic planning for future development.

This paper discusses the impact of the decline in the Asian economy on Australian contractors and describes the development of Australian construction contracting in Asia. In particular, the paper will focus on three leading Australian contractors who are involved in a number of large projects in the region and how their profit levels have been impacted.

Key words: Asia, Australia construction, profile, strategies, impact

Introduction

Australian construction contractors have recognised that, collectively, Asia represents one of the largest markets in the world. Prior to the economic decline, the construction industry in Asia was valued at more than $US400 billion and some Australian contractors were optimistically predicting that this market could represent as much as 50% to 70% of their overall turnover within the next ten years. The ASEAN group of countries alone represented a market valued at more than twice that of the Australian annual $US25 billion construction industry, which is a significantly small proportion of total work available in the Asia Pacific region. (ABS, 1998). Consequently, the prosperity of the Asian construction industry is of major significance for Australia, and all countries in the Asia-Pacific region.

As a result of the decline most levels of construction related businesses have experienced varying degrees of pressure on their profitability and future company growth.

Generalisations made about how the economic decline in Asia has impacted on foreign contractors, however, may be misleading as the impact has not been uniform for all contractors. The variance of impact on contractors may be attributed to a combination of:

1. The specific characteristics of the Asian construction industry.
2. The firms individual characteristics.
3. The firms overall strategic planning.
The primary objectives of this paper are to:

1. Investigate the development of Australian contractors in the Asian region.
2. Evaluate the strategies implemented by leading Australian contractors operating in Asia.
3. Evaluate the impact and outcomes resulting from the economic decline on construction in Asia.

These primary objectives will be addressed by focusing on three of the leading Australian contractors operating in Asia.

Development of Australian construction in Asia

Australian construction companies have been operating in the Asia region since the early 1970's, contributing to some of the largest projects in the region including the $US1.0 billion (approx.) Nam Theun Dam in Laos and the Honk Kong Airport. Lend Lease were one of the first into Asia commencing in 1973 with residential housing in Singapore. Leighton contractors commenced there Asian operations in 1975 trying to capture work generated by the growth in the legal and banking sector and John Holland moved into Asia in 1978 with major civil works projects in Indonesia.

Major Australian building materials companies were also entering operations in Asia, including Boral, CSR, James Hardie and Pioneer. Boral Australia moved into Asian in 1973 becoming a market leader in the Indonesian concrete and plasterboard market. In 1994 CSR became on of the largest concrete suppliers in China by entering in a $US30 million (aprox). joint venture for the construction of four large pre-mixed concrete plants in the northern China to replace traditional site mixed concrete (Forday, 1994). In 1996 James Hardie formed a joint venture with Jardine Davies Inc to build a $US35 million (approx.) fibre cement plant in the Philippines to replace the traditional use of plywood and gypsum plasterboard in framed construction, where there are concerns about durability (Duncan, 1996).

In contrast to South East Asia, Australia contractors and suppliers movement into Japan has been limited to domestic construction. Although more than 1.4 million (approx.) homes are built in Japan every year, 10 times the size of the Australian market only about 100 Australian homes were built in Japan in 1997. This is despite the removal of many trade barriers and substantial Australian and Japanese government assistance to potential investors and exporters. Many Australian contractors had the perception that the Japanese market was too difficult to penetrate. In 1996 the Australian Government criticised the

Building Standard Law of Japan, which requires different design criteria for timber-framed and steel-framed houses and did not recognise Australian plumbing standards. The Australian government is aiming for joint recognition of housing regulations (Duncan, 1997).

Table 1 provides a summary of some of the major projects in which Australian contractors have been involved:

Table 1. Major Australian projects in Asia

Contractor	Project Type	Country	Value $US(Approx.)
Transfield	Nam Theun dam	Laos	$1.0 billion
Concrete Constructions	RCBC tower	Philippines	$35 million
	KLCC fitout	Malaysia	$60 million
	PDB serviced apartments	Malaysia	$150 million
	Baiyoke Tower	Thailand	$110 million
Lend Lease	Apartments	Indonesia	$90 million
	Apartments	Singapore	$40 million
Multiplex	Four seasons hotel	Singapore	$90 million
	Landside Hotel	Malaysia	$50 million
Leighton	Aviation fuel service facility	Hong Kong	$170 million
	Aqueducts	Hong Kong	$60 million
	Residential towers	Hong Kong	$165 million
John Holland	Water privatisation	Malaysia	$220 million
	Friendship bridge	Vietnam	$30 million
	Mining	Indonesia	$65 million
Cordukes	Housing	Thailand	$15 million

Profile of three leading Australian contractors

Investigation of profiles of the leading Australian contractors operating in Asia will provide an indication of their development in the region. The three leading contractors that have been selected for this investigation are:

1. Leighton Contractors
2. Concrete Constructions
3. Lend Lease

1. Leighton

The Leighton Group, operating through its subsidiaries Leighton Asia and Thiess contractors, has been the single largest Australian contractor operating in Asia with $US350 million (approx.) of work annually since 1994. This represented 15% (approx.) of the Groups overall international turnover. Turnover in Asia was mainly resulting from contracts in Hong Kong where Leighton began its operations in 1975 (Leighton Annual Report, 1997).

Leightons largest contract in Hong Kong is the $US170 million (approx.) Aviation Fuel Service Facility at Chek Lap Kok airport. In 1996 70% (approx.) of Leightonís projects in Hong Kong was from infrastructure development, with total work in Honk Kong accounting for 75% (approx.) of its total revenue in Asia. Leighton Asias other projects in Hong Kong include: reclamation in Kulwin, underwater pipelines, housing, advanced earthworks for the Lantau Link, architectural services, museums, hospitals, the construction of one section of a 12.5km highway on Lantau Island and the design and build of two aqueducts involving the construction of an 11.9-kilometre, steel-lined treated-water aqueduct and a one-kilometre raw-water aqueduct. The aqueducts project also includes site formation works, construction of access roads and mechanical and electrical building services works and is due for completion in 2000. Leighton has also has built a close relationship with the Australian supplier, Pioneer, which supplies them with ready-mixed concrete (Ballantyne, 1997).

The opportunities in Hong Kong were seen as far greater by Leighton than in other parts of Asia. Other Australian contractors tended to by-pass Hong Kong for the less developed construction sectors in South-East Asia. Leighton also operated throughout Eastern Asia including, China, Malaysia, Thailand, Vietnam, and the Philippines where it had about 10 projects under way in 1997. Work outside Hong Kong accounted for 25% (approx.) of the overall work in Eastern Asia. Prior to the decline Leighton Asia employed 5,000 people in the region with 100 expatriate (approx.) staff based in Hong Kong, 80% (approx.) of whom were Australian (Callick, 1997).

2. Lend Lease

The Lend Lease Corporation is an Australian based international property and financial services group which has been operating in Asia since 1973. Lend Lease used the recession of the early 1990's in Australia to establish and consolidate joint ventures with significant Asian corporations. Having commenced Asian operations in Singapore, Lend Lease expanded to Jakarta, Bangkok, Kuala Lumpur, and Shanghai with the help of strategic partnerships in each location. In 1997, Lend Lease had a $US140 million (approx.) turnover in Asia which represents 5% (approx.) of its total international turnover (Lend Lease Annual Report, 1997).

In 1997, Lend Lease had six development projects in Asia with the majority of activity in Singapore and Indonesia and smaller operations in China, and Thailand. The largest of Lend Leases projects to date was a $US40 million (approx.) investment in a joint venture with Indonesian company Sinar Mas. The 50-50 joint-venture PT Lend Lease Graha Indonesia (LLGI) commenced in 1994 to provide integrated project management and design and construction of industrial, manufacturing and commercial projects. LLGI was involved in the "Grand Cempaka Apartments" project in Jakarta and was participating in the next stage, for Construction of a retail commercial development, building 500 shops and offices (Lend Lease Interim Report, 1997).

Also, in 1997 Lend Lease undertook a joint venture to open a chain of serviced apartments throughout the Asia-Pacific region with the Singapore-based Oakwood International Ltd, a subsidiary of the United States-based R&B Realty Group. This was a 5 year plan aimed at opening about 10,000 apartments in 34 properties throughout major cities including Shanghai, Beijing, Singapore, Hong Kong, Sydney, Manila, Tokyo and Bangkok. The joint venture, was to be 50% owned by Lend lease. Other developments in Singapore included Admiralty Industrial Park (Smith, 1997). Lend Lease is now also expanding into funds and assets management (Lend Lease Annual Report, 1997).

3. Concrete Construction

Concret Constructions has been operating in Asia since 1987. Commencing in Guam where they extended and refurbished the $US70 million (approx.) Hilton Hotel they moved into Thailand in 1992, followed by Indonesia, Malaysia and Vietnam where they constructed the Australian Embassy in Hanoi. Total exposure for Concrete Constructions was $US200 million (approx.) in 1997 (Korporaal, 1994).

The largest of Concrete Constructions project in Asia to date has been involvement in the $US120 million (approx.) Baiyoke Tower II in Bangkok. The tower was to be Asia's tallest reinforced concrete structure reaching a height of 460 metres, 8m taller than the Petronas Towers in Kuala Lumpurís and claimed as the worldís tallest building. The building in the heart of Bangkok was to symbolise the economic modernisation of Thailand, and involved a number of Australian consultants and contractors (Sexton, 1997).

Concrete Construction has also been involved the Kuala Lumpur City Central (KLCC) fitout of the concert hall and art gallery valued at $US60 million (approx.), the PDB 39 storey serviced apartments also in Malaysia and most recently the $US200 million (approx.) RCBC tower in the Philippines.

Strategies of Australian contractors in Asia

The focus on long term development in Asia has insured that the leading contractors have developed and implemented key strategy for successful operations in Asia.

The general strategies implemented by the leading Australian contractors may be summarised as:

1 Focus on specific clients.
2 Focus on specific project types.
3 Undertake joint ventures.
4 Give the company strong local presence.
5 Focus on work which it they had significant expertise.

The implementation of the strategies to focus on specific clients and project types has been best implemented by Leighton, who focused its bidding on long-term infrastructure projects that were supported by the government. The Hong Kong Governments select list of pre-qualified contractors was targeted by Leighton who recognised this as a key for successful operations. Leighton focused on undertaking the measures required for becoming a select contractor, for which they were successful, and could thus tender for all Government related work. The objective by Leighton on pursuing this strategy was that it would see them operating in a market where: i) the tendering process was transparent, ii) guarantee that the lowest tenderer, who conforms to the conditions, would win the job; iii) ensure involvement in long term projects; iv) the projects were underwritten by Government and iv) contracts were written in Hong Kong dollars pegged to the US currency and thus insulated from the currency instability.

The strategy to undertake joint ventures has been implemented by most of the leading contractors. Lend Lease has entered several joint ventures wilth local partners, including Sinar Mas in Indonesia, Technology Parks in Singapore, and Mitsubishi, LG Corporation and Li & Fung in China. Concrete Constructions achieved their objective of joint ventures with local companies including the Central Group of Companies in Thailand, the Low Yat Group in Malaysia, Pt. Duta Graha in Indonesia and the Hanoi General Construction Company in Vietnam. Concretes also undertook the acquisition local companies with the Malaysian construction group I & P Fletcher in 1994. Concrete Constructions also considered joint ventures with other Australian contractors including John Holland and Transfield. The aim being to maximise the economies of scale achievable in terms of technology, expertise and resources (Korporaal, 1994). Leighton Asia also undertook joint venture projects with other international contractors including Japanese contractor Kumagai Gumi Co. Ltd. in 1997 to design and build two aqueducts in Hong Kong.

The strategy of giving the companies a strong local presence has be achieved with most leading contractors establishing head offices and involving locals as employees and directors. Contractors recognised the need to provide the Asian industry with a positive signal that they were an organisation with a long-term interest in the region and that they were not just there because of the recession experienced in Australia during the early 1990ís. The appointment by Concrete Constructions, in 1994, of leading Hong Kong businessman and a director of Hong kong and Shanghai Banking Corporation Mr Paul Cheng to the board of the Concrete Constructions Group and its subsidiary Concrete Constructions Investments (Asiapac) (Harley, 1994), is a typical example of Australian contractors commitment to giving their companies a strong local presence.

The strategy of contractors focusing on work which it they had significant expertise has been implemented by most of the leading contractors. Concrete Constructions were able to focus on work in which they had significant experience with in Australia namely commercial CBD and high-rise buildings. The 39 storey PDB serviced apartments in Malaysia, the RCBC tower in the Philippines and the Baiyoke tower in Thailand are all examples of high rise construction in which Concrete Constructions has been focusing its expertise in. Similarly, Leighton have focused their expertise on civil and infrastructure construction for which they have significant experience in Australia. Lend Lease has significant expertise in both construction development and property investment and have developed this portfolio in Asia as part of their global development and investment strategy. An example is its $US10 million (approx.) investment in developing a produce markets-type project on 160 hectares north of Bangkok in early 1997 (Lend Lease Annual Report, 1997).

Impact on Australian contractors

The impact of the Asian crisis has had a varied impact on the overall performance of individual companies and the Australian construction industry in general. The impact on Australian contractors may be classified into the following two categories:

1. Direct costs
2. Opportunity costs

The direct costs, in financial terms, been minimal for larger contractors with construction activity in Asia to date representing, at best, only about 10% -15% of the overall company turnover with most only being exposed to between 5%-10%. This is reflected by the value of larger contractors share price remaining stable and no adjustment to their credit ratings at this stage.

Leighton, although having to write down $US6 million (approx.) from bad debts and currency losses, still achieved an operating profit of $US14 million (approx.) in 1997 with the region actually growing by approximately 25%. for a total work value of $US480 million (approx.). However, Leighton has lost three building contracts in Thailand. and two goldmining contracts in Malaysia and the Philippines. In Indonesia, its second biggest Asian market, and where most of the group is contracts were written in $US the impact has been minimal (Leighton Annual Report, 1997). The Honk Kong Governments plans to continue its significant spending on infrastructure over the next five years will see Leighton continue to operate in Asia. Leighton is continuing to develop its construction methods for quicker and cheaper housing and was recently successful in a $US140 million (approx.) bid to construct a major housing project for the Hong Kong Housing Authority. The project will include construction of six 41-storey residential towers, a five-storey car park and a kindergarten (Chandler, 1998).

Lend Lease were also able to achieve a profit of $US15 million (approx.) from its Asian operations in 1997. This resulted mainly from the sale of some of its Singaporean investments. However work has stopped on their biggest Asian project at the Cempaka industrial park in Jakarta which was under review with provision of $US5 million (approx.) being made against the project. The Thai Markets Stages II and III are also under review with provisions of $US4.5 million (approx.) made against the project. Lend leases exposure to the currency crisis was particularly significant in Indonesia and Thailand where the exchange rate fluctuations were $US40 million (approx.) in 1997. Lend Lease maintained a fully hedged position during the period and as such protected its equity position and profits from the impact of currency fluctuations, with the exception of the Thai Baht from 1 July 1997 to 8 July 1997 due to Bank of Thailand currency trading restrictions (Lend Lease Annual Report, 1997).

Despite ambitions of increasing their work to $US450 million (approx.) by 2000 Concrete Constructions have not survived the Asian crisis as well as their contemporaries. They lost $US2.4 million (approx.) in 1997 winding up an operation in Guam with larger losses pending following the resolution of disputes over work on the Baiyoke Tower in Bangkok. Although not directly attributed to the economic decline the Baiyoke Tower still had a final payment of $US35 million (approx.) unpaid after Concrete Constructions removed its crane from the top of the building in 1996. Concrete Constructions claimed that payments were consistently late throughout the contract period and rejects suggestions it removed the crane as a tactic to extract payment. Concrete Constructions had been prepared to leave the crane on top of the building so the contractor selected to build the communications tower did not have the considerable expense of erecting another. But after delays on the decision to award the broadcast tower contract, and with no date for when work might start, Concrete Constructions removed the crane. In 1997 an independent private assessor, Project Planning Services, has declared that Concrete Construction had not filled its obligations to the client Thai tycoon and managing director of the Land Development Co, Mr Panlert Baiyoke. Mr Panlert rejected suggestions that the economic downturn had affected his ability to pay (Sexton, 1997). However, despite these losses Concrete Constructions will stay in Asia hoping to capitalise on their experience of being one of the few contractors in the region with the technical and managerial skills of high rise construction.

The financial impact has been more significant on medium sized contractors as well as building materials companies who, with high over head costs and lacking the diversification in project numbers, or company size and structure to support a portfolio of development and investment that would absorb the short to medium down turn experienced in the Asian market. For example, mid-sized contractor Cordukes Ltd focused its operations mainly in Thailand and was unable to sustain its operations after the significant decline in the Thai Baht. Cordukes experienced a 94% (approx.) decline in profit and a s a result was forced to sell its half-share in the operations to its joint-venture partner (Allen, 1997). CSR had experienced an after tax loss of $US96 million (approx.) in it Asian operations which also included plasterboard facilities in Indonesia. It was also consolidating its operations in Taiwan into a single site (Hoyle 1997).

The most significant impact on opportunity costs, and that most difficult to measure in financial terms, are those associated with the momentum and critical mass that contractors have forfeited in return for consolidation and caution in bid selection. Contractors are now focusing on actual work rather than greedily rushing into costly bids for speculative work. Reduction in the amount of speculative work has resulted in a more competitive market and a decrease in costs for locals. Furthermore, the opportunity cost lost from the possible down sizing of operations associated with the economies of scale and local presence that was being constructed through joint ventures and local offices will be difficult to quantify and have a significant impact on future development.

Direct cost on the Australian construction industry has come from Asian inability to honour commitments to construction and investment in Australia. Many Asian investors have had to repatriate investments in Australia to meet commitments at home. This has resulted in a significant impact on the forward planning of work for which a number of leading Australian contractors had been committed, e.g. Multiplex.

Further direct costs are those associated with the potential decline in privately funded infrastructure construction in Australia such as mining. The Asian downturn has caused some mining projects which are under consideration to remain that way for longer than they otherwise would. Projects which have been impacted include the Rio Tinto approved $US405 million (approx.) project to develop the 15m tonne-a-year Yandicoogina iron ore mine in the Pilbara as well as a host of several other possible new iron ore projects in WA (Howarth, 1997). This has resulted in a significant impact on West Australian economic growth of 7.25% in 1997 with the majority of money in construction of mining operations being channelled through major Australian construction (Kitney 1997).

Conclusion

There are several key outcomes from the economic decline in Asia on construction that need to be considered:

1. Identification of weaknesses.
2. Evaluation of existing strategies and policies.
3. Future opportunities.

The recent economic decline in Asia has given us the opportunity to evaluate the strategies implemented by Australian contractors over the past 25 years of development in the Asian construction market. These initial 25 years may be considered as the first phase of development in a region where, despite the current down turn, development will increase well into the next century.

The down turn has given contractors the chance to re-evaluate their position in the Asia market and examine strategies required for successful operations. This will enable contractors to implement directions for the future so as to manoeuvre themselves into a position where they may capture a market that represents collectively one of the largest in the world. Successful operators of the future will be the contractors that have identified the areas where they have comparative advantages and develop corporate structures that facilitate economies of scale and maximise factors such as proximity, construction technology, human and physical resources, management and culture.

The world construction markets are becoming increasingly polarised into regional areas within the global economy. Asian governments need to develop a construction industry which provides transparency and accountability to encourage development and investment in the future. The current downturn in a number of Asian countries provides the opportunity for construction industry reform and adoption of regulation and policy that may reduce the problems encountered by internal crisis or external shocks

Large scale infrastructure projects will be a key to growth and recovery in Asian and the World Bank will continue to support this sector and is predicting that infrastructure investment into Asia will be significant. Although stalled in the short term, funds will be injected into the region at some stage in the near future, contractors need to be in a position to move into the market when it re-emerges.

References

Allen, L (1997) Big builders still confident of Asian projects. *Australian Financial Review,* 30 October.

Australian Bureau of Statistics (1998) Year Book Australia.

Callick, R (1997) Why Leighton likes being there. *Australian Financial Review,* 28 May.

Chandler, (1998) Leighton in $225m HK housing deal. *Australian Financial Review,* 16 June.

Ballantyne, T (1997) Australian firms play big role in new airport. *Australian Financial Review,* 29 April.

Duncan, C (1997) Builders chase Japanese contracts. *Australian Financial Review,* 20 February.

Duncan, C (1996) James Hardie forms Philippines joint-venture, 26 November.

Forday, G (1994) CSR in large China concrete venture. *Australian Financial, Review* 22 February.

Harley, R (1994) Appointment cements concretes Asia push. *Australian Financial Review,* 21 August.

Howarth, I (1997) Start for $705 million iron ore project. *Australian Financial Review,* 18 September.

Hoyle, S (1997) Asia misreading $96 million writedown. *Australian Financial Review,* 15 October.

Kitney, D (1997) Asia crisis threat to W.A. resources boom. *Australian Financial Review,* 4 November.

Korporaal, G (1994) Concrete constructions looks east. *Australian Financial Review,* 6 April.

Leighton Annual Report (1997).

Lend Lease Annual Report (1997).

Lend Lease Interim Report (1997).

Lend Lease Annual Report (1996).

Sexton, J (1997) SE Asia edifice complex unravels. *Australian Financial Review,* 31 October.

Smith, F (1997) Lend lease moves into serviced apartments. *Australian Financial Review,* 21 January.

Williams, L (1994) The need for a competitive edge. *Sydney Morning Herald,* 3 March.

A COMPARISON OF CLIENT AND CONTRACTOR ATTITUDES TO PREQUALIFICATION CRITERIA

ANTHONY MILLS
Department of Architecture, Building and Planning, University of Melbourne, Australia.
a.mills@architecture.unimelb.edu.au

MARTIN SKITMORE
School of Construction Management and Property, Queensland University of Technology, Australia. RM.Skitmore@qut.edu.au

Abstract

Contractor prequalification is concerned with assessing the likelihood of contractors meeting client and project requirements. The criteria used in this assessment have been developed in a largely idiosyncratic manner to date and with little or no consultation with the contractors affected. As a result, contractors are faced with a variety of calls for information by prequalifiers, the collection of which can be quite costly. This is leading to expensive duplication of effort by contractors in providing what is often similar information but in different formats. Furthermore, previous research has shown that the benefits of the information to prequalifiers is uncertain - many prequalifiers analyse the information in only a cursory manner. What is needed is some form of cost-benefit analysis to be carried out which will establish a common set of criteria for all to use.

As a precursor to this, the research described in this paper compares the different attitudes of both prequalifiers and contractors to prequalification criteria commonly in use in the Australian building industry. This was carried out via a postal questionnaire involving 49 contractors and 15 prequalifiers across Australia. The respondents were divided into three groups; (1) contractors doing work for mainly private sector clients, (2) contractors doing work for mainly private sector clients, and (3) construction prequalifiers (clients).

The results show that both clients and contractors have divergent opinions on the importance and value of the criteria in use. The possible reasons for these differences are discussed and the likely implications for future research in the topic.

Keywords: Benefits, costs prequalification, criteria, perceptions, standardisation.

Introduction

Contractor selection is one of the most important aspects of project management decision making. In the face of multiple and conflicting client and project goals (e.g., time, cost, quality and risk), this necessarily involves the consideration of several criteria for selection. The criteria that clients most often use has been the subject of several studies in an attempt to compile a comprehensive rank ordered list of "universal criteria" for general use Hatush and Skitmore (1997a).

Until now, it is the clients' views only that have been solicited in the search for such a list. Liston (1994) however, suggests that the criteria for selection should also include those which contractors also believe to be indicators of good performance. There are several obvious reasons for this:

- Contractors, being often more experienced than clients in such matters, may be in a better position to judge the relevance of potential performance criteria.

- Universal criteria may provide contractors with a more consistent basis upon which to tender or negotiate for work and a better basis for marketing their abilities CIDA (1995). As such, contractors are partial stakeholders in the process and thus, it can be argued, are entitled to have some input in the type of criteria used.
- Multiple criteria contractor selection is known to be a very subjective process *Russell et al* (1992a) Holt *et al* (1994) Liston (1994) Drew and Skitmore (1993) Hatush and Skitmore (1997a) and therefore not always fair to the contractors under consideration
- Most clients are still using *ad hoc* criteria Holt (1994a) Hatush and Skitmore (1997b) Holt *et al* (1994a) which does not give contractors confidence that the system is sufficiently well considered.
- The criteria used are client oriented CIDA (1995) and may therefore more reflect the client's predisposition more than the likely performance of contractors. They may even be centered on the individual client representative's own personal prejudices or political ambitions.
- It is known that construction managers have different views to public clients on the subject Russell *et al* (1992)
- The criteria measures are often fuzzy and imprecise Holt *et al* (1994b)
- Even public and private sector clients have different criteria preferences Russell *et al* (1992)

Of course, it can be argued that the client is footing the bill, *he who pays the piper should call the tune*. Anecdotal evidence suggests that, if each client adds a little to the project acquisition costs of successful (and unsuccessful) contractors the net result will be an increased industry overhead which must ultimately be passed back to the clients in the form of increased industry price levels.

What is needed is some form of contractor selection process that produces the best cost-benefit ratio. In this situation of course, the costs and particularly the benefits are difficult to estimate. Nevertheless, it is clear that it is the total costs and total benefits to all stakeholders that are at issue.

As a starting point, it is necessary to extend the existing list of selection criteria by considering the contractors' viewpoint. Asking contractors for their opinion on the importance and usefulness of the criteria gives some measure of benefits and costs, albeit mainly to the contractors. The next stage is to compare these contractors' views with the clients. If they are of a like-mind, then the problem is greatly simplified as there will be a homogeneous consensus group. If they are not so like-minded, then it may be necessary to start to find ways of incorporating these differences into the process.

Previous research

As mentioned above, despite the several studies in contractor selection criteria, very few have considered non-client stakeholder views to date. Russell *et al* (1992) analysed the attitudes of three types of client organisations: public owners, private owners and construction managers with results that "... indicate a significant statistical difference among public owners or construction managers, while public owners and construction managers responded similarly." The only other study to include non-clients was that of CIDA, who developed prescriptive criteria that were "subject to a broad industry consultation CIDA (1993) and therefore can be assumed to incorporate some degree of stakeholder views.

There have been several studies into the importance of criteria in the prequalification decision including; Liston (1995), Russell (1992), Holt *et al* (1994b), CIDA (1995) and Hatush and Skitmore (1997a). Each author developed a list of criteria that they considered were the most significant decision making factors. After exhaustively compiling an aggregated list of all possible criteria it was discovered that in many instances considerable overlap occurred between the criteria used by different authors. In addition, many of the criteria used by other researchers were based on local conditions, and were therefore, not appropriate to the Australian construction industry. Consequently it was decided that the CIDA (1995) model represented the most relevant and comprehensive set of criteria, and this has been used in this research.

Data collection

The research instrument was a postal questionnaire based on the CIDA criteria which was sent to groups of contractors and clients. Firstly, a pilot study undertaken comprising three domain experts who where contacted and asked to examine the layout, order and intelligibility of the questionnaire. In addition, the questionnaire was sent to an expert on survey design for evaluation. All comments were then incorporated into the final questionnaires.

The survey comprised 39 questions (coded B301-B339) relating to prequalification sub-factors. Each of the sub-factors can them be aggregated into one of the nine major Decision Factors. Respondent were asked to express their opinion of importance of each criteria on a Likert scale of 1 to 7.

The final questionnaire was sent to each individual with a covering letter and a stamped/self addressed envelope. Questionnaires were sent out to 158 companies in the construction industry throughout Australia. There were a total of 64 returned questionnaires giving a response rate of 41%.

After preliminary analysis of the data the number of useable questionnaires for analysis was 24 No. Contractors undertaking mainly *private* sector work, and 25 No. Contractors undertaking mainly *public* sector work, 15 No. public sector *clients*. The response rate was considerably higher than most postal questionnaires, which normally only attract return rates of between 20-30%, in addition according to Moser and Kalton (1971) if the response rate is less than 30% the results may be considered subject to non-response bias. However, in this case the response rate was higher and the each of the significant groups were well represented, therefore the opinions of all sections of the entire population are considered to be adequately reflected.

Most of the respondents to the questionnaire occupy senior management positions within their firms. If contractors are considered, most firms (96%) have a turnover of greater than AU\$ 1M, and 43% exceed AU\$ 5M. If clients are considered, all clients have capital works budgets that exceed AU\$ 50M. Therefore it was suggests that all respondents are in a position to have an understanding of the prequalification process and the subsequent issues involved.

Analysis and results

The results (Table 1) Mean Score and Rank by Group, shows that *Details of past projects* is the most important factor in prequalification decision making by all groups, and that *Success of completed projects, Past project time performance*, and *Bank reference* also seem to be important considerations by all groups in the survey. However, the table also indicates that in many instances each group has quite different views about the importance of some factors. For instance *Company organisation/history* was ranked second by *public* contractors and sixth by *private* contractors, but only twenty ninth by *clients*. This suggest that there may be a range of factors that have significantly differently levels of importance to each of the stakeholders in the prequalification process.

The objective was to find criteria where the importance is significantly different between each group of respondents. A Discriminant Analysis was undertaken on the 39 prequalification decision sub-factors, for the three groups of respondents; i.e. Private contractors, Public contractors and Clients. If differences exist, it will indicate which group has different responses to the criteria used for prequalification.

Discriminant Analysis is a statistical process that identifies variables that are important for distinguishing among groups, and which can then be used to develop a procedure for predicting group membership of new cases whose group is undetermined. Norusis (1994). The concept underlying discriminant analysis is a fairly simple, combinations of the independent, or predictor, variables that can be formed into a linear function, this then serves as the basis for classifying cases into one of the groups.

The interpretation of the discriminant weights, or coefficients, is similar to that of multiple regression analysis. The value of the coefficient for a particular predictor depends on the other predictors included in the discriminant function. The signs of the coefficients are arbitrary, but they indicate which variable values result in large and small function values. The relative importance of the variables can be obtained by examining the absolute magnitude of the standardised discriminant function coefficients. Generally, predictors with relatively large standardised coefficients contribute more to the discriminating power of the function, as compared to predictors with smaller coefficients.

However, as Malhotra (1993) states "if multicollinearity in the predictor variables exits, there is no unambiguous measure of the relative importance of the predictors in discriminating between the groups". When there is a high degree of correlation between some of the independent variables in the data, interpretation of the results is difficult. This is because one variable will assumed all the discriminating power of the other correlated variable. As a result interpretation of the results should be done with considerable caution.

The bivariate correlation was undertaken on all variables in the study, and the results showed that there were a number of variables that were highly correlated (Table 2 Questions having Multicollinerity) This showed that some of the questions in the survey seemed to elicit similar responses especially when they were grouped under major headings. For instance, the responses to questions B337, B338, and B339 were highly correlated, all questions refer to non-compliance or breaches of the law, and are grouped under the Decision Factor of Legislative Compliance. Consequently, it seems reasonable to delete questions B338 and B339, the leave the effect of Legislative Compliance to be taken up by question B337. The variables that remain are known as surrogate variables, and they represent the combined effects of both variables.

The following variables have been removed from the Discriminant Analysis due to the effects of excessive multicollinearity.

Table 1 Mean Score and Rank by Group

Ref	Decision Factor	Private	Rank	Public	Rank	Clients	Rank
B301	Company organisation/history	5.46	6	6.04	2	4.67	29
B302	Details of past projects-track record	6.04	1	6.16	1	6.20	1
B303	Current load	5.35	8	5.28	7	5.47	12
B304	Current directors	4.50	24	4.92	14	4.27	36
B305	Current management & administration	4.96	13	5.20	9	4.67	29
B306	Employee qualifications	4.75	16	5.08	12	5.00	22
B307	Major plant & equipment	4.04	37	3.84	37	4.00	39
B308	Success of completed contracts	5.58	4	5.76	3	6.13	3
B309	Geographic location of project	4.13	34	4.32	27	4.93	23
B310	Directors statement	4.36	27	4.56	19	4.16	37
B311	Asset and liabilities	5.01	12	4.92	14	5.86	4
B312	Profit & loss statement	5.08	11	4.56	21	5.79	7
B313	Movement of assets for year	4.35	28	4.32	27	5.29	17
B314	Cash flow forecast	4.68	17	4.36	24	5.29	17
B315	Bank reference	5.57	5	5.40	5	5.86	4
B316	Credit reference	5.35	8	5.16	10	5.71	8
B317	Turnover history	4.65	18	4.56	19	5.43	13
B318	QA certification	4.17	33	4.36	24	4.73	26
B319	Actual quality achieved in past	5.88	3	5.12	11	5.33	15
B320	Type of quality program	4.25	30	4.21	31	4.73	26
B321	OH&S key personnel	4.96	13	4.72	16	4.53	32
B322	Actual safety level achieved	5.46	6	5.32	6	5.67	9
B323	Type of safety program	4.21	31	4.60	18	5.20	20
B324	Past project time performance	6.00	2	5.60	4	6.20	1
B325	Management level utilized on past projects	4.58	21	5.04	13	5.67	9
B326	Reason for variance of time & cost in past	4.63	20	4.72	17	5.53	11
B327	Scheduled performance of past projects	5.33	10	5.24	8	5.80	6
B328	Human resources management process	4.54	23	4.30	29	4.80	24
B329	Labor relations statistics over last year	4.00	39	4.34	26	4.47	33
B330	Compliance with labor legislation	4.58	21	4.26	30	4.60	31
B331	Company training program	4.42	25	3.96	35	4.40	34
B332	expenditure on skill formation	4.08	36	3.80	39	4.13	38
B333	Skill formation policy & strategy	4.33	29	3.84	37	4.33	35
B334	No. of claims on previous projects	4.80	15	3.94	36	4.73	26
B335	Explanation of previous claims	4.03	38	4.01	34	5.20	20
B336	No. of claims referred to arbitration/litigation	4.21	31	4.44	22	5.33	15
B337	Record of conviction/non-compliance with law	4.65	19	4.16	32	5.40	14
B338	Reason for convictions/non-compliance with law	4.10	35	4.05	33	5.27	19
B339	Procedures to avoid futures breaches of law	4.39	26	4.42	23	4.80	24

Table 2 Questions having Multicollinerity

Decision Factor	Removed	Remain
Financial Capacity	B311	B312
Time Performance	B227	B226
Skill Formation	B332, B333	B331
Claims History	B335	B334
Legislative Compliance	B338, B339	B337

The purpose of the research is to discover if there are differences in the views of contractors and clients, the discriminant function appears to clearly separate the groups which suggests that clear differences exist. When the discriminant analysis was run on the three groups, and the results (Table 3) show that it is effective in separating the groups. The Eigenvalues of 2.74 (Function 1) and 2.2975 (Function 2) indicate that they are good discriminators.

Table 3 Canonical Discriminant Functions

Function	Eigenvalue	% Variance	Wilks' Lambda	Significance
1	2.6081	59.43	0.099966	.0010
2	1.9770	42.13	0.359619	.0362

The formula for discriminant analysis is similar to a simple linear equation, it is sometimes tempting to interpret the magnitude of the coefficients as indicators of the relative importance of the variables. However, it is far better to use the standardised coefficients which have been recalculated to a mean of zero (0) and standard deviation of one (1). Tabachnick and Fidell (1996)

Table 4 Standardised Canonical Discriminant Function Coefficients

Variable	Function 1	Function 2	Variable	Function 1	Function 2
B301	-1.1795	.41965	B319	.03752	-.66546
B302	.16908	.23790	B320	.36537	.32810
B303	-.44236	-.73869	B321	-.79597	-.47617
B304	.09922	.39629	B322	.62876	.70904
B306	-.59064	.33293	B323	.43838	.30237
B307	.35345	.23040	B324	.10095	-.53772
B308	.19616	-.05513	B325	.20470	.37983
B309	.65359	.82692	B326	.15336	.87836
B310	-.43416	-.65758	B328	-.01557	.70584
B312	.60191	-.91982	B329	-1.00897	.08808
B313	.19860	.54883	B330	.19098	.00733
B314	-.76854	-.61885	B331	-.25818	-.88376
B315	-.76854	-.61885	B334	-.04964	-.88128
B316	.39996	-.56647	B336	-.00227	.74814
B317	27079	-.21508	B337	.30611	.06073
B318	-.07087	.80000			

The actual sign (+/-) of the standardised coefficients are arbitrary, the negative coefficients could just have well been positive if the other signs were reversed. By looking at the groups of variables that have coefficients of different signs, it is possible to determine which variable values result in large or small function values. Thus, large positive coefficients will tend to increase the function score, and large negative coefficients tend to decrease the function score. The results Table 4 standardised canonical discriminant function coefficients shows that B301 (-1.17956) had the largest absolute value for Function 1, and B312 (-.91982) had the largest value for Function 2

It is also possible to determine the effectiveness of the analysis by observing the spatial separation of the groups based on discriminant functions. The Territorial Map (Figure 1) forms a 'star' pattern, with each group (1-*private*, 2-*public*, 3-*clients*) separated from the others. If Function 1 (x-axis) is considered, the centroids of the two groups of contractors (1-*private*, 2-*public*) are close, while the centroid of the clients (Group 3) is well separated. This indicates that Function 1 reflects the divergence of views between *contractors-clients* groups. Similarly, function 2 (y-axis) also shows that the centroids between the *public* and *private* contractors are well separated, while the client centroid is in between. This indicates that Function 2 is used to identify the discriminating variables between the *public-private* contractor groups. The Territorial Map (Figure 1) shows that the groups have their centroids clearly separated from the others, and there is no overlap between the membership of the groups. This again suggests that the respondents to the survey have distinctly different views on the importance of various prequalification criteria.

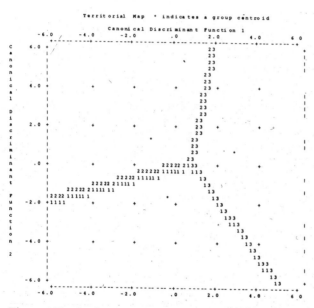

Figure 1 Territorial Map

Once the discriminant scores were computed the cases in the data was assigned to a particular group, this is then compared with the actual group membership which is already is known, and the accuracy of the classification is determined. Table 5 Classification of results shows that the cases are reasonably accurately classified by the above two functions (i.e. 92.2%), the discriminant functions clearly identify the groups based on the responses in the survey.

Table 5 Classification of results

Actual Group	No of cases	Predicted Group A	Predicted Group B	Predicted Group C	Percentage Correct
A (Private)	24	21	1	1	87.5%
B (Public)	25	1	23	1	92.0%
C (Clients)	15	0	0	15	100%
% Correct					92.2%

Table 6 Discriminating criteria between clients and contractors

Variable Code	Decision Factor	Decision sub-factor	Function 1 Coefficient *
B301	Technical Capacity	Company organisation/history	-1.17956
B329	Human Resources Management	Labor relations statistics over last year	-1.00897
B321	Occupational, Health & Safety	OH&S key personnel	-.79597
B314	Financial Capacity	Cash flow forecast	-.76854

* Standardised Canonical Function Coefficients

Discussion

Clients and contractors have different views on the importance of prequalification information. The results (Table 7) shows criteria that best discriminate between the views of clients and contractors. It can be seen that *company organisation/history* has the largest coefficient and therefore represents the most widely different view. For instance, from Table 1, both *private* (Rank 6) and *public* contractors (Rank 2) believe that *company organisation/history* to be an important technical indicator. The *clients* on the other hand, rank this criteria one of the least important (Rank 29) suggesting that a companies organisation and history are almost irrelevant in making prequalification decisions. Another divergent criteria *OH&S key personnel* also follows a similar pattern, clients consider it to be relatively unimportant (from Table 1) while contractors give it a higher ranking and therefore believe it to be an important prequalification decision making factors.

Table 7 Discriminating criteria between private & public contractors

Variable Code	Decision Factor	Decision sub-factor	Function 2 Coefficient *
B312	Financial Capacity	Profit & loss statement	-.91982
B331	Skill Formation	Company training program	-.88376
B334	Claims History	No. of claims on previous projects	-.88128
B326	Time Performance	Reason for variance of time & cost in past	.87836
B309	Technical Capacity	Geographic location of project	.82692

* Standardised Canonical Function Coefficients

The divergence of views amongst contractors is shown by reference to the coefficients in Function 2. The results highlights the fact that they may not be a homogeneous collection of firms. Scrutiny of the standardised coefficients (Table 7) indicates that *profit & loss statement, company training program* and *No. of claims on previous project* are some of the best discriminators of the attitudes of *private* contractors. If the ranks in Table 1 are also considered it shows that these factors are ranked relatively highly by *private* contractors but relatively lowly by *public* contractors. This may represent the factors that characterise *private* contractors, and it may be reasonable to suggest that private clients expect these types of characteristics from their contractors.

Summary and conclusions

Universal criteria may provide contractors with a more consistent basis upon which to tender or negotiate for work, and a better basis for marketing their abilities CIDA (1995). As such, contractors are partial stakeholders in the process and are entitled to have some input in the type of criteria used. Also the benefits of the prequalification process as a whole may improve if the selection of contractors is based on criteria that contractors themselves believe are important.

The outcomes of this research was to demonstrate that the views of contractors are quite different to those of clients. This suggests that, if the views of all prequalification stakeholders were solicited, the end result would be quite different to that which occurs in practice at present.

Multiple criteria contractor selection is known to be a very subjective process Russell *et al* (1992) Holt *et al* (1994) Liston (1994) Drew and Skitmore (1993) Hatush and Skitmore (1997a) and therefore may not always be fair to the contractors under consideration. The criteria used are client oriented CIDA (1995) and may therefore more reflect the client's predisposition more than the likely performance of contractors. They may even be centered on the individual client representative's own personal prejudices or political ambitions.

If contractors are not engaged in the process it is possible that criteria which is important to the prequalification decision are being overlooked by clients. Most clients are still using *ad hoc* criteria Holt *et al* (1994a) Hatush and Skitmore (1997b) Holt *et al* (1994a) which does not give contractors confidence that the system is sufficiently well considered. This research supports that view and suggests that the criteria that is ranked highly by contractors and low by clients are candidates, this includes; *company organisation/history*, and *OH&S key personnel*.

It is already known that construction managers have different views to public clients on this issue, Russell *et al* (1992), this research has shown that the contractors are a heterogeneous collection of firms that vary in their opinions of the importance of prequalification criteria. The *private* contractors consider the factors like; *profit and loss statement, No. of claims on previous projects* to be important, *public* contractors on the other hand do not put credence on these factors when making prequalification decisions. This may be because the preferences shown by *public* and *private* contractors represents their perception of their client's expectations.

Acknowledgements

The authors would like to thank the Office of Building Policy, Department of Infrastructure, Victoria and the Office of Housing, Victoria. And all industrial collaborators for providing support and information.

References:

CIDA(1993) *The Australian Construction Industry Pre-Qualification Criteria for Contractors and Subcontractors*, Construction Industry Development Agency, Commonwealth of Australia.
CIDA(1995) *Prequalification Criteria for Contractors: The Australian Construction Industry*, Construction Industry Development Agency, Commonwealth of Australia.
Drew, D., Skitmore, R. M. (1993) 'Prequalification and Competitiveness', *OMEGA International Journal of management Science*, 21, p. 363-75.

Hatush, Z., Skitmore, M (1997a) 'Criteria for contractor selection', *Construction Management and Economics,* 15, p. 19-38.

Hatush, Z., Skitmore, M. (1997b) 'Evaluating contractor prequalification data: selection criteria and project success factors', *Construction Management and Economics,* 15, pp. 127-147.

Holt, G., Olomolaiye, P., Harris, F (1994) 'Incorporating project specific criteria and client utility into the evaluation of construction tenderers', *Building Research and Information,* 22, 4, p. 214-221.

Holt, G., Olomolaiye, P., Harris, F. (1994a) 'Factors Influencing U.K. construction clients' choice of contractor', *Building and Environment,* 29, 2, p. 241-248.

Holt, G., Olomolsiye, P. Harris, F. (1994b) 'Evaluating prequalification criteria in contractor selection', *Building and Environment,* 29, 4, p. 437-443.

Liston, J.W.(1994) *Prequalification of contractors,* Physical Infrastructure Centre.

Liston, J.W. (1995) 'Prequalification of contractors', *Transactions of Multi-Disciplinary Engineering, Institution of Engineers, Australia,* 1, p. 17-21.

Malhotra, N.K. (1993) *Marketing Research: An applied orientation,* Prentice Hall, New Jersey, USA.

Moser, C.A., Kalton, G. (1971) *Survey methods in social investigation,* Heinemann Education,

Norusis, M.J. (1994) *SPSS Professional Statistics 6.1,* SPSS Inc., Prentice Hall, Chicago, Illinois.

Russell, J.S., Hancher, D.E. Skibniewski, M. J. (1992) 'Contractor prequalification data for construction owners', *Construction Management and Economics,* 10, p. 117-135.

Tabachnick, B., Fidell, L. (1996) *Using Multivariate Statistics 3 ed.,* Harper Collins, USA.

QUALITY ASSURANCE FOR PROFITABLE BUILDING SERVICES PROCUREMENT

LAM, K.C.

Department of Building Services Engineering, The Hong Kong Polytechnic University
Hung Hom, Kowloon, Hong Kong. bekclam@polyu.edu.hk

Abstract

Procurement of highly serviced buildings is a formidable task in today's construction industry. Designers and contractors must satisfy the needs of their clients and produce high quality buildings, which are value for money. To achieve this goal, the building professions should consider the use of Total Quality Management (TQM) and joint management team for quality attainment and profitable building services procurement. Also, the management of building services can be improved by the use of an appropriate procurement path under particular circumstances.

Keywords: Building services procurement, joint management, team.quality, total quality management

Introduction

Notwithstanding some good initiatives have come about in the last decade and are well placed to make valued contributions to the management of the design and construction of mechanical and electrical (M&E) services in the post-Latham era. The procurement of complex buildings is still fraught with expensive and complex problems of inadequate coordination of building services, poor design, over-engineering, inefficient construction and installation of M&E services, and difficult maintenance of buildings and their services. These problems are detrimental to the success of projects. Obviously, there is a need for better management of services from design to construction to address all these avoidable problems.

This paper is derived from the author's research project that aims to identify the most appropriate procurement method for highly serviced buildings in order to satisfy the clients' need for good quality building.

Building services account typically for 15-25% of the total cost of a new building and for highly serviced buildings this can be as high as 50-60%. Nowadays, building services has become very complex and forms an indispensable part of a building. Research results have indicated that procurement of highly serviced buildings is a difficult task in today's construction industry. But, designers and contractors must satisfy the needs of their clients and produce high quality buildings which are value for money. To achieve this goal, the building team cannot go on or do things as before, and it is the duty of the building professions to improve their performances by using Total Quality Management (TQM) and

Joint Management Team (JMT) that can encourage total involvement of all participants in the building process.

For quality building services, Total Quality Management is an appropriate answer. Its successful implementation can achieve better briefing, integration and coordination of mechanical and electrical services from design to construction. Furthermore, buildings can be brought on stream much quicker even with tight time scales and closely defined budgets. When people and resources involved in a project are properly managed through the joint management team, total project success becomes possible.

The research findings have indicated that the latest thinking regarding quality assurance, TQM/JMT, fair contract and proper selection of procurement path have profound influence on conceptual planning, design and construction of buildings, particularly for the complex and highly serviced buildings like hospitals, large hotels and intelligent sky-scrapers, etc. Research results have also indicated that reengineering the present management structure and attitude of working in the building industry is needed for better building performance and contractor's profit.

Quality management in building services

Building consists of three major parts, namely; a building element, a structural element, and a building service element. Building services are the only active component in an otherwise passive shelter. When a building is put into use, services have to perform day in, day out for the life of a building, and tend to affect everyone in the building. Building services are dynamic and undergo many changes in order to cope with users' needs during the useful life of a building. Obviously, to fully satisfy the customers' requirements, building services must be totally planned and managed during the design and construction processes so that all mechanical and electrical services will:

- integrate with other elements to form a perfect 'whole' (i.e. building) and be properly planned and coordinated for proper and smooth installation;
- perform satisfactorily in terms of maintenance of environmental comfort and provision of optimal convenience to users;
- operate effectively and use energy efficiently and in an environment-friendly manner;
- be cost effective in terms of life-cycle costs; and
- be designed with flexibility, adaptability, workability, reliability, manageability and safety in mind.

Building services account typically for 15-25% of the total cost of a new building. For some highly serviced buildings the cost can be as high as 50-60% (O'Hea, 1996) of construction cost. Despite the importance of the M&E services, traditional practices still fail to fully manage the services design and installation. This is unfortunate, as the result is often the unnecessary waste of time and money.

Obviously, to satisfy and benefit clients, it is necessary to improve the present management of design and construction process by implementing TQM.

The building services industry

Building services is not a single entity. It is made up of a diverse number of active participants, namely:

- the clients and end users who are served by the building services industry and who have an influence on it;
- the designers who design and integrate the building and the building services;
- the manufacturers/suppliers who design, produce and supply the components and equipment that constitute the service systems;
- the contractors who translate the designer's concepts and innovations into reality and who also plan, coordinate and install the services; and
- the facility manager who maintains the building.

As the building services consist of many organizations, it is clear that for best project results, a team approach is needed to carry out all these interrelated activities. They have to be undertaken by each of these five parties as an integrating unit as:

- the client's contribution can influence the services designer's design and other subsequent activities;
- the designer's design can influence physical installation of building services and maintenance aspects;
- the contractor's working and expert advice can affect the decision-making process of the design and management of services installation;
- services design and contractor's decisions on practicality, constructability and cost can influence the manufacture of M&E services, products or components; and
- services design and installation can have a significant impact on the management of the maintenance of M&E services, and thus the usefulness of a building.

Everyone should seek to identify what his or her customers require. This should also be coupled with the idea that everyone has a customer both within and outside an organization. Apparently, the implications of this 'customer' concept (as in a TQM) in relation to a construction project are vast and dramatic. Recognizing the inherent complexity of the building services design and the complex relationships between the five parties stated above, it is quite possible that things can go wrong if there is a lack of structured quality management for design and construction. Clearly, the knock-on effect created in one activity can have significant impact on subsequent activities or events, e.g. inadequate design produces unsatisfactory coordination, installation and maintenance of services and these can seriously affect the quality of a building.

Management of building services

Based on the author's research, it has been identified that the procurement of complex and highly serviced buildings (i.e. large hospitals, hotels, infrastructures and large office complex developments) is always fraught with expensive and complex problems of inadequate coordination of building services. These problems are detrimental to the success of any building projects, and therefore must be resolved by rational and professional project management from design to construction.

By their very nature M&E services cannot be designed and installed independently as this vital element has to be fully integrated systematically. Furthermore, services have to interrelate with other elements in a building. Hence, a high level of coordination is required. Surely, close working with other building designers and contractors is necessary to produce an integrated building in which all the three elements are fully planned, systematically organized, and combined and brought to fruition as required by the client, (i.e., a perfect building with good design, high quality, short completion time and last but not least, reasonable cost). As buildings grow in size and complexity, building services are becoming increasingly complex, and the traditional methods of working together are more difficult to facilitate effective management of the building process.

Problems in building services design and installation

The author's research results have indicated that the following factors, if not managed professionally and effectively in terms of quality assurance, can cause poor quality in design and construction of highly serviced buildings.

- Incomplete briefing with consequent incomplete design;
- Multi-headed client problems and longer briefing period;
- Sophisticated technology for highly serviced buildings demands a much higher level of integration of diverse and intricate building services and other elements of a building; continuing coordination of services within the building process;
- Clear and fair contract with unambiguous allocation of risks and responsibilities is not always considered essential and given to the contractors;
- Difficult management of the Multi-Temporary Organization; and
- Inappropriate procurement strategy with unsatisfactory project performance and team working.

Obviously, all these factors must be tackled and managed by all members in the design and construction teams, otherwise, many contractual/technical/management/financial/human/ quality problems could emerge. All these problems are detrimental to the success of building projects, and customer satisfaction will not be achieved. This constitutes to the poor image of the construction industry and the building industry will be condemned for the poor performance and building professionals are likely to blame one another for causing the problems. Indeed, clients are also likely to be criticized by the industry too as they may also be responsible for some of the faults (i.e. poor brief/information, frequent changes, inadequate fees and acceptance of tender by cost instead of quality). It is clear that everyone bears the responsibility and must respond to the requirements of quality building. As TQM focuses on continuous improvement and customer satisfaction, it can be applied to construction projects.

Details of quality problems affecting coordination of services

The aforementioned problems have been examined by Barton (1978), Michie (1981), Kwok (1988), Pasquire (1994), Lam, et. al (1996), Gibb (1995), and many other researchers. In the context of building services, the problems can broadly be divided into three categories, namely:

- Technical Problems - these include inadequate briefs, inaccurate or inadequate details of design concepts, lack of integrated / coordinated designs, incomplete design / information, mistakes / discrepancies, impractical / complicated designs, unclear design / construction responsibilities, poor installation and difficult maintenance;
- Management Problems - these include inadequate management in clients' own organizations, inadequate management of design and site supervision, inadequate management of services installation and building construction, inadequate management of multi-organizations, poor communication and site supervision and lack of professionalism on the part of project participants; and
- Human Relationships - these can best be related to the effectiveness of team working and cooperation. Adversarial relationships with a 'them and us' attitude and differentiation of the operating units are critical factors.

It is easy to blame client for driving down prices, and contractors for cutting their own throats to win work, but the M&E design profession must also carry its share of the blame. On those projects studied by the author, inaccurate / inadequate M&E drawings / details and poorly coordinated building services, and poor management of design and construction contributed significantly to the coordination problems. This points out that teamwork for integrated design and quality assurance is missing.

The first two problems are all related to quality of design and excellence in management. They can only be tackled by commitment to avoiding problems using quality management for design and construction processes which encompasses the following requirements:

- For design-related problems - Quality client's brief, adequate resources; good design management; completely integrated and coordinated design; constructability; complete design information; clear design and construction responsibilities; and continuous design review, checking, and feedback for improvement;
- For coordination problems - Clear responsibilities, adequate time and resources; completely coordinated services; building drawings and information; contractors site coordination; combined services drawings by contractors; good management and supervision of coordination of services; and assistance from the design team and continuous review;
- For management problems - Effective project management; team working spirit; effective communication and better interaction between client and designers, and designers and contractors; and systematic yet flexible procedures; professional approach and continuous review; and
- For procurement problems - Use of the most appropriate procurement system is a contributory factor in achieving project success as this will influence the degree of

management of a project and coordination of services from design to construction by clients, designers and contractors.

As seen from the above, all the identified problems can be resolved by the implementation of quality management i.e. getting things right the first time and every time but with teamwork – Client, Designers, Builder and Contractors.

Human issues are difficult to overcome as the state of legal independence between participants and their traditional methods of working together are often an obstacle to adopting a team environment. To improve working relationship, a cultural change away from the "them and us" attitude participants traditionally adopt in the construction industry is required. Obviously, for quality building to succeed, we must begin with a return to basics. This starts with the formation of a coherent project team, one in which customers' interests are the interests of all involved. A new paradigm in Total Quality Implementation must exist, the participants within the project organization need to satisfy the needs of the next participants in the line who use their output.

Construction projects are becoming much larger and increasingly complex, and each of the design and construction process incorporates a chain of activities which incorporates many people and organizations. Design and construction faults or problems could emerge if there is a lack of quality management between the client, designers, contractors and the manufacturers. The effects of this deficiency often result in building defects; poor quality of work and consequently high maintenance costs; high levels of variations and uncertainty of cost; and often late completion of work and a high incidence of claims. Griffith (1990), Griffith and Sidwell (1995), Kumaraswamy (1996), Colin, Langford and Kennedy (1996), Lewis and Atherley (1996) and other researchers have identified similar problems in buildings which are related to failure in quality of design and construction, and their findings are also applicable to building services. From the author's analysis, problems can be found to stem from:

- Clients - inadequate brief; lack of information or decisiveness; unrealistic time, cost and quality targets; too many changes; and unclear responsibility and risk allocation; and inadequate cost for coordinate of services;
- Designers - inaccurate or inadequate details of design information; incorrectly specified or misused materials or components; inadequate coordination between client/designers; lack of design empathy for construction; inadequate contract document; inappropriate contract type, inadequate site supervision; poor interaction between client/designers and contractors; and lack of professionalism of project participants; and
- Contractors - unrealistic tender pricing; inadequate planning and management; lack of competence of project participants, inappropriate contractor selection, inappropriate form of contract, inadequate coordination between designers and contractors, poor communication; inadequate supervision; poor installation; and low quality of testing and commissioning.

All these problems are detrimental to the success of building projects and must be overcome and avoided by giving due consideration to each of the problems identified in term of adequacy, appropriateness, practicability, constructability, simplicity, effectiveness, value analysis, coordinated project documentation, cooperation, effective management and the use of good engineering. For quality building services, an overall quality management system must be traceable across the total building process and should assure quality from top to bottom, i.e. client to suppliers. In this respect, Total Quality Management (TQM) provides a good environment for achieving this objective. Brian Moss in his CIBSE presidential address (Building Services July 1992), has advocated TQM and he also contends that TQM is capable of producing far reaching results, better quality, higher profitability, growth and a better working spirit. TQM is different from QA in that it is a corporate philosophy based on customer satisfaction, rather than a supplier procedure (Drummond, 1992). Customer satisfaction and continuous improvement are the fundamental goals of TQM. Building construction is a 'people' activity demanding a high level of personal skills from conception of the design and throughout construction with particular requirements in terms of problem solving. Clients provide input to the design process. The design process provided by the designers gives input to the construction process. Suppliers provide the raw materials necessary to construct the facility. This amalgamation of people, equipment, and processes forms a system to achieve the objective of the building process. If this system is to work at its best, there should occur a sense of teamwork among all parties associated with the process.

The integrated team would of course achieve the management of the following quality issues:

- Quality design based on an appropriate quality system can improve integration and coordination of building services and buildings;
- Efficient management of construction and coordination due to availability of complete design, systematic project information and more efficient and effective means of Coordinated Project Information (CPI). All technical and management issues can be monitored and reviewed systematically;
- Improved quality and constructability;
- Fewer delays and disruptions due to more effective planning, better design, team working and "right first time"; and
- A systematic checking process for design and construction process so that problems can be identified and corrected much earlier.

Clearly, a strategy of TQM promoting a systematic and disciplined approach mechanism would have a direct bearing upon known coordination problems.

Constraints in using TQM/JMT

In reality, the coordination of all the separate management processes which go to make up TQM is difficult to achieve in building construction due to the nature of the works. Difficulties can arise as a result of:

- complexity of the layers of internal and external customers involved in a single project;
- lack of compatibility between members of the various design and construction teams;
- inappropriate choice of procurement arrangement;
- conflict of interests and differences in objective;

- the state of legal independence between participants;
- widely diverse nature of the building services industry, the naturally occurring (time, resources, market, etc.) and imposed constraints (i.e. codes and standards etc.);
- differences in culture and the variances in patterns of behaviour; and
- as an advertisement (registered firm status) rather than a stepping stone toward better quality.

Nowadays, most building services and buildings are much more complex. Clients are more commercially minded, more discerning and more demanding of better value for money, faster construction, a higher standard of services performance and provision, and generally have higher expectations of quality. These changes in M&E services requirements, together with changes in procurement methods, shortening of design periods and the increase of specialist subcontractor design; tight fees and low profit margins, high overheads and risks, keen competition, etc. have all resulted in complications in the design, procurement and quality control of M&E services. To implement TQM successfully, the building services industry should reengineer and emphasize teamwork and partnering (short or long term) between participants, and use the most suitable procurement path to accomplish the requirements of the client. Indeed, the CIBSE/DoE Conference More Efficient Construction, (1996) has voiced the following statements:

- get the job right in the first place; work as a single team with a common goal; use of construction management; better and complete design and use of specialist contractor's knowledge (by David Arnold);
- collaboration and harmonisation between the consultant and specialist (by Martin Davis);
- designers should produce designs and specifications that do not limit the methods or approaches to the installation and its processes (by John-Deal); and
- the reluctance on the part of designers to get into detail probably causes the lion's share of disruption, delay and cost escalation on any project (by Colin Stoker).

Similarly, the Working Group II of the Construction Industry Board (Building Services, Jan. 1996) also sees the following as the principal areas (needed in a TQM) for action: change the industry culture; better brief; design and construction processes work as one; foster teamwork and partnership; make quality the main requirement of all elements of the design and construction processes; knowledge and constructability; life-cycles; design and maintenance considerations; and quality and value must not be ignored in the pursuit of lowest price, etc.

As seen from the above, all these comments need to be seriously considered and incorporated for quality attainment of building services. However, one cannot impose QA or TQM into contract, and TQM must come from the members of the team as a corporate strategy.

Correlation between procurement system and quality

Apparently, there is a correlation between the type of procurement adopted and the quality to be achieved. Though Burt (1978) does not consider the impact of a contracting arrangement on quality to be critical, he agrees that in particular circumstances one arrangement may offer advantages over another. Griffith and Sidwell (1995) view this issue differently as they consider procurement has a profound influence upon quality and therefore procurement system must be carefully chosen. Holness and Osborne (1996) also concur that the various procurement methods available have different implications on quality but contend that the choice of contracting arrangements is an integrated part of quality attainment, but not the determining factor.

The author's research has indicated that the procurement path is a crucial factor in attaining project success. The studies also reveal the importance of working in harmony and the need for good management of design and construction. However, effective management of design and construction will rely on good team working which are ingredients of a TQM. Hence, the author would strongly argue that quality attainment can be influenced by the building procurement path, as each procurement strategy would have a particular impact on the working of the building team and the performance of the client, designers and contractors under a particular project environment and characteristics.

TQM/JMT and project performance

The author's research results based on case studies have clearly identified that good quality of management of design and construction is the most important factor in attaining quality buildings. All results suggest that the use of an appropriate procurement path for particular set of project characteristics and environment, together with the most suitable management strategy would have a significant effect on quality attainment. Obviously, when teamwork is treated as a first priority, (as called for in TQM), the non-traditional procurement methods would in theory give better results. However, this does not automatically mean the need for a design and build method of procurement, although this method has a distinct advantage over others in applying quality management to its process (e.g. better brief, simplified contractual arrangements, integrated design and construction, improved communication, increased operational efficiency, single project team and harmony in design and construction). It is true that good quality could also be achieved through other methods of procurement together with good team-working spirit, and through the existence of a good team of professional designers and contractors.

Use of TQM/JMT for coordination of services

The procurement of complex buildings is always fraught with expensive and complex problems of inadequate coordination of building services. Coordination problems are very complex in nature. To add to the complexity, all the identified coordination problems do have knock-on effects on one another. However, TQM and JMT with the most suitable procurement strategy can overcome the problems. The author has identified many case studies that the following TQM/JMT attributes are conducive to successful coordination of services.

- client must consider quality and price, not lowest price;

- project success depends on effective management of coordination of services within a construction process. The services designer must design the building services fully in a way that equates to the approach adopted by other members of the design team including proper detailed coordination;
- coordination of services is not entirely a technical issue. Good management of design and installation of services is very important;
- effective coordination of services on site is a function of high level of building team working and this requires effective project management especially for a temporary multi-organization;
- client characteristics, project characteristics and allocation of risks and responsibilities will influence project success and management of coordination of building services;
- though a selected procurement path has a direct bearing on the success of a particular project, no single method of procurement can be suitable for every project all of the time. If success is to be achieved, selection of the most appropriate organization(s) for design and construction for each project is essential;
- all parties must fully committed to the need for coordination of services right from the start of a project and cultural change in the construction industry is necessary;
- coordination must start in the early stages by all parties; and
- building process cannot go on as before, the industry should achieve first class customer satisfaction.

Conclusion

Coordination of services has been identified as one area, among many, in which the building services industry performance is inadequate. Therefore, clients are still not satisfied with the products that the professional produce, and all parties involved in a project are also frustrated in the coordination chaos. To achieve the best project success and quality attainment, it is considered essential to have a correctly selected procurement strategy together with a good team of designers and contractors all implementing a Total Quality Management system for their design and construction. This not only results in a better building, but the entire building team will benefit more.

Nevertheless, implementation of TQM/JMT is not easy. Because of the widely diverse nature of the building industry, the naturally occurring and imposed constraints, the extent of required knowledge and expertise, the difference in objective classification, the difference in culture, and the variations in patterns of behaviour. They all can adversely affect the implementation of TQM/JMT. Despite all these difficulties, TQM/JMT is a MUST if project success is to be pursued by all. It should stress that the implementation of TQM/JMT does not guarantee or insure against project failure, but provides a better approach to coordinating building services. TQM/JMT is process-oriented and not so much result-oriented. If the processes are right, the results (i.e. quality of coordination of services) are likely to follow.

Services may account for as much as 50% of project cost. It is vital that building services design and installation are adequately designed and coordinated from design through construction and that if this is done, significant improvements in project performance will be realised. Besides, all parties involved will be satisfied as they all can achieve their goals and earn their rewards.

There has been a growing awareness of the importance of the teamwork approach in construction industry. We also know this involves cultural change away from the way participants traditionally work in the construction industry in order to achieve Total Quality Management. Of course, TQM requires a joint management team approach. When all members of the building team are properly managed and equipped through the quality teamwork approach, the construction process should be improved, and consequently, the resulting product – the building and its M&E services – would satisfy the requirements of all customers – client and his building team.

References

Barton, P.K. (1978), Co-ordination of Mechanical and Electrical Engineering in Building Production, IOB Special Paper, U.K.

Burt, M.E. (1987), A Survey of Quality and Value in Building, BRE Publications, Watford.

CIBSE/DoE Conference on "More efficient construction", *Building Services Journal*, January 1996 (p.p. 24-26).

Conlin, J.T., Langford, D.A. and Kennedy, P. (1996), The Relationship Between Construction Procurement strategies and Construction Contract Disputes, CIB (p.p. 360-371).

Drummond, H. (1992), The Quality Movement, Kogan Page, London.

Gibb, A.G.F. (1995), Maintain control or delegate responsibility? The design development dilemma at construction interfaces, ARCOM 95, U.K. (p.p. 202-211).

Griffith, A. (1990), Quality Assurance in Building, Macmillan, England.

Griffth, A and Sidwell, A.C. (1995), Constructability in building and Engineering Projects, Macmillan, England.

Holness, N.A. and Osborne, A.N. (1996), An Evaluation of Quality as a Sympton of Inter-organisational Conflict in the Construction Industry, ARCOM 96, U.K. (p.p. 347-357).

Kwok, P.K. (1988), Coordination of building services in tall buildings, Building Journal Hong Kong China, p. 98-104.

Kumaraswamy, M.M. (1996), Is Construction Conflict Congenital ?, ARCOM 96, U.K. (p.p. 190-199).

Lam, K.C., Gibb, A.G.F. and Sher, W.D. (1996), Impact of building procurement methods on coordination of Building Services, ARCOM 96, U.K. (p.p. 150-159).

Lewis, T.M. and Atherley, B.A. (1996), Analysis of Construction Delays, CIB (p.p. 60-71).

Michie, A. (1981), Integration and Coordination of Building Services and its Relationship With Project Management, BSER&T (p.p. 15-26).

Moss, B.C. (1992), Satisfying the user, Building Services Journal, July. (p.p. 33-35)

O'Hea, J. (1996), A vision for the Millennium, Building Services Journal, July, 1996 (p.p. 24-25).

Pasquire, C. (1994), Early Incorporation of Specialist M&E Design Capability, CIB/Proceedings (p.p. 259-267).

Working Group II of the Construction Industry Board, Building Services Journal, January, 1996 (p.p. 28).

CONSTRUCTABILITY-THE IMPLICATIONS FOR PROJECT PROCUREMENT IN THAILAND

CHANDRA BHUTA and SUJAY KARKHANIS

School of the Built Environment, Victoria University of Technology, PO Box 14428, MCMC, Melbourne, VIC 8001, Australia. ChandraBhuta@vut.edu.au

Abstract

The paper presents some findings of a study undertaken by the principal author in order to perform a constructability comparison between the construction industry in Thailand and Australia. The results of the study provide information on the level of usage and awareness of constructability in these two countries. These results may help the Thai construction industry to become more efficient and become globally competitive. The study was prompted by similar studies undertaken in Australia and the United States on constructability. However, none of these studies attempted to compare the level of usage and awareness of constructability in context of the Thai construction industry. The study is important for the Australian construction industry since it has large stakes in once booming Thai construction industry. An understanding of the construction practices in Thailand and vice versa is very vital for a successful business partnership between the two countries.

As a first step towards this understanding, a study was conducted through a survey questionnaire sent to more than seventy-five construction organisations in Thailand and Australia. The questionnaire was designed to help in identifying the reasons for use or otherwise of constructability concept in the building industry, to evaluate the extent of its implementation and barriers to the implementation, to gain a general understanding of the industry view on constructability, and to examine whether constructability can be used to improve an overall efficiency. A response rate of more than 50% was achieved. An analysis of the study sample from Thailand indicated that awareness of constructability was poor among Thai construction industry. The survey also indicated that the organisations aware of constructability believed this awareness to be beneficial to the building industry. The use of constructability in the decision-making process was quite significant in Australia as compared to Thailand.

Keywords: Australia, constructability, industry survey, Thailand

Introduction

The term 'Constructability' implies an optimum use of construction knowledge and experience in planning, design/engineering, procurement, and field operations to achieve overall project objectives. A number of studies/reports published on performance measurement in the construction industry had highlighted the importance of constructability factors in successful delivery of a project [Ramahi & Bhuta (1994)]. The

World Bank funded construction projects around the world reported an average cost overrun of 65% and time overrun of 68% [Thompson & Perry (1993)]. On the contrary, the Royal Commission of New South Wales in Australia reported cost and time overruns of in the range of 20-25% on construction projects undertaken in Australia. It is not unreasonable to assume that most of the World Bank funded projects would be located in developing countries. Thus the difference in cost and time overruns between a developed country (in this case Australia) and developing countries is quite significant. Some proportion of this cost and time overrun may have been influenced by constructability factors. The use of constructability factors has long been researched by many researchers in developed countries like the United States and Australia. However, there seem to be a perceptible lack of research on constructability factors in the developing countries.

The paper presents some findings of a study undertaken by the principal author [Chaimukda & Bhuta (1997)] in order to compare constructability practices between a developing country and Australia. Thailand was a natural choice since it happened to be Chaimukda's home country. Also Thailand has close economic ties with Australia. The comparative study provided information on the level of usage and awareness of constructability in a developed country and a developing country. The findings of the study may help the construction industry in a developing country to become more efficient and globally competitive. The study was prompted by similar studies undertaken in Australia and the United States on constructability. However, none of these studies attempted to compare the level of usage and awareness of constructability in context of the construction industry in a developing country. The study is also important for the Australian construction industry since it has large stakes in the construction industry of a number of developing countries around the world. A mutual understanding of the constructability practices is very vital for a successful business partnership between developing countries and Australia.

Literature review

One of the earliest research on improvements in constructability factors was conducted by Proctor & Gamble Ltd in the United States in late 1970s. The research emphasis was on modularisation and pre-assembly in construction and developed a construction checklist to facilitate its application. In late 1970s the American construction industry was on decline in its cost-effectiveness and quality. In 1980 the well known the Construction Industry Institute (CII) was established to conduct research on the improvement of cost-effectiveness of the American construction industry. The Constructability Task Force, one of research arms of the CII, conducted some pioneering work and published a number of research papers [CII (1986), CII (1987), CII (1987)]. Many researchers within CII [Tatum (1987), Tatum et al (1987), O'Connor et al (1987), O'Connor (1988)] identified ways to improve constructability. Tatum, among other things, identified constructability factors that may be most crucial during the concept phase of a project. He also emphasised the role of prefabrication, pre-assembly, and modularisation in its contribution towards improvements in constructability. O'Connor's team identified constructability factors for engineering, procurement, and field operations.

The advent of research on constructability issues in Australia has been fairly recent. During the period 1991-93, the Construction Industry Institute, Australia (CIIA) collaborated with the CII in the United States to develop constructability principles appropriate to the Australian context. Since then a number of researchers have worked in Australia to understand the constructability factors for project procurement in Australia [Griffith (1996)].

Ogunlana and Promkuntong [Ogunlana (1996)] had conducted a research in Thailand to investigate the causes of construction delays in developing countries, especially the fast growing economies in south- east Asia. They identified a number of issues affecting construction delays including the implications of constructability factors for project procurement in Thailand. However, none of the studies conducted hitherto attempted to compare constructability practices between a developing country and a developed country.

Aims of the study

The Australian construction industry has large interests in the south-east Asian countries and as such this study is primarily focused on south-east Asia. As a first step towards the understanding of the constructability practices a study was conducted through a survey questionnaire sent to construction firms in Australia and Thailand. The specific aims of the survey were:

- Examine and determine the current status within the construction industry (Thailand and Australia) on constructability factors for project procurement.
- Identify and evaluate the reasons for under utilisation of constructability concepts in the construction industry.
- Test whether use of constructability factors can enhance project outcome.

The survey sample was selected randomly, however, some criteria were set to obtain a sample relevant to the study as follows:

- Constructions firms who have participation in construction management/project management.
- Contractors/Sub-contractors operating nationally and overseas.
- Clients having construction interests nationally and overseas.

Structure of the survey

The survey questionnaire was structured in three parts viz. constructability concepts, use of constructability in projects, and barriers to the use of constructability. The first two parts of the survey questionnaire examined the following:

- Firm's understanding of the constructability concepts
- Objectives of using constructability
- Benefits of using constructability factors for project procurement

- Stages and level of usage of constructability factors in project procurement

Under the third part of the questionnaire the following areas were examined:

- Company information including its area of operation and turnover
- Barriers to use of constructability factors in construction

Response rate

The survey questionnaire was sent to more than Eighty construction firms in Australia and Thailand. The response rate was 48% (13) and 57% (34) for firms in Australia and Thailand respectively. The value within the brackets represent the number of responses. It can be seen that the percentage response rate for Thailand and Australia is fairly even [Figure1].

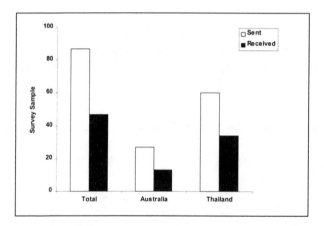

Figure1 Response rate

Type of respondents

The survey questionnaire had asked respondents to identify their role as follows:

- Owner
- Owner Operator
- Developer
- Builder
- Consultant
- Other

Most types of firms were represented in the final response, although there was some skewing towards Consultant and Builder/Developer type of firms [Figure 2]. The total number of responses exceed the number of respondents because some respondents belonged to more than one category.

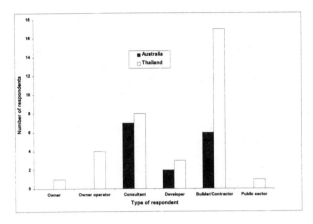

Figure 2 Type of respondents

Analysis of the survey questionnaire

Constructability awareness

As expected the survey analysis brought out a definite contrast in constructability awareness between the construction industry in Thailand and Australia. While the term 'Constructability' was not new to respondents in Australia; more than two-third of the respondents from Thailand were only partially aware of this concept [Figure 3]. A lack of research in constructability may be a possible cause of its lack of field application in Thailand.

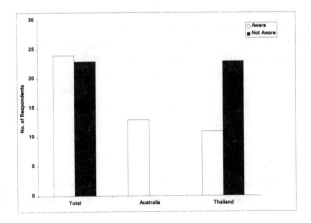

Figure 3 Constructability awareness

Constructability and type of project

The new projects including commercial, and engineering account for more than three-quarter of the projects for which constructability factors are a major consideration. Again, the Australian respondents were considerably ahead of their counterparts in Thailand [Figure 4]. It is clear from Figure 4 that the commercial projects are more likely to be subjected to constructability review than not. However, an interesting thing to observe is the equality between the Australian respondents and their counterparts in Thailand on the application of constructability factors for the residential projects.

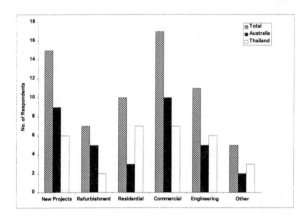

Figure 4 Constructability and type of project

Constructability and stage of project

It is a well-known in the construction industry that the activities in the initial stages of a project are likely to have most impact on the ultimate cost of a project. Use of constructability in the early stages of a project is likely to pay more than its application at the later stages. The Figure 5 shows that most of the respondents understand this concept. Nonetheless, use of constructability at all stages of a project is vital for its successful application. In Australia, the Construction Industry Institute, Australia (CIIA) guidelines on constructability recommend application of constructability at the planning and design stages of a project.

Constructability of a project

It was very heartening to know that constructability of an asset was considered very important by many respondents, both in Australia and Thailand. The survey analysis indicated that more than two-third and about a quarter of respondents from Australia and Thailand respectively believe that applying constructability during construction can reduce construction costs. The respondents were asked to indicate their understanding of the concept of constructability [Table 1].

A similar nature of response was received for the benefits of constructability of a project.

Australian respondents indicated a substantial use of constructability factors in decision making process. However, the same was not true for their counterparts in Thailand. This may be partly due to lack of proper planning in the Thai construction industry. Nonetheless, the Thai construction industry needs to improve its usage of constructability in decision making in order to be competitive.

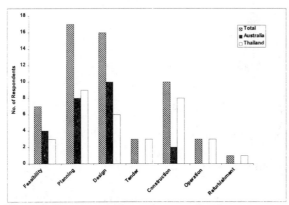

Figure 5 Constructability and stage of project

Table 1. Constructability of a project

Understanding	Australia	Thailand
Reduce construction costs	9 (69%)	8 (24%)
Improve construction during planning and design	12 (92%)	8 (24%)
Select construction methods	8 (62%)	7 (21%)

Other factors
   - Contracting strategy
   - Project objectives
   - Organisation structure
   - Lack of design excellence
	8 (62%)	15 (45%)

Reasons for the use of constructability

The reasons for the use of constructability on projects is influenced by the following:

- Project characteristics
- Project delivery systems
- Project participants
- Cost constraints

The main reason influencing the respondents on use of constructability for a project was the project characterisitics [Figure 6]. Surprisingly the cost constraint was at the bottom of the list indicating that cost is not considered to be the most important factor inspite of contrary belief.

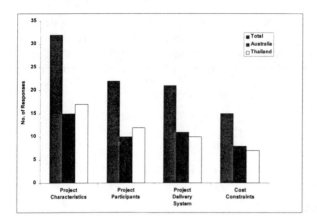

Figure 6 Factors influencing the use of constructability

Objectives in dealing with constructability

The prime objective for majority (more than 80%) of the respondents in dealing with constructability was reduction in the construction cost and improve the construction process. This is not surprising since the basic aim of constructability is to facilitate efficient construction process. Nonetheless, many respondents had additional objectives in dealing with constructability viz. improving the design phase, minimise on-going maintenance cost etc.

Barriers to the use of constructability

None of the respondents from Australia perceived any barriers to the use of constructability. This not unexpected, since all respondents from Australia use constructability on their projects. However, the respondents from Thailand seem to have a considerable number of barriers for the implementation of constructability on their project. Table 2 lists the barriers perceived by respondents from Thailand.

The Table 2 clearly indicates the lack of understanding of constructability concept along with the lack of understanding of the benefits of constructability. These two are major impediments for the application of constructability in the Thai construction industry.

Future of constructability

The respondents were asked if they sense constructability growing in the building industry. More than three-quarter of the respondents in Australia and Thailand felt that constructability in the building industry was growing. This view was shared almost equally by respondents from both the countries.

Table 2. Barriers to the use of constructability

Barriers	Number of respondents
Availability and reliability of data	11 (32%)
Reliability of the construction technique	3 (9%)
Lack of understanding of constructability	16 (47%)
Insignificant benefits of constructability factors	14 (41%)
Tight budgets	14 (41%)
No system to incorporate constructability factors into the design process	8 (24%)
Most decisions do not need constructability factors	2 (6%)
Other	8 (24%)

Conclusions

The survey set out to investigate and report on the following issues:

- Examine and determine the current status within the construction industry (Thailand and Australia) on constructability factors for project procurement.
- Identify and evaluate the reasons for under utilisation of constructability concepts in the construction industry.
- Test whether use of constructability factors can enhance project outcome.

The principal idea behind the survey was to provide industry professionals with the knowledge of the level at which constructability is being used by the construction industry in Australia and Thailand.

Only a small percentage of respondents from the Thai construction industry seem to be aware of the constructability factors and its benefits in construction cost reduction. On the other hand the respondents from the Australian construction industry seem to be well aware of the concept of constructability and its potential to provide cost savings. There is general lack of understanding of the constructability factors amongst respondents from Thailand which may be partly responsible for their lack of usage of constructability factors in decision making. The application of constructability factors for project procurement is mainly evident in the commercial construction. The Australian construction industry uses constructability factors on almost all types of projects and are far ahead of their counterparts in Thailand. However, for residential projects the application of constructability is on equal terms.

As far as the barriers to the application of constructability are concerned the Australian construction industry sees no problems. On the other hand the respondents from Thailand perceive a number of problems of which the lack of understanding of constructability concept seems to be the important one. Other problems mentioned include lack of reliable data, lack of monetary funds to support constructability reviews, and lack of a proper system to implement constructability. In spite of all this the future is very promising. More than eighty percent of the respondents from both the countries felt that constructability factors may be considered for most of their future projects.

The application of constructability is crucial for the avoidance of cost and time overruns, especially for the developing country like Thailand. The experience and expertise of the Australian construction industry can make a significant contribution in implementing constructability for its peers in Thailand.

References

Chaimukda, N and Bhuta, C (1997) Constructability comparison between Thailand and Australia. Masters thesis, Victoria University, Melbourne (unpublished).

Griffith, A and Sidwell, A (1996) Constructability in Building and Engineering Projects. Anthony Rowe Ltd, Chippenham, Wiltshire.

O'Connor, J, Rusch, S, and Schulz, M (1987) Constructability Concepts for Engineering and Procurement. *Journal of Construction Management and Engineering*, 113, pp 235-248.

O'Connor, J (1988) Constructability Improvement during Field Operations. *Journal of Construction Management and Engineering*, 112, pp 69-82.

Ogunlana, S and Promkuntong, K (1996) Construction delays in a fast-growing economy: Comparing Thailand with other economies. *International Journal of Project Management*, Vol 14, No 1, pp 37-45.

Ramahi, H and Bhuta, C (1994) Quality performance measurement in the construction industry. Masters thesis, Victoria University, Melbourne (unpublished).

Tatum, C (1987) Improving Constructability During Conceptual Planning. *Journal of Construction Management and Engineering*, 113, pp 191-207.

Tatum, C, Vanegas, J and Williams, J (1987) Constructability Improvement Using Prefabrication, Preassembly, and Modularisation. Stanford University, California.

The Construction Industry Institute (CII) (1986) Constructability: A Primer. United States.

The Construction Industry Institute (CII) (1987) Constructability concept file. United States.

The Construction Industry Institute (CII) (1987) Guidelines for implementing a constructability program. United States.

Thompson, P and Perry, J (1993) Engineering construction risks: A guide to project risk analysis and risk management. Thomas Telford, London.

Author Index